The War of the Soups and the Sparks
reveals how science and scientists work.
Valenstein describes the observations
and experiments that led to the discovery
of neurotransmitters and sheds light on
what determines whether a novel concept
will gain acceptance among the scientific
community. His work also explains the
immense importance of Loewi, Dale, and
Cannon's achievements in our understanding of the human brain and the way
mental illnesses are conceptualized and
treated.

ELLIOT S. VALENSTEIN is emeritus
professor of psychology and neuroscience
at the University of Michigan and the
author of *Blaming the Brain: The Truth
About Drugs and Mental Health* and *Great
and Desperate Cures.*

THE

WAR

OF THE

SOUPS

AND THE

SPARKS

———

THE

WAR

OF THE

SOUPS

AND THE

SPARKS

The Discovery of Neurotransmitters
and the Dispute Over How Nerves
Communicate

———

ELLIOT S. VALENSTEIN

Columbia University Press
NEW YORK

Columbia University Press
Publishers Since 1893
New York   Chichester, West Sussex
Copyright © 2005 Columbia University Press
All rights reserved

Library of Congress Cataloging-in-Publication Data

Valenstein, Elliot S.
The war of the soups and the sparks : the discovery of neurotransmitters and the
dispute over how nerves communicate / Elliot S. Valenstein.
p. cm.
Includes bibliographical references and index.
ISBN 0-231-13588-2 (cl. : alk. paper) — ISBN 0-231-50973-1 (elec.)
1. Neurotransmitters—History. 2. Neural transmission—History. I. Title.
QP364.7.V34 2005
612.8—dc22

2005042103

Columbia University Press books are
printed on permanent and durable acid-free paper.
Printed in the United States of America

c 10 9 8 7 6 5 4 3 2 1

This book is dedicated to two institutions and two people. To the City College of New York, for providing a returning World War II veteran with the opportunity to make a fresh start. At CCNY, I first learned what scholarship really meant. There professors encouraged even undergraduates to question everything and everyone, including themselves. My professional career ended up at The University of Michigan, where for more than 35 happy years I have been continually stimulated, challenged, and helped by my colleagues.

This book is also dedicated to the memory of Professors William C. Young and Walle J. H. Nauta. In graduate school, Bill Young taught me by his example how to select research problems worth pursuing and to appreciate that every manuscript can be improved with more thought and still another draft.

After completing my Ph.D., I went to work at The Walter Reed Institute of Research. Because of the crowded conditions at the Institute, I shared an office with Walle Nauta, at the time probably the leading neuroanatomist in the world. A constant stream of eminent neuroscientists came to consult with Walle, and he always insisted that I sit in on their discussions about brain and behavior. Even though it was often not true, he always made me feel like I had something to contribute. No one could have had a more stimulating environment to grow intellectually.

CONTENTS

ILLUSTRATIONS

PREFACE

During the past fifty years I have seen many advances in our understanding of how the brain works. None of these advances, however, has had a more revolutionary impact on our ideas about the brain than the discovery that nerves secrete chemical neurotransmitters when communicating with other nerves and the muscles they innervate. Yet few people, including most neuroscientists, know much about how neurotransmitters were discovered, the fierce and lengthy dispute about their very existence, or the scientists involved and the social and political events that affected their lives and work.

I first became interested in how neurotransmitters were discovered when I was writing the book *Blaming the Brain*. After I became aware that all the early drugs used to treat mental illness had been discovered accidentally, it occurred to me that such discoveries could not have happened any other way.[1] So little was known about brain chemistry in the 1950s that it would not have been possible to predict the physiological or psychological effects of any of these drugs. Moreover, because neurotransmitters were not thought to exist in the brain, for a number of years it was not even possible to offer a reasonable explanation of what the drugs might be doing there even after their effects were discovered.

When I finished *Blaming the Brain*, I started to look into the history of the discovery of neurotransmitters, to satisfy my own curiosity. I was soon captivated by what I found to be a fascinating story. When I talked to my friends about what I was uncovering, it became clear that very few knew much, if anything, about this history. Here was a little-known and fascinating account of an important subject. I decided to make this history the focus of my next book.

Initially I concentrated on learning how the evidence for neurotransmitters had been acquired. As I learned more about the scientists involved, however, I felt that it was essential to include in my book something about their lives, how their scientific interests developed, and the different ways they responded to evidence suggesting that chemical substances were involved in transmission of the neural impulse. I have therefore included considerable information about the lives of many of the scientists and more extensive biographical details about Henry Hallett Dale, Otto Loewi, and Walter Bradford Cannon, the three most central to this history. The first two shared the Nobel Prize for the discovery of neurotransmitters, and Walter Cannon might have shared the prize with them had he not been persuaded to adopt and defend a controversial theory.

Political events greatly affected the lives and work of a number of the scientists important to this history. During World War I, many changed the direction of their research in order to contribute to the war effort. For some, the disruption was much greater during the period leading up to World War II. Otto Loewi, for example, already a Nobel laureate, was arrested by Nazi storm troopers and thrown in jail. Many German scientists were dismissed from their positions because of the Nazi "racial" policies, and the more fortunate were able, often with the help of colleagues and money from the Rockefeller Foundation, to join laboratories abroad. There an impressive number made major contributions to science, some of which are relevant to this history. I have included several accounts of these events because the historical context is necessary for understanding what transpired.

An important part of this history involves the episode called the "War of the Soups and the Sparks," a dispute over whether nerve impulses are transmitted chemically or electrically. The dispute was mainly between pharmacologists, who had uncovered the first evidence of chemical transmission, and neurophysiologists, experts on the nervous system, who dismissed these new findings and remained committed to electrical explanations of neural transmission. Although initially there were good scientific reasons to question the possibility of chemical transmission, the controversy was sustained and fueled by the competing interests of the two disciplines.

There is always the question of where to begin the history of any topic. It seemed to me that the logical place to begin this story is the period around 1900. It was then that, after several decades of controversy, the "neuron doctrine" was finally accepted. This doctrine asserted that the nervous system is comprised of nerve cells separated from each other. Although the instruments that eventually made it possible to see the gap between nerve cells would not be available for another fifty years, there were good arguments, although these were disputed, that at least a functional gap existed between the terminals of neurons. This controversy, and how it set the stage for investigating whether the gap was bridged electrically or chemically, is the subject of the first chapter.

Also important in setting the stage for the eventual discovery of neurotransmitters were early investigations of the effects that different drugs have on visceral organs. Many new drugs had become available during the second half of the nineteenth century, and the effects of these drugs were commonly tested on visceral organs such as the heart, lungs, blood vessels, intestines, and glands, as well as on skeletal muscles. These studies provided the basis for the observation that some drugs produced the same effects on these organs as did stimulating the nerves that innervate them. Chapter 2 describes how we learned about the so-called autonomic nerves, which innervate visceral organs, and how this research paved the way for early speculation that a chemical process might somehow be involved in transmission of a nerve impulse.

Chapter 3 describes how some gifted British scientists came to speculate that a chemical process was involved in neurotransmission. Their speculations arose mainly from the observation that adrenaline seemed capable of mimicking the effects of sympathetic nerve innervation at all sites. Although this line of research came tantalizingly close to proving the existence of chemical neurotransmitters, it was virtually ignored, for reasons that will be explored in that chapter.

The early speculation that neural transmission could involve a chemical process was ignored for about fifteen years. During that period, however, Henry Dale began to investigate the properties of a number of drugs that evoked or blocked visceral responses, and in the process he accumulated much of the basic information that was essential for eventually proving that autonomic nerves secrete chemical neurotransmitters. Chapter 4 describes Dale's life, education, research style, and personality, as well as the circumstances that led him into this particular line of research.

World War I interrupted any further progress on the question of possible chemical transmission. From 1914 to 1918 Dale, like most British scientists, was heavily committed to war-related research. It was not until 1921, then, that new interest was suddenly aroused about the possibility of the existence of neurotransmitters, or neurohumoral secretions, as they were called at the time. At that time Otto Loewi reported that he had obtained evidence that the nerves controlling heart rate secrete neurohumoral substances; his data were derived from a simple experiment that had occurred to him in a dream. Chapter 5 describes Loewi's background, his close friendship with Dale, and the experiences that may have inspired his dream. Also described in chapter 5 is the fierce opposition to Loewi's conclusion and the way he responded to the challenge.

Dale, who was preoccupied with other interests and administrative responsibilities after the end of World War I, remained on the sidelines during much of the controversy over Loewi's work. Dale was persuaded, however, by Loewi's evidence that neurohumoral transmissions regulate heart rate, and he began to investigate whether other parts of the nervous system might also utilize neurohumoral secretions. With the help of some brilliant young collaborators and a technique brought to his laboratory by Wilhelm Feldberg, one of the many scientists forced to flee Nazi Germany, Dale was able to prove that virtually all peripheral nerves secrete chemical neurotransmitters. This part of the story, which led to Loewi and Dale sharing the Nobel Prize, is told in chapter 6.

Paralleling the work of Loewi and Dale in Europe was that of Walter Cannon, who inadvertently stumbled on other evidence that the sympathetic nervous system secretes adrenaline. Unfortunately, Cannon was persuaded to adopt a theory about these neurosecretions that became controversial and eventually was proven wrong. Although Cannon was nominated for the Nobel Prize a number of times, this controversial theory may have been responsible for his not being selected to share the prize with Loewi and Dale. Cannon was unquestionably the most eminent physiologist in the United States. He was a remarkably interesting man, not only because of the broad theoretical concepts he introduced but also because of his willingness to become actively involved in social and political issues. Chapter 7 describes Cannon's life and work.

Opposition to the notion of chemical transmission seems to have intensified after the awarding of the Nobel Prize. Opponents might have been willing to concede that neurohumoral secretions were adequate for innervating visceral organs, but they argued that this process would be much too slow to produce the swift responses manifested by skeletal muscles. Moreover, there were not many people willing to entertain even the possibility that nerves in the brain communicate by secreting neurotransmitters.

The dispute was exacerbated by the fact that the techniques of the pharmacologists and the neurophysiologists differed so greatly that the evidence each side generated tended to be ignored by the other. Beyond that, the dispute about chemical transmission seems to have been partly fueled by the resentment of some neurophysiologists at the intrusion of pharmacologists into a field where they were the experts. This "War of the Soups and the Sparks" is described in chapter 8.

When their Nobel Prize was awarded in 1936, Otto Loewi and Henry Dale were over sixty, as was Walter Cannon. Their most productive research years were behind all of them, but they remained active in supporting the research of others and the causes they believed in. A description of the final years of these three remarkable, but very different, scientists is presented in chapter 9.

Even after resistance to the theory of chemical transmission in the peripheral nervous system had significantly diminished, opposition to the notion of chemical transmission in the central nervous system, especially in the brain, continued. It was much more difficult to obtain evidence of neurohumoral secretions in the brain, as the neurons there are much more tightly packed, making their connections much harder to study. Moreover, there were no clear responses evoked by neurons in the brain that were comparable to those of visceral organs responding to signals from nerves in the autonomic nervous system. Convincing evidence of chemical transmission in the brain had to wait for the development of a number of anatomical, chemical, and physiological techniques and instruments that cumulatively made it possible to gradually acquire compelling evidence that neurotransmitters were secreted by brain neurons. Over time, attitudes evolved from a general skepticism about there being any neurotransmitters in the brain to the view, widely held today, that there may be as many as one hundred different chemicals that act as neurotransmitters in the nervous system. These later developments form the subject of chapter 10.

A few comments about the value of history in general, and this historical account in particular, are presented in a brief epilogue.

NOTES

1. The initial psychotropic drugs discovered in the 1950s include chlorpromazine, the first antipsychotic drug, which was marketed in the United States as Thorazine; iproniazid, the first antidepressant, marketed as Marsilid; and the first anti-anxiety drugs: meprobamate, marketed as Miltown and Equanil, chlordiazepoxide, marketed as Librium, and diazepam, marketed as Valium. How these were accidentally discovered is described in chap. 2 of E. S. Valenstein, *Blaming the Brain: The Truth About Drugs and Mental Health* (New York: Free Press, 1998).

ACKNOWLEDGMENTS

I am most fortunate to work in a field where so many colleagues will take the time to respond unselfishly to requests for information, advice, and comments. I want to acknowledge the following individuals and institutions for their help.

Horace W. Davenport, professor emeritus of physiology at the University of Michigan, shared his experiences of working with Walter Cannon at Harvard and made a number of helpful comments on that chapter.

Edward Domino, professor emeritus of pharmacology at the University of Michigan, read an early version of the manuscript and made me aware of issues that needed to be explored.

Robert D. Myers, professor emeritus of pharmacology, East Carolina University, shared recollections of his experience working with Wilhelm Feldberg at the Mill Hill Laboratory.

Richard Hume, professor of molecular, cellular, and developmental biology at the University of Michigan, provided helpful information about the evolution of neurotransmitters.

Professor Solomon Snyder, distinguished professor of neuroscience, pharmacology, and psychiatry at Johns Hopkins University, commented most helpfully on my manuscript and shared over a pleasant lunch some of the history behind his major contributions to our understanding of neurotransmitters.

Professor Steven J. Cooper, head of psychology at the University of Liverpool, called to my attention several publications relevant to this history and shared his knowledge of the prevailing attitude of English physiologists toward the field of pharmacology during the first half of the twentieth century.

Jonathan Leo, professor of anatomy, Western University of Health Sciences in Pomona, California, and Dr. Tilli Tansey of the Wellcome Trust Center for the History of Medicine at University College London brought useful articles to my attention.

Professor Kent Berridge, my colleague in biopsychology at the University of Michigan, made many thought-provoking comments on several revisions of the entire manuscript.

Professor Michael Myslobodsky, of Tel Aviv and Howard University, drew the figure depicting the leech muscle preparation for detecting acetylcholine. Professor Myslobodsky, Paul Valenstein, and Max Valenstein provided assistance in reproducing the figures for publication.

I want also to acknowledge the Nobel Archives for providing me with copies of nominating letters and evaluations of candidates for the Nobel Prize. Maria Holter translated some of the Nobel archive documents from Swedish to English.

The Rockefeller Foundation provided information from their archives on the work of the foundation in helping scientists and intellectuals persecuted by the Nazis.

The University of Michigan research librarians helped me locate numerous old and obscure references.

I am also indebted to the staff at Columbia University Press for being enthusiastic about the book from the outset. My editor, Robin Smith, made many constructive suggestions that improved the book. Ron Harris and Ann Young coordinated the various stages of production of the book and kept everything running smoothly.

Elliot S. Valenstein
Professor Emeritus, University of Michigan

CHAPTER 1

Setting the Stage

The Neuron Doctrine and the Synapse

If there exists any surface or separation at the nexus between neurone and neurone, much of what is characteristic of the conduction exhibited by the reflex-arc might be more easily explicable. . . . At the nexus between efferent neurone and the muscle-cell, electric organ, which it innervates, it is generally admitted that there is not actual confluence of the two cells together, but that a surface separates them. . . . In view, therefore, of the probable importance of this mode of nexus between neurone and neurone it is convenient to have a term for it. The term introduced has been synapse.

—Charles Scott Sherrington (1906)[1]

It is now recognized that, with only a few exceptions, there is a physical gap between nerve cells and between them and the muscles and glands they innervate. Over much of the last half of the nineteenth century, however, eminent anatomists argued over whether a gap existed between nerve cells or whether instead nerve fibers formed a continuous network with no separation between them. The controversy over the existence and the nature of the gap had to be resolved before the question of whether nerves communicated chemically or electrically could come to the fore. By 1906, as the above quote by Charles Sherrington indicates, it was recognized that there was at least a functional, if not a physical, separation between nerves and the muscles they innervate. How this agreement came about is described briefly in this chapter.[2]

The discovery and gradual improvement of the compound microscope between 1800 and 1850 had made it possible to examine plant and animal tissue in much greater detail. This paved the way for Theodor Schwann (1810–1882) and Matthias Schleiden (1804–1881) in Berlin to propose the "cell theory," which stated that all plant and animal tissue is made up of separate cells, each surrounded by a membrane.[3] By the 1850s most biologists had accepted the theory that the cell was the basic unit of living tissue, but it was not agreed that the cell theory applied to the nervous system. Although nerve cells had been described, several anatomists maintained that the nervous system functioned through a network of very fine fibers that were not separated from each other. Moreover, some anatomists did not believe that the nerve fibers were actually connected to the nerve cells. Although not the first to advance the idea, Wilhelm von Waldeyer [Hartz] (1836–1921) in Berlin proposed the "neuron doctrine" in 1891; this stated that the nerve cell and its fibers constitute the basic unit of the nervous system. However, neither the techniques for staining nerve cells nor the power of the compound microscopes available at the time was sufficient to determine with certainty whether the fibers from different neurons were contiguous or separated by a gap.

The situation had begun to change after 1873, the year Camillo Golgi (1843–1926) at the University of Pavia introduced a method of staining nerve cells by impregnating them with silver nitrate.[4] The silver made nerve fibers stand out clearly as black against a background of a different color (the "black reaction"). Moreover, because only a small percentage of nerve cells were stained in this process, individual neurons could be followed more easily in their entirety.[5] Despite the advantages of the Golgi silver stain, it was still not possible to see whether the fine terminals of neurons were actually touching or whether a minute gap existed between them. (The gap between neurons was not actually seen until the electron microscope became available in the 1950s.)

This question of whether nerve fibers from different cells were connected was in dispute throughout much of the last decade of the nineteenth century. Ironically, although Golgi and the Spanish anatomist Santiago Ramón y Cajal (1852–1934) both used the Golgi stain, they reached different conclusions about the existence of a gap. Golgi became the leading protagonist for the "reticular theory," which held that the nerve fibers formed a contiguous network, or reticulum.[6] Cajal, on the other hand, took the lead in summarizing the evidence for the neuron doctrine, which maintained that a gap exists between the terminals of nerve fibers and both other nerve cells and the muscles they innervate.[7]

Cajal was persuaded by several lines of evidence that a gap existed. By studying neurons in embryonic and immature animals, Cajal was able to show that developing nerve fibers first grow out of the nerve cell bodies, growing next to muscles and to other nerve cells only at later stages. He also drew on the observation that when a nerve fiber is cut, interrupting its connection to the cell body, the fiber degenerates, but the degeneration stops at the boundary of the next cell.[8]

Cajal also concluded that transmission between nerve fibers is unidirectional. Nerve impulses, Cajal argued, do not spread diffusely in all directions, as would seem to be the case in a nerve net. He called this unidirectionality the "law of dynamic polarization," and he described the interface between neurons as acting as a valve controlling one-way traffic. He argued that the generally short and heavily branched fibers called dendrites are receptive in that they receive information from other cells and convey it to the cell bodies. The fibers called axons, which are generally longer and with fewer branches than the dendrites, conduct away from the cell toward other nerve cells or muscles. This conclusion was supported by the observation that the dendrites in sensory nerves, like the optic and olfactory nerves, have their dendrites connected to the eyes and nose, respectively, while their axons extend into the brain. In contrast, the axons of motor neurons extend into the muscles they innervate. Cajal proposes that where the axons of one neuron meet the dendrites of another neuron, transmission is always from axon to dendrite and never the other way around.

By 1906 the majority of neuroanatomists supported the neuron doctrine. That year, Golgi and Cajal received word from Stockholm that they were to share the Nobel Prize "in recognition of their work on the anatomy of the nervous system." The Nobel Committee acknowledged Golgi as a "pioneer of modern research into the nervous system." Today Golgi is recognized as one of the founders of neuroscience. In addition to the histological stain that bears his name, Golgi discovered a basic structure within the nucleus of a cell that is now known as the "Golgi apparatus." He also discovered the "Golgi tendon organ" and a type of cell in the cerebellum of the brain that now bears his name. Golgi made other contributions as well, including the first full description of the course of malaria in humans.[9] In the citation for Cajal, the Nobel award described him as the person most responsible for "giving the study of the nervous system the form it has taken in the present day."

Cajal and nearly everyone else were taken aback when Golgi used his Nobel lecture, "The Neuron Doctrine, Theory and Fact," to launch an attack on the neuron doctrine. He stated that he was still "far from being willing to admit the idea of individuality and of functional independence of each nerve element," and went on to assert that the neuron doctrine was "generally recognized as going out of favor." Not only was the lecture inappropriate for the occasion, but Golgi ignored much of the evidence and arguments that Cajal and others had collected over the previous decade. It was a strange performance by a man who had deservedly received the highest recognition for his many contributions, but who apparently could not accept that he might not always have been right.[10] Golgi's Nobel lecture had no detectable impact on the growing acceptance of the neuron doctrine, although the reticular theory lingered on for several years, supported by a small group of defenders.[11]

It was the pioneering neurophysiologist Charles Scott Sherrington (1857–1952), later Sir Charles, who introduced the term "synapse" to describe the gap between neurons and between neurons and the muscles they innervate. Sherrington, who would later share the 1932 Nobel Prize in Physiology or Medicine with Edgar Adrian, had studied spinal reflexes and how antagonistic muscles are reciprocally inhibited to allow for smooth movement. Even though he could not see it, Sherrington recognized that there must be some kind of separation between neurons. He was quite familiar with Cajal's work and had even persuaded his own mentor in physiology, Michael Foster (1836–1907), to invite Cajal to give the 1894 Croonian Lecture to the Royal Society.[12] Not only had Cajal's work persuaded Sherrington that there must be at least a functional gap at the end of axons, but from his own work he recognized that the delay in transmission at the junction between neurons or between a neuron and a muscle could only be explained by the existence of a gap. In his 1906 book *The Integrative Action of the Nervous System*, Sherrington noted that the existence of a physical gap between a nerve and the muscle cell that controls the electric organ used in some fish to shock prey had already been demonstrated.

Sherrington first used the term "synapse" in 1897.[13] He been asked by Michael Foster to write a separate volume on the nervous system for a revision of Foster's classic *Textbook of Physiology*. In working on the volume, Sherrington found that he needed a word to refer to the junction point between neurons. With the help of a friend, a Greek scholar, he adopted the term "synapse" from a Greek word meaning "to clasp." Sherrington found this word preferable to other alternatives that seemed to imply a closer physical bonding than he thought existed between neurons.[14]

Prior to the pioneering work of the men discussed above, early work with curare had in fact suggested that the point at which neurons and muscles meet has special properties. The renowned French physiologist Claude Bernard (1813–1878) had started to experiment with curare around 1840 after receiving a poison arrow coated with this substance from a friend, who had obtained it from a South American native.[15] When Bernard thrust the arrow into the thigh of a rabbit the animal became paralyzed, although its heart was still beating. Bernard was fascinated by the action of the drug and began to investigate. The work of others had demonstrated that an animal injected with curare could be kept alive, although only by putting the animal on an artificial respirator, because the drug disrupted breathing by blocking all movement of the chest muscles.

In a series of cleverly designed but simple experiments, Bernard was able to show that curare neither paralyzed the muscle nor blocked nerve conduction. He demonstrated, for example, that muscles were still able to respond to direct electrical stimulation even after they were injected with curare. In another experiment, Bernard isolated the sciatic nerve and its attachment to the leg muscle. He then immersed the sciatic nerve in a bath of curare solution, but left its point of attachment to the muscle outside the bath. When he stimulated the sciatic nerve the muscle responded normally, indicating that the curare did not interfere with nerve conduction. However, if the point of attachment between the nerve and the muscle was placed in the curare solution, stimulation of the nerve no longer made the muscle contract. Although Bernard never published his conclusions, comments in his notebook indicate that he realized that curare must act on the junction between the nerve and the muscle and that therefore there must be some kind of independence between them which would explain the action of curare.[16]

Sherrington knew about Claude Bernard's work because John Langley, one of his mentors, had studied curare and had noted that it as well as nicotine prevented muscles from responding to nerve stimulation. In one experiment, Langley dipped a thread in either nicotine or curare and carefully applied it directly at the point where nerve and muscle meet; in this way he confirmed Bernard's observation that this was the only place these drugs were effective in preventing the nerve from stimulating the muscle.

Although Sherrington did not discuss how communication might take place across the synapse, by giving the gap a name he had helped to focus attention on the special properties of this region. The emphasis on the synapse eventually led to the question of whether nerve impulses are transmitted across the synapse electrically or chemically.

It has become almost mandatory in historical accounts of ideas about chemical transmission to cite the German electrophysiologist Emil Du Bois–Reymond (1818–1896). In his 1877 textbook, Du Bois–Reymond commented that there were two possible processes by which a nerve might stimulate a muscle:

> Of known natural processes that might pass on excitation, only two are, in my opinion, worth talking about: either there exists at the boundary of the contractile substance a stimulatory substance in the form of a thin layer of ammonia, lactic acid, or some other powerful stimulatory substance, or the phenomenon is electrical in nature.[17]

Du Bois–Reymond, professor of physiology at the University of Berlin, was a brilliant man.[18] He designed ingenious recording instruments (galvanometers) for detecting electrical discharges of nerves and is generally considered the person responsible for creating the field of electrophysiology. Although Du Bois–Reymond may have considered the idea that a chemical change was a component of muscle contraction, he did not consider the possibility that transmission between the nerve and muscle is chemical.[19]

The possibility that chemical mediators might be involved in synaptic transmission would not be seriously entertained for another forty years. It emerged slowly, from the studies of the effect of various drugs on visceral organs and on the skeletal muscles innervated by spinal nerves. Skeletal muscles, which are also called "striated" (or in older literature, "striped") muscles, are attached to the skeleton and move the body in space. In contrast, the muscles of most, but not all, visceral organs, such as the intestines, lungs, and blood vessels, are called "smooth" (or in early literature, "plain") muscles.[20] How the study of the effect of drugs on visceral organs and skeletal muscles contributed to the discovery that nerves secrete chemical substances is introduced in the next chapter.

CHAPTER 2

The Autonomic Nervous System

Testing Drugs on Visceral Organs and Skeletal Muscles

The known physical characters of drugs are insufficient to account for the effects they produce . . . in consequence I consider that there is a chemical combination between the drug and a constituent of the cell—the receptive substance. On the theory of chemical combination it seems necessarily to follow that there are two broad classes of receptive substances; those which give rise to contraction, and those which give rise to inhibition.

—John Newport Langley (1921)[1]

During much of the nineteenth century, pharmacology primarily involved cataloging information about the use of drugs to treat illness. Much of this information was passed down without any valid testing of the effectiveness of the drugs or much understanding of how they exerted the effects claimed for them. The drugs, which were mostly crude extracts from plants, were known collectively as *materia medica*, and those who conveyed this field's knowledge to medical students often had the title of professor of *materia medica*, not pharmacology. Around the turn of the century, however, pharmacology began to evolve into an experimental science investigating how drugs exerted their effects and what they could reveal about normal physiology.

Many new and interesting drugs had become available during the latter half of the 1800s, and their effects could most conveniently be discovered by observing how visceral organs and skeletal muscles responded to them. The action of drugs was commonly assessed by noting the changes they evoked in blood pressure, heart rate, salivation, tears, sweat, glandular secretions, pupil size, and vasoconstriction and dilation, as well as in the contraction and relaxation of both smooth and skeletal muscles. Often drugs were tested to see how they affected the capacity of a nerve to produce its normal response on these organs. It was learned, for example, that stimulating the vagus nerve typically slowed heart rate and that some drugs could block, while others could enhance, this response.

In order to understand how the various drugs exerted their effects, it was necessary to learn more about the nerves that innervate the visceral organs. The visceral organs are innervated by the part of the nervous system referred to as the autonomic nervous system. It was originally called the "involuntary" nervous system. The terms "autonomic" and "involuntary" were adopted because visceral responses are to a great extent automatic and self-regulatory and are not capable of voluntary control.

Much of the basic knowledge of how the autonomic nervous system controls visceral organs was acquired during the period from 1890 to the mid-1920s, mainly through the research of Walter Holbrook Gaskell (1847–1914) and John Newport Langley (1852–1925), who were colleagues at Cambridge University. A contemporary of Gaskell and Langley wrote that perusing any study of the autonomic nervous system created before their work would be "like reading a description of circulation before Harvey."[2] Their contributions did not end with their detailed description of the anatomy of the autonomic nervous system. Their experiments on how drugs affect the capacity of the autonomic nerves to evoke visceral responses also contributed greatly to the understanding of the physiology of that system.

Both Gaskell and Langley had distinguished themselves earlier as students of Sir Michael Foster, professor of physiology at Cambridge University. Gaskell was five years older than Langley, and he succeeded Foster as head of the department. Langley later succeeded Gaskell in that role. History has been kinder to Langley, but Gaskell is no less deserving of remembrance. Ernest Starling said of him:

> In all his work Gaskell was never content with the mere record of a new fact. The ever-recurring question was 'What does this mean?' and the search after the meaning of phenomena led him to wide generalizations, parts of which may not stand with future investigation, but all of which has served and will serve as a beacon light in revealing problems and showing the way of research to those working in allied branches of the subject.[3]

Henry Dale, who was a student of both men, said of Gaskell that he "was certainly the most stimulating teacher of advanced students that I have ever encountered." He described Gaskell's lectures as "an exciting intellectual adventure." Dale was not equally complimentary about Langley's lectures, describing them as a dry recitation of facts. Langley may have been a dry lecturer, but he was a brilliant researcher, and he attracted outstanding students to his laboratory, among them Charles Sherrington, Thomas Elliott, and Henry Dale—all major contributors to this history.

Of the two, Gaskell was more concerned with the anatomy of the autonomic nervous system. Yet, perhaps surprisingly, given that anatomy is usually considered an observational science, Gaskell was more inclined to theorize and speculate than was Langley. Although today Langley's contributions are better known to physiologists than Gaskell's, that was not always the case. After Langley's death in 1925 it was noted that there was a tendency to "belittle the value of Langley's addition to the field" and to "represent them as counting for not more than the completion of a scheme of which Gaskell had conceived the framework and laid down the fundamental principles."[4] Perhaps revealing as much about himself as about Gaskell, Langley wrote:

> In reviewing Gaskell's work one can not fail to be struck by the carefulness and accuracy of his observations. But the bent of his mind lay in the direction of generalization. A fact once definitely ascertained was never viewed by him as an isolated phenomenon, it was used as a basis for formulating the general rule. If he sometimes generalized too hastily, it was but the defect of his virtue.[5]

Gaskell's interest in the evolution of what he called the "involuntary nervous system" enabled him to have a unique perspective on the way the system is organized. His authoritative text on the origin of vertebrates, which was published in 1908, described how the study of the innervation of muscles in a wide range of invertebrates had helped in investigating the innervation of both "voluntary" and "involuntary" muscles in vertebrates.[6] He explained that there is little or no voluntary control over the "involuntary" muscles of visceral organs.

Gaskell studied serial sections of the canine spinal cord, and he described the nerve fibers that leave the cord at different levels. He stained the nerve fibers with osmic acid, and in a landmark study published in the latter part of the nineteenth century, he traced the neural pathways that innervate most visceral organs. This was tedious work, but knowledge of these pathways was not only basic to understanding the physiology of visceral organs, but essential for the research that eventually discerned neurohumoral secretions at autonomic synapses.

Gaskell determined that two neurons are involved in innervating visceral organs. The first, now known as the preganglionic neuron, originates in either the spinal cord or the brain stem, and its axon generally terminates in a cluster of nerve cells called a ganglia. As explained below, the location of these ganglia varied with different parts of the "involuntary nervous system." Axons from the second neurons, now called postganglionic autonomic nerve fibers, continue on to innervate the visceral organs.

Gaskell divided the "involuntary nervous system" into three parts based on the location of the first nerve cell in the pathway: the bulbar (or cranial) pathway, the sympathetic pathway, and the sacral pathway. The "bulb" refers to the posterior part of the brain, called the hindbrain or medulla oblongata; the thoracic (chest) and lumbar (lower back) region of the spinal cord comprised their sympathetic division; and the sacral division involved the lower part of the spinal cord.[7] Today Gaskell's bulbar and sacral parts are combined and called the parasympathetic division. Gaskell also described how the anatomical arrangement of the sympathetic division differed from that of his two other parts (see figure 2-1 below).

Gaskell recognized that almost all visceral organs are innervated by nerves from both the sympathetic division and what is now called the parasympathetic division and that their effects are usually opposed. Thus stimulating the vagus nerve (part of the parasympathetic division) slows heart rate while the sympathetic nerve that innervates the heart accelerates heart rate. In his 1881 Croonian Lecture to the Royal Society, Gaskell described the different rhythms of the heart, and in later reports he described how the cardiac nerve in frogs and crocodiles contains both the vagus nerve and the sympathetic nerve, which in mammals are separate. He showed that stimulating the cardiac nerve could produce either acceleration or slowing of heart rate depending on which nerve fibers were primarily activated. This work, as will be seen later, was important in explaining the results obtained by Otto Loewi that ultimately provided the first direct evidence that nerves secrete neurotransmitters.

Gaskell also provided a way to understand some of the pharmacological observations that eventually led to the discovery that sympathetic nerves secrete an adrenaline-like substance. His embryological and comparative anatomical studies were extensive, including such invertebrates as leeches and other annelids (segmented worms). Although Gaskell had no way of knowing it at the time, the response of the leech muscle would be used in a sensitive bioassay that made it possible to prove that parasympathetic nerves secrete the chemical acetylcholine. Gaskell also discovered that sympathetic nerves arise from the same primordial cells as the cells in the adrenal medulla that secrete adrenaline. This made sense of John Langley's later observation that adrenaline seems capable of mimicking the responses evoked by stimulating sympathetic, but not parasympathetic nerves. Gaskell summarized his studies of the neural pathways that innervate visceral organs in his 1916 book *The Involuntary Nervous System*.[8] The book was published posthumously; Gaskell finished the final revisions only days before he died.

While Gaskell's contributions were primarily, but not exclusively, anatomical, John Langley is remembered mainly for his physiological studies, although he also contributed to study of the anatomy of the autonomic nervous system. Langley introduced the names preganglionic and postganglionic fibers and substituted the term "autonomic nervous system" for Gaskell's "involuntary" nervous system. Langley chose the word "autonomic" because it implied "a certain degree of independent action, but exercised under control of a higher power."[9]

For more than thirty years Langley was the editor of the *Journal of Physiology*, published in London. The journal was founded by Michael Foster in 1878, but when it fell seriously in debt and was threatened with extinction, Langley was able to obtain the money that would enable him to purchase the controlling share of the stock and establish himself as editor. He took his editorial responsibilities seriously, and his critical comments on manuscripts submitted to the journal were often anticipated in fear and not infrequently resented. As the editor of what was widely regarded as the leading physiology journal in the world, Langley exerted a great influence on the field.

In his more anatomically oriented research, Langley used drugs to trace the neural pathways that innervate the different visceral organs. He "painted" nicotine on nerves with a fine artist's brush and carefully soaked up any excess with blotting paper. As noted in chapter 1, in some experiments Langley soaked a piece of sewing thread in a drug so that he could then carefully "paint" the drug onto different parts of an autonomic pathway. He found that nicotine first stimulated and then blocked transmission across sympathetic ganglia. If he stimulated a preganglionic nerve after applying nicotine at the ganglia where this nerve connected to a postganglionic nerve, the visceral response was blocked. The nicotine was apparently effective at the synapse in the ganglia because the visceral response could be evoked by stimulating the postganglionic fiber. By observing where nicotine blocked the response normally evoked by stimulating

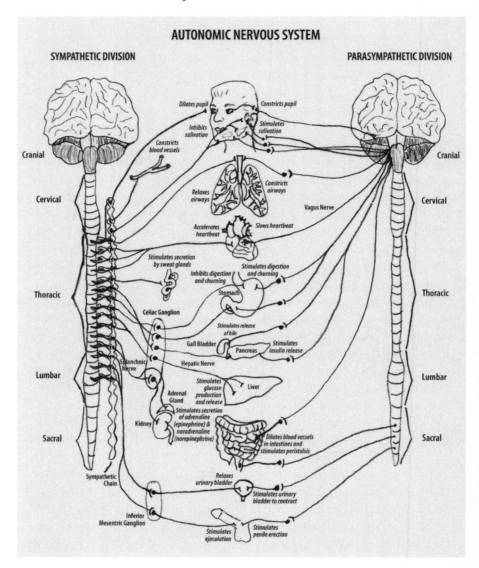

Figure 2.1   Basic organization of the sympathetic and parasympathetic divisions of the autonomic nervous system. The innervation of visceral organs generally requires two neurons. A preganglionic neuron originates in either the spinal cord or the brain stem and connects (synapses) with a postganglionic neuron that innervates the visceral organ. The preganglionic sympathetic neurons originate in the thoracic and lumbar regions of the spinal, and generally synapse close by in a sympathetic chain. An important exception involves the innervation of the adrenal gland, which is done directly by the splanchnic nerve, a preganglionic sympathetic neuron. The preganglionic parasympathetic neurons originate in either the brain stem or the sacral region of the spinal cord, and synapse either close to or within the visceral organ they innervate.

different presynaptic neurons, Langley was able to determine where different synapses occurred.[10]

He used many different drugs in studying the autonomic nervous system, including muscarine, pilocarpine, atropine, and curare. Many of these drugs were extracts obtained from plants initially brought to Europe from Africa, South America, and Asia.[11] Some of the drugs, as will be described, acted selectively on either the sympathetic or the parasympathetic division of the autonomic nervous system, by either enhancing or blocking the responses normally evoked by stimulating nerves. These experiments by Langley and others played a crucial role in setting the stage for the eventually discovery that nerves secrete chemical substances.

Around 1900, Langley began to experiment with adrenaline. Earlier, George Oliver and Edward Schäfer in England had undertaken the first systematic studies of the effects of adrenaline. Schäfer headed the Physiology Laboratory of University College London. Oliver was a practicing physician with a talent for designing and building instruments for measuring physiological responses, and he often tested them on members of his own family. For example, he created a device for measuring the diameter of blood vessels located close to the surface of the skin. When Oliver's son swallowed a small amount of an adrenal gland extract, the diameter of his arteries decreased and his blood pressure increased dramatically. (One can only wonder about Oliver's judgment: the experiment seems particularly rash, given the fact that there had been several reports of animals dying after receiving adrenal extracts.)

Oliver was excited by this observation of adrenaline's effects, and he traveled to London to talk with Edward Schäfer, whom he had known when they were students together.[12] When he arrived, Oliver found Schäfer busy measuring the blood pressure of a dog, and he persuaded him to inject some of the adrenal extract he had brought with him. Within minutes of the extract's injection in a vein, the dog's blood pressure exceeded the capacity of the mercury device being used to measure it. Thus began a series of collaborative investigations on adrenaline. In articles published in 1894 and 1895, Oliver and Schäfer summarized the responses of different visceral organs to the adrenal extract by stating that they were identical to those seen following stimulation of sympathetic nerves.[13]

Prior to Oliver and Schäfer's publications, attempts at investigating the function of the adrenal gland had been frustrating. It was known that the adrenal gland was essential, because all animals died if the gland was removed. Although it had been proposed that the gland might be essential for removing toxic substances from the blood, this hypothesis did not seem right—every animal injected with the adrenal gland extract available at the time died within a twenty-four-hour period rather than at variable lengths of time. Although Oliver and Schäfer's extract was crude, it apparently was an improvement over that used by previous investigators. They had tried various dilutions using alcohol and glycerine and both fresh and dried adrenal gland tissue. They also boiled and filtered the diluted solution and found that animals survived the injection, provided quite dilute solutions of the adrenal preparation were used.[14]

Oliver and Schäfer also studied responses obtained from extracts from the cortex (the outer part) as well as the medulla (the core) of the adrenal gland. They reported that only extracts from the adrenal medulla increased heart rate, caused blood vessels to contract, and increased the tone of skeletal muscles. In 1885 they concluded a long series of experiments with the statement that "like the thyroid gland, the suprarenal capsules [adrenal gland medulla] are to be regarded, although ductless, as strictly secreting glands."

Note on the History of Adrenalin(e), Noradrenalin(e), Epinephrin(e)

Interest in the properties of adrenal extracts increased around 1900, and several refinements occurred during the next few years. In 1899 John Abel at Johns Hopkins University in Baltimore produced a more refined extract, which he named "epinephrin," now spelled "epinephrine" in the United States. Jokichi Takamine, a Japanese industrial chemist working with Abel, further purified Abel's extract by using large amounts of ammonia, and he obtained a patent for both the process and the trademark, Adrenalin. When Takamine joined Parke Davis & Co. in Detroit in 1903, the company marketed the purified extract from oxen and sheep adrenals as "Adrenalin." By 1905, two different chemists had independently synthesized adrenaline.[15] Nevertheless, extracts from the adrenal gland continued to be used for a number of years in physiology and pharmacology laboratories. This was a source of confusion caused by the fact that, unknown to the researchers, most adrenal gland extracts contained various mixtures of both noradrenaline (a related chemical that was later found to be the neurotransmitter used by sympathetic nerves) and adrenaline.

In the early literature adrenaline was spelled adrenalin (with and without the final "e" and usually without being capitalized) and also adrenin, suprarenin, and later epinephrine (also with or without the final "e"). To avoid confusion, "adrenaline" will be used throughout this book, except in quotations, where the original spelling will be retained.[16]

In 1898, only a few years after Oliver and Schäfer's publications, Max Lewandowsky, a Russian researcher working in Germany, reported that an adrenal extract made the pupils dilate and the eyeballs protrude, effects similar to those obtained following stimulation of the sympathetic nerves that innervate the eyes.[17] John Langley extended Lewanowsky's observations by providing additional examples of the correspondence between the effects of adrenaline and sympathetic nerve stimulation.[18] This included, among other responses, an increase in blood pressure, inhibition of sphincter action in the stomach and intestines, inhibition of bladder function, pupillary dilation, and contraction of the uterus. Langley concluded that "in many cases the effects produced by the adrenal extract and electrical stimulation of the sympathetic nerve correspond exactly." However, he was also aware, as early as 1901, that in some instances the responses to adrenaline and sympathetic stimulation were not identical.[19]

Langley also reported that the adrenal extract duplicated only the effects of the thoracolumbar division, or sympathetic division, of the autonomic nervous system—stimulating the vagus nerves and other nerves in the craniosacral (parasympathetic) division of the autonomic nervous system generally produced effects opposite to those obtained by the adrenal extract.

The effects produced by supra-renal extract are almost all such as are produced by stimulation of some one or other sympathetic nerve . . . the effect of supra-renal extract in no case corresponds to that which is produced by stimulating normal conditions of the cranial autonomic or sacral autonomic nerve.[20]

These observations would eventually lead to the first speculation that adrenaline-like chemical substances might normally be involved in the innervation of visceral organs by sympathetic nerves.

At the time there were different hypotheses about the way adrenaline and other drugs evoked responses from visceral organs. Despite Lewandowsky's conclusion that adrenaline acted directly on eye muscles, others argued that it acted directly on the innervating nerves. Langley showed that adrenaline had the same effect on different muscles even after the sympathetic nerve had been cut and had degenerated. He was, however, at a loss to explain this result and simply concluded that this finding is "out of reach of any immediate hope of explanation."

Langley had observed that at some sites sympathetic nerves cause muscles to contract while at other sites they cause muscles to relax. Even collateral branches of the same sympathetic nerve might produce opposite responses on different visceral organs. Similarly, adrenaline produced different responses on different visceral organs, but generally mimicking the responses evoked by the sympathetic nerve. Langley was also aware that many other drugs did not have the same effect on all visceral organs, and he concluded that these drugs must trigger the release of an endogenous chemical substance located in muscles. Langley called these endogenous chemicals "receptor substances," which he conceived of as endogenous chemicals that actually produce the response. He speculated that the "receptor substances" were inactive until transformed by a neural impulse or by a drug that combined with them, and he proposed that differences in these endogenous chemical substances explained why the same drug or neural impulse does not produce the same response at all sites.

Langley's concept of "receptor substances" was a major step in the advancement toward our present ideas about drug–receptor interactions. He proposed that the effectiveness of drugs depended on their ability to bind with and transform endogenous chemicals in order to make them active. Langley concluded that there must be two types of receptor substances, one capable of producing the excitation that causes muscles to contract, and the other an inhibitory substance that makes them relax. His conception of receptor substances was drawn partly from chemistry, where two substances can combine to produce a third substance with new properties.[21] While Langley's conception of how receptor substances became active is different from our modern conception of receptors, it was an important step forward in that it recognized that there was a binding between a drug and receptor and that there are different receptor substances for the same drug.[22]

In his studies of how drugs like curare can prevent nerves from evoking a response, Langley came close at one time to suggesting that nerves secrete chemical substances. He wrote that the way curare blocks the innervation of skeletal muscles seems "to require that the nervous impulse should not pass from nerve to muscle by an electrical discharge, but by a secretion of a special substance at the end of the nerve."[23] However, when Langley spoke of the "secretion of a special substance at the end of the nerve" he did not mean to suggest that it was the nerves that secreted these chemical substances, for we know that he believed the substance was secreted by the muscle at the region where it was connected to the nerve.

Langley's concept of "receptor substances" was not accepted at the time, and a number of eminent physiologists and pharmacologists criticized the concept on various grounds.[24] It was not until the early 1930s that the receptor concept began to gain wider acceptance. This was to a large extent due to the work of Edinburgh pharmacologist Alfred Joseph Clark, which enabled him to determine that drugs are generally effective only on the membrane that surrounds a cell. Moreover, by a brilliant quantitative analysis, Clark was able to conclude that drugs do not act on the entire cell membrane, but on discrete receptor units embedded in the cell membrane.[25]

Much of the later work on adrenaline in Langley's laboratory was done in collaboration with Thomas Renton Elliott (1877–1961). Elliott is commonly credited with being the first person to hypothesize that sympathetic nerves secrete adrenaline when innervating visceral organs. Although Elliott clearly advanced us toward recognizing that nerves secrete the chemical substances now called neurotransmitters, whether he himself actually made this suggestion is an interesting question that is discussed in the next chapter.

CHAPTER 3

An Idea Ahead of Its Time

The First Hint at the Existence of Chemical Neurotransmitters

Adrenaline might then be the chemical stimulant liberated on each occasion when the impulse arrives at the periphery.

—Thomas Renton Elliott (1904)

Thomas Renton Elliott (1877–1961) is commonly given credit for being the first to suggest the existence of chemical neurotransmitters.[1] That Elliott was a remarkably astute scientist and scholar there can be no doubt, but the question of whether Elliott actually concluded that nerves secrete chemical substances is more ambiguous than usually reported.

After completing the academic requirements for the medical degree in 1900 at the age of twenty-three, he decided to get involved in research before starting the clinical training. He soon demonstrated his potential for research and was awarded the prestigious Coutts-Trotter fellowship, which enabled to work with John Langley in the physiology department at Cambridge. Elliott left Cambridge in 1906 in order to complete his medical education. It was during the period 1900 to 1906 that Elliot completed the major portion of the work that led him to suggest that adrenaline was released at sympathetic nervous system synapses. What he meant by that, however, has generally been misunderstood.

When Elliott started at Cambridge, Walter Gaskell had recently stepped down as professor and head of the physiology department and John Langley had succeeded him. Langley assigned Elliott the task of extending his own observations on the similarity of the effects produced by adrenaline and by sympathetic nerve stimulation. This became the focus of Elliott's work. Not wanting to be quick to publish, he systematically explored the action of adrenaline on different visceral organs and glands in different animal species and compared the responses to those obtained by stimulating the innervating sympathetic nerves. In doing this, Elliott went well beyond confirming his mentor' s earlier observations.

Elliott presented his results at a meeting of the Physiological Society in May 1904. He first reviewed Langley's evidence that adrenaline acts directly on the smooth muscles and not on the innervating nerves. He also noted that the effects of adrenaline on the muscle actually increase after cutting the nerve. This important observation, which was later confirmed and extended by others, is now called "denervation supersensitivity," a phenomenon that will be discussed further in the chapter on Walter Cannon, who used it to detect the release of adrenaline. Elliott reported that at virtually all visceral organs the effect of adrenaline is identical to that produced by stimulating the innervating sympathetic nerve, and he concluded his presentation with the often-quoted statement that "Adrenaline might then be the chemical stimulant liberated on each occasion when the [sympathetic nerve] impulse arrives at the periphery."[2]

Elliott had not only extended Langley's observations by providing many more examples of the similarities in the effects of adrenaline and sympathetic nerve stimulation, but he also suggested that adrenaline might normally be released at the synapse. However, a careful reading of what Elliott actually wrote does not support the assumption that he had proposed that it is the sympathetic nerves that secrete the adrenaline. Although he hypothesized that adrenaline might be released whenever a sympathetic nerve impulse reached the muscle he seems to have concluded, in agreement with Langley, that it was the organ stimulated that did the secreting, not the nerve.

Elliott followed up his presentation at the Physiological Society with a 68-page publication entitled "On the Action of Adrenalin." The paper contains an impressive amount of data demonstrating that adrenaline is effective only on smooth muscles and glands innervated by sympathetic nerves. He also speculated that synapses at the terminals of parasympathetic nerves and between all pre- and postganglionic nerves, as well as between spinal motor nerves and skeletal muscles, might share a biochemical similarity, although the substance is not adrenaline.

The cranial or sacral [divisions of the parasympathetic branch of the autonomic nervous system] on the other hand together with all the preganglionic "synapses" are rather related biochemically to the junctions of the skeletal [spinal] nerves with striped [skeletal] muscles.[3]

Elliott had based this speculation on the fact that muscarine, nicotine, and some other drugs, but not adrenaline, either blocked or facilitated transmission at all of these synapses. This was a most prescient speculation, as it is now known that one substance, acetylcholine, is secreted at all these sites. At the time, however, there was little interest in acetylcholine, and none of the drugs effective at these sites was thought to be natural substances in the body, as adrenaline was.

Among the conclusions Elliott reached in his 1905 paper was that

> In all vertebrates the reaction of any plain [smooth] muscle to adrenalin is of a similar character to that following excitation of the sympathetic (thoraco-lumbar) visceral nerves supplying that muscle. The change may be either to contraction or relaxation. . . . A positive reaction to adrenalin is a trustworthy proof of the existence of sympathetic nerves in any organ. . . . Sympathetic nerve cells with their fibres, and the contractile muscle fibres are irritated by adrenalin. The stimulation takes place at the junction of muscle and nerve.[4]

Elliott did not conclude in any of his publications that sympathetic nerves secrete adrenaline. It is possible that John Langley may have discouraged him from further speculation in print as, according to Henry Dale, Langley was himself "impatient of speculative theory" and advised his students to "make accurate observations and get the facts. If you do that the theory ought to make itself."[5]

Elliott's speculation had little effect on others at the time, and Langley did not even mention the possibility that sympathetic nerves might secrete adrenaline in his 1921 textbook on the autonomic nervous system.[6] This, ironically, was the same year that Otto Loewi reported the results of an experiment demonstrating that nerves actually secrete chemical substances.

Henry Dale, whose role in this story will be discussed in the following chapter, was a close friend of Elliott, and he may have been partly responsible for discouraging Elliott from pursuing his ideas at the time. Dale told Elliott about several of his own observations where the effects of adrenaline and sympathetic nerve stimulation were not identical. And well they might not have been, as we now know that it is not adrenaline but noradrenaline (norepinephrine), a closely related substance, that is actually secreted by sympathetic nerves in most mammals. Despite having demonstrated a number of similarities between the effects of adrenaline and nerve stimulation, Langley reported that in no case is "there complete correspondence between the action of a drug and the effects of nerve stimulation."[7]

Elliott left Langley's laboratory in 1906 and begin his clinical training at University College Hospital in London. He completed the medical degree in 1908, with multiple honors, and was appointed to the clinical staff there.[8] Elliott also taught in the medical school, and, with Thomas Lewis, he introduced changes in medical education that encouraged the staff as well as the students to become involved in clinical research based on the experimental biological sciences.

Although he was heavily involved in clinical work and teaching, Elliott found time for additional research on the sympathetic nervous system. Invited to give the Sidney Ringer Memorial Lecture in 1914, he spoke about the adrenal gland's relation to the sympathetic nervous system. Leaning heavily on the work of Walter Gaskell, Elliott reviewed the embryological evidence demonstrating that the secretory cells in the adrenal medulla, the chromaffin cells, arise from the same primordial cells (the ectodermal sympathoblasts) from which postganglionic sympathetic neurons originate.

> We have seen how the ganglion cell [postganglionic sympathetic nerve] and the adrenalin cell are both derived from what is almost a common cell with power to transmit a nervous impulse or to excrete adrenalin.[9]

Elliott speculated that sympathetic nerves at one time had the capacity to produce and secrete adrenaline as do the chromaffin cells of the adrenal gland, but during the course of evolution they lost this ability.

> Their present anatomical separation may be an index of a differentiation of functions which was *once held* [emphasis added] by the two in common, when the adrenalin liberation was a part of the nervous impulse.[10]

Elliott speculated that the terminals of the postganglionic sympathetic nerves might have acquired the capacity to store the adrenaline secreted by the adrenal gland.

> It is conceivable that as the nervous cell developed its peculiar outgrowths for the purpose of transmitting and localizing the nervous impulse, it might lose its power of producing adrenalin and come to depend on what could be picked up from the circulating blood and stored in its nerve endings. Removal of the glands would cut off this source of supply, and paralysis of the nerves would result sooner or

later in that territory where the nerves had been functionally most active and had consumed their stores.[11]

Elliott pursued the idea that adrenaline secreted by the gland was stored at the nerve-muscle synapse. He found, however, that a week after adrenalectomy some sympathetic nerves still had the capacity to evoke a response.[12] In the end, Elliott concluded that circulating adrenaline from the adrenals is stored at most sympathetic synapses, but he was noncommittal about whether it is stored in the muscle or the nerve. He considered this "a question of secondary interest" because he regarded the place where the two meet—the myo-neural junction—as a separate unit.[13] Although Elliott considered it possible that adrenaline might be involved in neural transmission even if a membrane separated the nerve from the muscle, Langley concluded that this would make chemical action at the synapse less likely, "for it would involve the secretion of a substance from the nerve endings."[14]

Elliott continued to follow the later developments that eventually proved the existence of chemical neurotransmitters, but he did no further work on the problem. He wrote nothing more about the adrenal glands or the sympathetic nervous system after 1914. He preferred to guide and advise others doing research without getting directly involved in the work. Nevertheless, Elliott's early research on the sympathetic nervous system and his contributions to medical education were so highly regarded that in 1913, although not yet thirty-six, he was elected to the Royal Society.

When World War I started in 1914, Elliott was stationed in France with the Medical Corp. He rose to the rank of colonel and was awarded the distinguished service order and made a commander of the order of the British Empire for his research on the treatment of war-related injuries. Elliott was very much involved in coordinating research on the causes and treatment of wound shock, which at the time was responsible for the loss of many lives. In this role, Elliott met the Harvard physiologist Walter Cannon, who was serving with the American forces in France in 1917. The two became good friends and remained in contact, mostly through correspondence, throughout their lives. Later, as will be described, Cannon became aware of Elliott's early work on adrenaline and the sympathetic nervous system, and this influenced his own work on the same problem.

In 1953, seventeen years after Henry Dale had shared the Nobel Prize with Otto Loewi for proving the existence of neurohumoral secretions, he expressed his debt to Elliott in a dedication that read:

> To T. R. Elliott,
> Who had so much to do with the
> beginning of these adventures
> and, long after they have ended,
> is still my counselor and friend.[15]

Elliot's ill health in later years eventually forced him to retire, but he continued to provide thoughtful and helpful advice to all who sought him out. He died in 1961 at the age of eighty-three.[16]

A year after Elliott's 1905 publication demonstrating a strong relationship between adrenaline and the sympathetic nervous system, Walter Dixon (1871–1931), a leading figure in British pharmacology, described a study that suggested that a humoral factor was secreted by the vagus nerve, the parasympathetic nerve that innervates the heart. Dixon was one of the first in Great Britain to have the title of professor of pharmacology, a position he had held at King's College, London. When he transferred to Cambridge University, Dixon assumed responsibility for building a school of pharmacology committed to research. He was well aware of what was going on in John Langley's physiology laboratory, and he certainly was familiar with Elliott' s research on adrenaline and the sympathetic nervous system, as he was thanked by Elliott in a footnote for providing help with the manuscript.

Dixon was highly respected as a researcher and teacher. The young Henry Dale published one of his first papers with Dixon, and he always regarded him as a friend and a respected elder colleague. Dixon's interests varied widely, and he tended to initiate the investigation of a chemical substance and then switch his interest to another substance almost immediately. He was one of the first to investigate the properties of mescal and other hallucinogens, and he tried several of these drugs on himself. After trying mescal he reported:

> When sitting with closed eyes, balls of red fire pass slowly across the field of vision. Later these changed to kaleidoscopic displays, with ever-increasing colours, or revolving wheels of colour being arranged in a definite pattern, which constantly changes. Only seen with closed eyes. After-images are prolonged.[17]

Dixon was also interested in addiction, and he studied the action of such drugs as alcohol, morphine, cocaine, cannabis, and tobacco. Later he served on the League of Nations Committee on Drugs of Addiction and also on the league's Committee on the Standardization of Drugs, which was headed by Henry Dale. From his varied experience with drugs, Dixon had formulated a general theory of drug action. He proposed that all drugs interact with endogenous substances to form a new chemical substance that is responsible for the effects produced. Dixon wrote that the reason "few drugs exert a similar effect upon all tissues" is that the drugs interact with different endogenous substances. This idea, if not identical to, was clearly very close to Langley's theory of "receptor substances," described in the previous chapter.

Much of Dixon's research was guided by his belief in the importance of endogenous substances for understanding not only the action of drugs but also the action of nerve impulses. In 1906 he reported the results of an experiment that appeared to demonstrate that chemical mediation was involved at the synapse between the vagus nerve and the heart. It had been known since 1845 that stimulating the vagus nerve slowed heart rate, and by 1900 frogs were commonly being used in physiology classes to demonstrate the action of the vagus nerve and to illustrate how various drugs affected that nerve's capacity to slow heart rate. It was also well known that while adrenaline accelerated heart rate, pilocarpine, muscarine, and some similar alkaloid drugs slow heart rate in the same manner that stimulating the vagus nerve did.

At the time there were many new drugs available, most of which were extracts from plants known to the natives of the various countries then being explored by Europeans. Muscarine, for example, was extracted from the poisonous mushroom *Amanita muscaria*, while pilocarpine, which causes salivation when chewed, was obtained from South American shrubs of the genus *Pilocarpus*. The various effects of muscarine in particular were studied, and this led to the realization that it mimicked not only vagal stimulation but most of the effects produced by stimulating other parasympathetic nerves. It was also known that the drug atropine, likewise obtained from a plant, blocked the effects of both muscarine and parasympathetic nerve stimulation.[18] While it was not known how muscarine exerted its effect, Langley had demonstrated, as he had done with adrenaline, that it did not act on the axon or nerve fiber itself. These observations led to the speculation that some natural alkaloid substance might play a role in mediating parasympathetic nerve effects.

At a meeting of the Therapeutic Society in London held in December 1906, Dixon reported the results of an experiment on the vagus nerve. A brief report of the study was published the following year in a relatively obscure magazine. The title of both the presentation and the published report, "On the Mode of Action of Drugs," reflected Dixon's broader interest in how drugs interact with endogenous chemical substances. He described the reason for undertaking the investigation:

> When a muscle contracts, when a gland secretes, or a nerve ending is excited, the cause in each case may be due to the liberation of some chemical substance, not necessarily set free in the circulation, as in the case of secretin, but more likely liberated at the spot upon which it is required to act. In order to test the validity of this reasoning, I investigated the action of the vagus nerve upon the heart.[19]

What Dixon did was to extract a substance from the heart of a dog, the donor animal, before and after stimulating the vagus nerve. It was a crude experiment even at the time.

> Animals were killed by pithing; they were bled and the vagus nerves were then placed on the electrodes and excited for half an hour. The heart was next extirpated, placed in boiling water for ten seconds and extracted with alcohol.[20]

The fluid extracted from the boiled heart underwent further processing. As Dixon described the process, the extracted fluid was first

> evaporated until dryness and then taken up again with 100 per cent alcohol. This was again evaporated off again on the water bath, and a few drops of normal saline solution added. The solution so obtained was found to have the power of inhibiting the frog's heart.

Dixon wrote that fluid extracted from a nonstimulated heart also slowed heart rate, but to a lesser degree.

> [The non-stimulated heart] also gave a supply of this inhibitory substance, but in a smaller degree than in the excited heart. I interpret these experiments to mean that *some inhibitory substance is stored up in that portion of the heart* [emphasis added] to which we refer as a "nerve ending," that when the vagus is excited this inhibitory substance is set free, and by combining with a body in the cardiac muscle, brings about this inhibition.

It seems clear that Dixon believed the inhibitory substance was stored in the area of the heart where the vagus nerve joins it. Although a neural impulse is required to release that substance, Dixon never suggested that it was secreted by the nerve. He also concluded that drugs like muscarine that inhibit heart rate must also "act by liberating the inhibitory hormone," whereas drugs like atropine, which block the action of both muscarine and vagal stimulation, "either prevent the liberation of the hormone, or saturate the substance in the end organ upon which it acts." What Dixon concluded from his experiment was made even clearer by the recording secretary who summarized his 1906 presentation:

> Professor W. E. Dixon gave an account of his experiments on vagus inhibition. He was of the opinion that the heart contains a substance—"pro-inhibitin," which as a result of vagus excitation is converted into a chemical body—"inhibitin." This substance, combining with the heart muscle, results in cardiac standstill.[21]

Dixon concluded that drugs and neural impulses both liberate endogenous substances:

> Drugs act, as far as we can judge, by influencing the tissues in exactly the same way as the physiologist affects the tissues when he stimulates a nerve. In any case the analogy between the activity of a tissue, produced on the one hand by exciting a nerve electrically and on the other by the administration of some drug, is so close as to warrant the conclusion that the ultimate effect is produced by the same mechanism in both cases.

Dixon's observation of the action of drugs like muscarine on structures innervated by parasympathetic nerves was analogous to Thomas Elliott's much more extensive investigation of the action of adrenaline on structures innervated by sympathetic nerves. Because Dixon knew Elliott and his work very well, this tends to support the conclusion that Elliott never meant to imply that sympathetic nerves secrete adrenaline. At the very least, Elliott never clearly committed himself to that position at the time.

From the perspective of what is known today, it is difficult to account for the results Dixon reported. While we now know that the vagus nerve secretes acetylcholine to inhibit heart rate, it is highly unlikely that Dixon extracted this substance from the heart he was working with. Acetylcholine is very unstable, and the elaborate processing Dixon used to extract the substance from the stimulated heart makes it highly unlikely that any acetylcholine present would have remained active. The synaptic physiologist Hugh McLennan later wrote about Dixon's methodology that it is:

"inconceivable that this substance [acetylcholine] was present in Dixon's extract."[22] There may well have been some substance present in the heart after it was stimulated for thirty minutes that could explain the slowing of heart rate Dixon observed, but it is unlikely to have been acetylcholine. It is not useful to speculate further about Dixon's results as neither he or anyone else ever attempted to replicate the experiment. Dixon never published any actual data. He presented no numbers indicating how much the heart rate decreased or how reliably he could reproduce his results. He simply stated that the extracted fluid inhibited the frog's heart.[23]

Dixon, commenting on the lack of influence of his earlier work, later noted that "the birth of the chemical transmission era was perhaps premature and the theory on which our discussion was centered, aroused little comment at the time, and its youth and adolescence were equally discouraging."[24]

After Dixon died in 1931, Henry Dale wrote that "the late W. E. Dixon was at the time almost alone in proposing that the vagus nerve, secreted a chemical substance responsible for slowing the heart." Dale later stated that "it was beyond doubt that Dixon, following Elliott's suggestion concerning adrenaline, had at that early date a conception of the general nature of the mechanism which later evidence has completely justified."[25] It is true that Dixon had "a conception of the general nature of the mechanism": he concluded that the chemical substance mediated the effect of the nerve impulse; but he did not conclude that this substance was secreted by the nerve.

Dale's statement about Dixon's contribution was made in 1934 when he and Otto Loewi were well on the way to proving that parasympathetic nerves secrete acetylcholine. Dale appears to have been overly generous in crediting Dixon and Elliott with the origin of the idea that autonomic nerves secrete chemical substances. While both Elliott and Dixon proposed that chemical substances are involved in sympathetic and parasympathetic nerve action, neither concluded that such a substance is secreted by nerves. Dixon was particularly clear in his belief that the substance liberated came from the stimulated heart muscle, not the nerve.

As noted above, Dale was originally skeptical, if not opposed, to Elliott's idea that adrenaline is released (irrespective of its origin) at sympathetic nerve terminals. Dale's skepticism was based partly on his observation (with George Barger) that several other amines were more potent than adrenaline in reproducing sympathetic effects and also on his observations that the effects of adrenaline and sympathetic nerve stimulation were not always identical. These observations led Dale and Barger to conclude in 1910 that:

> To suppose that such bases [amines] and sympathetic nerve impulses alike owe their action to the liberation of adrenine seems to us to create additional difficulties for the conception.[26]

The problem of how the vagus nerve causes slowing of heart rate was not picked up again until 1920, when Otto Loewi began his seminal studies that eventually proved that the nerves slowing heart rate secrete acetylcholine. Ironically, at the same meeting of the British Medical Association where Dixon presented his results on cardiac slowing, Reid Hunt, an American pharmacologist working at the time with Paul Ehrlich in Frankfurt, reported that acetycholine is the most powerful substance known for lowering blood pressure, a well-established parasympathetic response.[27] Hunt concluded that

> [Acetycholine] is a substance of extraordinary activity. In fact, I think it safe to state that, as regards its effect upon the circulation, it is the most powerful substance known. It is one hundred times more active than choline, and hundreds of times more active than nitroglycerine; it is a hundred times more active in causing a fall in blood-pressure than is adrenaline in causing a rise.[28]

Because acetylcholine was not known to be a natural substance found in animals, it did not occur to either Dixon or Hunt that their reports might be related. However, about a decade later, Hunt's report did influence Henry Dale, who at the time was demonstrating that acetylcholine was much more potent than muscarine or any other known substance in mimicking all the effects of parasympathetic nerve stimulation. The story of how Dale came to investigate acetylcholine is the subject of the next chapter.

CHAPTER 4

Henry Dale

Laying the Foundation

It would hardly have been possible, indeed, for me to make a conscious choice of pharmacology as the main subject of my future activities, at the time when my formal studies came to an end with my graduation, first in science and then in medicine. For I had not, so far, encountered pharmacology as a subject either of study or research, and had no basis for a judgment of the kind of interests and opportunities which it might offer.

—Henry Hallett Dale (1961)[1]

More than anyone else, Henry Hallett Dale (1875–1968) is responsible for the discoveries that provided the foundation necessary for proving that nerves secrete humoral substances. However, this was not his intention at the time he was working, and it took the speculation of others to provoke him to look at the problem.

Dale was born in London in 1875, the third child of seven. He later said that despite a concerted effort by a relative to prove otherwise, no family members engaged in any kind of scientific work. His father managed the pottery branch of a large firm, as did his paternal grandfather. Dale's maternal grandfather, Frederick Hallett, came from a family of farmers, but he became a skilled cabinetmaker and owned a small factory that made fine furniture.

Dale began his education at a neighborhood school in Crouch Hill, North London, that was run by two ladies. At the age of eight, he was sent to a nearby private school. The vice principal of the school, a man named Edward Butler, was the author of several popular books for naturalists and was an authority on an order of insects. Dale attributed his early interest in natural science to Butler's influence, particularly to his insistence that the way to prove that you really understood something was to explain it to someone who knew nothing about the subject. He must have been a superb teacher, for Dale in later years related happy memories of being kept after school for half an hour or more while Butler prodded him to improve an explanation he had written. Butler would say: "Now my boy, you are going to stay with me till you have written that, so that I, as the most stupid person imaginable, cannot possibly misunderstand it." Dale apparently enjoyed the challenge of rewriting an explanation again and again, until finally he elicited a smile and a pat on the back from Butler with a "Now you've got it; I can't misunderstand it and I can't improve it. Off you go." Dale later reported that he would walk home, late for tea but brimming over with the happiness of achievement. Later, he attributed to this training the reason why his work submitted to the *Journal of Physiology* was often accepted without corrections from the editor, John Langley, who was widely known for his "savage" criticism of manuscripts.[2]

Dale was an excellent student. At the age of fifteen he won second prize, and the following year, first prize honors, in the examination of general proficiency given by the College of Preceptors. Despite this show of talent, it was understood within his family that he would leave school and enter his father's office. However, during his last year at the Tollington School he did so well on an examination that he was offered an additional tuition-free school year. During that year, Dale took an examination in twenty-one different subjects given by the London chamber of commerce and won a prize of 20 pounds, not an insignificant amount of money at the time. He was nevertheless still expected to leave school, until a fortunate circumstance arose. It seems that his father, while on a trip north, boasted about his son's achievement to the headmaster of the Leys School in Cambridge. This led to Dale being offered the opportunity to take a competitive examination late and to be judged against those who had already taken the exam. Dale was awarded one of the three scholarships offered and entered the Leys School in September 1891, remaining there until 1894.

At the Leys School, Dale's teachers recognized his inclination toward natural science, and they had him concentrate in that area with the goal of obtaining a scholarship in science at Cambridge. This encouragement gave Dale, for the first time, the ambition to do something besides "occupying a stool in some city office." Although Dale later regretted that he might have specialized too early, he was definitely not a "specialist drudge." He edited the *Leys Fortnightly*, the school magazine, and was awarded a prize for an article he wrote for the magazine and another prize for translating a passage from Virgil into English prose.

After passing an examination in the science courses at the age of seventeen, Dale was permitted to substitute physiology, a more experimental field, for biology. He was fortunate in being guided in physiology by the science master Alfred Hutchinson, who had his students work through Michael Foster's definitive four-volume *Textbook in Physiology*, published in 1891. Some supplementary readings of more recent work were also assigned. Dale was thus exposed at a young age to what was new in physiology, such as Minkowski's research on the relation of the pancreas to diabetes and the thyroid gland to myxoedema, work not yet included in Foster's textbook. Dale was receiving excellent preparation to impress his future mentors at Cambridge.

In the fall of 1894, Dale entered Trinity College, Cambridge, with a small scholarship and some additional support in natural science, which included physiology, chemistry, and physics. He impressed his instructors and won a Major Foundation Scholarship in the spring of 1896. The freedom at Trinity College was enormously attractive to Dale, and he later counted the six years spent there as being among the happiest of his life. The scholarship students at Trinity formed a small, congenial group of friends, twelve of whom formed the University Natural Science Club. The group met in members' rooms, with the host responsible for providing refreshments—sardines on toast was a common fare—and for reading a paper or leading a discussion on some current topic of scientific interest. At one such meeting they discussed the newly discovered X-rays, and at another meeting there was a demonstration of these rays with a Crooke's tube borrowed from the Cavendish Laboratory.[3] At still another meeting, one of the members set up what must have been one of the earliest demonstrations of "electromagnetic (wireless) transmission," wherein a bell was rung by a remote signal sent from the Cavendish Laboratory at the exact moment prearranged.

At the end of his third year at Trinity, Dale succeeded in getting a B.A. degree by passing both parts of the special physiology examination a year earlier than usual. He was anxious to do this because he wanted to keep up with the other members of the natural science group, most of whom were a year ahead of him and had started to eat at the Bachelors' table. Dale was given a part-time job teaching biology three times a week to a small group preparing for an examination. His father was pleased with this, for he considered it a step toward Dale's successful career as a schoolmaster, which would justify the expenses he had incurred in sending his son to Leys and Trinity. Dale, however, felt that he had no special talent to help him arouse interest in elementary science in boys who in most instances had no aptitude or inclination for the subject. He did, however, have one boy in his class, Peter Laidlaw, who later became a valued collaborator and a Fellow of the Royal Society.

As teaching undergraduates did not appeal to Dale, he never applied for or held a university appointment. He later wrote that he recognized that many regard teaching as not only personally rewarding but as a way of generating ideas that may facilitate one's own research, but he added:

> I do not believe that one could safely assume a uniformity of its influence for all people; and I am inclined to believe that for me it was, on balance, of advantage to be free from it at that period.[4]

When he completed his examinations at Cambridge, his teachers took it for granted that Dale would stay on and become a candidate for a college fellowship. However, the annual support of 100 pounds was not sufficient to live on, and he felt that he could not ask his father for any more help. Financial support had been a continuous concern for Dale throughout his course of study. He had hoped to receive the Coutts-Trotter studentship at Trinity because it brought with it a living stipend, but it was awarded to Ernest Rutherford. The competition between the two must have been fierce, as both were clearly qualified. Rutherford received the Nobel Prize in chemistry in 1908, although he is generally thought of today as a physicist because of his work on electromagnetic fields and the nucleus of the atom. Dale went on to supplement his inadequate stipend with whatever "demonstratorship" and private tutoring he could obtain. Rutherford was appointed professor at McGill University the following year, and Dale then shared the Coutts-Trotter studentship with another student. John Langley had wanted Dale to receive it alone, but another professor had nominated the Honorable R. J. Strutt (the future 4th Lord Raleigh), necessitating a compromise. Dale was allowed to keep his modest college stipend and this made it possible for him to subsist.

At Cambridge Dale had been instructed by both John Langley and Walter Gaskell. Langley, while brilliant, was exacting, hypercritical, and reluctant to speculate or theorize, while Gaskell was a teacher who described experimentation to test hypotheses as an exciting adventure. Dale later wrote

I believe, in retrospect, that the opportunity of daily contact with men of such contrasted attitudes, each of them in the highest rank as an exponent and practitioner of his own conception of scientific research, had an educative value much greater than we recognized, at the time when we enjoyed it.

Dale left Cambridge in 1900 to complete the requirements for his medical degree at St. Bartholomew's Hospital in London. He was awarded another fellowship, the Schuster Scholarship in anatomy and physiology, which supported him during the two years spent at "Old Bart's." Dale neither enjoyed nor felt comfortable in the clinical environment. He thought that the medicine he was exposed to had little scientific basis and was frustrated by staff members who tended to back up their statements by simply referring to their experience or else "made obviously foolish pretensions to giving it a basis of experimental science." Dale could not help contrasting the "oracular authority" of the clinical staff with the give-and-take exchanges with his teachers at Cambridge where even "the very great W. H. Gaskell, seemed to be ready for discussion on an assumed basis of equality with the humblest of us, and eager to encourage us in frank and critical enquiry. "[5]

After completing his clinical training, Dale, who still needed to be concerned about support, was given an opportunity to apply for the George Henry Lewes Studentship. This fellowship, which was for Cambridge men only, had been established by the novelist George Elliot. Being awarded the fellowship helped Dale make the career choice between clinical medicine and research. He chose research even though the level of support was modest and then only guaranteed for a few years.

In 1902 Dale was given the use of a research room in Professor Ernest Starling's laboratory at University College, London. Starling and his brother-in-law, William Bayliss, had just discovered secretin, and Dale was given the task of studying whether this hormone affected the production of insulin. Dale studied the histological changes in the pancreas produced by prolonged stimulation with secretin. He made some interesting observations that seemed to indicate that some pancreatic cells might be converted to the insulin-secreting islets of Langerhans cells, but the validity and true meaning of this observation was never established.

More important for Dale's scientific career than his preliminary studies of secretin and insulin was the fact that working in Starling's laboratory provided him the opportunity to meet Otto Loewi, who was working in pharmacology in Marburg, Germany. Loewi, who was close to the same age as Dale, was spending a brief time in England in order to gain some firsthand experience with pharmacological research in Great Britain. This meeting of Dale and Loewi, which is described in the next chapter, formed the beginning of a close friendship and a sharing of scientific interests, which ended only with Loewi's death in 1961. In 1936 they would share the Nobel Prize in Physiology or Medicine.

Not long after Dale and Loewi met, Dale was given the opportunity to spend several months (October 1903 to February 1904) in Paul Ehrlich's laboratory in Frankfurt-am-Main. On the way to Frankfurt, Dale stopped to visit Loewi in Marburg, and the two of them traveled together to Frankfurt. Ehrlich, a bacteriologist and immunologist, was at the time the director of the Royal Institute of Experimental Therapy (Königliches Institut für Experimentelle Therapie) in Frankfurt, had theorized that many physiological processes result from the interaction of a chemical substance with a preformed receptor. Ehrlich applied this theory to the field of immunology, where he argued that antibody formation results from the binding of antigens to specific chemical configurations on cell surfaces, which he called "side chains." According to Ehrlich's theory, this binding, or recognition of the antigen by side chains, resulted in the release of antibodies into the bloodstream. Ehrlich's ideas had evolved from earlier research on staining techniques in which he had developed a similar concept of a specific affinity between the stain and the target tissue.[6] Later, when Ehrlich began to concentrate on chemotherapy, he emphasized that the chemical constitution of therapeutic drugs have a special affinity for the cells of the pathogenic organisms against which they must act. This was the origin of the concept of the "magic bullet," the name given to the arsenic drug that Ehrlich later developed for treating syphilis.

This idea of specific binding was closely related to Langley's concept of "receptor substances," and Ehrlich and Langley were well aware of each other's work. Ehrlich eventually adopted Langley's terminology and began to use the term "receptor substance" rather than "side chains." Dale, having been a student of Langley, was familiar with this history and was pleased when Starling helped arrange for him to spend some time with Ehrlich.[7] Starling also had an interest in the concept of receptor substances because it had obvious implications for the affinity that exists between a hormone and its targets. Dale later described his first contact with Ehrlich:

Everyone who visited Ehrlich at that time received a brief and cordial welcome before being immediately plunged into a turbulent stream of excited descriptions of Ehrlich's latest scientific findings and theories, profusely illustrated by diagrams in dye on any available surface, so that the visitor, even if his own interests and work lay in a related field of scientific research, soon felt he was losing the ground from under his feet, and there was nothing for it but to submit resignedly to his flood of words.[8]

Dale did not spend enough time with Ehrlich to enable him to accomplish much, and he later wrote that: "I had no results of my own worthy to record, but I have been glad to have had contact with Ehrlich's stimulating mind and personality."

In 1904, after Dale had returned to London, he accepted a position as a pharmacologist at the Wellcome Physiological Research Laboratories, located at Herne Hill, a suburb of London. Ernest Starling had recommended him for the position when Henry Wellcome asked him to suggest some candidates. Dale's friends tried to persuade him not to take the position, arguing that working for a pharmaceutical company would be a "dead end," amounting to "selling his scientific birthright for commercial pottage." At the time, most pharmaceutical companies did research only directly related to developing drugs and little, if any, basic research. However, Henry Wellcome was quite persuasive in reassuring Dale that he would be free to follow his own scientific interests and that he himself had established the laboratory in order to make contributions of "permanent scientific value."

Dale carefully weighed all the factors. He felt that his academic prospects were not promising, because he had not done anything especially noteworthy. The Wellcome position would provide him with his own laboratory and the opportunity, as he put it, "to make my own mistakes." Dale later admitted that he was also attracted by the offer of a "marrying income." For many years, he had lived on a bare subsistence level with income from various fellowships. Having a "marrying income" must have been on his mind, for in November 1904, shortly after accepting the position, Dale married his first cousin "Nellie" (Ellen Harriet Hallett). It turned out to be a most fortunate move, as within two years Dale was appointed director of the Wellcome Laboratories and at the end of the ten years (1904–1914) spent there, he was recognized as a major figure in pharmacology and physiology.

Henry Solomon Wellcome (1853–1936)

It is impossible to know what Dale would have accomplished had it not been for the support and opportunities that Henry Wellcome provided. Wellcome was a remarkable man and while not a scientist himself, his enormous importance in nurturing scientific work, preserving medical history, and supporting exploration in various fields was later recognized when he was elected a Fellow of the Royal Society.

Although Henry Wellcome headed the Burroughs Wellcome and Company, a British pharmaceutical firm, he was born in the United States and was an American citizen. He later acquired British citizenship. Wellcome was born in a log cabin 125 miles from Milwaukee. His father was an itinerant missionary who traveled in a covered wagon through Wisconsin and Minnesota. When Henry was eight years old the family moved to a small settlement in the Blue Earth County of Minnesota. When the settlement was attacked by a Sioux tribe, Henry helped to hold off the Indians by casting rifle bullets and assisting in caring for the wounded. Nevertheless, he had a lifelong sympathy for the plight of the Indians and later contributed a considerable amount of money to support an Alaskan Indian tribe. Henry Wellcome later wrote a book about these people, *The Story of Metlakatla,* which was published in 1887.

Henry Wellcome's early education was achieved in a log house frontier school. He developed an interest in chemistry and pharmacy, and at the age of fifteen he started a three-year period as a drug clerk in Rochester, Minnesota. There he was befriended by Dr. William Mayo, the father of the Mayo brothers, who later founded the now famous Mayo Clinic. William Mayo, who was a friend of Henry Wellcome's uncle, advised the young Henry to pursue his interest in chemistry and pharmacy by getting additional schooling. Henry took the advice and started at the Chicago School of Pharmacy, later switching to the Philadelphia School of Pharmacy and Chemistry. After graduation, he worked as a pharmacist for various New York firms, traveling around the United States and South America. In Peru, he learned about cinchona cultivation and the method of producing quinine from the bark of that tree, and this gave birth to a life-long interest in plants that had medicinal properties.

In 1880, at the age of twenty-seven, Wellcome decided to settle in England, where his ancestors had come from. He was soon devoting his enormous energy to developing Burroughs Wellcome and Company, started with his partner Silas Burroughs. The company manufactured fine chemicals, alkaloids, and medicinal products. Following Burroughs' untimely death, the entire enterprise was placed completely in Henry Wellcome's hands. The firm was a success almost from the outset, and the company's reputation grew after the company was the first to market drugs in a compressed form under the proprietary name "Tabloids." These pharmaceutical pills were a great financial success, and Wellcome then had more time to travel and pursue his many scientific and intellectual interests.

Around 1895, Wellcome began to develop the research side of the company, and he established several laboratories for that purpose. One laboratory was devoted to pursuing his interest in the active ingredients in plants that had medicinal properties. There was also a research laboratory that studied serum therapy. Although these laboratories were clearly pursuing research that potentially could advance the commercial interests of the company, at the time Wellcome was unusual in giving researchers a relatively free hand to work on problems that had little or no immediate promise of producing marketable drugs. The investigators were also free to publish the results of their investigations, a policy that helped attract competent researchers.

Wellcome developed an interest in tropical diseases. He loved to travel to out-of-the-way places and was in Khartoum soon after General Kitchener had conquered the Sudan. At that time Khartoum was a hotbed of infectious diseases—malaria, dysentery, and other tropical illnesses. In 1903 he founded what would become the world-famous Wellcome Tropical Research Laboratories at Gordon Medical College in the Sudan, which played a major role in improving the health of people in that region.

Wellcome also had a strong interest in the history of medicine, and for a number of years he collected everything he could find about the practice of medicine from the earliest times to the present. In 1913 this collection became the core of the newly established Wellcome Historical Medical Museum in London. Wellcome's traveling and his interest in archaeology, ethnology, and mythology provided the opportunity

(continued)

for him to visit relics and excavation sites in Ethiopia, Palestine, and other locations. He did some independent investigations himself and funded the archeological expeditions of others. Some of the findings from these expeditions were brought back to England and eventually placed in different museums in that country.

Wellcome visited Africa many times and also supported a number of explorations to the "dark continent." He was an active member and officer of the Royal African Society and established the Wellcome Gold Medal given by that organization. He also funded a number of other medals and prizes, including several in the United States and England for work on the history of medical subjects. In 1905 he established a hospital dispensary in Mengo in Uganda. He also supported missionary projects and had a long friendship with the explorer Henry M. Stanley, who found the (never really lost) Scottish missionary, David Livingstone. Stanley later described his travels in his highly successful two volumes, *In Darkest Africa*, published in 1890. In 1931 Wellcome established the Lady Stanley Memorial Hospital in Makona, Uganda.

His interest in tropical medicine brought Wellcome into contact with General William Gorgas, who helped establish the role of the mosquito in spreading malaria and yellow fever among workers building the Panama Canal. Wellcome established the Gorgas Memorial Laboratories in Panama.

Toward the end of his life, Wellcome arranged for the many museums and laboratories he had established to continuously draw support from the profits of Burroughs Wellcome and Company. Wellcome lived simply, mainly in hotels, and spent little of his money on personal luxuries. He was a retiring man who lived a somewhat lonely life. He could exhibit a stubborn self-confidence in his own judgment, which often proved better than the "best advice" he was given. He had one son from a marriage that was subsequently dissolved, but in his will he left almost all of his wealth to science and medicine. He established the Wellcome Trust Fund, which continues to this day to provide major support for medical research in Great Britain as well as funds for fellowships and education.

Henry Dale, who headed the Wellcome Trust Fund for a number of years after he retired from active research, wrote in a letter to the *Times* (August 1, 1936), after Wellcome's death, that Henry Wellcome told him that "he chose to spend his wealth in supporting research as another man might chose to spend his on a racing-stable." Dale continued by noting that "the whole of his power of supporting research, which was the real aim of his ambition, came to him from the continued and increasing success of his business. . . . He knew, of course, that every public association of his name with what was scientifically worthy of regard might indirectly enhance the reputation and prosperity of his business: but that would give him added power to support research, much of it in fields which had no obvious relation to his business interests."

Soon after Dale started his new position, Henry Wellcome mentioned to him "that when he could find the time without interfering with plans of his own, it would give him a special satisfaction if he could clear up the problem of ergot as the pharmacy, pharmacology, and therapeutics of that drug being then in a state of confusion." This was not a completely disinterested suggestion, for several pharmaceutical companies at the time were exploring the feasibility of using an extract from the ergot fungus in obstetrics. Ergot had been used by midwives for centuries but was not commercially available in a standardized dose. Wellcome had recently learned that Parke Davis, a rival firm, had used a bioassay to standardize an ergot preparation they were planning to market. It was difficult for Dale, who was just starting out in a new position, to ignore Wellcome's suggestion, no matter how mildly it was proposed, and he soon became involved in testing the properties of various compounds that were extracted from the ergot fungus.

History of the Ergot Fungus

Ergot, which is a fungus that grows on some grains, especially on rye, has a long history in folk medicine. An Assyrian tablet from 600 B.C. refers to a "noxious pustule in the ear of grain," and one of the sacred Pharsee books refers to "noxious grasses that cause pregnant women to drop the womb and die in childbed." Both of these references are thought to be to the ergot fungus.

The Greeks and Romans did not usually eat rye, and there are no written references to ergot in their literature. Although there do not

(continued)

seem to be any written accounts of the use of ergot for medicinal purposes during the Middle Ages in Europe, there is other evidence that
some of its properties were known earlier. Ergot poisoning was recognized, and there are reports of seizures caused by consuming ergot.
Epidemics caused by ergot poisoning were described, and these were
characterized by gangrene of the feet or hands so severe that a limb
might become dry and black as charcoal, sometimes falling off without
any blood loss. The burning sensation from the limbs was called St. Anthony's Fire, because of the belief that the shrine of that saint could
provide relief from the excruciating burning sensation and even a cure.
An epidemic in 1670 left the first recorded evidence that ergot poisoning was recognized as the cause of St. Anthony's Fire. More recent reported outbreaks of St. Anthony's Fire include one in Russia in 1926,
another in Ireland in 1929, and still another in France in 1953.

It is now known that ergot spores are either deposited by insects or
carried by the wind onto grain, usually rye, where they germinate and
gradually consume the grain. These spores eventually harden into a
purple curved body, called the sclerotium, that is the commercial
source of ergot.

Ergot was being used as an herb in obstetrics before it was identified as the cause of St. Anthony's Fire. Midwives had been aware of it
for centuries before any physicians adopted it. Ergot produces uterine
contractions, and there are clear records of its being used as early as
1600 to induce labor in pregnant women. There are also several later
reports of the use of ergot by physicians. A letter in a New York medical journal in 1818 described a physician's experience with ergot,
which was called *pulvis parturiens*: "It expedites lingering parturition
and saves to the accoucheur a considerable portion of time, without
producing any bad effects on the patient. . . . Previous to its exhibition
it is of the utmost consequence to ascertain the presentation [i.e., to
determine the orientation of the child in the uterus] . . . as the violent
and almost incessant action which it induces in the uterus precludes
the possibility of turning. . . . If the dose is large it will produce nausea
and vomiting. In most cases you will be surprised with the suddenness
of the operation; it is, therefore, necessary to be completely ready before you give the medicine. . . . Since I have adopted the use of this
powder I have seldom found a case that detained me more than three
hours."[9]

In the United States the use in obstetrics of some form of ergot preparation had been increasing. When in 1824 the incidence of still-born infants was noted to be on the rise in some eastern cities, the Medical Society of New York called for an inquiry. A report issued in 1824 found ergot responsible and commented sarcastically that ergot should be called *pulvis ad morte* rather than *pulvis ad partum*. The report recommended strongly that ergot be used only to control post-partum hemorrhage, not to hasten parturition. However, even when Henry Dale started to do research on ergot, the accoucheurs were using a manufactured product called "Liquid Extract of Ergot," a watery preparation given to induce labor.

As it turned out, the ergot fungus proved to be a treasure trove of active pharmacological substances, and their investigation set the direction of most of Dale's research. George Barger, a Wellcome chemist whom Dale had previously known at Cambridge, had extracted several substances from ergot, presumably also with the prompting of Henry Wellcome. These extracts needed to be tested for their biological action. Among the substances Barger had extracted were histamine (not yet identified by this name), acetylcholine, tyramine, and other amines that mimicked many of the effects of adrenaline and sympathetic nervous system stimulation. Histamine, tyramine, and acetylcholine are not normally found in fresh ergot, but rather are products of putrefaction. Tyramine, for example, is also found in spoiled meat, and the acetylcholine present in ergot was due to bacterial contamination caused by *Bacillus acetylcholini*, the same bacillus responsible for fermenting sauerkraut. The acetylcholine found in ergot is relatively stable, as ergot does not contain any cholinesterase, the enzyme that rapidly degrades or inactivates acetylcholine in the body. Although at the time none of these extracts were known to be natural constituents in animals, they were being studied because of their physiological effects. In a number of respects the history of histamine and acetylcholine study closely parallel each other in ways other than their derivation from ergot. They both were synthesized earlier and were later found to have similar effects in stimulating the uterus. Neither histamine or acetylcholine were known to be present in mammals until the 1920s.

Two additional extracts from ergot, ergotamine and ergotoxine, proved enormously valuable for testing the presence of adrenaline, for they were found to block the effects of adrenaline and sympathetic nerve stimulation on blood pressure. Dale later described, with some humor and modesty, how this property of ergotamine was discovered. When he had just begun working at the Wellcome Laboratories he was asked to perform some perfunctory tests on one of the ergot extracts supplied by George Barger. He felt unprepared for the task and later wrote, "Pharmacological research was for me a complete novelty, and I was, frankly, not at all attracted by the prospect of making my first excursion into it on the ergot morass."[10]

Dale began by testing ergotoxine, a substance closely related to ergotamine, on the arterial pressure of a cat. This test, he later wrote, was one of the few "within reach of my limited competence." As he was finishing this work, a sample of dried adrenal gland extract was sent up to him to test. He used the same cat to make the test and found that the adrenal extract did not produce the expected increase in blood pressure. At first Dale thought this odd result must have been due to his incompetence, but when he once again tested some adrenal extract on cats previously injected with ergotoxine he got the same result. It turned out that ergotoxine not only blocked but also reversed the effects of adrenaline and sympathetic nerve stimulation. Dale had inadvertently found the first adrenaline blocking agent, and this later became a basic pharmacological tool for testing for the presence of adrenaline.

Ultimately Dale did find a crude extract that caused the uterus to contract, but the substance was not identified and characterized until the early 1930s, when three different laboratories reported obtaining a purified ergot extract that when taken by mouth caused vigorous uterine contractions.[11] However, by this time Dale was no longer at the Wellcome Laboratories, and his interests had turned to other matters.

Henry Wellcome soon recognized Dale's keen judgment and administrative ability, and within eighteen months he made him director of the Wellcome Laboratories. This enabled Dale to obtain the help of several chemists, in addition to his friend George Barger, and he could call on a number of people to assist in studying the properties of the various ergot extracts. With the collaboration of Peter Laidlaw, the undergraduate student Dale had taught at Cambridge, Dale demonstrated that the extract, which was later identified as histamine, produced a marked decrease in blood pressure. At first this was a puzzle, for they had observed that this substance constricted blood vessels in the uterus. Vasoconstriction should have produced an increase, not a decrease, in blood pressure. The explanation came when Dale found that the extract produced a marked vasodilation and an increase in permeability of the capillaries of the uterus.[12] Dale concluded that even though the substance caused vasoconstriction in some blood vessels, the vasodilation of the capillaries had the greater effect and was therefore responsible for the overall lowering of blood pressure.

This early work on histamine prepared Dale to appreciate later the possibility that the lowering of blood pressure that occurred during shock might be due to the dilation of the capillaries and leakage of blood through the capillary walls. During the First World War, Dale did research on a histamine animal model of "secondary wound shock," a sometimes fatal drop in blood pressure experienced by soldiers with battle wounds. At that time, however, it was not known that histamine was a natural constituent in the body. Reflecting much later on how close he had come to discovering the role of histamine in shock, Dale wrote: "I have been quite shocked on reading anew one passage of our discussion to discover how we seemed to have gone almost out of our way even to preclude this possibility from consideration."[13]

Dale's research on histamine was a significant contribution in the eventual discovery of the importance of this substance in allergic reactions and shock, but it is his research on the pharmacology of other substances that is most relevant to the present story. Dale found several amine substances, such as tyramine, that could reproduce at least some of the effects of adrenaline and sympathetic nerve stimulation. With the help of George Barger, Dale found noradrenaline to be the most potent substance in mimicking sympathetic responses. Today noradrenaline (more commonly called norepinephrine) is recognized as the neurotransmitter secreted by most sympathetic nerves, but at the time it too was not known to be a natural substance in the body and Dale thought of it as only an interesting synthetic compound.[14] Reflecting back to this time, Dale acknowledged his general hesitancy to speculate when he commented: "I failed to jump to the truth, and I can hardly claim credit for having crawled so near and then stopped short of it."[15]

Dale was sent another ergot extract for routine testing around the year 1910. The substance was found to produce a profound inhibition of heart rate. In fact, the first time Dale used it, he thought he had killed the cat he was using as a subject, because he could not detect any heart rate or blood pressure. Although the substance was like muscarine in many of its effects, it was more potent and, unlike muscarine, which is quite stable, it was a labile ester that rapidly became inactive. Dale recalled that Reid Hunt and his assistant René Taveau had produced acetylcholine from choline and had reported in 1906 that its effect on circulation "is the most powerful substance known."[16] With the help of Arthur Ewins, a chemist in the laboratory, Dale was able to prove that this ergot extract was acetylcholine, and they found that it not only slowed heart rate, but also reproduced many other parasympathetic effects, such as increasing salivation and causing the esophagus, stomach, intestines, and bladder to contract.

In a letter to Thomas Elliott, Dale wrote:

> We got the thing [acetylcholine] out of our silly ergot extract. It is acetylcholine and a most interesting compound. It is much more active than muscarine, though so easily hydrolysed that its action, when it is injected into the blood stream, is remarkably evanescent, so that it can not be given over and over again with exactly similar effects, like adrenaline. Here is a good candidate related to the rest of the autonomic nervous system, I am perilously near wild theorizing.[17]

However, Dale did not do any "wild theorizing" and did not even hypothesize, at least in any publication, that acetylcholine might be secreted by parasympathetic nerves. He also ruled out the possibility that it would have any application as a therapeutic drug: " Acetylcholine occurs occasionally in ergot, but its instability renders it improbable that its occurrence has any therapeutic significance."[18]

In regard to any possible physiological significance of acetylcholine, Dale wrote:

> The question of possible physiological significance, in the resemblance between the action of choline esters and the effects of certain divisions of the autonomic nervous system, is one of great interest, but one for the discussion of which little evidence is available. Acetylcholine is, of all the substances examined, the one whose action is most suggestive in this direction.[19]

Although Dale had observed that the effectiveness of acetylcholine in reproducing parasympathetic responses surpasses adrenaline's ability to mimic sympathetic responses, he noted an important difference, that adrenaline is found in the body while acetylcholine is just a drug, although one with most interesting properties: "There is no known depot of choline derivatives, corresponding to the adrenine [adrenaline] depot in the adrenal medulla, nor indeed, any evidence that a substance resembling acetylcholine exists in the body at all."

Dale found that acetylcholine and muscarine had the same effect at a number of sites. There were other sites, however, where muscarine was not effective in mimicking acetylcholine, although at these sites low doses of nicotine did mimic acetylcholine.[20] When it was found that atropine blocked transmission at the muscarinic sites, but not at the nicotinic sites, Dale designated acetylcholine active sites as either "muscarinic" or "nicotinic," a terminology still used today.[21]

Dale also introduced other terminology that was widely adopted and is still in use. He had found that many ergot extracts could mimic either sympathetic or parasympathetic nerve stimulation, and he and George Barber introduced the terminology "sympathomimetic" and "parasympathomimetic" to describe any drug that mimicked the effects of these two divisions of the autonomic nervous system:

> We are dealing with a range of compounds which thus simulate the effects of sympathetic nerves not only with varying intensity but with varying precision. In some cases the points of chief interest in the action of a compound are those in which it differs from the action of adrenine. A term at once wider and more descriptive than "adrenine-like" seems needed to indicate the type of action common to these bases. We propose to call it "sympathomimetic," a term which indicates the relation of the action to innervation by the sympathetic system, without involving any theoretical preconception as to the meaning of that relation or the precise mechanism of the action.[22]

A similar argument was advanced to justify use of the term "parasympathomimetic."

In his characteristically cautious manner, Dale wrote:

> The fact that its [acetylcholine's] action surpasses even that of adrenine [adrenaline], both in intensity and evanescence, when considered in conjunction with the fact that each of these two bases reproduces those effects of involuntary nerves which are absent from the action of the other, so that the two actions are in many directions at once complimentary and antagonistic, gives plenty of scope for speculation.[23]

Dale here describes acetylcholine's action as "immediate" and "intense" but also "extraordinarily evanescent." He suggested that the short duration of acetylcholine's action and suggested that this was due to an enzyme, an esterase, in the body that rapidly broke it down into acetic acid and an inactive choline. Dale wrote:

> The possibility may indeed be admitted of acetylcholine, or some similar active and unstable ester, arising in the blood and being so rapidly hydrolyzed by the tissues that its detection is impossible by known methods. Such a suggestion would acquire interest if methods for its experimental verification could be devised.[24]

Apparently, Dale did consider at this point that there might be some natural substance in the body that resembles acetylcholine, but because it is broken down so rapidly he knew of no way to recover any of it. Dale would have to wait two decades before Wilhelm Feldberg would bring the leech muscle technique to his laboratory, which would make it possible to detect the small amounts of acetylcholine secreted by nerves. This advance is described in chapter 6, along with the events that led to the awarding of the Nobel Prize for his proving the existence of neurohumoral secretions at different peripheral synapses.

From the perspective of what we now know, it seems remarkable that Dale did not speculate further about the possibility that parasympathetic nerves secrete acetylcholine. Several factors explain why Dale did not make this connection. As Dale later wrote, they had at the time "no evidence at all that acetylcholine was a constituent of any part of the animal body."[25] Much later, when he was reflecting on his failure in 1914 to make the theoretical leap, he wrote that "by this time, both Elliott and I seem to have become shy of any allusion to the 'chemical transmission' theory, which had originated ten years earlier."[26] It does not, however, seem to have been a question of "shyness," but rather that Dale was simply not thinking at the time about the possibility that autonomic nerves secreted chemical substances.

Dale was by temperament reluctant to speculate much beyond what the evidence in hand could firmly support. He was cautious and perhaps prone to give too much weight to exceptions that might make any generalization risky. Dale was later described by Lord (Edgar) Adrian as more likely "to apply the brake than to be the first in the gold rush, but the gold he has found will keep its value."[27]

In 1914 Dale was elected to the Royal Society in recognition of his many contributions to pharmacology and physiology, and he was also appointed to the National Research Committee (now National Research Council). Dale was only twenty-nine when he started working at the Wellcome Lab-

Figure 4.1   Henry Hallett Dale in his thirties, and in his early sixties.

oratories, and the support he received for his research was unusual for a person of that age. He was able to take good advantage of this support and was enormously productive. There were, however, aspects of the position at the Wellcome Laboratories that became a source of irritation for him.

Dale had come to expect that from time to time he would be asked to assess the biological properties of some new compounds even though they were not relevant to his own interests, but such requests were becoming more frequent and burdensome. The director of Burroughs Wellcome's Chemical Research Laboratories had been sending over large numbers of compounds for routine screening. Determining whether a compound had any interesting physiological properties could be quite time-consuming, but when Dale objected, the director informed Henry Wellcome that his work was being held back by Dale's failure to cooperate. Although Dale was able to use the conflict to gain additional staff, the incident was an unpleasant reminder that despite Henry Wellcome's assurances that he would be completely free to pursue his own scientific interests, his time and laboratory resources were increasingly being drawn upon to support the commercial interests of the company.

Around this time, another difficult incident took place, concerning the appropriate terminology to be used in referring to the adrenal gland extract. Dale generally called the substance adrenalin and sometime adrenaline, as did most pharmacologists and physiologists. But after Parke, Davis & Company in Detroit patented a purified crystal under the proprietary name "Adrenaline," some executives at the head office told Dale, who was submitting a paper for publication, that he should not use a patented name belonging to a rival company. Dale resented the interference and clearly said so. He had always used the term and insisted on his right to continue using it. The executives were close to Henry Wellcome, and Dale was not sure how the dispute would be resolved. In the end, Wellcome supported Dale, but the experience was unpleasant, and Dale began to consider offers of other positions.

Several possibilities presented themselves, but he was most attracted by the offer of the directorship of the biochemistry department of the newly formed Institute of Medical Research. He did not want to leave the Wellcome Laboratories, however, until he clarified what support he could count on in the position, and he also wanted to know who would be appointed director of the new institute. Moreover, because Dale did not consider himself a biochemist, he wanted to be able to create a position for a chemist skilled in synthesizing new compounds. Dale's strength was his ability to develop techniques for determining the physiological effects of various compounds on various organs and then draw inferences about the normal physiology of these organs. He did not have the chemist's ability to extract and synthesize potentially interesting compounds. Dale, who always seemed to know what and whom he needed to move his research to the next level, insisted that George Barger be offered a position in his new laboratory. After that was settled, Dale accepted the position at the Institute for Medical Research in Hampstead, a region in north London.

World War I broke out one month after Dale started at the institute, and he, like most scientists in Great Britain, now devoted his time to the war effort. Dale was well prepared for the task by virtue of his early studies on what turned out to be histamine, and as already noted, he did research during the war on the causes and treatment of "secondary wound shock." This work, which will be discussed more fully in chapter 7, brought him in contact with others working on shock, including the American physiologist Walter Cannon.

During the war Dale also served on a committee working on ways to standardize the packaging and labeling of drugs. Drug standardization had become an acute problem with the outbreak of the war, because Germany had been the largest supplier of the drugs used in Great Britain and this source was now cut off. He helped to establish international standardization of drugs by getting agreement on the tests for specifying the quality and potency of drugs produced in different countries. Later, in 1925, he chaired a conference on this subject in Geneva for the Health Organization of the League of Nations.

When the war ended in 1918, Dale, who was recognized as one of the leading pharmacologists in the world, began to be pursued by institutes in the United States. He was invited to give the Herter Lectures at Johns Hopkins in 1919. At the time this was considered to be the most prestigious lectureship in medical science in the United States. After his lecture, Dale was offered a university chair and a position as head of the renowned department of physiology at Johns Hopkins. During the same trip to the United States, Dale visited New York as a guest of Simon Flexner, director of the Rockefeller Institute. Flexner had started to court Dale in London, and he offered him a position as head of the projected department of pharmacology at the Rockefeller Institute. Dale declined this offer as well as the position at Johns Hopkins. The facilities and support at the Rockefeller Institute would have far surpassed anything that Dale had at the Institute of Medical Research in Hampstead, but he turned the position down because he felt, "that when my own country was just emerging from the strain and impoverishment, inflicted by the whole four years of the war, it was not time to leave untried the prospect and opportunity which I had accepted there, in favor of the one offered in the U.S.A."[28]

Dale's laboratory at the Institute for Medical Research was always referred to as F4, being the fourth laboratory on the first floor of the Institute. It was a large communal laboratory with storage cabinets packed with equipment from floor to ceiling. On one end was the equipment for smoking the paper used to record responses on kymograph drums as well as vats of the varnish used to preserve the kymographic records worth saving.[29] Located close to a chemical balance was a note that read "Near enough is not good enough." The note had been placed there by the head technician, who had been with Dale for many years and no doubt knew his commitment to precision.

Dale created a friendly, open laboratory where all felt free to question any one else's ideas, and interactions produced collaboration, not competition. Everyone worked close to each another, so that if one's experiment wasn't working or there was a necessary waiting period, it was natural to pay attention to how your neighbor's experiment was progressing. This physical arrangement facilitated collaborative research as did the lunch and tea rooms shared by all working in the institute.

By 1920, Dale had contributed enormously to our knowledge of the pharmacological and physiological properties of many different compounds that mimicked sympathetic or parasympathetic effects and the various drugs that enhanced or inhibited these effects. He was reluctant, however, to speculate, at least in print, on the possibility that any of these compounds were secreted by autonomic nerves. He had, however, laid the foundation for others, particularly for Otto Loewi, who was willing to speculate about the significance of the experiment that had first occurred to him in a dream.

CHAPTER 5

Otto Loewi

An Inspired Dream and a Speculative Leap

The background of a man's achievement is usually the most interesting part of his life story. An advantage of autobiography may be that the author reports more competently than others on his inner experience and its effects; yet, even so, fiction may come into play because a retrospective report may not always truly reflect the past as it happened.

—Otto Loewi (1960)[1]

In fact, pharmacology started from physiology, used its methods and has as its main goal the revealing of physiological function from the reaction of living matter to chemical agents. The very conception to use drugs as tools in the study of function oriented me to pharmacology.

—Otto Loewi (1954)[2]

By 1914 Henry Dale had established that acetylcholine was the most potent substance known capable of mimicking parasympathetic effects. Moreover, he had also shown that noradrenaline (norepinephrine) was much more potent than adrenaline (epinephrine) in reproducing sympathetic effects. He had the two most important pieces of the puzzle in hand, but he did not risk speculating about what the whole picture might look like and he never hypothesized that these two substances might be secreted by nerves. It is true that it was not known that acetylcholine and norepinephrine were natural substances, but it is also true that Dale, by his own later admission, was especially cautious about speculating and theorizing. In later life he seems to have considered John Langley's aversion to theorizing a shortcoming, but whether it was because of the example Langley set or because of his own personality, Dale seems to have been similarly inclined. Of course, setting high standards for the evidence necessary to support a theory is not a flaw. Theorizing in science can be either inspired insight or a seductive trap that can cause one to waste years chasing shadows down blind alleys. The problem is that it is usually impossible to know in advance which outcome it will be.

It is also true that Dale's research was interrupted at a critical point by the outbreak of World War I. From 1914 to 1918 Dale was committed to doing research related to the war effort. After the war ended in 1918, Dale continued his wartime studies of histamine and anaphylactic shock, but there was a fifteen-year interval before he once again did research on acetylcholine. Probably the most important reason Dale did not put together all the evidence he had discovered is that the possibility that nerves secrete chemical substances was not an issue at the time. The evidence might have been right in front of him, but he did not see it because he was not thinking about it. This situation would be changed by Otto Loewi, a man with a very different temperament.

Otto Loewi (1873–1961) was born in Frankfurt into a Jewish family of successful wine merchants. In his youth he spent summers in the Haardt Mountains in southwestern Germany, where his father owned an old manor house surrounded by a large garden and vineyards. Loewi had a classical German gymnasium education with nine years of Latin and six years of Greek. He read original texts in those languages and became well schooled in classical antiquity. Although he performed well in the humanities, by his own admission Loewi did poorly in the sciences, especially in physics and mathematics.

In his youth Loewi developed what would become a lifelong enthusiasm for art. Trips to Belgium had stimulated his interest in the early Flemish painters, and he planned to study art history at the university. His father, however, wanted him to do something more practical and persuaded his son to study medicine. Loewi agreed and started medical school in Strassbourg. He was often bored, however, and would skip classes to hear lectures on the arts. As a result, he barely passed the third-year examination and decided to take a year off in Munich, ostensibly to prepare for the final year of medicine. Munich had a lot to tempt him, and Loewi could not resist spending considerable time at the opera, art galleries, and museums.[3]

Loewi must have done some work in Munich, because he successfully completed his final year in medicine after returning to Strassburg. He did the required thesis in pharmacology under Oswald Schmiedeberg, a renowned pharmacologist. The thesis involved studying the effects of arsenic, phosphorus, and other substances on the isolated heart of the frog. The topic had attracted him, in part, because of his reading of Walter Gaskell's 1882 Croonian Lecture on the vagus nerve's innervation of the frog's heart. As noted earlier, the isolated heart of a frog had for some time been commonly used by physiologists and pharmacologists to study the effects of drugs on heart rate, and this preparation would be used by Loewi in the experiment that eventually proved that nerves secrete chemical substances.

One of Loewi's friends in Strassburg was Walther Straub. Later their stays in Marburg overlapped.[4] Straub had developed what became known as the "Straub Cannula" for collecting fluid from the isolated heart of a frog. The technique involved removing a frog's beating heart and tying off all the blood vessels except the aorta. A glass cannula was inserted through the aorta into the ventricle of the heart and then filled with saline. The cardiac nerve in the frog contains both the vagus and the so-called accelerator (or sympathetic) nerve, and this was left attached when the heart was removed from the body. Changes in heart rate were recorded on a kymograph drum before and after the nerve was stimulated or a drug was introduced into the cannula. Loewi used this technique in a number of experiments prior to using it in the critical 1921 experiment, which initiated his investigation of neurohumoral secretions and led to the awarding of the Nobel Prize he would share with Henry Dale.

After completing his medical education in 1896, Loewi acquired additional training in biochemistry during the year following graduation from medical school. He then spent a frustrating clinical year as an assistant doctor in the city hospital in Frankfurt. At the time, there was a high mortality rate among patients with tuberculosis and pneumonia, and Loewi was frustrated by the lack of any effective treatment. He decided not to practice medicine and began to consider a research career. He was familiar with the research of Claude Bernard and others, who had used curare to infer the existence of a specialized neuromuscular site, and he was attracted by the possibility of using drugs to uncover physiological principles.[5]

Although he had done little research other than what was required for his medical degree, Loewi managed to convince Hans Horst Meyer, an eminent German pharmacologist, to take him on, and he started work in Meyer's laboratory in Marburg in 1898.[6] Loewi spent much of the next six years studying glucose metabolism, nutrition, and diabetes, but he also did some research on kidney and heart function. Loewi's most notable work during this period involved obtaining evidence that animals could synthesize proteins from amino acids. Prior to that time, it was generally believed that the diet of animals had to contain whole proteins in order for them to maintain a nitrogen equilibrium. He had read a paper published by a colleague who had developed a simple procedure for decomposing proteins into their degradation products.[7] This gave Loewi the idea of trying to maintain animals on a diet made up from these degradation products. Although the animals did not like the diet, he found a way to maintain them in good health and thereby proved that animals are able to synthesize the proteins they need to sustain life.[8]

Loewi wished to widen his knowledge of physiological techniques, and with Meyer's help it was arranged for him to spend several months in 1902 visiting physiologists in England, mostly in Ernest Starling's laboratory in University College London. Loewi found Starling's enthusiasm contagious and was very much attracted by his leadership style and personality, which combined modesty with scientific erudition. This style and personality contrasted favorably with the behavior of some of the more authoritarian German professors Loewi had known. Of this experience, Loewi later wrote in his autobiographical sketch:

> When I first saw Starling's laboratory, I was surprised by the striking contrast between its extremely limited space and primitive equipment and the high level of the work that had come out of it. Like everybody else, I was charmed at first sight by Starling's appearance, his expressive features, his shining eyes. I soon recognized that he was the most dynamic and contagiously enthusiastic man I had ever met, full of ideas and optimism, as well as very critical of other people's and his own achievements. He possessed a serenity and simplicity, humility and a kind of naïveté that is characteristic of so many men of genius. He made the atmosphere of his laboratory warm, informal, and stimulating.[9]

Loewi was equally impressed by William Bayliss, Starling's brother-in-law and frequent collaborator. It was an exciting time, as Starling and Bayliss had just discovered the hormone secretin and how it is released into the circulation from the duodenum. The work with secretin had led Starling and Bayliss to introduce the term "hormone," which they defined as a substance secreted into the bloodstream and distributed to other parts of the body, where it often exerted profound effects.

It was during this visit that Loewi first met Dale, who, as was noted in the previous chapter, was working in Starling's laboratory. Dale found

Loewi to be a charming, witty man and an engaging conversationalist, and their lifelong friendship began at this time. Loewi had wide interests, and he made many friends during this brief visit. One of the goals Loewi set for the visit was to improve his English. However, Dale later reported that he was impatient with being corrected, saying: "I have not the time to learn English correctly, I just wish to speak it fast."[10]

The following year Loewi visited England again. Humorous reports of his improvised English had spread around the physiology community. When Loewi heard that Sir John Burdon Sanderson, Regius Professor of Medicine at Cambridge, wanted to meet him, he felt honored. Later, when he found out that what Sanderson really had wanted was to experience a sample of his English, he took it with good humor. He enjoyed and was immediately comfortable with the informality of English physiologists and especially their uninhibited style in exchanging ideas with students as well as colleagues. This was quite different from the tradition in Germany, where the "Herr Doktor Professor" was rarely challenged.

On this second trip to England, Loewi spent time with John Langley and Thomas Elliott in Cambridge. Dale later wrote that the possibility of the existence of neurohumoral secretions may have been planted in Loewi's mind during this meeting with Elliott. There is no record of their conversation, but years later Dale described a conversation he had had with Loewi at that time:

> Loewi was in Langley's department at Cambridge; and I remember that when Loewi came back to London, before he returned to Marburg, he and I had dinner together one evening and he told me how much he enjoyed the discussion with Elliott and what a high opinion he had formed of his promise. The two had met and interchanged ideas; and the idea of chemical transmission from nerve endings might surely have had some mention between them, though we cannot expect either of them to have any memory now of such talk about it, a half century later. What we do know is that Loewi, later in the same year, threw off the suggestion that muscarine might be liberated to transmit vagus effects.[11]

Dale was referring to the fact that, in 1904, a year after Loewi and Elliott met in Cambridge, Loewi had remarked to Walter Fletcher, a colleague of Dale, that when the vagus nerve inhibited heart rate a chemical like muscarine might be involved.[12] It is impossible to know whether Loewi's comment had been influenced by Elliott's work. It could have been an independent thought as Loewi was well aware that Oswald Schmiedeberg, his thesis mentor, had demonstrated more than thirty years earlier that muscarine had the same effect on the heart as vagus nerve stimulation.[13] Loewi's remark, however, was only a casual observation that he did not pursue and apparently completely forgot about until he was reminded of it in 1929.

In 1905, when Hans Meyer accepted a position as professor of pharmacology at the University of Vienna, Loewi followed him. Loewi and Meyer continued to do collaborative research and co-authored several articles on the relative potency of a number of synthetic amines, especially "arterrenol," which is another name for noradrenaline (norepinephrine). At the time Loewi, like Dale, thought that noradrenaline was just an interesting drug, not a natural substance.

In Vienna, Loewi also collaborated on several studies with Alfred Froelich, who was already well known for the syndrome that bears his name. The Froelich Syndrome is an endocrine disorder characterized by delayed puberty, undersized testes, and obesity. Loewi and Froelich studied the effects of various drugs on autonomic nerves. In one important study they demonstrated that prior treatment with cocaine increases the response to adrenaline, a phenomenon now known as "cross-sensitization." This cocaine-enhanced response to adrenaline is quite specific and, as will be explained in a later chapter, could be incorporated into a technique for detecting the presence of small amounts of adrenaline. Loewi also worked with the drug eserine (physostigmine), which enhances the response to parasympathetic nerve stimulation.[14] These studies would later be useful in proving that nerves secrete chemical substances, but at the time this was not something that Loewi was thinking about.

In 1908 Loewi married Guida Goldschmiedt, the daughter of a professor of chemistry in Prague, and the following year they moved to Graz in Austria, then part of the Austro-Hungarian Empire, where Loewi had been appointed professor and head of pharmacology. Graz is an old and picturesque city built on a hill. Although it was not as cosmopolitan as Vienna, Graz's university and medical school were highly respected and the city had a good opera company and excellent theater offerings. Loewi was happy there. He had a large laboratory and excellent students and collaborators. He and his wife lived in Graz for thirty years, and three sons and a daughter were born to them there. In 1938, however, as will be described in a later chapter, he had to leave Austria after the Germans annexed that country.

Figure 5.1 The young Otto Loewi, whose passion was art (left); and as an eminent fifty-six-year-old pharmacologist visiting Boston in 1929 (right). Courtesy of the Wellcome Trust.

Loewi was a lively lecturer, popular with students and highly respected by the medical faculty, which elected him dean in 1912. He continued to work on carbohydrate metabolism and experimental diabetes, but he also did research on the action of digitalis on the heart and published several papers on how calcium and other drugs affect the vagus nerve's capacity to slow heart rate. This work further prepared Loewi to pursue his seminal experiment, which eventually proved that the vagus nerve secretes acetylcholine. The idea for this experiment occurred to him in a dream at a time when he was not consciously thinking about neurohumoral secretions. Later in his life Loewi remarked that he had occasionally thought about the possibility of neural secretions, but because he saw no way of attacking the problem: "it entirely slipped my conscious memory until it emerged again in 1920." Loewi described the experience of his dream on many different occasions, always in a similar manner:

The night before Easter Sunday of that year I awoke, turned on the light, and jotted down a few notes on a tiny slip of thin paper. Then

I fell asleep again. It occurred to me at six o'clock in the morning that during the night I had written down something most important, but I was unable to decipher the scrawl. The next night, at three o'clock, the idea returned. It was the design of an experiment to determine whether or not the hypothesis of chemical transmission that I had uttered seventeen years ago was correct. I got up immediately, went to the laboratory, and performed a simple experiment on a frog heart according to the nocturnal design.[15]

Loewi's version of these events seems to be somewhat more dramatic than what actually occurred. Dale, who remarked later that he was one of the first to hear Loewi's account of "the remarkable story of the dream," recalled that Loewi had originally told him that when he awoke on the second night at 3:00 A.M. he made careful notes so that he would have no trouble deciphering his thoughts the next morning.[16] Loewi's more dramatic later version is that he went directly to the laboratory at 3:00 o'clock in the morning and that by "five o'clock the chemical transmission of the nervous impulse was conclusively proved."[17]

Moreover, Loewi's recollection that the dream occurred "on the night before Easter Sunday" does not seem to be accurate. He performed the experiment in late February, perhaps extending into early March, and he sent the manuscript describing the experiment to the journal later in March. The article is marked as having been received in the editorial office of *Pflügers Archiv* on March 20, 1921. However, in 1921 Easter Sunday was on March 27, a week after the manuscript was received by the editor.[18]

While these discrepancies do not subtract from the importance of the experiment, they would seem to reflect the artistic bent of Loewi's temperament, which might have made it difficult for him to resist making the description of what transpired even more dramatic than it actually was. Dale, in contrast, would probably never have strayed from the facts, no matter how inconsequential. It may also be true that Loewi's different temperament played a role in allowing him to risk drawing a conclusion from quite preliminary and inconclusive experimental results.

The experiment inspired by the dream was also described by Loewi on many occasions:

The hearts of two frogs were isolated, the first with its nerves, the second without. Both hearts were attached to Straub cannulas filled with a little Ringer solution. The vagus nerve of the first heart was stimulated for a few minutes. Then the Ringer solution that had been in the first heart during the stimulation of the vagus was transferred to the second heart. It slowed and its beats diminished just as if its vagus had been stimulated. Similarly, when the accelerator nerve was stimulated and the Ringer from this period transferred, the second heart speeded up and its beats increased.[19]

Loewi reported that he used the hearts from fourteen frogs of two different species (*Rana esculenta* and *R. temporaria*) and from four toads.[20] As already noted, in these amphibians the heart is innervated by a mixed nerve; the vagal-sympathetic trunk contains the sympathetic nerve as well as the vagus, a parasympathetic nerve. For that reason, electrical stimulation of the nerve trunk sometimes produced an acceleration and sometimes a slowing of heart rate, and it was not possible to predict which would occur.[21] Loewi initially gave the name "Vagusstoff" to the humoral substance that slowed heart rate, and he called the substance responsible for accelerating heart rate "Acceleransstoff."

It was certainly an exaggeration to state, as Loewi did many times, that the initial experiment had "immediately and conclusively" proven "that the nerves do not influence the heart directly, but liberate from their terminals specific chemical substances." In fact there was a protracted and sometimes acrimonious dispute over Loewi's data and what, if anything, his experiment had proved. It took almost a decade before Loewi's conclusions were widely accepted.

Many physiologists were simply unwilling to believe that nerves produce their effects by secreting chemical substances, let alone that different nerves secrete different substances. Others reported that they were unable to replicate Loewi's results.[22] There are many reasons why the initial attempts to replicate Loewi's results may have failed. First, as it was later learned, the "Vagusstoff" turned out to be acetylcholine, a neurotransmitter that is, as Dale hypothesized in 1914, rapidly degraded by an enzyme now known to be cholinesterase. Loewi was most fortunate in using frogs and toads in his experiment and particularly in working with these cold-blooded animals during February and early March. In the colder months, acetylcholine is said to be more stable because there is less cholinesterase available. Had Loewi's initial experiments been done during the warmer months, it is unlikely that the experiment would have demonstrated any effect at all. This factor may have been partially responsible for the failure of others to replicate his results. It is also now known that frog species differ in the amount of cholinesterase available, and this may have contributed to the failure to confirm Loewi's initial report. Loewi also pointed out in 1924 that successful results required the selection of hearts that are maximally responsive.[23]

Moreover, Loewi pipetted the fluid from the stimulated heart and then often delivered it back to the same heart, although he did use a second heart in some tests. Thus, in much of his initial work, the same heart served as both donor and recipient.[24] This meant that Loewi had to withdraw some fluid from the ventricle of a heart that had been stimulated and wait several minutes until the heart rate returned to normal base levels. There are good reasons for questioning whether the little acetylcholine released could have remained active over this period of time. Even Loewi admitted later that it was remarkable that the initial experiment worked:

If I had carefully considered in the daytime I would undoubtedly have rejected the kind of experiment I performed. It would have seemed likely that any transmitting agent released by the nervous impulse would be in amount just sufficient to influence the effector organ. It would seem improbable that an excess that could be detected would escape into the fluid which filled the heart.[25]

Another problem that may have been responsible for a failure to replicate Loewi's results was the physical disturbance produced when fluid is pipetted back into a heart. It was found that even the small changes in hydrostatic pressure introduced by dripping fluid into the cannula could produce changes in heart rate, even if there is no active substance in the solution. The recipient heart sometimes slowed down when the perfusate came from a heart beating at a normal rate or when the vagus nerve had not been stimulated.[26]

Considering all that could have gone wrong, Loewi was clearly lucky that his initial experimental procedure produced any interpretable result. Actually, the data presented in the first publication was not at all convincing, as it consisted solely of a few not very persuasive kymographic tracings of changes in heart rate. Moreover, different results were obtained with the two frog species and the results with the toads were also different.[27] A more cautious scientist would probably have collected much more additional evidence, and ruled out alternative explanations, before submitting a paper proposing such a bold and controversial hypothesis.

Loewi did include the phrase "First Communication" in the title of the article, indicating that more information was being collected. Indeed,

Loewi soon published additional papers, each with more evidence on the same effect. By 1926 Loewi had published eleven communications in a numbered series all bearing the same introductory title: "Concerning a humoral transfer of a heart nerve effect" (in German).[28] He also wrote another six articles related to different aspects of the same effect during this period. It was as though the criticism he received as well as the recognition of the potential importance of the work had provided an energy and focus to Loewi's work that exceeded anything evident before this.

There were so many criticisms of Loewi's claims that in 1926 he was asked to respond to his critics by demonstrating the experiment at the twelfth International Physiology Congress held in Stockholm. By this time Loewi had apparently removed many of the sources of difficulty in replicating his results, but he was nevertheless somewhat hesitant about accepting the invitations, remarking that, "Like most experimenters, I had experienced time and again that experiments before a large audience often fail although they never did in the rehearsals."[29]

Nevertheless, Loewi was successful in demonstrating his results eighteen times during the course of the Stockholm meeting. As Loewi later told William Van der Kloot, his colleague in pharmacology at New York University, he "had been obliged to stand at one end of the room and simply give instructions, so that the possibility of his secreting some chemical under his fingernails and dropping it on the preparation could be eliminated."[30] One of the changes in his procedure that made it easier to replicate his initial results was the addition of physostigmine (eserine) to the solution to prevent inactivation of the "Vagusstoff" (acetylcholine).[31] Ulf von Euler, who was at the congress in Stockholm, later wrote that Loewi's demonstration was what first aroused his interest in neurohumoral transmission.[32] Later, von Euler, who spent part of 1930 and 1931 in Dale's laboratory, proved that the sympathetic neurotransmitter is noradrenaline (norepinephrine), not adrenaline (epinephrine).[33]

Although R. H. Kahn, a physiologist in Prague, reported in 1926 that he had confirmed Loewi's results using an improved technique, there continued to be a number of reports of failure to replicate the effect.[34] Loewi's successful demonstrations in Stockholm had by no means convinced all his critics. In 1927, for example, Louis Lapicque, a renowned Parisian neurophysiologist, declared that the idea of humoral transmission of nerve impulses was "unthinkable."

In 1932 W. A. Bain in Edinburgh modified the apparatus used by Kahn and was able to obtain even more robust evidence supporting Loewi's con-

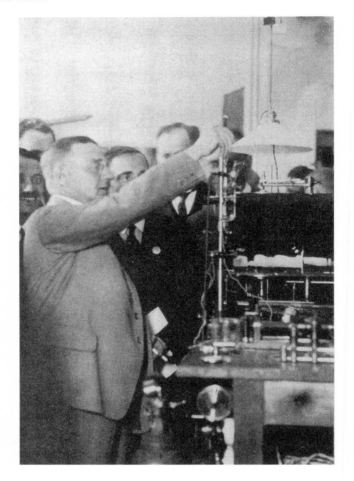

Figure 5.2   Otto Loewi demonstrating neurohumoral transmission in 1926 at the International Physiology Congress in Stockholm. Picture reproduced (without credits) in B. Holmstedt and G. Liljestrand, Readings in Pharmacology (New York: Macmillan, 1963), p. 194.

clusion about neurohumoral secretions. In the introduction to his paper, Bain described the shortcoming in Loewi's method that might have accounted for the difficulty in replicating his results:

> Many who have attempted to repeat Loewi's fundamental work on the frog-heart have found it difficult to get confirmative results, and some have failed to get any results at all. . . . Others again, while ob-

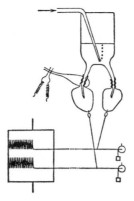

Figure 5.3 Apparatus used by R. H. Kahn. This was an improvement over the technique used by Otto Loewi and produced much more reliable evidence of neurohumoral transmission.

taining a percentage of positive results, are unable for various reasons, to accept Loewi's conclusions. . . . Loewi, in his experiment, collected the fluid from a vagus-stimulated heart and applied this to the same or to another heart. He considered that it was important in such experiments to apply the smallest possible quantity of fluid in the nerve-stimulated heart, his idea being that the smaller the quantity of fluid in contact with the heart the more concentrated would be the resulting neuromimetic fluid, and thus the more definite the effects on the heart to which it is applied.

Nevertheless it would appear that in Loewi's method a considerable amount of the vagus substance formed during stimulation must be destroyed before it can be applied to another heart. . . . From this consideration it appeared possible that better results might be obtained by a method in which the irrigating fluid was passed somewhat rapidly through the heart in the hope that the "vagus substance," almost as soon as it is formed, would pass into the irrigating fluid and thus away from at least one of the factors operating for its destruction.[35]

Loewi never published a diagram of the technique he used, and in most later accounts of his work the improved technique introduced by Bain (see fig. 5-3) is reproduced, with the implication that it is the apparatus Loewi had originally used.

The same year that Bain replicated Loewi findings with his improved technique, Leon Asher of Basel wrote that one had to be delusional to believe that Loewi's experiment proved anything:

> Only the failure of the necessary experimental critique and a little "autistic" thinking by definition of Bleuler could allow one to consider experiments which in the present situation do not prove anything, much less than absolutely proving the existence of a vagus hormone.[36]

Later that year Asher was finally convinced that nerves do indeed secrete inhibitory and excitatory substances. Although he reluctantly agreed that Loewi was right, Asher felt compelled to add that "it was just a piece of luck," as Loewi's experiment had not proven anything. Much earlier, in 1917, Asher had tried an experiment somewhat similar to that of Loewi's, but it failed to prove anything and he may have been reluctant to acknowledge Loewi's success where he had failed.

Early on Loewi had recognized that there were two possible origins of the humoral substances responsible for slowing and speeding up heart rate:

> On the one hand, they may originate directly from the effect of nerve stimulation independent of the type of cardiac activity. . . . From a different viewpoint, there is also the possibility that these substances are only products of the specific type of cardiac activity which is released by the nerve impulse; under such circumstances, therefore, the identification of their action with the nerve stimulus would be only accidental, so to speak.[37]

Loewi was referring to the persistent problem of trying to determine whether the humoral substance detected originated in the innervated organ (in this instance the heart muscle) or from the nerve.

In his second publication in the series on neurohumoral substances, also published in 1921, Loewi attempted to rule out the innervated heart as the source of the chemical substance. When he paralyzed the donor heart with a high dose of nicotine, for example, he found that it was still possible to transfer the "Vagusstoff" after the vagus nerve was stimulated. Although the evidence did not completely rule out the possibility that the "Vagusstoff" had come from the heart, Loewi was apparently satisfied that he had shown that it was, at least, not dependent on any change in heart rate.

In the same second publication in the series, Loewi began the process of identifying the chemical nature of "Vagusstoff."[38] He was well aware that Henry Dale's 1914 publications on acetylcholine had provided a good basis for suspecting that it was the most likely candidate for being "Vagusstoff," and he began to accumulate evidence that this was the case. He showed, for example, that of all the known substances that could mimic the action of the vagus nerve—muscarine, pilocarpine, choline, and acetylcholine—only acetylcholine satisfied all the pharmacological tests.[39] Choline had been a possibility until he demonstrated that not enough was present in the "Vagusstoff" to produce the effect observed. Moreover, he noted that choline, even in high concentrations, has only a relatively weak action on the heart and hardly ever arrests it in diastole.

In regard to the site of action of "Vagusstoff" and "Acceleranstoff," Loewi raised the long-standing question of whether the action of acetylcholine is on the heart muscle or back on the nerve that secreted it. He answered this question, much as Langley had done many years earlier in another context, by demonstrating that "Vagusstoff" had the same effect even after he had severed the vagus nerve and it had degenerated. Loewi concluded that neither vagal nor sympathetic substances act upon the nerve, but directly on the effector organ, in other words, the heart.

As already noted, Loewi had, with his colleague Emil Navratil, demonstrated that physostigmine (eserine) increases the response to "Vagusstoff."[40] It had previously been shown that this drug potentiates the slowing of the heart caused by vagal stimulation. Since it was known that eserine (physostigmine) inhibits the esterase enzymes that inactivate acetylcholine, Loewi and Navratil postulated, as had Dale much earlier, that the heart probably contains an endogenous esterase, which they called "cholinesterase."[41] In the eleventh paper in their series, Loewi and Navratil wrote:

> We proposed in our tenth paper that the administration of vagal substance or acetylcholine produces only a very short-lasting effect on the heart since both were speedily metabolized. We therefore investigated whether the long duration of action of vagal substance and acetylcholine, which is seen when physostigmine or egotamine are given beforehand, was due to an inhibition of their metabolism. These experiments showed that the metabolism of vagal substance and acetylcholine by heart extracts, was indeed inhibited by physostigmine and

ergotamine. This is further evidence for the analogous behavior of the two substances [i.e., "Vagusstoff" and acetylcholine].[42]

It was presumed that some enzyme like cholinesterase must be responsible for metabolizing acetylcholine, and in 1930 Loewi reported finding traces of cholinesterase in the heart.

While eserine (physostigmine) enhanced the slowing effect following stimulation of the vagus nerve, atropine blocked this response. The atropine and eserine were specific in their effects, as they had no influence on the acceleration of heart rate produced by stimulating sympathetic nerve fibers. Loewi and his colleagues found, for example, that eserine did not enhance the heart-rate response to muscarine or any of the other candidates being considered, except for choline, which had already been ruled on other grounds. All of these tests had provided strong support for the conclusion that acetylcholine was mediating the slowing of heart rate produced by vagal stimulation.

Loewi also demonstrated that the drug ergotoxine, an alkaloid substance that Dale had earlier shown would block and even reverse the effect of adrenaline, prevented the acceleration of heart produced by sympathetic stimulation.[43] Even though the results with ergotoxine suggested that the Acceleransstuff is adrenaline, Loewi hesitated to commit himself. He did conclude that some amine substance closely related to adrenaline, if not adrenaline itself, was probably involved.[44]

Loewi had taken a risk in initially claiming that he had demonstrated the existence of neurohumoral secretions based on what was initially a rather weak experiment, which lacked many controls that were needed to make it convincing. However, after being challenged, he focused all work in his laboratory on collecting supporting evidence and eventually proved that he was right, but it took about a decade before he had convinced the hardened skeptics.

Although neurophysiologists eventually conceded that Loewi was correct in his conclusion that neurohumoral secretions regulated heart rate, many regarded this as a special case that did not apply to other peripheral nerves, let alone to the central nervous system. Through 1933, even Loewi was reluctant to extend the idea of neurohumoral transmission, probably regarding it, as most neurophysiologists did, as too slow a process for transmission at most synapses. Loewi was particularly skeptical of the possibility of neurohumoral transmission at either the autonomic ganglion or at the synapse between spinal motor nerves and skeletal muscles. He expressed this reservation in his Harvey Lecture in 1933, when he stated that: "I personally do not believe in a humoral mechanism in the case of striated muscle."[45]

This was at the very time Dale and his colleagues had started to collect evidence that acetylcholine might be involved in innervating skeletal muscles. Dale once described Loewi's reluctance to accept the possibility that acetylcholine is secreted by spinal motor nerves as "an attitude of almost obstinate skepticism." Dale wrote that Loewi seemed almost frightened by the idea:

> Loewi seems to have taken alarm, for the time being, at the thought of such a possible extension of his discovery, and to have gone to the length of establishing his own alibi by a public disclaimer of belief in chemical transmission at the motor nerve endings.[46]

In 1935, however, Loewi made it clear in his Ferrier Lecture to the Royal Society that he was now convinced that acetylcholine was secreted at these additional sites.[47] This evidence, collected in Dale's laboratory, led to Dale sharing the Nobel Prize with Loewi, and its discovery is described in the next chapter.

CHAPTER 6

The Road to the Nobel Prize

It was not until 1929, after an interval of some fifteen years, but then in a new atmosphere of generally awakened interest in its possible significance that my direct participation in experiments with acetylcholine was awakened.

—Henry Dale (1953)[1]

World War I began only a few months after Henry Dale had left the Wellcome Laboratories to start working at the Institute for Medical Research. As did most of the leading British scientists, he spent much of the next four years working on war-related problems. When the war ended, Dale was increasingly burdened with administrative responsibilities and committees like the one he headed for the League of Nations on drug standardization. He was, however, able to continue his research on histamine and to make important contributions to our knowledge of this substance.

As noted in chapter 4, Dale and his colleagues had described the capacity of the substance later identified as histamine to produce a profound vasodilation and lowering of blood pressure. Histamine was not, however, clearly known to be present as a natural substance in the body of animals. There had been some earlier reports of finding a suspected histamine-like substance in the body, but it was generally thought to have been there as a result of some bacterial contamination. In 1927 Dale and his colleagues isolated a substance from fresh samples of the liver and lungs of animals that produced a profound and immediate vasodilation.[2] The same year others clearly established that this substance was histamine. It had previously been called the "H-substance" because it produced effects similar to histamine, but it was not clear if the two substances were identical. Dale summarized much of this work in the three Croonian lectures he delivered in 1929.[3] Finding that histamine was a natural substance in the body may well have stimulated Dale to start thinking about whether acetylcholine was also a natural substance in animals.

Following his major paper on acetylcholine published in 1914, Dale did not do any research on that substance for fifteen years. Although he certainly was aware of Loewi's series of publications on neurohumoral transmission through much of the 1920s, it was not until 1929 that Dale's interest in acetylcholine was reawakened. He subsequently made clear that it was Loewi's research on neurohumoral transmission that: "had given such ideas for the first time an experimental reality. . . . And thus it was not until 1929, after an interval of some fifteen years, but then in a new atmosphere of generally awakened interest in its possible significance that my direct participation in experiments with acetylcholine was awakened."[4]

In 1929 Dale wrote to a friend: "I am more and more convinced that the thing [acetylcholine] is there to be found, if only we can overcome the technical difficulties."[5] That same year Dale and Harold Dudley found acetylcholine and also histamine in the spleen of the ox and the horse, and later in human placenta.[6] How they were able to do this has been described by the neuroscience historian Stanley Finger:

> Dale and the chemist Harold Dudley went to a local slaughter-house and collected spleens from horses and oxen that had just been killed. They then minced the spleens, soaked the material in alcohol, filtered it, and manipulated it in other ways. Although seventy-one pounds of minced horse spleen yielded just one-third of a gram of acetylcholine, Dale and Dudley were thrilled. For the very first time, acetylcholine had actually been obtained from an animal organ.[7]

Dale and Dudley had picked the spleen to study because they had previously found that an unidentified substance extracted from this organ produced effects similar to acetylcholine. In their article, they pointed out that prior to this work, acetylcholine had been considered "only as a synthetic curiosity, or as an occasional constituent of certain plant extracts. It appears to us that the case for acetylcholine as a physiological agent is now materially strengthened by the fact that we have now been able to isolate it from an animal organ and thus to show that it is a natural constituent of the body."[8]

Dale and Dudley reviewed all the similar effects produced by acetylcholine and parasympathetic stimulation. Even though they had not found acetylcholine in nerves, they wrote cautiously that finding it to be a natural constituent of the body increased the possibility that a "substance indistinguishable from acetylcholine by its action" has a "physiological role" to play.

By 1930 Loewi's conclusion that the vagus nerve secretes acetylcholine when innervating the heart had started to gain acceptance. However, evidence that acetylcholine is secreted at other sites had not yet been demonstrated. This started to change as a result of research by several investigators. In 1931, for example, Erich Engelhart, working in Loewi's laboratory, reported finding acetylcholine in the aqueous humor of the eye after stimulating the oculomotor nerve. The aqueous humor was later shown to be capable of evoking a vagus-like slowing of a tortoise's heart.[9] The next year, Canadian investigators reported that when they stimulated the chorda tympani nerve on one side of the body of a cat, causing the salivary gland on that side to secrete saliva, the salivary gland on the opposite side also secreted saliva after a delay even though it was denervated.[10] It was presumed that the delay was necessary to allow diffusion of the released acetylcholine from one side to the other through capillaries. Similar reports from others began to provide evidence that not only is acetylcholine secreted by the vagus nerve, but it might be more generally involved in mediating responses evoked by other parasympathetic nerves.

In 1930 Dale and John Gaddum began to investigate whether chemical mediation is involved in activating skeletal muscles. Exploring a phenomenon studied earlier by Charles Sherrington and others before him, they confirmed that a denervated skeletal muscle contracted if a nearby parasympathetic nerve, not connected to the muscle, was stimulated.[11] Dale and Gaddum hypothesized that the denervated muscle might become more sensitive to acetylcholine and that therefore a small amount released by stimulating a parasympathetic nerve could reach the muscle through the blood and cause it to contract. They demonstrated that a skeletal muscle contracted when even a small amount of acetylcholine is injected into its vascular system.[12] Although this demonstrated that skeletal muscles can respond to circulating acetylcholine, it did not prove that these muscles are normally stimulated that way or that the nerves secrete acetylcholine.

Several prominent neurophysiologists criticized Dale and Gaddum's suggestion that skeletal muscles normally contract in response to "the peripheral liberation of acetylcholine" by spinal nerves.[13] They argued that it had not been proven that spinal nerves actually secrete acetylcholine, and they pointed out that acetylcholine might come from some other organ, as it had recently been found in organs as diverse as the placenta and the cornea as well as the spleen. Some neurophysiologists were willing to concede that acetylcholine might play an ancillary role in the response of skeletal muscles, but they insisted that the normal mode of innervation was electrical.

What was needed to bolster the argument was a method for detecting the small amounts of acetylcholine that were presumed to be secreted at nerve terminals. Neither Dale nor Loewi had developed such a technique. However, the leech muscle preparation developed by Bruno Minz in Wilhelm Feldberg's laboratory in Berlin proved to be just the method needed at this time. Minz had first described the technique in 1932, and Feldberg subsequently used the method in several studies that demonstrated that acetylcholine was secreted by the vagus and other nerves in mammals.[14] Several of these experiments were done in collaboration with Otto Krayer, who later became head of pharmacology at Harvard.[15] Krayer had developed a good technique for drawing venous blood, and Feldberg had the leech muscle preparation. Together they demonstrated that after the nerve was stimulated the blood drawn from the vein surrounding that nerve terminal contained acetylcholine. It was this technique that Feldberg brought to Dale's laboratory in 1933, and during the next three years it played a major role in proving that acetylcholine is secreted at many different peripheral synapses. The political events that were responsible for Feldberg leaving Germany and working in Dale's laboratory are an important part of this history.

Wilhelm Feldberg (1900–1993) was born in Hamburg into a Jewish family. His father and uncle had gone into business selling women's clothing, and they expanded this into a highly successful department store. This financial success made it possible for the family to indulge their interest in the arts, and Feldberg's sister later became a successful artist. Because of the family's wealth, Feldberg was later able to select positions based solely on his interest, regardless of whether any salary was included. Although he did keep abreast of the family business, it was an older brother who took it over while he studied medicine in Heidelberg and Munich, and completed the medical degree at the University of Berlin.

Feldberg received his medical degree in 1925, but, like both Dale and Loewi, he preferred research to practicing medicine, and he joined the laboratory of Professor E. Schilf at the University of Berlin. That same year he married Katherine Scheffler, the daughter of Karl Scheffler, a noted art historian. Katherine had studied literature and anthropology and worked as a translator, but she also showed a lively interest in her husband's research and career, an interest that was responsible for Feldberg first meeting Dale.

Feldberg met Dale only a few months after he and Katherine were married. Feldberg's family was close-knit, and it was expected that he would spend much of every weekend with them. However, Katherine recognized that this was interfering with her husband's work. Together, they decided that the least offensive way to terminate the family weekend visits was to get away for awhile. Feldberg's mentor, Schilf, had previously translated Langley's book *The Autonomic Nervous System* into German and was able to arrange for Feldberg to spend 1925 in Langley's laboratory in Cambridge.

Feldberg had been in Cambridge for only six months when Langley died unexpectedly. Dale then invited him to spend the remaining time in his laboratory in London. Feldberg later described how both Langley and then Dale urged him.to take great care before publishing any experimental results. Langley's advice was to repeat experiments many times: "if you obtain a result in five consecutive experiments, but not on the sixth, you must do another twelve experiments." Dale's advice was to concentrate on perfecting the experimental methodology: "Feldberg, you must work like an astronomer. Prepare for weeks, for months, if necessary for years, until your method is working to perfection, then do one experiment, perhaps two—and publish the results."[16]

Not only did Feldberg profit from the time he spent with English physiologists and pharmacologists, but the period away was successful in enabling him to discontinue the "required" weekend family visits and to spend the time instead doing research. Feldberg studied the action of histamine, and in 1930 he coauthored with Schilf a 600-page book on the pharmacology and physiology of histamine. The book was dedicated to "H. H. Dale, In Dankbarkeit und Verehring" (In gratitude and admiration).[17]

In 1932 Dale and Feldberg met again in Germany. The occasion was a meeting of the German Pharmacological Society in Wiesbaden. The meeting was followed by the Internisten Kongress, where Dale was the invited lecturer. Dale attended both, and Feldberg later described their meeting:

> At the meeting of the Pharmacological Society (Otto Loewi was there, too) I gave a communication on the release of an acetyl-choline-like substance from the tongue into the blood during stimu-

lation of the chorda-lingual nerve, illustrating how easy it had become to detect the released acetylcholine once its destruction was prevented by an intravenous injection of eserine and the venous blood was allowed to pass over the eserinized leech muscle preparation. Another communication on the leech muscle preparation itself was given by [Bruno] Minz who at the time was working with me in the Physiological Institute of the University of Berlin. Dale was greatly interested in our communications.[18]

Dale was anxious to hear more about the leech technique, and he arranged to have lunch with Feldberg. At a lull in their conversation, Dale asked Feldberg what he thought about Hitler, whose Nazi party was rapidly gaining strength in Germany. Feldberg replied: "Sir Henry [Dale had been knighted that year], you need not worry, he will never win, and if he should, he will cook with water only," using the German expression that meant nothing will come of it. Dale, who paid more attention to political events, replied: "Feldberg, you had better stick to your experiments."

Only months later, on January 30, 1933, the Nazis took over the government after the Reichstag fire incident, and the first anti-Semitic laws were promulgated shortly thereafter. Feldberg was working in Paul Trendelenburg's Institute of Physiology in Berlin. In April of that year, while in the midst of an experiment, he was called to the director's office and told that he had to leave the institute by noon that day. Feldberg later described his first reaction as "idiotic," as he only thought to ask: "But what about my experiment today?" The director consented to allow him to finish the experiment, but made it clear that he had no choice and Feldberg had to leave permanently by the end of the day.

Feldberg called his wife to inform her about what had happened, and she volunteered to come to the laboratory to help him finish the experiment. It was after midnight before they were ready to leave. Feldberg later wrote that his departure would have been totally ignored except for two visiting Japanese colleagues who had heard of his dismissal and were waiting patiently at the door for his departure. The Japanese colleagues bowed silently as they left and bowed once again when Feldberg and his wife turned back.[19]

During 1933 roving bands of brownshirts were roughing up Jews in broad daylight, and 100,000 Germans, Jews, and political leftists were put in hastily built camps or taken into police custody on fabricated charges. At the time it was still possible for Jews to leave Germany, providing they left most of their assets behind. The Rockefeller Foundation had begun to provide grants to institutions that would offer positions to eminent scientists and intellectuals who had lost their positions because of the German Nazi government. The grants were barely adequate, but they could be renewed for up to three years.[20]

Feldberg had heard about the Rockefeller Foundation program, and in May he sought out Dr. Robert Lambert, assistant director of the Rockefeller medical science division in Europe, who was in Berlin at the time.[21] Feldberg's name was not on his list, but Lambert said it sounded familiar and after searching through his diary he declared: "Here it is. I have a message for you from Sir Henry Dale whom I met a fortnight ago in London." Lambert reported that Dale had said that if by any chance he should meet Feldberg in Berlin, and he had been dismissed, to tell him that Dale would like him to come to London to work with him there. Lambert's report to the Rockefeller Foundation indicated that he had had a conversation with Feldberg on May 24 and was "very favorably impressed with his personality and with the importance of his investigations." He also reported that although Feldberg had some private means, the Germans would permit him to take very little money out of the country. Feldberg had requested an annual stipend of 300 pounds, but Dale had informed Lambert that a family needed a minimum of 600 pounds to live in London.

Dale had written several letters to Lambert and others at the Rockefeller Foundation about Feldberg, as well as on behalf of other German scientists he knew were in difficulty. In a letter dated May 3, Dale wrote to Lambert, mentioning his high opinion of Feldberg and adding that: "of all the people I have had in my own laboratory, he is the one who has most directly continued to work in the field to which his association with me first introduced him. From my point of view, there is nobody whom I would more willingly have working with me."[22]

Technically Feldberg's situation did not meet the Rockefeller Foundation's conditions for support. Feldberg had taken a position as a private docent at the Institute of Physiology because it allowed him to devote full time to research with no teaching or other nonresearch duties required. However, because he had no salary it meant that he had not technically been fired and had not, therefore, met this Rockefeller Foundation requirement for support. Feldberg not only did not have a salary, but even paid for his assistant and the research expenses from his own money. It took several letters from Dale to finally work out an arrangement whereby the Rockefeller Foundation granted funds to support a specific research project that included money in the budget for a subsistence wage for Feldberg.[23]

Feldberg left Germany, arriving in England on July 7, 1933. Feldberg's wife, Katherine, who was not Jewish, had insisted that he leave Germany immediately, while she remained behind in Berlin with their two children to pack their personal belongings. The first question Dale asked Feldberg after welcoming him to England was: "Feldberg, what do you think now about Hitler?" The only answer he could think of was: "Sir Henry, can I help it that history made a mistake?"

Bruno Minz, who had taken the lead in Feldberg's laboratory in developing the leech muscle technique for detecting acetylcholine, was not as fortunate as Feldberg in finding a place to continue his work. Minz, who was also Jewish, had to flee Berlin. He had obtained a temporary position in Paris. However, he eventually had to flee Paris also, ending up in Algeria, where he found work as a physician in a military hospital. Attempts to find a position for him in the United States were unsuccessful. After the war Minz was able to pursue his research career at the Sorbonne in Paris, where he made a number of important contributions.[24]

During his first weeks in London, Feldberg was in constant fear that his wife and children might not be permitted to leave Germany. There were stories of passengers being taken off the trains at the Dutch frontier. Katherine managed, however, to get on a boat in Holland. Hours before the boat was due in England, even before the custom officials had arrived, Feldberg was at the dock, pacing up and down. Finally, the boat arrived and Katherine and the children disembarked. An immigration official, who had been observing Feldberg's pacing and his agitated state, said to Katherine as he handed back her landing permit: "Mrs. Feldberg, you must never again leave your husband alone."

Feldberg later commented that the immigration officer's remark was typical of the compassion that most English had for the refugees from Germany. This compassion was typical of many, but not all. While English scientists generally welcomed the persecuted German scientists into their laboratories as colleagues, many English physicians perceived German doctors as competitors, and the major medical organizations opposed their admission. Employment of the refugee physicians was generally restricted and existed only on a limited scale. Lord Dawson, the president of the Royal College of Physicians, for example, sent a memorandum to the home office in 1933 stating "that the number of German physicians who could teach us anything could be counted on the fingers of one hand."

This was at a time when German medicine was arguably the best in the world.

The Feldbergs rented a two-room furnished flat only a ten-minute walk from the institute at Hampstead. This enabled Feldberg to spend the maximum amount of time working in the laboratory. Katherine would generally walk with him to the institute in the morning, wish him good luck on the experiment scheduled for that day, and assure him that he need not worry if he had to work late. Feldberg would typically call Katherine when he was ready to leave, and they would meet halfway.

One day, when Feldberg and his wife had stopped to admire some lobsters in a fish market, Lady Dale walked by and asked what they were doing. Katherine explained that her husband loved lobsters and whenever an experiment had gone well in Germany, they would celebrate with a lobster dinner. When Henry Dale arrived at his office the next morning, he told Feldberg that from now on "lobster experiments would be celebrated in [his] house." Sometimes, after Dale had been away attending a meeting, he would come to the laboratory in the evening and ask: "A lobster experiment today?" Feldberg's presence obviously brought about a change in Dale, who for some time had been pulled away from the laboratory by administrative responsibilities. The potential of the leech preparation, however, had motivated him to return to the laboratory. Lady Dale told Katherine that she was so pleased they had come, because her husband was once again excited about doing experiments.

There is no doubt that Feldberg made a major contribution to the ongoing research in the laboratory, but Dale also helped many German scientists get positions in other laboratories in Great Britain besides his own. He was aware of the danger of the Nazi movement before most were, and he was genuinely concerned about those who were being persecuted.

Hitler's Gift

During the first year after the Nazis came to power, 2,600 scientists left Germany.[25] As these included some of the most eminent scientists, it had an enormous impact on science in Germany, a country where science had been preeminent. From 1901, the first year the Nobel Prize was awarded, to 1932, German scientists had received one-third (33 out of 100) of the prizes in science. Great Britain received 18, and

United States scientists were awarded only 6. Those figures were turned on their head after the war ended.[26]

Twenty of the refugees from Hitler's regime later received the Nobel Prize, and fifty who emigrated to Great Britain became fellows of the Royal Society. Twenty-five percent of German physicists left the country, and the majority of the scientists working on the atomic bomb were refugees from the Nazis, not all of them German and not all of them Jewish. Among the scientists working on synaptic transmission and neurochemistry who left Germany between 1933 and 1937 were Hermann Blaschko, Edith Bülbring, Wilhelm Feldberg, Bernard Katz, Otto Krayer, David Nachmanson, and Marthe Vogt. There were many more who left countries other than Germany because of the Nazis. Among these were Otto Loewi and Stephen Kuffler, both of whom left Austria after the Germans annexed that country. Kuffler, who worked for awhile on synaptic transmission with John Eccles and Bernard Katz in Australia, became a member of the National Academy of Sciences in the United States, and a foreign member of the Royal Society. A few examples that illustrate Dale's consideration for others are described below.

Although the father of Edith Bülbring (1903–1990) was German, her mother was a Dutch Jew. Bülbring was working in Berlin, doing research with Professor Friedmann, a noted authority on infectious diseases. Friedmann was Jewish, and in 1933 he was dismissed from the hospital along with all the other Jewish doctors. Bülbring, who had a German name and was registered as a Protestant, was safe for a time, but one day she was summoned by the hospital authorities and told that because of her Jewish mother she could no longer work there. At the time her sister was taking a trip to England and Edith joined her there, in part to visit Professor Friedmann, who was working in the National Institute for Medical Research, which Dale was then heading.

While Bülbring was visiting Friedmann, Dale engaged her in a long conversation about her work and then asked her whom she would like to work with in England. Actually, her plan was to try to secure a position in Holland, but when Dale put the question to her, she replied spontaneously that she would like to work with him. Dale said that he already had fifteen people from the Continent working at the institute, but he would make a call to see what could be arranged. Dale

(continued)

was able to arrange a position for Edith Bülbring in a new pharmacology laboratory in London. She later moved to Oxford University, where her important contributions to smooth muscle physiology were recognized with many awards and honors, including election to the Royal Society.

The story of Marthe Vogt, who was born in 1903 and is still alive in 2004, living in California, illustrates that of a number of scientists who left Germany not because they were Jewish, but because they could not tolerate the Nazi regime. She was the daughter of the renowned neuroanatomists Cecile and Oskar Vogt. At a time when it was quite unusual for a woman to obtain a medical degree, Marthe Vogt had both the M.D. and Ph.D. degrees. While the Vogts were not Jewish, they had many Jewish friends and colleagues. Oskar Vogt had helped the Soviet government establish a brain research institute in Moscow and had advised Soviet physicians on the preservation of Lenin's brain. All of these activities and contacts made the Nazis suspect Oskar and Cecile Vogt, and they were discharged from their position at the Max Planck Institute. They might have been arrested had not the powerful Krupp family, several of whom had been patients of the Vogts, interceded on their behalf.

By 1935 Marthe Vogt had become head of the chemical division of the Kaiser Wilhelm Institüt für Hirnforschung (Institute for Brain Science), but had made up her mind to get out of Germany. With Rockefeller Foundation support, Dale was able to accommodate her in his laboratory. In 1936 Marthe Vogt, with Dale and Feldberg, was the first to prove that spinal motor nerves secrete acetylcholine.[27] After the war, Marthe Vogt and Feldberg traced the distribution of acetylcholine in the brain. Their report on the location of acetylcholine in the brain raised the possibility that this substance might also be a brain neurotransmitter. Marthe Vogt later moved to Cambridge University and then to the pharmacology department of Edinburgh University. In 1954 she published a paper on the regional distribution of sympathin (a name used at the time for adrenaline and noradrenaline together) in the brain, providing more suggestive evidence that brain neurons might secrete neurotransmitters.[28] Before she retired, Marthe Vogt received many awards from physiological and pharmacological societies around the world and she was also elected a fellow of the Royal Society.

Otto Krayer was another German who was not Jewish but left Germany because of the Nazi regime. Krayer was a relatively young pharmacologist at the University of Berlin and had collaborated with Feldberg in a study demonstrating that the mammalian vagus nerve secretes acetylcholine. After Hitler came to power, Krayer was offered the position of chair of pharmacology in Düsseldorf to replace Philipp Ellinger, who had been dismissed because he was Jewish. Krayer's ethical principles would not allow him to accept the position under those conditions, and he wrote a letter explaining why he thought such dismissals were wrong. This took a lot of courage, and Krayer was dismissed from his position in Berlin.

Krayer was not able to emigrate immediately, but with help from a Rockefeller Foundation fellowship he obtained temporary employment at University College in London. He refused to return to Germany when the fellowship period ended, and he accepted a position at the American University of Beirut. When Reid Hunt was approaching retirement as professor of pharmacology at Harvard, Walter Cannon, professor of physiology there, recommended Krayer for the position. Dale also wrote in support of Krayer, who was given the position. Krayer became enormously popular with students and faculty at Harvard, where he built a leading department of pharmacology.

Immediately after arriving in England, Feldberg set up the leech muscle preparation for detecting acetylcholine. The untreated leech muscle is not especially reactive to acetylcholine, but if eserine (physostigmine) is added to the bath, its sensitivity is increased more than a million fold.[29] Essentially, the technique involved perfusing leech muscle suspended in a saline-eserine bath with blood drawn from a vein that drained the region surrounding the nerve terminal under study. A strain gauge attached to the leech muscle was used to detect any contraction that occurred when the blood was added to the bath.

As is the case with any biological technique, there were some "tricks" that had to be learned to make it reliable. First of all, because the muscles from all leeches are not equally sensitive to acetylcholine, the right strain of leeches had to be used.[30] It was also necessary to adjust the amount of eserine (physostigmine) added to the saline solution, because every muscle

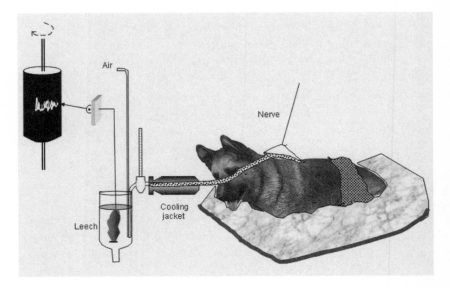

Figure 6.1   The leech muscle technique introduced by Bruno Minz and Wilhelm Feldberg. The response of the leech muscle is extremely sensitive to the presence of acetylcholine. Drawing by Professor Michael Myslobodsky.

has a different baseline sensitivity to acetylcholine. Moreover, the experimental animal (usually a cat or dog) must be injected with eserine and an anticoagulant such as heparin. The eserine protected the acetylcholine collected in the blood from degradation by cholinesterase, while the anticoagulant prevented the blood from clotting. Finally, before the blood was injected into the saline-eserine bath it had to be passed through a cooling jacket to adjust its temperature to a level suitable for the cold-blooded leech.

Blood was withdrawn from the vein before and after the nerve under study was stimulated. The stimulated nerve was suspended on a thread that lifted it away from the surrounding tissue. This precaution prevented the surrounding tissue from being stimulated and possibly being the source of any acetylcholine detected. When all of these conditions were met, the leech muscle could respond to a dilution of only one part of acetylcholine in 500 million parts of the bath solution. By systematically diluting solutions containing an unknown amount of acetylcholine and comparing their effects to standard solutions of acetylcholine it was even possible to achieve a reasonable quantitative estimate of the amount of this substance present in any solution.[31]

In less than three years, between 1933 and 1936, Feldberg published twenty-five experimental papers, with Dale and with various members of the exceptional group of colleagues working in the laboratory.[32] It can be argued that Dale would not have shared the Nobel Prize with Otto Loewi in 1936, had it not been for Feldberg's contribution. However, Feldberg always minimized his own role, saying that he "may have brought a key capable of opening some doors, but it was Sir Henry and John Gaddum [another of Dale's collaborators] who knew what doors needed to be opened." This was too self-effacing, as even before he had joined Dale's laboratory Feldberg had used the leech preparation in Germany to demonstrate that acetylcholine is secreted by different nerves. Not only had he demonstrated that the vagus nerve secretes acetylcholine in mammals, but he and Otto Krayer showed that the lingual nerve, which innervates the tongue, also secrets acetylcholine. Moreover, Feldberg and Minz had used the leech muscle preparation to demonstrate that the splanchnic nerve secretes acetylcholine when innervating the adrenal medulla.[33] This finding was important because the splanchnic nerve is a sympathetic preganglionic nerve that innervates the adrenal medulla without terminating on a postganglionic fiber, and served as an early hint of what was later proven, namely that all preganglionic nerves—sympathetic as well as parasympathetic—secrete acetylcholine. Even before joining Dale's laboratory in 1933, Feldberg was well on the way to demonstrating that acetylcholine is secreted by many autonomic nerves. The importance of Feldberg's many contributions was later recognized, and in 1947 he was elected a fellow of the Royal Society.[34]

The first experiment with the leech preparation in Dale's laboratory demonstrated that acetylcholine is secreted by other branches of the vagus nerve, in addition to the branch innervating the heart. After Dale and Feldberg reported in 1934 that acetylcholine is secreted by all the branches of the vagus nerve, including the one innervating the stomach, Dale proposed that all branches of a nerve secrete the same chemical substance.[35] It was the neurophysiologist John Eccles who first referred to this as "Dale's Principle" or "Dale's Law." It was accepted as a basic principle of neural functioning until relatively recently, when it had to be modified after the discovery that neurons are capable of secreting several different neurotransmitters.

Dale and Feldberg reported an exception to what seemed, at the time, to be the general rule that postganglionic sympathetic nerves always secrete an adrenaline-like substance. They found that the postganglionic sympathetic nerve that innervates the sweat glands on a cat's paws secretes acetylcholine.[36] To avoid confusing the pharmacological and anatomical classification, Dale recommended that autonomic synapses should be referred to as either cholinergic or adrenergic, a designation still used today. Dale summarized the conclusion he reached from the many studies he did with Feldberg and other collaborators in his laboratory in the following statement:

> We can say that postganglionic parasympathetic fibres are predominately and perhaps entirely "cholinergic" and that postganglionic sympathetic fibres are predominately, though not entirely, "adrenergic."[37]

Feldberg and Gaddum demonstrated that acetylcholine is secreted at sympathetic ganglia synapses as well as at parasympathetic ganglia.[38] To do this, they drew blood from the vessels surrounding the superior cervical ganglion region, where some pre-and postganglionic sympathetic nerves synapse. The technique for doing this had recently been described by A. W. Kibjakow, a Russian working in Kazan.[39] By drawing blood before and after stimulating preganglionic sympathetic nerves, Feldberg and Gaddum could demonstrate that acetylcholine is secreted by the nerve. This demonstration was especially significant because it showed for the first time that neurotransmitters were secreted not only to innervate smooth muscles and glands, but also at synapses between neurons. Up to this point there had remained the possibility, although the evidence was piling up against it, that the acetylcholine detected might come from the innervated muscle rather than the nerve. But at the ganglionic synapses there was no muscle involved. Despite this demonstration, the concept of neurohumoral transmission at synapses between two neurons was particularly resisted by neurophysiologists.[40] Dale later commented that John Eccles "continued for some time to dig his toes in firmly in opposition to any suggestion of a cholinergic transmission at synapses in ganglia, still maintaining that it must be electrical. One could hardly avoid a suspicion that he must still be unconsciously influenced by the somewhat closer analogy between these peripheral interneuronal synapses and those of the central grey matter [the spinal cord and the brain]."[41]

Dale and his colleagues began to investigate whether neurohumoral transmission exists outside the autonomic nervous system. Although neurophysiologists offered many reasons why chemical mediation is too slow for the innervation of skeletal muscles, there was, as mentioned in the last chapter, some indirect evidence that spinal nerves might secrete acetylcholine.[42] The leech muscle preparation made it possible in 1936 to demonstrate much more persuasively that spinal nerves also secrete acetylcholine.[43]

A few years earlier, George Lindor Brown, a neurophysiologist, had joined Dale's laboratory. He was the only neurophysiologist to work in the laboratory, and the circumstances of his being there are somewhat ironic. Brown had been working in John Eccles' laboratory when Dale heard him present a paper at a meeting in Oxford. Dale was impressed with Brown and invited him to join his laboratory. Using essentially the same techniques he had been using in Eccles' laboratory, Brown stimulated spinal motor nerve with a brief, weak electric pulse. Under this condition, the innervated muscle received only a single neuronal impulse and it responded only feebly with a short-lasting contraction. However, when eserine was applied at the point of the synapse, a vigorous, repetitive firing of the muscle fibers occurred.[44] Since eserine facilitates only the response to acetylcholine, this demonstration provided additional evidence that acetylcholine was being used as a chemical transmitter by spinal motor neurons. Because these neurons originate in the spinal cord they are considered central nervous system neurons, and therefore the demonstration provided support for the idea that neurohumoral transmission exists outside the autonomic nervous system.

The controversy continued, but by 1936 Dale and his collaborators had extended Loewi's work by providing persuasive evidence—if not completely convincing to everyone—that neurotransmitters were secreted at all peripheral synapses, not just at the synapse between the vagus nerve and the heart. They had provided evidence both that all autonomic preganglionic nerves secrete acetylcholine and, although its significance was still contested, that the spinal motor nerves innervating skeletal muscle also secrete acetylcholine. With only a few exceptions, postganglionic parasympathetic synapses were shown to be cholinergic, whereas postganglionic sympathetic synapses were demonstrated to be adrenergic.

In 1936 Loewi and Dale shared the Nobel Prize in Physiology or Medicine for demonstrating neurohumoral transmission. The citation for the award was written by Gorän Liljestrand of the Royal Karolinska Institute:

You, Professor Loewi, first succeeded in establishing proof of such transmission and in determining the nature of the effective substances. This work was, in part, built up on earlier research to which you, Sir Henry, made an essential contribution. The results were consolidated and complemented in many important respects by you and your collaborators. You and your school have also greatly extended the range of the new conception by later discoveries. Through these various discoveries, which have stimulated research in innumerable parts of the world, therefore demonstrating once again the international character of science . . .

Liljestrand got it just about right. It was Loewi who had made the speculative leap based on his initial experiment. He was guided toward this leap, at least partly, by the evidence that Dale had collected earlier. Loewi, who was at first hesitant to extend his conclusions beyond the nerves regulating heart rate, was convinced by the evidence from Dale's laboratory to change his mind. In his 1936 Nobel Lecture, Loewi stated that: "The total result of the many different, resultant investigations on various organs can be summarized by saying that up until now no single case is known in which the effect of the stimulation of the parasympathetic nerves was not caused by the release of acetylcholine."

In a tribute to Dale on his eighty-fifth birthday, Loewi reminded his audience that in 1933 he had said that: "I personally do not believe in a humoral transmission in the case of striated muscle." He continued by noting that: "Within the next year Dale demonstrated the chemical nature of transmission from spinal nerves to striated muscles. It was a discovery that made me happy, which may seem strange in view of the sentence just quoted. But it was not strange at all, for the fact stood forth that physiology owed and still owes to Dale the knowledge that the transmission of impulses from *all* peripheral, efferent nerves to the effector organs is of a chemical nature. "[45]

Liljestrand also wrote in the Nobel citation that, in addition to contributing to our understanding of synaptic transmission, neurohumoral transmission was beginning to have practical implications. He noted: "Certain observations made during recent years point to practical consequences which will be of value in combating a number of pathological conditions." Liljestrand probably had in mind some work that followed the observation of Mary Walker at the Alfege Hospital in Greenwich, England. Walker had reported that hypodermic injections of physostig-

Figure 6.2   Sir Henry Dale and Professor Otto Loewi in Stockholm in 1936 on the occasion of the awarding of their Nobel Prize. Courtesy of the Wellcome Trust.

mine (eserine) had "a striking though temporary effect" in increasing muscle strength in a patient suffering from *myasthenia gravis,* and it also improved the patient's swallowing and respiration. Walker and others had been encouraged to administer physostigmine because its well-established ability to block the action of cholinesterase might potentiate acetylcholine's ability to activate skeletal muscles.[46]

An examination of the nominations of Loewi and Dale for the 1936 Nobel Prize for Physiology or Medicine reflects the political climate of the time.[47] Normally recipients of the Nobel Prize receive many nominations from their native country. However, by 1936 the Nazi Party in Austria had become quite active and overt anti-Semitism was growing. Although Otto Loewi had been the professor of pharmacology at the University of Graz for more than twenty-five years, he received only two nominations from Austria and none from his own university. There were no nominations from Germany, his native country. One of the nominations from Austria was from Loewi's friend Ernst von Brücke, a professor at Innsbruck, but the letter consisted of only a single paragraph.[48] The second nomination, also from Innsbruck, was submitted by A. Jarisch, who nominated both Loewi and Henry Dale.

In contrast to the few nominations Otto Loewi received from Austria (and the total absence from Germany), Henry Dale was nominated by many eminent scientists in Great Britain. The list of British scientists who wrote in support of Dale's nomination included Sir Charles Sherrington and Sir Edgar Adrian, both of whom were Nobel Laureates. There were also nominating letters from A. V. Hill, F. G. Hopkins, J. H. Burn, and George Barger, a former collaborator.

It was actually Henry Dale who submitted the longest, most detailed, and most persuasive letter nominating Otto Loewi for the Nobel Prize. Dale described the early speculation by Thomas Elliott and Walter Dixon that chemical substances might mediate sympathetic and parasympathetic innervation, but noted that "direct experimental evidence of such process was lacking, and it was nothing more than an interesting and unfruitful speculation." Dale described how this was changed by Loewi's work:

> The position was entirely changed by a series of brilliant observations, made by the simplest and most direct methods, which Otto Loewi began to publish in 1921, and continued to describe, in a succession of some fifteen publications by himself or his immediate pupils, during the decade from 1921–1931. . . . This ideally simple observation at once established the truth of a conception which had seemed to be beyond the reach of experimental verification. Loewi's researches with his pupils, however, with the same simple technique, soon led him far beyond this mere demonstration of transmission of nervous effects by the release of chemical agents.

Dale went on to describe in detail the evidence Loewi had collected to prove that it is acetylcholine that is secreted by the vagus nerve and to support the conclusion that a substance "probably related to adrenaline" is secreted by the sympathetic nerve innervating the heart. Throughout this highly persuasive letter nominating Loewi, Dale included several reminders, modestly presented, of his own contributions to this work. Dale made reference to his early work that had laid the foundation for Loewi's work:

> I had myself, in 1914, pointed out that acetylcholine reproduced the peripheral effects of parasympathetic nerve impulses, with a fidelity resembling that which adrenaline had earlier been shown to reproduce those of sympathetic nerve impulses, and suggested that the relation was similar in both cases.

And when Dale described Loewi's discovery of cholinesterase in heart muscle, he noted that he had anticipated the existence of such an esterase enzyme in 1914. Dale also noted in his letter that the importance of Loewi's work had grown as a result of the demonstrations of active neurohumoral substances at all peripheral nerves, including spinal motor nerves, adding that, "Since a large part in this more recent development has been played by work in my own laboratory, I feel a special confidence in claiming that it has been made possible and has owed its original stimulus to the discovery made by Loewi in 1921."

In nominating Loewi for the Nobel Prize and documenting his own contributions along the way, Dale had developed a strong argument for the prize to be shared between the two of them. There were several other letters that nominated both of them for the prize, but none as effective as the one Dale himself had written, although he never explicitly suggested that the prize be shared.[49]

The awarding of the Nobel Prize did not end the controversy between the proponents of electrical transmission and those of chemical transmission. By this time most neurophysiologists were willing to concede that neurohumoral transmission might occur at autonomic nervous system synapses, but they opposed the idea of chemical transmission in the central nervous system. Some opposed the theory of chemical transmission not only in the brain but also at the spinal motor neuron synapses with skeletal muscles. This continuing controversy, known as the War of the Soups and the Sparks, is the subject of chapter 8.

But first it is necessary to discuss a relatively neglected side of the argument for neurohumoral transmission. Although it was the similar effects produced by adrenaline and sympathetic nerve stimulation that first raised the possibility that the neural impulse was chemically mediated, it was the work on acetylcholine that provided the most compelling evidence for neurohumoral transmission. Otto Loewi had concentrated his efforts on proving that his "Vagusstoff" was acetylcholine and did not pursue to the same degree his strong suspicion that the "Acceleranstoff" was probably adrenaline. Even in 1935 Loewi hesitated to call the sympathetic transmitter adrenaline: "In spite of analogies, however, and although personally I am convinced of the identity, I do not feel justified as yet in assuming that the sympathetic transmitter is adrenaline, and I will therefore call it 'the adrenaline-like' substance."[50]

When Henry Dale resumed his research in this area after having put it aside for fifteen years, it was the role of acetylcholine that he and his collaborators explored. In the meantime, the American physiologist Walter Cannon stumbled on evidence that adrenaline-like substances, which he came to call sympathin, were serving as chemical transmitters at sympathetic nerve synapses. Cannon's many contributions to physiology and the story of how he came close to sharing the Nobel Prize with Loewi and Dale are described in the next chapter.

CHAPTER 7

Walter Cannon

A Near Miss by America's Most Renowned Physiologist

As a matter of routine I have long trusted unconscious processes to serve me. . . . [One] example I may cite was the interpretation of the significance of bodily changes which occur in great emotional excitement, such as fear and rage. These changes—the more rapid pulse, the deeper breathing, the increase of sugar in the blood, the secretion from the adrenal glands—were very diverse and seemed unrelated. Then one wakeful night, after a considerable collection of these changes had been disclosed, the idea flashed through my mind that they could be nicely integrated if conceived as bodily preparations for supreme effort in flight or in fighting.

—Walter Cannon (1965)[1]

Although the American pharmacologist Reid Hunt first reported that acetylcholine was the most potent drug known for lowering blood pressure, the possibility that it was secreted by parasympathetic nerves was not pursued until fifteen years later, when Otto Loewi began to work on the problem. And even afterward it was primarily European pharmacologists who extended Loewi's research. A search through the *American Journal of Physiology* from 1926 through 1929, for example, reveals only one publication on acetylcholine (and it was judged to be "insignificant").[2]

There was one American, however, who came close to being the first to prove that a humoral substance might be secreted by nerves. This was the Harvard physiologist Walter Bradford Cannon (1871–1945), the foremost physiologist in the United States at the time. Cannon had stumbled on evidence that sympathetic nerves secrete an adrenaline-like substance even before Otto Loewi published his first tentative evidence on "Vagusstoff." Walter Cannon might have shared the 1936 prize with Otto Loewi and Henry Dale had he not adopted a controversial theory about the nature of these secretions.[3] The life and work of this remarkable man are an essential part of this story.[4]

Cannon had an enormous influence on the field of physiology in his time. He was a prolific and creative experimentalist with a manual dexterity that enabled him to keep animals in excellent health even after difficult and extensive surgery.[5] Perhaps most remarkable about Cannon was his ability to see how disparate experimental results could be integrated into broad physiological principles, which had great influence not only on physiologists, but also on those in many other fields. He practiced what he called "synthetic or integrative physiology," and many of the concepts he introduced, like "homeostasis," have been so widely adopted by the general public that many have forgotten who first introduced them. He was also an inspiring and much-revered teacher, and during his lifetime more than four hundred students and postdoctoral fellows, from many countries around the world, worked in his laboratory.[6] He had a friendly, open manner that enabled him to work with people in all walks of life. Colleagues in other countries often commented that his direct and unpretentious manner reflected what was best in the American character. In addition to all this, Cannon was a concerned citizen who did not hesitate to get involved in neighborhood problems as well as affairs of the world.

Cannon was born in 1871 in Prairie du Chien, Wisconsin, and was of Scotch-Irish ancestry. Prairie du Chien was the site of Fort Crawford, where the army physician William Beaumont conducted some of his seminal observations on digestion in the stomach. These observations were made on a subject named Alexis St. Martin, a French-Canadian fur trapper whose stomach was left externally exposed after an accidental gunshot injury. Cannon's initial research was on gastric motility and digestion, and he took pleasure in this connection to Beaumont, often citing Beaumont's early observations in his own publications.

Walter Cannon's father, Colbert Cannon, worked for the Chicago, Milwaukee, and St. Paul Railroad, first as a newsboy and later as a supervisor of newsboys. He married in 1870, and Walter Cannon was born a year after the marriage. Colbert Cannon switched jobs several times, trying farming in Minnesota before settling in St. Paul, where he eventually became the chief clerk in the car service department of the St. Paul and Pacific Railroad. In 1881, when Walter was only ten years old, his mother died of pneumonia. His father then married Caroline Mowrer, who was devoted to her husband's children and they to her.[7]

Colbert Cannon was a clever and inventive man and an independent thinker. He was promoted to the position of "car accountant" after he devised a novel graphic method to keep track of the company's freight cars. The promotion enabled the family to move to a larger and more attractive house in St. Paul, where Colbert used one room as an office in order to realize his long-standing dream of practicing medicine. He had no formal training in medicine, but this was not unusual among physicians at the time. He got what he knew of physiology, anatomy, and therapeutics from books, and he treated his friends and neighbors without charge. He placed hot cups on the chest of patients to treat tuberculosis, applied electricity to various body parts, injected herbal substances into the buttocks, and rubbed the body with coconut oil to treat other ailments. He had many opinions about healthy living, such as wearing flannel and chamois cloth chest protectors in the winter, and he embarrassed his children by not allowing them to wear the pointed shoes that were stylish at the time. Colbert also had strong opinions about food. Meat was not served in the house, and the family meal often consisted of nuts, distilled water, Graham bread, and Postum.

Although many of Colbert's ideas were unfounded, he had an active and curious mind and was constantly engaged in trying out something new. Despite bouts of depression, he was a good father, building dollhouses for the girls and skis and a huge bobsled with a steering wheel for Walter. He bought a tool chest for Walter and worked beside him building toys. He strongly believed that education should make children self-sufficient, and he placed a brooder and an incubator in the family dining room and taught his children how to raise chickens. There was little decoration in the dining room except for a bookcase of encyclopedias, which were often used to settle family disputes. He bought so many books and magazines that he put a strain on the family budget. Despite Colbert's interest in education, however, he took Walter out of school at the age of fourteen, believing that working would provide a better education than "loafing in school." At the time Walter didn't resent this, but after spending two years working at a job he found dull, he convinced his father that he should return to school.

Walter Cannon started high school when he was almost eighteen, the age most children graduate. The St. Paul High School had an excellent reputation, and many of the students went on to university, most to the University of Minnesota. Cannon had become interested in the debate on evolution between Thomas Huxley and Bishop William Wilberforce, and he read voraciously about the subject. Finally, as he later wrote, "my inner turmoil drove me to the confession that I no longer held the views accepted by members of the congregational church which I had joined."[8] When the minister of the local congregational church called Walter in for a talk, he took the worst possible approach, appealing only to authority. He asked Walter what right he had to question the great church scholars. Walter responded predictably, resigning from the church and angering his father, the deacon and a Sunday school teacher at the same church. However, Walter received support from a young Unitarian minister, a graduate of the Harvard Divinity School, who had recently been appointed to the high school staff.

Although Cannon's rebellion was the subject of much gossip around town, he was by no means ostracized. He was voted the "most popular boy" in high school, was elected class president, served as editor of the high school newspaper for a year, and was one of the top students, winning many honors. He was also editor of the class yearbook. The faculty chose him to give the commencement address, and Cannon used the opportunity to talk about the importance of tolerance for different opinions in science and religion. His commencement address, "Don't Be a Clam," was awarded a prize as the best student essay.

Cannon wrote well and he gained the respect and support of his English teacher, Mary Newsom, the daughter of a newspaper editor in St. Paul. She often invited Cannon to her home to talk about books and ideas, encouraging his independence and convincing him that he had to go to college. On a trip east, she wrote to Cannon bubbling over with enthusiasm about her visit to Harvard. Convinced that Harvard was the best place for Walter, Mary Newsom recruited teachers at the high school to write letters supporting his admission. During the early summer of 1892 Cannon took the Harvard entrance examinations in elementary Greek, German, French, and trigonometry. Then, with his father's help, he obtained a summer job as assistant to the paymaster of the Great Northern Railroad. Walter traveled widely with the paymaster, which afforded him an opportunity to see the extent of the plains, the mountains out west, and Indians paddling birch bark canoes in Idaho. Walter's lifelong respect for "working-class" people is foreshadowed in his diary description of the men building the railroad around Spokane:

There were two gangs of men—each gang seized a rail and laid it in contact with the last rail—the nails were placed a uniform distance apart, the car was jerked forward—two more rails were laid—and so the work went on. . . . Men worked as if their lives depended on accomplishing as much as possible each second. The great clouds of choking dust blackened their eyelids and covered their lips and made their voices hoarse. The laborers were great hard men with their hair and face-lines and clothing filled thick with dirt.[9]

Near the end of the summer, Walter received word that he had been admitted to Harvard. The family could ill afford the expense of sending him there, so it helped that he was offered a $250 scholarship. He still had to watch his finances closely; the Harvard catalogue mentioned that $372 would just barely cover the yearly cost of tuition, room, board, and incidental expenses. He found a room to rent for $57 for the year and ate at the Foxcroft Club, which was subsidized by Harvard. The meals were *a la carte*, with such charges as one cent for butter and ten cents for two eggs. Cannon figured out how to get the most nutrition for the least money. He also was able to earn some money tutoring students.

Cannon had decided on a career in medicine. His first-year classes, selected for him by his high school Latin teacher, included English rhetoric and composition, French, German, chemistry, botany, and zoology. It was a heavy load, but he maintained close to a straight A record. Mary Newsom wrote advising him to "Go to the Professors. You are not wasting their time if they are true teachers of men." She also sensed that he might be missing female companionship and wrote that: "You must not go through your college course without a woman's companionship and influence."[10] Cannon was well aware of the differences in social classes between most of the students at Harvard and the residents of Cambridge. He wrote a description of some of his New England "Brahmin" classmates for a class assignment:

The two came swinging down Brattle Street with their brilliant black shoes, prominently creased trousers, long overcoats flaring from the waist downward, new yellow gloves, and heavy canes, all of the newest and most improved style. Their faces smooth and clear cut looked striking under their shiny silk hats. As the two glided lithely along they met a little mucker who stopped short on seeing them. He stared at them with wide-open eyes and mouth as they approached and when they had passed he turned and watched them swagger on.[11]

Cannon began to attend the Boston Symphony concerts and public lectures on social issues. His essays written for English classes began to express his views on broader social issues, rather than just commenting on the different social classes around Cambridge. By the time he was a junior, Cannon was taking an increasing number of electives in addition to science courses. He took a course on the Italian government and law and was particularly impressed by a psychology course taught jointly by William James and Hugo Müstenburg.

The zoology faculty considered Cannon one of their most promising students, and he was invited to spend a summer in Alexander Agassiz's laboratory in Newport, Rhode Island. He there began a warm friendship with Herbert Spencer Jennings, a graduate student who had done his undergraduate studies and a year of graduate work at the University of Michigan. Jennings would later become well known for his classic studies on the behavior of the paramecium and other lower organisms.[12] Jennings wrote to a friend about spending time with a Mr. Cannon, from Minnesota, who was going into medicine and who had: "a decided leaning toward sociological matters . . . a very fine fellow, whole-souled and genuine—no veneer or falsity, like the easterners, many of them."[13] Jennings and Cannon remained lifelong friends.

Cannon later commented that Charles B. Davenport, George H. Parker, and William James were the teachers who influenced him the most. With Davenport he did his first investigation of a biological phenomenon—the orientation of protozoans to a light source. He was Parker's assistant for two years, and the two became good friends. He described William James' lectures as "fascinating in the freshness and constant unexpectedness of his ideas." Once when walking home with James, Cannon expressed an inclination to study philosophy. James turned to him and said, "Don't do it. You will be filling your belly with the east wind."[14]

By his senior year, Cannon's professors were treating him as a junior colleague, and he was included in gatherings at their homes. In one diary entry he noted that he had been at Professor Shaler's home and spent a pleasant evening listening to Shaler and Lloyd Morgan trade animal stories. In 1896 Cannon graduated summa cum laude from Harvard University. He wrote to Johns Hopkins Medical School, considered at the time the country's best medical school, inquiring if he could get part-time employment while attending. Cannon's letter was never answered, and so in 1896 he enrolled in Harvard Medical School.

Medical education at Harvard, as well as at other schools, was in a state of transition at that time. In the 1850s Harvard's medical school had offered only two years of lectures to students, many of whom had little background in the sciences. Admission to the medical school did not even require a university diploma. There were no laboratories for the faculty or the students. Although there were several calls to reform the medical school, not much happened until Charles William Eliot became university president in 1869. Eliot, who had been a chemistry professor, turned out to be a remarkably effective and innovative leader. During his forty years as president of Harvard, Eliot transformed what was a small parochial Unitarian college into a great university. He introduced many changes in the curriculum and allowed undergraduates to select courses, and after he raised salaries he was able to recruit an outstanding faculty. Eliot recruited Henry Pickering Bowditch as an assistant professor of physiology by providing him with a research laboratory and the opportunity to participate in reforming medical education at Harvard.[15]

Bowditch, a graduate of Harvard University and Medical School, had studied in France with Paul Broca and Jean Charcot.[16] When he accepted the position at Harvard he purchased physiological equipment in Europe at his own expense before sailing to Boston. Later, when he became dean of the medical school, he started to build the physiology department and to make research a part of the mission of the medical school and the hospital. He also helped establish the *American Journal of Physiology*, which published its first issue two years after Cannon entered the medical school. Prior to that time, American physiologists were dependent on European journals to publish their work.

Early in his first semester Cannon decided that it wasn't necessary for him to attend all the lectures, and he asked Bowditch if he might do some research with him. At the time Bowditch was excited about the potential of the recently discovered Röntgen-rays (X-rays), and he suggested that Cannon and Albert Moser, a second-year medical student, explore the usefulness of X-rays to study the passage of food through the digestive tract. This suggestion would completely change Cannon's career as at the time he was interested in neurology. Through an unpredictable path, the work on gastric physiology led Cannon to study how emotions influenced the sympathetic nervous system and the secretion of adrenaline.

Cannon and Moser's first project was to try to use X-rays to study deglutition (swallowing) by following an opaque substance as it moved through the esophagus. Cannon obtained a card of small pearl buttons normally used on clothing, and these were placed far back in the dog's throat, forcing it to swallow. After some trial and error, they were able to observe the button as it passed through the esophagus to a point just above the diaphragm. They repeated the experiment on roosters, geese, frogs, and cats and constructed special restraining boxes for each of these animals.

Cannon and Moser then switched from buttons to bismuth subnitrate, a radio-opaque substance, which they placed in a gelatin capsule that they mixed in with food. Bismuth subnitrate was being used at the time to treat various gastrointestinal disorders. This procedure worked so well that when the American Physiological Society held their meeting in Boston, Bowditch arranged for a group to come to the medical school to see Cannon and Moser's demonstration of food passing through the alimentary canal, including peristaltic waves and how the consistency of different foods determined the rate of descent.[17]

All this research activity had left no time for Cannon to work to support himself, and just when it appeared that he might have to give up research, he received the news that the faculty had awarded him a scholarship. Cannon then decided to undertake an independent study of peristalsis in the stomach, as Moser had become too busy with clinical training. This was more difficult than studying movement of food through the esophagus, but was important because so little was known about the role of stomach movement in digestion. Bowditch had once remarked that butchers knew as much as physicians about the motor activity of the stomach.

Cannon gradually built up a collection of colored tracings he made of stomach movement at different stages after the food entered. Much of this work was done on cats, but he also drew pictures obtained from humans while they were standing, sitting, or prone, and breathing normally or holding their breath. This work aroused some interest in the possibility that the technique might be used to identify malignancies and other stomach pathology that might uniquely distort gastric motility. His publication in the *American Journal of Physiology* describing this work evoked complimentary reactions from physiologists in Europe and the United States.[18]

In his third year of medical school, Cannon rotated through various clinical services. Third-year students were required to take charge of six pregnant women from before they went into labor through convalescence of the mother after delivery. This involved visiting homes of the poor, usually recent immigrants living in South and West Boston. While many of the students found this an unpleasant task, Cannon enjoyed the opportunity to have contact with people of different nationalities and customs. He wrote about his experiences to an older friend back home, who replied that: "It is very-broad minded of you to approve of the old women and to admit that they know anything about babies. I never knew a young doctor before to admit it."[19]

Cannon managed to continue some research on gastric motility while doing the clinical rotations, and during this period he made his first observations on the physiological changes associated with emotional states—a topic that would in time become the focus of his research. He had noticed that when a cat changed from being relaxed to struggling, stomach movements ceased and the outline of the stomach widened in the pyloric region. He began to stress cats in different ways, and he observed that gastric motility always ceased until the stress was removed. The same changes occurred when a cat was in a state of rage.

His research was going well, and Cannon was enjoying the clinical experience with patients, but he was unhappy with the clinical classes. The lectures were often delivered in a dry manner, allowing no time for discussion. He had some friends in law school and was impressed with their lively debates about the cases analyzed in class. Cannon persuaded a young instructor in neurology to hand out a description of a patient from his private practice and to give students a week to study the case before discussing it in class. It proved so successful that the instructor adopted this way of teaching his entire class.

Encouraged by the instructor, Cannon published an article in which he proposed that the "case method" of teaching should be used to supplement lectures. He wrote that: "With a good leader . . . the underlying pathological condition, the disturbed physiology, the therapeutic action of the drugs employed, could constantly be brought forth to give the cases a rational explanation and to teach the students the deeper insight which vision through general principles affords."[20]

Cannon received many complimentary responses to his article, including one from William Osler, who was generally regarded as the leading authority in clinical medicine. A few months later, Cannon, still a medical student, was invited, along with some distinguished professors of medicine and education and Harvard's president, Charles Eliot, to participate in a special meeting on medical education sponsored by the Boston Society for Medical Improvement. Cannon was coming to be known as a person with many different talents. Before starting his final year in medical school, he was invited to teach the undergraduate Harvard and Radcliffe College class in vertebrate anatomy, a course he had assisted George Parker in teaching while still an undergraduate.

As he approached graduation from medical school, Cannon was offered a full-time instructorship in zoology. The chairman of physiology also wrote to Charles Eliot about the possibility of offering Cannon an appointment in physiology. The letter praised Cannon as an unusually promising person who "understands that an investigation is only well-begun at the stage most students think it is finished." Cannon was offered a position in physiology, although he was not yet sure that he really wanted a career in research. He wrote to Cornelia James, the woman he would marry a year later, that he had accepted a position as instructor in physiology at Harvard and that this might be the beginning of a new career— "Who can tell?"

When Cannon started in the physiology department in July 1900, he was given a small laboratory, which enabled him both to continue work on how emotions affect gastric motility and to explore several new lines of research.[21] A number of academics began to visit at this time to learn more about his "case method of teaching." Cannon's reputation was growing among the faculty administrators, and he was also popular with students, who asked him to give a special lecture on the power of suggestion in medicine. Charles Eliot remarked at a faculty meeting that "he regarded Cannon as a valuable person to hold onto," and even though the physiology department was reducing the number of instructors, Cannon was rehired. Nothing motivates like success, and Cannon was becoming convinced that he wanted a career in research.

When Cannon was a senior in Harvard College, he had begun going out with Cornelia James, a first-year student at Radcliffe, who also hailed from Minnesota. They knew each other from St. Paul's High School, where he had been four years ahead of her. They became quite close in Cambridge and were engaged before he graduated from medical school. When it was announced in the physiology department that he was getting married, one of the staff remarked that "a young married is a young man marred." However, Cannon later wrote that "throughout my married life my wife has been my best, my most helpful and most devoted counselor and companion." Cornelia was a remarkable woman with an independent spirit and strong views about the importance of getting involved in work of social importance. Later in her life she wrote several novels, one of which was quite successful.[22]

After their marriage, Walter and Cornelia enjoyed a strenuous honeymoon of canoeing, camping, and mountain climbing. While in Glacier National Park they were told that Goat Mountain at the head of Lake MacDonald had never been climbed, and they decided to try. It was a rugged climb and they had several close calls—once when they barely escaped from a huge rock that broke loose, just missing them, and another time when Cannon was stuck for quite a while in a crevice on a steep surface of the mountain. When they finally reached the bottom it was almost dark. They met a couple of geologists doing some surveys for the government. The geologists were so impressed with the tale of their climb that they officially named the peak Mount Cannon.[23]

Cornelia often influenced her husband to become involved in various social causes. It didn't take much persuasion, as he was similarly inclined. The Cannon household became a meeting place not only for young academics, but also for groups concerned with improving the schools and the lot of the mostly Irish immigrants living in East Cambridge. Although they had some early difficulty having children of their own, they eventually had a son and four daughters. Much later Henry Dale wrote after visiting the Cannons in their old, colonial-style house in Cambridge that he had been welcomed into a household glowing with happiness and affection and that anyone who had not seen Cannon as a father had "missed a very important and attractive side of his character."[24]

Cannon's reputation as a physiologist was growing rapidly, and in 1902 he was offered a position at Western Reserve University. Harvard did not want him to be lured away, and he was thus promoted to the rank of assistant professor. Three years later Cannon was offered a professorship at Cornell University. At the time he was having a disagreement about teaching methods with William Porter, a senior professor responsible for much of the administration of the physiology department. Cannon presented his disagreements to Henry Bowditch and Charles Eliot. It was feared that Cannon might leave if the problem couldn't be resolved. Cannon was very popular with students, and some of them collected signatures on a petition supporting him. Bowditch planned to step down as chairman of physiology and, after consulting with others, he decided that while Porter was a good man, Cannon "is the best man in our school, a man of very unusual merit and tact as well, who values even the slightest good work of others . . . If we lose Cannon we've made a real loss." Charles Eliot appointed a committee of senior medical professors to interview both Cannon and Porter, and in the fall of 1906 it was decided that Cannon should succeed Bowditch as chairman and be given the George Higginson Professorship in Physiology.[25]

Expressing his appreciation for the promotion, Cannon wrote to Charles Eliot: "I shall do my best to show my loyalty and devotion to her [Harvard]"—whereupon Eliot appointed him to a number of committees. Cannon never served on any committee in a perfunctory manner. He often broadened his responsibilities, as for instance when he chaired Harvard's committee on the use of animals in research. He collected information on existing practices around the country and used this information to write a small manual of rules on the care of laboratory animals. Cannon circulated the manual to the deans of seventy-nine medical schools, and many adopted his guidelines.

At the time, concern over the use of animals in experiments was heightened by a growing antivivisection movement. Cannon wrote that in a democracy the public should be involved in the issue, and he proposed establishing an auxiliary board of laymen to support medical research. He wrote a pamphlet describing how animal research benefited humanity and joined others in attempting to block proposed legislation that would have restricted animal experimentation. The American Medical Association asked Cannon to head a committee for the protection of medical research. Among others who served on the committee were Reid Hunt, the pharmacologist who had first reported on the potency of acetylcholine; Simon Flexner, the director of the Rockefeller Institute; and Harvey Cushing, the noted neurosurgeon, who became a close friend of Cannon. He received numerous letters expressing appreciation for the work he did in standardizing rules to govern animal experimentation and for opposing the antivivisectionists. Harvey Cushing wrote to him: "You fully deserve the grateful acknowledgement of the entire profession. It scares me to think that someone else other than yourself might have been made chairman of our committee."

Cannon was pained when his friend and former teacher, William James, disagreed with his defense of animal experimentation. James had a longstanding abhorrence of animal experimentation that may have stemmed from animal experiments he had observed and from the demonstrations he had helped prepare when he was an assistant in anatomy. James had written a letter to the *New York Post* in 1909, which was subsequently circulated by the Vivisection Reform Society of New York. James argued that medical and scientific men acted like a "trade union or corporation that could not be trusted to be either truthful or moral in setting up rules that might condemn any of their members and who would not accept that treatment of animals might be somebody else's business, including a God in Israel to whom he owes account."

Cannon's response was that there were already laws against cruelty to animals and that several callous employees in animal laboratories had been dismissed during the previous year. He also argued that special legislation to control medical research was unnecessary and likely to be restrictive. William James was not convinced, and both his letters to "my dear Cannon" and his public statements continued to express his opposition to what he believed were ineffective recommendations. Actually, James' position on vivisection was more nuanced than many inferred from the *New York Post* letter.[26] He did not believe that vivisection or even inflicting pain on animals was always wrong, but he felt that those doing it could not be trusted to police themselves. As James' health was failing at the time Cannon decided not to press the disagreement with his former teacher any further. William James died in 1910, not long after this exchange with Cannon. Cannon also disagreed with William James on the physiological basis of emotions, but he softened his critique of what is now called the James-Lange theory of emotions until well after William James had died.[27]

Cannon's research was going well. While Pavlov in Russia and William Bayliss and Ernest Starling in England were well known for their studies of gastric and pancreatic digestive juices, Cannon was getting to be well known for his research on the mechanical factors in digestion—the grinding and churning of food in the stomach and intestines. His interest in the clinical implications of his research motivated him to design an apparatus to amplify the sounds of movement coming from the digestive tract of cats. This brought Cannon's research to the attention of gastroenterologists and to surgeons who were interested in exploring the potential of this technique to detect complications following stomach surgery.[28] Cannon summarized his research on gastrointestinal movement in the book *The Mechanical Factors of Digestion*, published in 1911.

Medical students and even undergraduates were attracted to Cannon, and there was a steady flow of them knocking on his door looking for an opportunity to work in his laboratory. He accepted a number of them, and he had a constantly evolving list of projects for them to choose from. He seemed to think about physiology all the time, and colleagues joked that he shouldn't take any vacations because he always returned with more ideas than could possibly be pursued. The file he kept of potential research projects included such notes as:

> "Peptic ulcers from simultaneous stimulation of vagus and
>   sympathetic?"
> "Factors affecting the growth of hair?"
> "Milk secretion as affected by fright?"
> "Effect of secreted adrenalin on bronchioles"
> "Catnip in relation to sex? Spayed cats?"[29]

Cannon's influence on students was often lasting, as the experience of Albert Hyman illustrates. Hyman was a freshman at Harvard College, and Cannon was his faculty advisor. They would sometimes meet in the laboratory. One time when Hyman was seeking some advice, Cannon was in the process of trying to restore normal heart rhythm in a cat after he had induced ventricular fibrillation. Although Cannon never succeeded in finding a reliable method of restoring normal rhythm in a fibrillating heart, he did stimulate Hyman to think about the problem and two decades later to find a way to do just that. Hyman later described how he had naively asked Cannon why it was so difficult to get the heart beating again:

> Dr. Cannon looked at me with a wistful smile, just one of the many components of an unusually expressive face which was beloved by all of his students and associates. "Young man," he said, "that is indeed a very important question. All I can say at this moment is that I am glad that you asked it; and if the seeds of inquiry have been properly implanted by this incident, perhaps some day you may be able to produce the answers."[30]

Despite all that Cannon was involved in—research, writing, lecturing, teaching, administrative responsibilities, and committee work—he also found time for community projects. Cornelia may have first had the idea to use Harvard facilities to improve the Cambridge public school system, but Cannon soon became involved. When Harvard's division of education did not approve the plan, Cornelia wrote to her mother that the professors selfishly wanted a private school only for their own children. She then convinced her husband to use the front porch of their home for a model kindergarten and first-grade class. Cannon helped build the playground equipment in their backyard, and this was used by neighborhood children as well as those enrolled in the school. The Cannon's living room also be-

Figure 7.1   Walter Bradford Cannon in his laboratory. Courtesy of the National Library of Medicine.

came a meeting place for discussing how to upgrade the Cambridge public schools.

In 1908 Cannon was searching for a strategy to study how emotional states influenced gastrointestinal motility when he took on a graduate student whose interest would radically change the direction of his research. Roy Hoskins had completed an undergraduate degree and some graduate training in the biological sciences at the University of Kansas before coming to Harvard as a doctoral candidate in physiology. This was a new degree in the medical school and Hoskins was Cannon's first Ph.D. student.

Hoskins arrived with a strong commitment to working in endocrinology.[31] Cannon, however, felt that he himself knew little about the endocrine glands and tried to persuade Hoskins to work on digestion. When Hoskins resisted, Cannon suggested he study the thyroid gland because he suspected that some thyroid conditions might be caused by emotional states.[32] Hoskins agreed, but he ended up working mostly on the adrenal gland. When Hoskins completed his dissertation and accepted a position at Starling Medical College in Ohio, Cannon encouraged him to investigate whether adrenaline secretions are influenced by emotional states. Within a year after leaving Harvard, Hoskins informed Cannon that he had found adrenaline in the blood of an animal after it had become enraged. Cannon proposed that he and Roy Hoskins coordinate their research on adrenaline secretion during emotional states.

Working with Daniel de la Paz, a postdoctoral fellow from the Philippines, Cannon compared the adrenaline contents in the blood of calm and stressed cats. He stressed the cats by exposing them to barking dogs. Blood was drawn from the cat when it was quiet and after it was stressed by the dogs.[33] To determine whether adrenaline was present in the blood, Cannon used a method developed at the Rockefeller Institute by Samuel Meltzer and his daughter, Clara Meltzer Auer. They had found that in a cat with its superior cervical sympathetic nerves cut, even a small subcutaneous injection of adrenaline caused the pupils to dilate. Denervation was thus found to increase the sensitivity of the iris to adrenaline. Using this technique to detect the presence of adrenaline, Cannon was able to show that adrenaline is released into the blood whenever a cat is stressed, and in 1911 he published a paper entitled "The Emotional Stimulation of Adrenal Secretion."[34]

Cannon confirmed these results by using a still more sensitive bioassay for detecting adrenaline. It had been reported that strips of a cat's intestinal muscle continued to contract rhythmically if left in a solution of blood, but the contractions were inhibited by adrenaline even when it was diluted to one part in twenty million. Cannon and De La Paz were able to confirm that adrenaline was released into the blood following stress, but they were completely perplexed when they found adrenaline in the blood of a stressed cat after the adrenal glands had been removed.[35] This was in January 1911—Cannon had inadvertently stumbled on the first indication that an adrenaline-like substance was secreted by some organ other than the adrenal medulla. However, the possibility that this substance might be secreted by sympathetic nerves was not something Cannon was thinking about at the time. After ruling out other explanations that occurred to him, Cannon decided that the adrenaline must have come from a small part of the adrenal gland that was inadvertently left in the animal.

Cannon began to think about all the different physiological changes that were evoked by adrenaline and that these all seemed to occur during stress. Then, as Cannon later wrote: "One wakeful night, after a considerable collection of these changes had been disclosed, the idea flashed through my mind that they could be nicely integrated if conceived as bodily preparations for supreme effort in flight or in fighting." On January 20, 1911, Cannon wrote in his scientific diary: "Got idea that adrenals in excitement serve to affect muscular power and mobilize sugar for muscular use—thus in a wild state readiness for flight or fight!"

It also occurred to Cannon in 1911 that prolonged stress might induce different pathological states. This was before there was any serious interest in what came to be called psychosomatic medicine.[36] Cannon had diabetes in mind when he reported that stress produces an increase in sugar in the blood and urine. The article concluded with the statement that, "In the light of the results here reported the temptation is strong to suggest that some phases of these pathologic states are associated with the strenuous and exciting character of modern life acting through the adrenal glands."[37]

Cannon then explored several novel ways of demonstrating how excessive adrenal activity might cause, or at least exacerbate, pathological states.[38] Working together, Cannon and Hoskins stimulated the sympathetic nerve that innervated the adrenal medulla and found that with prolonged stimulation the gland seemed to become exhausted, losing its capacity to secrete adrenaline.

Cannon explored other ways that adrenal secretions might prepare animals for emergencies, and he investigated whether adrenaline might help stop bleeding. Working with Walter Mendenhall, a teaching fellow in the physiology department, Cannon found that both stimulating the splanchnic nerve, which causes the adrenal medulla to secrete adrenaline, and injecting adrenaline shortens blood clotting time. They also found that blood clotting time decreased when a cat became angry or enraged. Cannon also found that adrenaline delayed muscle fatigue, and he wrote in his diary: "Much excited by the possibility of adrenaline activation explaining 'second wind.'"[39]

During the winter of 1913–1914 Cannon gave several popular lectures on his new ideas in the Boston area. Over two hundred people came out in a heavy rainstorm to hear his public lecture at the Harvard Medical School. An imaginative writer of detective stories later described how a doctor Walter Cannon of Harvard could determine if any adrenaline was present in a speck of blood left at a murder scene by observing the reaction of a strip of animal intestines to the blood. A New York newspaper described the increase in blood sugar levels during stress under the byline: "Man is Sweetest When he is Angry." Cannon was making adrenaline a household word.

In 1913 Cannon used the Carpenter Lecture at the New York Academy of Science and the Lowell Lecture in Boston to describe his ideas on the emergency theory of adrenal function. Earlier that year, he had tested the urine of the Harvard football players following the Harvard-Yale game. He found high levels of sugar in the urine (glycosuria) not only in those on the field, but also among members of the team sitting on the bench. War clouds were gathering over Europe, and Cannon talked about sports being the "physiological equivalent of war" and a possible substitute outlet for the fighting instinct. The lectures were well received and much discussed, and the *Journal of the American Medical Association* reprinted excerpts from his Carpenter Lecture. Cannon summarized this work in the book *Bodily Changes in Pain, Hunger, Fear, and Rage,* published early in 1915.[40]

The war in Europe began on July 28, 1914. Cannon, like many people in the United States, including President Woodrow Wilson, had taken a neutral stance. When Harvey Cushing organized a volunteer surgical unit to work in a British hospital, Cannon refused to sign up. However, when the luxury liner *Lusitania* was sunk by a German submarine in May 1915, Cannon, along with much of the country, began to actively side with the British and French.

The National Academy of Sciences established the National Research Council in 1916 to look into the threats caused by the war, and Cannon agreed to chair a committee for developing research programs to investigate medical problem of soldiers. He wrote to Charles Sherrington to determine what problems British physiologists were working on, and he drew up a list of several topics, such as surgical shock, irritable heart syndrome, and fatigue, for the committee to study. He agreed to head the subcommittee on shock.

After President Wilson asked Congress for a declaration of war against Germany on April 2, 1917, Cannon enlisted, and in May he left for Europe as a first lieutenant with the Harvard Base Hospital Unit no. 5, which was attached to the British Army until American troops arrived in Europe.

When Cannon got to France, he found that the hospital in Boulogne was a depressing, dirty tent city. Although it was well behind the front lines, there were no facilities, animals, or equipment to do any research. One day Thomas Elliott visited the field hospital, and a lasting friendship between them began. They were aware of each other's work on adrenaline and the sympathetic nervous system.[41] Although Cannon had never treated a patient in shock, Elliott, who had the rank of colonel, was able to arrange for Cannon to work at a British clearing station that received many cases of serious shock.

At the clearing station Cannon became familiar with the course of blood pressure changes that occur with shock. He learned that there is "primary shock", in which wounded soldiers immediately experience a precipitous drop in blood pressure, and also a condition called "secondary shock," in which soldiers with wounds that are not necessarily severe sometimes have a delayed, but precipitous fall in blood pressure that could be fatal. Secondary shock sometime occurred even after soldiers had been sent to the rear.

When Cannon examined the blood of soldiers in shock, he found it to be acidic. He speculated that the acidity might cause blood to be stored in the capillary bed, a condition that could explain the decrease in blood pressure. Cannon tried to correct the acidosis by administering common cooking soda (sodium bicarbonate) to soldiers who appeared to be going into shock. After being given teaspoons of baking soda in solution some soldiers seemed to recover and one soldier with severe wounds, low blood pressure, and a rapid heart rate improved so dramatically after being given the baking soda solution intravenously that Cannon called it a "red letter day." He contacted William Bayliss, and they started to collaborate in exploring this possibility, although in the long run it proved to be a false lead. Dale, who was exploring the use of histamine to create an animal model of wound shock, was actually on the right trail, but at the time no one knew that histamine is a natural substance in the body.

The war ended on November 11, 1918, but Cannon, who then had the rank of major, had to wait until January 1919 before he could go home. A troopship was finally available to transport him back home, and when the *Manchuria* completed its eleven-day crossing of the Atlantic, Cannon spied Cornelia waiting to greet him at the dock in Hoboken. He had a warm reunion with his family, and within a few weeks he was back lecturing at the medical school and doing research in his laboratory.

As already mentioned, Cannon had become interested in exploring whether prolonged stress could induce various pathological states by excessive stimulation of the adrenal medulla. Cannon had earlier reported that the adrenal medulla could become exhausted and no longer capable of secreting adrenaline if it received excessive stimulation. The theory and the experimental results had not gone unchallenged, and a dispute began when George Stewart and Julius Rogoff of Western Reserve University reported that adrenaline levels remained relatively constant after prolonged stress and they argued that the adrenal gland did not become exhausted. Moreover, Stewart and Rogoff found that animals can cope with stress after the splanchnic nerve that innervates the adrenal medulla is severed.

Cannon answered that he had never claimed that adrenaline secretion was the only physiological response to stress, but rather that its effects supplemented and enhanced the activity of the sympathetic nervous system.[42] Arguments went back and forth. Stewart and Rogoff criticized Cannon's bioassay method for measuring adrenaline because it was not quantitative. Cannon responded that this criticism was equivalent to rejecting a carpenter's level because it did not measure the precise angle of inclination. He did, however, begin to use other ways of detecting the presence of adrenaline. As it was known that a heart deprived of its nerve connections became supersensitive to adrenaline, Cannon began to use the denervated heart to detect adrenalin. It was this methodology that eventually led Cannon to the conclusion that sympathetic nerves secrete adrenaline-like substances.[43]

In 1920 Cannon and Joseph Uridil, a postdoctoral fellow, undertook some experiments designed to answer a criticism by Stewart and Rogoff, but they ended up producing a totally unanticipated finding. Cannon had reported that stimulating the splanchnic nerve that innervates the adrenal medulla caused the denervated heart to beat faster, but only if the adrenal gland were intact. Stewart and Rogoff, however, reported that the denervated heart accelerated following splanchnic nerve stimulation even after the adrenal gland had been removed, and they attributed the acceleration to an increase in blood pressure.

Cannon and Uridil found that Stewart and Rogoff were correct. The denervated heart did beat faster during stimulation of the splanchnic nerve, even when the adrenal gland was removed. However, they were able to rule out increased blood pressure as the explanation. It was a puzzle for Cannon, and, after exploring other possible explanations, he concluded that splanchnic stimulation caused some sympathomimetic substance to be secreted by the liver. This conclusion was based on the observation that heart rate was not accelerated during splanchnic nerve stimulation if the hepatic nerve was cut.[44] Cannon had not at this point identified the sympathomimetic substance, but he considered it to be a hormone that was secreted from the liver into the vascular system and carried in the blood to the heart and to other organs, where it produced various sympathetic effects.

These results were published in 1921, the same year that Otto Loewi published his first evidence that neurohumoral secretions were involved in the regulation of heart rate. Had Cannon persisted in exploring why sympathetic nerve stimulation caused the denervated heart rate to increase, he might well have discovered then, as he did later, that sympathetic nerves secrete an adrenaline-like substance. At the time, however, Cannon was preoccupied with defending his "emergency theory" and not thinking about whether sympathetic nerves might secrete an adrenaline-like substance.[45]

The subject was not pursued any further until 1930, when Zénon Bacq came to work with Cannon. Bacq, a twenty-six-year-old physician from Belgium, began to prepare cats with denervated hearts and with the adrenals and livers isolated from their nerves. When they stimulated different sympathetic nerves in these animals they saw an increase in heart rate, but only after a delay of several minutes. The increase in heart rate reached an elevation of 25 to 30 beats per minute. While this was not as large as the 100 beats per minute increase that emotional excitement could produce in an intact cat, it was still a significant increase and needed to be explained. They also saw an increase in blood sugar even in animals that had the adrenal glands and liver isolated from their connecting nerves and from the vascular system.[46] It became necessary to consider other sources of what appeared to be a sympathomimetic substance that was secreted into the blood when sympathetic nerves were stimulated. By 1931 Cannon and Bacq had summarized their findings in several publications:

In a cat with heart denervated, adrenals and liver excluded from action . . . stimulation of the sympathetic supply to smooth muscles . . . causes, for about two minutes, a gradual increase in blood pressure, a gradual increase of heart rate, and a gradual increase of salivary secretion. The time relations of the increase and the slow return to the former state are, in each instance, similar to those reported for the denervated heart alone [reported in a previous paper] as a consequence of emotional excitement.[47]

They attributed these results to the secretion of some sympathomimetic substance that came from a source other than the adrenal medulla or the liver. Although on the basis of its effects Cannon suspected the substance was adrenaline, he suggested another name for it because of its different origin: "We would suggest that it be called provisionally *sympathin*, with the understanding that as knowledge of its character increases it may prove really to be adrenin [adrenaline] developed for local action in smooth muscle cells."[48]

Cannon viewed sympathin as a "hormone" that cooperated with and augmented the action of the sympathetic nervous system and the adrenal gland.[49] This was possible, Cannon argued, because sympathin, like adrenaline, is relatively stable and can remain active while traveling around the body in the blood. Because of the sympathetic nervous system's role in preparing animals to meet emergencies, Cannon considered it adaptive for this system to discharge as a whole, and he contrasted this with the parasympathetic nervous system, in which the anatomical arrangement made it possible to evoke separate responses.[50] Moreover, acetylcholine, the substance most capable of mimicking parasympathetic responses, is rapidly degraded and therefore cannot remain active long enough to travel to different organs.

In 1931 Cannon indicated that he believed it was the smooth muscles that were secreting sympathin when they were stimulated by sympathetic nerves.[51] Cannon felt that this conclusion was in agreement with Thomas Elliott's, although Elliott had never decided whether adrenaline is secreted by the muscles or by nerves. By this time, George Parker, Cannon's former teacher and close friend and colleague at Harvard, had concluded that nerves are capable of secreting humoral substances.[52] Parker drew this conclusion from his study of chromatophores, the pigmented cells that enable some crustaceans, fish, amphibians, and reptiles to camouflage themselves by changing their skin color.[53] Parker presented evidence that adrenaline produces the same effect on the chromatophores as stimulation of their innervating nerves, and he contrasted his views with those held by Cannon:

> The exact source of the several humoral substances thus associated with muscle has not excited much attention. In the most recent contribution to this subject, that by Cannon and Bacq, the smooth muscle is continually referred to as the element from which the humoral substance emanates. . . . It seems quite evident from the conditions under which the humoral substances of muscle are produced that these substances must originate either from the muscle cells or from the nerve terminals. . . . Because of the small size of the terminals, it would be natural to assume that the given substance was produced by the muscle, but if the most recent work on chromatophores has any meaning at all, it points most conclusively to the nerve terminals as the source of these materials. . . . In many respects this principle is

a special application of Bayliss and Starling's concept of hormones, though the distances traveled by the substances from their region of origin to those of application are often extremely short.[54]

In a 1933 article entitled "Chemical Mediators of Autonomic Nerve Impulses," Cannon wrote that Parker's conclusions might be right for chromatophores, but that he believed they did not apply to the sympathetic nervous system's innervation of smooth muscles.[55] Cannon argued that the very fine single nerve endings ("nerve twigs") that innervate each smooth muscle cell are unlikely to be able to release sufficient amounts of any chemical substance that would enter the blood and act as a hormone at distant sites.

Shortly before Zénon Bacq's fellowship ended, Arturo Rosenblueth arrived in the Physiology Department.[56] Rosenblueth, who would replace Bacq as Cannon's "first assistant," was a young Mexican physician, the eighth child of a Jewish immigrant from Hungary and a Mexican-American mother. Everyone who knew Rosenblueth considered him brilliant and multitalented. He was one of Mexico's top chess players and a pianist skilled enough to have considered a career on the concert stage. He was also fluent in English, French, and Spanish, and he had a good command of German. He had studied in France and had a German medical degree. When he returned to Mexico in 1927, he practiced psychiatry for a short period, but did not find it satisfying and applied for and received a Guggenheim fellowship to work in physiology at Harvard.

Cannon recognized that Rosenblueth was enormously gifted and invited him to join his laboratory. Cannon was fifty-nine at the time, not in the best of health, and, like any senior researcher with many different projects underway, he needed the assistance of younger scientists. Rosenblueth was delighted to be invited to work with the "great man." And he soon came to revere Cannon. Horace Davenport, who was in Harvard's department of physiology at the time, wrote that Rosenblueth "worshipped Cannon, and when he spoke of Cannon there was a tone of reverence in his voice. I once heard him refer to Bradford Cannon as The Son, as if Cannon was God. I suppose that Rosenblueth looked upon himself as the Holy Ghost, for he was not afflicted with disabling modesty."[57]

As Davenport implied, Rosenblueth could be arrogant, and it is not surprising that many resented him. Chandler McC. Brooks, who also worked with Cannon and knew Rosenblueth, described him as "too facile, brilliant, well-informed, and certain of his opinion for most of us to handle."

Cannon always thought highly of Rosenblueth, but a number of people who admired and loved Cannon resented the influence Rosenblueth seemed to have over him.

Rosenblueth was enormously energetic and, after observing some of Cannon and Bacq's experiments, he started to turn out papers for publication so fast that Cannon had difficulty examining them as closely as was his normal practice. Between 1931 and 1933 Rosenblueth published, either alone or with others, twelve papers, and by 1944 he had published seventy-four papers in the prestigious *American Journal of Physiology*, in addition to a book.

Rosenblueth also had considerable mathematical ability, and he peppered his publications with equations, logarithmic functions, and mathematical reasoning. In recalling his interactions with Rosenblueth, Horace Davenport, who once described Rosenblueth as Cannon's "dark angel," remarked that, "I can't say how profound was his mathematical skill, but it was certainly far beyond that of most of the mathematically illiterate physiologists of the day."[58] Rosenblueth's mathematical ability was certainly not all "smoke and mirrors," as he had a close collegial relationship and sometimes a collaboration with Norbert Wiener, who was certainly competent in mathematics. Wiener, the founder of cybernetic theory, regularly attended a roundtable discussion group organized by Rosenblueth, and later the dedication in his book on cybernetics read "to Arturo Rosenblueth, for many years my companion in science."[59]

In his first experiments in Cannon's laboratory, Rosenblueth demonstrated that prior administration of cocaine potentiated the response of the nictitating membrane to both adrenaline and sympathin.[60] The nictitating membrane is a thin membrane that is pulled over the eye for protection in some animals. The increased sensitivity produced by cocaine presented an opportunity for Cannon and Rosenblueth to compare the properties of adrenaline and sympathin.[61] Before 1933 Cannon had believed that the two substances were probably the same. He wrote that year: "Until a few months ago the evidence was fairly consistent that the substance set free from the heart on sympathetic stimulation is adrenin [adrenaline]."[62] However, Cannon went on to note that he and Rosenblueth had recently found that sympathin and adrenaline are not identical and that, moreover, there were two different types of sympathin.

The evidence for two different sympathins was based on experiments that compared the responses of different organs to adrenaline and sympathin after administration of ergotoxine, a substance Dale had discovered that would block the action of adrenaline.[63] They found that while ergo-toxine blocked adrenaline's capacity to increase blood pressure and heart rate, it did not block these same responses when produced by sympathin.[64] This convinced them that the two must be different. Moreover, they also observed differences in the responses to sympathin depending on from where in the body it was obtained.[65] From such observations, Cannon and Rosenblueth inferred that there were two different kinds of sympathin, an excitatory sympathin E and an inhibitory sympathin I.

At this point, Rosenblueth, employing his mathematical bent, intro-duced a degree of additional complexity to the two-sympathin theory. He had studied the responses of smooth muscles to increasing doses of adrena-line and concluded from the shape of the curves that at a certain point adrenaline must combine with some hypothetical substance to form a sec-ond substance, which he labeled AH (adrenaline plus a hypothetical sub-stance). As Cannon and Rosenblueth had concluded that there must be two different sympathins, there must be two hypothetical substances, one of which formed sympathin E when combined with adrenaline, and the other which formed sympathin I. Cannon had by this time changed his mind and now believed that adrenaline was secreted by sympathetic nerves. Accord-ing to this new theory, a sympathetic nerve impulse first releases adrenaline onto the smooth muscle, where it combined with one of two endogenous substances in the muscles to form either sympathin I or E.[66]

The two-sympathin theory was not well received even by Zénon Bacq, who had coauthored with Cannon the publication that introduced the term "sympathin." After he returned to Belgium, Bacq's research con-vinced him that the two-sympathin theory being promoted by Cannon and Rosenblueth was wrong, and he wrote a number of letters to Cannon explaining why he thought so. Bacq greatly admired Cannon, and, as he later commented, he was reluctant to start a long controversy with his "beloved teacher." He tried to persuade him not to persist in holding on to the two-sympathin theory, for which he held Rosenblueth responsible:

How and why was this unlucky hypothesis of the two sympathins E and I put forward? It can be said that Rosenblueth was very fond of theoretical speculations and mathematical analysis; all his work is impregnated with this spirit. When he wrote a paper or a mono-graph, he emphasized discussion, not the facts. Cannon lived closer to the facts of classical physiology; he constantly invented and per-fected new techniques. But neither of these two scientists had enough interest in biochemistry or pharmacology to avoid following the wrong path leading to the theory of the two sympathins. . . . I was struck by the fact that this interpretation was not only improbable but also useless, since nobody had any notion or even guess as to what was the chemical nature of these hypothetical substances E and I. . . . I showed that all the facts observed by Cannon and Rosenblueth could be easily explained if one accepted the idea that I is epinephrine [adrenaline] and E, norepinephrine [noradrenaline].[67]

Many others were also critical of the two-sympathin theory. Although Henry Dale was too diplomatic to express his reservations in a public forum, comments in several of his letters made it clear that he regarded the two-sympathin theory as unnecessarily complex and most likely wrong.[68] Nevertheless, this did not prevent Dale from giving Cannon credit for his evidence that sympathetic nerves secrete a chemical substance.

Cannon and Rosenblueth's 1937 monograph *Autonomic Neuro-Effector Systems* reviewed the arguments for the two-sympathin theory.[69] When the book was reviewed in the journal *Nature*, the reviewer "JHG" (probably John H. Gaddum, a colleague of Dale) described the book as "authoritative, "most valuable," and "impressive," but criticized as "unnecessarily complicated" the theory that different receptor substances in the smooth muscles combine with a substance secreted by the sympathetic nerves and then are released into the bloodstream

More than a few of Cannon's contemporaries believed that this opposition to the two-sympathin theory prevented Cannon from sharing the 1936 Nobel Prize with Loewi and Dale.[70] The theory was criticized when Cannon presented it at the International Congress of Physiology held in Leningrad in 1935. Zénon Bacq and Göran Liljestrand of Stockholm presided at the session in which Cannon delivered a lecture on this subject. The following year, the Nobel Committee considered Cannon's nomination for the Nobel Prize in Physiology or Medicine. This was the year the award was given to Loewi and Dale. Cannon's nomination was evaluated by Göran Liljestrand. He described Cannon's work on the emergency function of the "sympatico-adrenal system" as "prize-worthy," but he criticized the two-sympathin theory, mentioning the contradictory evidence presented by Zénon Bacq, who "was previously Cannon's colleague." Liljestrand concluded that because of the weakness of the two-sympathin theory, Cannon's "prize-worthiness" had decreased."[71]

Cannon's son Bradford later wrote that he never heard his father express any regret or disappointment over not sharing the Nobel Prize.[72] After the prize was awarded, Cannon reviewed the history of the field and the important contributions of Loewi and Dale without mentioning any of his own work. He concluded with the following observation: "Obviously a new realm in physiology is disclosed by these important researches, and it will be well to watch for further advances in the attractive realm in which Dale and Loewi have pioneered."[73]

Cannon sent congratulatory letters to Loewi and to Dale. Dale's reply was modest and diplomatic: "So often it must be the right impression, made at the right moment, which brings [the award] to a man no more deserving than many others. If all my friends who deserve it could receive the award, we should need more fortunes of many explosive millionaires to meet the demand."[74]

In 1939 Kálmán Lissák, a Rockefeller Foundation fellow from Hungary, joined Cannon. Lissák had developed a technique for extracting dialysate from postganglionic sympathetic nerves, which could then be analyzed to determine its physiological and chemical properties.[75] Cannon had earlier suspected that adrenaline was the mediating substance secreted by sympathetic nerves, and Lissák found that the extract from sympathetic nerves had the same capacity as adrenaline to increase heart rate and blood pressure and to cause the pupils to dilate, the nictitating membrane to contract, and the nonpregnant cat's uterus to relax. Moreover, all of these effects were potentiated by prior treatment with cocaine, just as with adrenaline.[76] Cannon was satisfied that the mediating substance in his theory was adrenaline, and that it was secreted by postganglionic sympathetic nerves. Cannon and Lissák concluded, "The results reported in the present communication are consistent with the view that sympathetic fibers liberate adrenaline at their terminals and that this agent, when it escapes into the blood stream, has been modified in such a manner that it has the peculiar actions of sympathin on remote organs in the body."[77]

The Cannon-Rosenblueth sympathin theory received two fatal blows at the end of the 1940s. The first came when Raymond Ahlquist (1914–1983), a pharmacologist at the Georgia School of Medicine, provided evidence that there are two different adrenergic receptors rather than two types of sympathin. Ahlquist had been interested in sympathomimetic drugs since the early 1940s, when he was involved with attempts to grow the plant *Ephedra sinica* in the western part of South Dakota. At the time this plant was the major source of the drug ephedrine and had to be imported from China. Although the project was abandoned when it became possible to synthesize ephedrine, Ahlquist continued to investigate the properties of different sympathomimetic drugs. Using rabbits, cats, and dogs, he investigated how sympathomimetic drugs affected blood vessels, intestines, pupils, uterus, heart, and the nictitating membrane. In a report published in 1948,[78] he compared the responses evoked by adrenaline, noradrenaline, and isoproterenol, the latter being a substance that is chemically close to, but not identical to, noradrenaline. From the ranking of the potency of these three substances, Ahlquist was able to conclude that there had to be two different receptors, which he called alpha and beta adrenergic receptors.[79]

In a tone that an editor who rejected the manuscript described as "audacious" and "irreverent," Ahlquist concluded that Cannon and Rosenblueth's two-sympathin theory, "widely quoted as a law of physiology, is no longer necessary." He went on to recommend that the use of the term "sympathin" "be discouraged."[80] Although references to sympathin continued to appear in the literature for several years, mention of the two-sympathin theory soon disappeared. It was ten years, however, before Ahlquist's proposal of two separate adrenergic receptors was widely recognized as being correct.

The other crushing blow to the two-sympathin theory came from the work of Ulf von Euler. Von Euler had spent six months in Henry Dale's laboratory during the early 1930s. When he returned to Stockholm he began to apply the bioassay techniques he had learned from Dale to study adrenaline and other natural substances that might be biologically active. He was able to obtain a much more purified extract from sympathetic neurons than Lissák had working in Cannon's lab. Using a new fluorescence technique developed by his Swedish colleague Nils-Åke Hillarp, von Euler was able to prove that sympathetic nerves secrete noradrenaline. Von Euler reviewed the state of the field before presenting the evidence that sympathetic nerves secrete noradrenaline.

> The pioneer work of Loewi and Cannon and their associates on the mechanism of sympathetic—or rather adrenergic—nerve action has revealed that stimulation of such nerves is accompanied by the liberation of some active principle with sympathomimetic properties. It was primarily assumed that the neurohumoral agent was identical with adrenaline, but this view seemed to need further consideration, however, in the light of the findings of Cannon and Rosenblueth

that stimulation of sympathetic nerves elicited remote actions which did not wholly conform with those of adrenaline. In order to reconcile the seemingly conflicting evidence, Cannon and Rosenblueth (1933, 1937) elaborated a hypothesis involving the primary liberation from the adrenergic nerve-endings of a mediator substance which should then combine with some constituent within the reacting cells under formation of the final active substances, which, on account of their supposed actions, were termed sympathin E (excitatory) and I (inhibitory). The assumption of these authors that the primary mediator should be identical with adrenaline seemed to be supported by the later work of Cannon and Lissák (1939).[81]

Von Euler provided convincing evidence that Cannon and Lissák were in error—the adrenaline that they had detected in sympathetic nerves is actually changed to noradrenaline by a process called demethylation before it is secreted. He concluded that "it seems most likely that the adrenergic transmitter is noradrenaline." At first von Euler concluded that some sympathetic nerves secrete adrenaline and that Cannon and Rosenblueth's sympathin E and I might correspond to noradrenaline and adrenaline. Later he seemed to agree with Ahlquist's conclusion that rather than two different sympathins, there are two different adrenergic receptors. The truth may be that the two issues are somewhat confounded and that while Cannon and Rosenblueth were correct in believing that there were two different substances involved, they were wrong in concluding that the nerves secreted a single substance that became two when combined with different substances in the smooth muscles. Ironically, von Euler found that Loewi had actually been correct when he tentatively concluded that it is probably adrenaline that is secreted by the sympathetic nerve that innervates the frog's heart. He found that this sympathetic nerve in frogs does secrete adrenaline, not noradrenaline, but that this is an exception and not the case in mammals.

For a while, von Euler used Cannon's term "sympathin" for noradrenaline, as a way of distinguishing it more clearly from adrenaline and because he was not in favor of introducing new names. However, "noradrenaline" (and "norepinephrine") were gradually adopted, and references to sympathin began to drop out of the literature. Cannon died in 1945, so it is not known whether he would have continued to defend the two-sympathin theory, but in 1950 Rosenblueth was still defending it.[82] He was, however, virtually alone in doing this. In 1953 Dale displayed his great skill of softening criticism by embedding it in a rather transparent compliment when he remarked that: "I can't suppress an impulse of admiration for the tenacity with which the last of Professor Cannon's distinguished collaborators, Professor Rosenblueth, in spite of all this recent evidence, still keeps the flag of the Sympathins, E and I, firmly nailed to the mast of his belief and his advocacy."[83]

Cannon continued to receive nominations for the Nobel Prize for his many different contributions until 1945, the year of his death. However, 1936 was the last time he was a serious candidate. Whether Cannon deserved to share the prize with Loewi and Dale certainly can be disputed. It is true that Loewi and Dale's evidence for chemical transmitters of neural impulses was more direct and compelling than the evidence Cannon presented and was not tainted by a controversial and what was ultimately shown to be an incorrect theory. Moreover, Cannon initially believed that sympathin is a blood-borne hormone triggered by sympathetic nerves rather than a neurotransmitter as we recognize it today.

It deserves to be noted that Cannon's experiments were closer to reproducing natural conditions in that they were often performed on unanesthetized mammals responding to meaningful stimuli. In contrast, the experiments of Dale and his collaborators were performed on anesthetized animals responding to nerve stimulation, while Loewi's experiments were performed on the isolated hearts of frogs and toads. Moreover, while Loewi and Dale concentrated their efforts mainly on acetylcholine, Cannon was studying the sympathetic nervous system and adrenergic substances.

In his 1932 review Dale made the point that Cannon and his coworkers were doing most of the work on chemical mediation of sympathetic nerves, although at the same time he conveyed his skepticism about the two-sympathin theory:

> Further progress in our knowledge of this chemical transmitter of the peripheral effects of true sympathetic nerves has come largely from Cannon's laboratory at Harvard. . . . Cannon supposes that the actual transmitter is a substance capable of producing either type of effect, as adrenaline does, according to the type of receptive substance which it finds, and combines with in the effector cell. He imagines that two types of such combination may occur, producing what he calls "Sympathin E" and "Sympathin I," which have aug-

mentor [excitatory] and inhibitor effects, respectively; and it is these combinations, he believes, which escape to some extent into the blood stream. It should be said, I think, that the behavior of the substance transmitting parasympathetic effects, concerning which more is known, provides no analogy for this conception. . . .

We may safely leave the details of chemical transmission of peripheral sympathetic effects to the further investigations of Cannon and his school, and return to the transmission of the peripheral effects of parasympathetic nerves.[84]

In 1953, when Dale commented on his own role in these events, he noted that although he and George Barger had found by 1914 that noradrenaline is considerably better than adrenaline at mimicking the effects of the sympathetic nerves, they did not pursue this observation because they thought that noradrenaline was just an interesting drug. He later wrote about these events after Cannon's death:

Doubtless I ought to have seen that nor-adrenaline might be the main transmitter. . . . If I had so much insight, I might even then have stimulated my chemical colleagues to look for nor-adrenaline in the body; but they would have certainly failed to find it, with the methods which were then available. . . . If I had taken the additional step, even in hypothesis, much trouble might, perhaps, have been saved in after years for my late friend, Walter B. Cannon: For the observed differences between the effects of adrenaline and those of "sympathin," as liberated by stimulation of sympathetic nerves, and especially of the hepatic nerves, which led Cannon and his associates to put forward the elaborate theory of the complex sympathins, E and I, were practically the same as those between adrenaline and noradrenaline.[85]

Neither Dale nor Bacq was correct when he assumed that noradrenaline was equivalent to sympathin E and adrenaline was equivalent to sympathin I, as later discoveries made it clear that the different responses were due to different receptors, not different neurotransmitters.[86] Cannon was, however, actually correct in concluding that the substance he had detected was not identical to adrenaline. Ulf von Euler—who would share the 1970 Nobel Prize for demonstrating that it is norepinephrine that is secreted by sympathetic nerves—put the question of priority in perspective when he wrote, "not before the classical series of experiments of Cannon and Rosenblueth did it become evident that sympathetic nerve stimulation led to the appearance of some factor other than adrenaline."[87]

It is perhaps ironic that the two-sympathin theory actually was correct with respect to the secretions of the adrenal medulla; it was later found that this gland secretes both norepinephrine (noradrenaline) and epinephrine (adrenaline). Cannon and Rosenblueth's theory, however, referred to secretions of sympathetic nerves, not the adrenal medulla.

Despite the evidence supporting chemical mediation of neural impulses, most neurophysiologists vigorously opposed the idea, except in the case of the innervation of some visceral organs. The vigor with which the idea of chemical transmission in the central nervous system was opposed is described in the following chapter.

CHAPTER 8

The War of the Soups and the Sparks

Even after the Nobel Prize was awarded to Otto Loewi and Henry Dale in 1936, most neurophysiologists did not accept neurohumoral transmission of the nerve impulse as a general principle, although many were willing to concede that chemical transmission might be adequate for the sluggish response of visceral organs. Neurophysiologists in general were convinced that only electrical transmission is fast enough to activate skeletal muscles, and for them the possibility that nerve impulses at brain synapses might be transmitted chemically was not worth thinking about. The neurophysiologist John Eccles wrote in 1936 that the "presumed chemical nature of the synaptic transmitter in the central nervous system . . . is almost entirely based on an extrapolation from the ACh [acetylcholine] hypothesis for sympathetic ganglia" and that the evidence that this could be applicable to the central nervous system was "almost negligible."[1]

Although the opposition to the principle of chemical transmission was based on reasonable arguments and conflicting lines of evidence, it is difficult to avoid the impression that many neurophysiologists resented the intrusion of pharmacologists into what had been their exclusive area of expertise. In describing the history of this period, Zénon Bacq observed that the eventual acceptance of chemical transmission deprived the neurophysiologists of "a vast field which passed into the hands of biochemists and pharmacologists."[2]

Neurophysiologists regarded their own data, which by the 1930s was being collected via multistage vacuum tube amplification and displayed on fast responding cathode ray oscilloscopes, as more meaningful, reliable, and scientific than the data pharmacologists collected with their bioassays and smoked-drum kymographs. There was also a tendency among some neurophysiologists to look down on pharmacologists, who spent their time investigating "spit, sweat, snot, and urine," and whose sometime association with the pharmaceutical industry was seen as "consorting with trade." It should be recalled that Henry Dale's friends advised him not to sell "his scientific birthright for commercial pottage" when he was considering the position offered by Henry Wellcome. In looking back to this earlier period, Zénon Bacq recalled that,

> Often in my younger years, during discussions with "true" physiologists, when a contradiction occurred between facts observed by physiologists on the one side and pharmacologists on the other, the choice was made in favour of the physiological observations because a pharmacological experiment, on principle, did not carry the same weight. This elite attitude of the physiologists implied that they considered the general intelligence of the "pharmacologists" of a lower level.[3]

Such nonscientific factors may have colored the dispute, but the arguments raised against chemical transmission were always framed around conflicting data and observations. Many neurophysiologists were convinced that chemical transmission is too slow to be the basis for the fast responses of skeletal muscles that only they could record with their technically more sophisticated electronic equipment. Although it was no longer possible for neurophysiologists to deny the evidence that acetylcholine is secreted by spinal motor nerves, they insisted that any role it played in neurotransmission was a secondary one, restricted to modulating the primary response to electrical transmission.

The list of neurophysiologists who refused to accept that acetylcholine plays a major role in innervating skeletal muscles included eminent scientists of the stature of Albert Fessard, Ralph Gerard, Rafael Lorente de Nó, John Fulton, Herbert Gasser, Joseph Erlanger, and John Eccles, the last three of whom were later awarded the Nobel Prize.[4] Erlanger, for example, citing his recording of the progress of a neural impulse along a damaged axon, stated that: "If an inactive stretch of fiber over 1 mm in length does not stand in the way of electrical transmission of the impulse, is it reasonable to maintain that the discontinuity at a synapse will stop such transmission."[5] A few neurophysiologists were more open to the possibility of chemical transmission. Detlev Bronk, for example, was clearly on the fence in 1939 when he wrote that: "I have no desire to defend either the acetylcholine hypothesis or the theory of excitation by circulating currents. . . . I would argue for a pluralistic theory."[6] In general, however, the two sides were far apart, and the dispute called the "War of the Soups and the Sparks" extended over two decades.[7]

John Eccles was not as willing as some others to concede a role for chemical transmission. In a letter dated February 11, 1939, Henry Dale wrote to Eccles that John Fulton had informed him that he (Eccles) was about to report "some experiments that strongly support the acetylcholine hypothesis." Eccles replied on March 1 that he could not imagine where Fulton had gotten that idea and added: "I fear your hopes of an early rapprochement will have suffered a disappointment before this [letter] reaches you . . . the fact is that I have become more antagonistic than ever to the humoral view."[8]

John Eccles is generally considered to have been the leading and most influential opponent of chemical transmission. Eccles was a native Australian who, after completing his medical degree with first class honors at Melbourne University in 1925, enrolled on a Rhodes scholarship at Magdalen College, Oxford. From 1927 to 1931 Eccles worked in Charles Sherrington's laboratory, receiving his doctoral degree in 1929. Much of his research, including his thesis, was on neural excitation and inhibition. Eccles was under consideration to succeed Sherrington, but when he was not chosen he returned to Australia in 1937 to head a small research unit in the pathology department of a hospital in Sydney.

In a letter dated September 16, 1937, Dale wrote to Eccles:

> Putting all chaff aside, I do want to take the opportunity to say how much I deplore your departure from physiology in this country. We have, in the last few years, had numerous opportunities of controversy, which on your side, and I hope on ours, has always been fair and good tempered. Our differences of opinion and interpretation, however, have not in the least weakened my admiration for the splendid work which you have been doing . . . I hope very sincerely that you will find conditions in Sydney such as to enable you to continue the work which you want to do. . . . If Sydney does not give you the opportunity that you want, you must come back to us.[9]

In Eccles' response to Dale on November 28 he noted that he had been in Sydney seven weeks and was beginning to feel at home. He wrote that that his "artillery" would soon be ready, but he expected that by that time "you will have constructed new lines of fortification for one to have a crack at. Of course at this distance we will miss the short range practice that we had at the Physiological meetings, but, if you don't make it too hot for me, I may venture over to England in a few years."[10]

In 1944 Eccles left Australia to become chair of physiology at the University of Otago Medical School in New Zealand. He had written earlier to Dale asking if he would recommend him for the position, and Dale wrote back that he would be very glad to act as a "referee" for him.[11] At Otago, Eccles and the philosopher of science Karl Popper spent much time together. His later books and theoretical articles on the brain, mind, and consciousness reflect Popper's influence. Eccles returned to Australia in 1952 as professor of physiology at the Australian National University in Canberra and remained there until 1966. These were Eccles' most productive years. During the years in Canberra, Eccles had an estimated seventy collaborators from twenty-two different countries, and wrote approximately four hundred scientific papers and three books. He attracted some outstanding collaborators, including Bernard Katz, a future Nobel Laureate, and Stephen Kuffler, who would later be elected to both the National Academy of Science in the United States and the Royal Society in Great Britain.

Eccles was elected to the Royal Society in 1941. He was knighted by Queen Elizabeth II in 1958, and in 1963 Sir John shared the Nobel Prize in Physiology or Medicine with Alan Hodgkin and Andrew Huxley.[12]

Prior to returning to Australia, while at Oxford, Eccles had regularly opposed the theory of chemical transmission, particularly at the synapse between spinal motor nerves and skeletal muscles. Despite this, a warm relationship existed between Eccles and the group associated with Dale. Feldberg later wrote that Eccles' opposition "had a most beneficial effect. We were not allowed to relax, but were forced to accumulate more and more detailed evidence in support of our theory"[13] Similarly, Dale described Eccles as an "ideal sparring partner," who honed the arguments supporting chemical transmission. Eccles apparently felt the same way, judging from his later description of an exchange he had with Dale at a Royal Society symposium held in 1937: "I learned there the value of scientific disputation—that it provides a great incentive to perfect one's experimental work and also to examine it more critically. Of course the critical appraisal is even more searchingly applied to the experiments of one's opponents."[14]

The debates between Eccles and Dale were always constructive and conducted in a friendly spirit, even though it often appeared to those who did not know them that they were about to come to blows. Sir Bernard Katz,

Figure 8.1   John Eccles in his neurophysiology laboratory.

who left Nazi Germany for England in 1935, described his first visit to Cambridge to attend a meeting of the Physiology Society that year.

> To my great astonishment I witnessed what seemed almost a stand-up fight between J. C. Eccles and H. H. Dale, with the chairman E. D. Adrian acting as a most uncomfortable and reluctant referee. Eccles had presented a paper in which he disputed the role of acetyl-choline as a transmitter in the sympathetic ganglion, on the grounds that eserine, a cholinesterase inhibitor, did not produce the predicted

potentiating effect. . . . When Eccles had given his talk, he was coun-
terattacked in succession by Brown, Feldberg, and Dale. . . . It did
not take me long to discover that this kind of banter led to no resentment between the contenders, it was in fact a prelude to much
fruitful discussion over the years and indeed to a growing mutual admiration between Dale and Eccles.[15]

Eccles and Dale often exchanged letters explaining their positions or
their reactions to each other's publications, and they customarily sent each
other copies of manuscripts before they were published. Their differences
were often softened by humor. On one occasion, for example, Dale and
Feldberg had just finished a vigorous tennis game when they met Eccles.
They had worked up quite a sweat, and Dale could not resist remarking to
Eccles that the sweat in their socks was undoubtedly dripping with acetylcholine. This was an allusion to Dale and Feldberg's demonstration that
the nerves that innervate the sweat glands on animal paws secrete acetylcholine, an exception for a sympathetic nerve. Eccles replied that before he
could accept that they would have to assay their socks.[16] Another example
is a 1937 exchange between Feldberg and Eccles. Feldberg was working in a
laboratory in Sydney at the time, and Eccles, who was in Canberra, sent a
message to him. The message was: "Acetylcholine is all wet." Feldberg immediately sent back a telegram: "Prefer wet acetylcholine to dry eddy
currents."[17]

In his 1937 review of the arguments for and against chemical and electrical transmission of nerve impulses to skeletal muscle, Eccles described
the two positions as being "in sharp contrast," with no apparent compromise possible.[18] His review of the evidence and arguments for the two theories was, however, balanced, and he acknowledged that there was as yet
"no conclusive evidence for or against either of them." He did conclude,
however, that acetylcholine could not be the main means of transmission,
although he acknowledged that it "may have a secondary action in raising
the excitability of the effector cells and in counteracting the onset of
fatigue."[19]

In December 1939 Cannon wrote a strong defense of chemical mediation of neural impulses in which he primarily addressed the opposing arguments of John Eccles and also John Fulton.[20] Cannon referred to those
defending the traditional electrical theory as "electragonists," while he
called those supporting chemical mediation "chemagonists," explaining
that "agonist" meant contestant or combatant. He noted that the electrag-

onists were willing to grant that acetylcholine mediated transmission at the synapse between the vagus and the heart, but denied any important role for that substance at the spinal nerve synapse on skeletal muscles: "they agree that acetylcholine is a deputy of nerve impulses at vago-cardiac synapses, but deny it that function for myoneural synapses."[21]

Cannon began by listing what both sides agreed on. They agreed, for example, that at parasympathetic synapses the perfusion of minute quantities of acetylcholine evoked the same responses as neural impulses. The electragonists also accepted the evidence, most of it collected by Dale and his colleagues, that acetylcholine was released at postganglionic parasympathetic synapses, at all autonomic ganglia, and also at the synapses between spinal nerves and skeletal muscles.

However, Cannon explained, neurophysiologists believed that acetylcholine was important only for evoking the slow visceral responses, and could not mediate the fast responses of skeletal muscles. Cannon noted that Eccles maintained that acetylcholine had only a "trophic influence" on skeletal muscle responses, increasing the response generated by the electrical impulse. Fulton was cited as describing the acetylcholine released as only a "byproduct of nerve metabolism."[22]

Cannon answered the argument that chemical transmission is too slow by pointing out that this was only an assumption, as "too little is known of the speed of chemical processes at synapses to justify categorical limitations." Moreover, he noted, the argument could be turned around by asking: "How can the electragonist explain the 0.2–0.4 millisecond delay that everyone agrees occurs at synapses?" This delay, Cannon argued, is five to eleven times greater than is required for electrical transmission across the synapse: "The relatively long delay in the [autonomic] ganglion is matched by a similar delay in the motor end plate [of skeletal muscles]. Here the electragonists have a real problem. The chemagonists, on the other hand, can readily account for the extra time as due to the requirements of an interposed chemical mediation."

Cannon then discussed several phenomena that he considered difficult, if not impossible, for the electragonists to explain. Among them was the fact that there are certain drugs that either block or enhance the response to nerve stimulation without altering the nerve impulse, but do modify acetylcholine effectiveness. He cited the example of curare, which blocked the response of skeletal muscles, but did not interfere with the nerve impulse. Curare did, however, raise the threshold of response to acetylcholine. Similarly, a small intravenous injection of eserine enhanced skeletal muscle responses to nerve stimulation although it did not alter the nerve impulse. Eserine, Cannon reminded readers, prolongs the action of acetylcholine by inhibiting cholinesterase. Such results, Cannon concluded, could be understood only by assuming that the nerve impulse is effective only to the extent that acetylcholine is active.

Cannon cited a number of other lines of evidence that presented problems for the electragonists but could be explained by the known characteristics of acetylcholine, cholinesterase, and various drugs. He then concluded:

> All these observations harmonize perfectly with the chemical theory of transmission and find no illumination whatever in the electrical theory. . . . Until the electragonists can display an instance of electrical transmission without acetylcholine at neuromuscular and neuroneuronal synapses their argument cannot be on the same footing as that of the chemagonists.[23]

As persuasive as Cannon's arguments may now seem, he did not change the mind of anyone committed to electrical transmission. At a 1939 symposium on the synapse, for example, when Dale remarked that nature would not have arranged for acetylcholine to be released at synapses just to fool physiologists, a neurophysiologist responded that the same could be said for the presence of action potentials that crossed the synapse. Eccles later characterized his position at the time: "My position was not that chemical transmission did not occur, but that it was a later slow phase of transmission, the early fast phase being electrical."[24] Eccles also claimed that the fast responses of muscles were not affected by eserine.

One of the problems that neurophysiologists had the most difficulty explaining was how at some sites a neural impulse produced excitation and at other sites it produced inhibition. Inhibition was clearly a problem for the electrical transmission theory. The various explanations proposed usually involved complex and often strained hypotheses of how wave interference, or "eddy currents" of different polarities, might produce inhibition.

Eccles had been attempting to explain how eddy currents could cross the synapse and, depending on their polarity, produce either excitation or inhibition. In 1947 he had a sudden inspiration that led to a new theory to explain inhibition of skeletal muscles. The theory was based on the properties of a small "interneuron," a neuron discovered with the aid of the Golgi stain. The cell was believed to have an unusual pattern of discharge and was found to exert an inhibitory influence on spinal motor neurons. It was initially called the Golgi cell, but today is generally called the Renshaw cell, after its discoverer, Birdsey Renshaw of the Rockefeller Institute for Medical Research.[25] Eccles' theory of inhibition was based on some properties this cell was believed to possess. He described how the theory had occurred to him in a dream: "On awakening I remembered the near tragic loss of Loewi's dream, so I kept myself awake for an hour or so going over every aspect of the dream and found it fitted all experimental evidence."[26]

Eccles wrote an article describing his theory, including a diagram purporting to explain how the "Golgi cell" could produce inhibition by inducing an anodal state at its point of contact with a postsynaptic neuron. The theory, which was published in the journal *Nature* in 1947, received a lot of attention and for several years it seemed to be gaining some experimental support. In the end, however, it was Eccles who proved the theory wrong, and he, to his credit, was the first to acknowledge it.

To measure inhibition, neurophysiologists today determine the voltage difference between the inner core of a neuron and its outer membrane. If the voltage difference is increased it is more difficult to excite (depolarize) the neuron. This is what constitutes an inhibitory state, or an inhibitory postsynaptic potential (IPSP). Just the reverse occurs when a neuron's state of excitability is increased and an EPSP is induced. Until 1951 the evidence for inhibition in the spinal cord had been collected with extracellular recording electrodes, which were relatively insensitive to voltage changes of the magnitude that had to be measured to detect inhibition in a single cell. In mid-1951 glass microelectrodes capable of penetrating and recording from single neurons became available. They were glass tubes filled with a saline solution and measuring only about one-50,000th of an inch. The electrodes made it possible to obtain much more precise measurements of the voltage changes that characterized inhibitory and excitatory states. The technique was rapidly adopted by Eccles.

The critical experiment was done one day in mid-August of 1951 by John Eccles, Jack Coombs, and Lawrence Brock. It was quite a day, because in the middle of the experiment, while Eccles attended to the cat they were using, Brock left the lab for the hospital to attend to Coombs' wife, who was delivering a baby girl. After the birth, they reassembled around the cat and by the end of a long day of recording voltage changes from the spinal motor nerves that innervate skeletal muscles, it became clear that Eccles' "Golgi-cell hypothesis of inhibition" was wrong and that only chemical transmission could explain their results. As they wrote in the publication,

The potential change observed is directly opposite to that predicted by the Golgi-cell hypothesis, which is thereby falsified. . . . It may therefore be concluded that the inhibitory synaptic action is mediated by a specific transmitter substance that is liberated from the inhibitory synaptic knobs and causes an increase in polarization of the subadjacent membrane of the motor neurone.

The striking contrast between the small presynaptic spike and the large [post-] synaptic potential further makes it appear improbable that the trans-synaptic flow of current would be adequate to evoke that large post-synaptic response. On the electrical theory of transmission it does not seem possible to provide an explanation for the great amplification observed. On the other hand, just such a large amplification is actually produced by the chemical transmitter mechanism at the neuromuscular junction, where the most probable explanation is that the acetylcholine liberated from the nerve terminals triggers off the sodium carrier mechanism.[27]

The "sodium carrier mechanism" had been discussed in recent publications by the British neurophysiologists Alan Hodgkin and Andrew Huxley, who had shown that the passage of sodium and other ions in and out of a neuron is what determines its state of excitability.[28] Hodgkin and Huxley also described the "sodium pump" mechanism, which increases the voltage differential across a neuron by pumping sodium out of the cell against an osmotic gradient. Based on this work, Eccles and his collaborators concluded:

The experimental observations on synaptic excitatory and inhibitory action require for their explanation two specific transmitter substances. The excitatory substance probably acts by stimulating the sodium carrier mechanism, while it is suggested that the inhibitory substance possibly acts by stimulating the sodium pump.[29]

Eccles sent Dale an advance copy of the manuscript, writing that "I hope that these modest offerings to the theme of chemical transmission will in some measure atone for my long delay in conversion."[30] Dale replied that he had derived great pleasure from reading the paper, adding: "Your new-found enthusiasm [for chemical transmission] is certainly not going to cause any of us any embarrassment." In a second letter, Dale wrote: "I do congratulate you all, not only upon the beauty of the observations recorded, but on the very attractively clear and concise account of them in the paper.[31] Dale later wrote that Eccles' change was like the conversion of Saul on the way to Damascus, when "the sudden light shone and the scales fell from his eyes."[32]

Although Eccles had capitulated in 1952 and accepted the chemical transmission hypothesis for spinal motor neurons, this did not abruptly end the opposition by others to the theory of chemical transmission at central nervous system synapses. A history of long and influential opposition had to be overcome. Throughout the 1940s and much of the 1950s, neurophysiologists had continued to insist that transmission was primarily electrical in the spinal cord and probably exclusively electrical at brain synapses. The neurophysiologists favoring the electrical hypothesis and the pharmacologists supporting the chemical hypothesis used such different techniques and obtained data so difficult to compare that each tended to ignore the other side's arguments. Neurophysiologists might talk in private about chemical transmission, but they usually avoided the subject in their publications. This led Dale to comment later that the subject of chemical transmission was treated "like a lady with whom the neurophysiologist was willing to live and consort in private, but with whom he was reluctant to be seen in public."[33]

John Fulton, the editor of the *Journal of Neurophysiology* and one of the most influential neurophysiologists at the time, was a strong supporter of the electrical transmission position, but sometimes he seemed to be hedging his bets. In a letter dated August 22, 1943, Dale, displaying his masterful sense of irony, wrote to John Eccles: "I am told that John Fulton, in a recent number of *Science*, has begun to balance himself more carefully than before on the top of the hedge, so that eventually we may find you all on the same, safe side."[34]

In 1949 John Fulton was still carefully on "the top of the hedge" while continuing to give more weight to the electrical transmission explanations. He had written in his influential textbook, *Physiology of the Nervous System*:

[Although] the theory of chemical mediation of nerve impulses appeared acceptable to many physiologists in the case of autonomic nerves acting on their effector organ, this concept, when applied to synapses and neuromuscular junctions, was less satisfactory and encountered increasing opposition. In addition to a great number of difficulties and contradictions, which were partly reviewed by John Eccles (in 1937) and have increased continuously since then, there are two main objections, the first being the time factor. This factor was of less importance in the case of the slowly reacting cells innervated by the autonomic nervous system. But the transmission of nerve impulses across the neuromuscular junctions and synapses occurs within milliseconds. No evidence was available that the chemical process can occur at the high speed required. . . . The idea of a chemical mediator released at the nerve ending and acting directly on the second neuron or muscle thus appeared to be unsatisfactory in many respects.[35]

In a later review of this history, Eccles wrote that at a symposium held in Paris in 1949 many participants conceded that acetylcholine mediated transmission at various peripheral synapses, "however, there was still fairly general agreement that central synaptic transmission was likely to be electrical." He also recalled that a great amount of opposition to the theory of chemical transmission at central nervous system synapses was still being expressed by Albert Fessard and others at a 1951 symposium in Brussels.[36]

In general, the possibility of chemical transmission hardly had any effect on the thinking of scientists, aside from those directly involved in studying synaptic transmission. For example, in their highly successful physiological psychology textbook published in 1950, Clifford Morgan and Eliot Stellar described synaptic transmission as occurring when the electrical changes in the presynaptic neuron induce a depolarization in the postsynaptic neuron. They did not mention chemical transmission as even a possibility.[37] The text presented, without comment or qualification, Eccles' earlier diagram, which attempted to explain inhibition as solely an electrical process. In later editions there was no reference to the fact that Eccles had changed his mind. Neither did Morgan and Stellar mention acetylcholine when describing the autonomic nervous system. Not only the Morgan and Stellar book, but virtually all physiological psychology textbooks of this period did not consider any of the implications of chemical neurotransmitters, if they mentioned the topic at all.

By 1953 some neurophysiologists had accepted John Eccles' evidence that transmission at the synapse between spinal motor nerves and skeletal muscle is chemical, but there remained many who were not persuaded that acetylcholine is the primary means of transmission at this synapse. At a 1953 Philadelphia symposium on chemical transmission, Ralph Gerard, a renowned neurophysiologist and the person credited with introducing the term "neuroscience," was given the task of summarizing the meeting. Those in attendance were primarily the Europeans who had provided the bulk of the evidence supporting chemical transmission in the periphery, including Otto Loewi and Henry Dale. Gerard began by remarking, "As the lone American in this distinguished galaxy and as one of the few physiologists amidst the pharmacological cohorts, the Committee may have wished me to serve as devil's advocate for an electrophysiological approach to neural functioning."[38]

Gerard accepted the challenge and presented a number of arguments for electrical and against chemical explanations. He argued that Eccles' 1952 experiment had not completely ruled out electrical explanations of inhibition in the spinal cord. Gerard raised the speed of conduction issue, stating that:

> No liberation or diffusion of chemicals could account for the ability of an impulse to jump over a millimeter of inactivable nerve fiber in a fraction of a millisecond. . . . Moreover, many, if not all of the phenomena of junctional transmission are neatly accounted for by the properties of eddy currents and the geometry of the junctional region.

The weakness of Gerard's arguments are now apparent, but at the time they provided support for those who refused to accept chemical transmission.

Ralph Gerard was not alone among neurophysiologists in contesting Eccles' evidence. At the same 1953 Philadelphia symposium, A. K. McIntyre of the University of Otago in New Zealand showed resentment that John Eccles had abandoned the neurophysiologists' position. McIntyre, who was a colleague of Eccles at Otago and had published an article with Brock, one of his collaborators, remarked that he found it: "particularly entertaining that so much of the newest evidence presented in favour of neurohumoral transmission mechanisms in the central nervous system comes, not only from an electrophysiologist, but from such a high voltage spark as J. C. Eccles."[39] McIntyre went on to state that: "A critical evaluation of the very interesting experiments of Brock, Coombes, and Eccles with intracellular recordings from motorneurones reveals that rather sweeping conclusions have been drawn from somewhat slender evidence."

While the nature of the innervation of skeletal muscles was still being disputed by neurophysiologists in 1953, the possibility of chemical transmission in the brain was rarely even discussed by that group. Wilhelm Feldberg, a pharmacologist, had been studying the effects of injecting various drugs into the ventricles of the brain, and he raised the question of the nature of synaptic transmission in the brain at the Philadelphia symposium. Feldberg stated that in his view synaptic transmission in the brain is chemical and "that there is no fundamental difference between the transmission processes in the central nervous system and those in the peripheral nervous system." Feldberg did make it clear, however, that the available evidence supporting either side of the dispute was not yet convincing.

> We are faced with the problem that, whatever our bias, we cannot state with certainty whether the transmission is chemical or electrical. But, according to which view we hold, our approach will be different. Anyone who takes it for granted that transmission is electrical, assumes the phenomenon can be fully dealt with by analyzing electrical changes in the central nervous system, and such an analysis forms the main subject of his research. On the other hand, the adherents of the chemical theory believe that a fruitful analysis of this kind requires accurate knowledge of the nature of the central synaptic transmitters.[40]

How the evidence accumulated to prove that neuronal transmission in the brain is chemical is the subject of chapter 10.

Henry Dale, Otto Loewi, and Walter Cannon were too old to contribute to this later part of the story. This book, however, is not only an account of scientific discoveries, it is also the story of the scientists involved—their personalities, how they worked, their friendships and disputes, and the social and political events that had an impact on them and their work. It seems only appropriate, therefore, to first describe the final years of the three principal scientists who established the foundation for the revolutionary changes in our understanding of how the nervous system works. That is the subject of the next chapter.

CHAPTER 9

Loewi, Dale, and Cannon

The Later Years

Otto Loewi, Henry Dale, and Walter Cannon had each passed his sixtieth birthday before 1936, the year the Nobel Prize was awarded for the discovery of chemical neurotransmitters. Their major scientific accomplishments were behind them, but they continued to make contributions, each in his own way, and under very different social and political circumstances.

OTTO LOEWI

Otto Loewi was sixty-three in 1936. He had remained active in research at the University of Graz, but in less than two years his life was completely disrupted by political events. The Austrian Nazi party had grown in strength after Hitler came to power in Germany in 1933 and was particularly active in the city of Graz. Nazi demonstrations were increasingly frequent, and anti-Semitism was becoming more virulent. Although most Austrian newspapers had expressed pride in November 1936 that an Austrian professor had been awarded the Nobel Prize, some newspapers, particularly the clerical press, according to the *New York Times,* buried the story on the back pages because Loewi was Jewish.[1]

Two years later, in 1938, the German Army marched into Austria and was greeted by crowds of cheering people lining the streets. Loewi has to have been aware of what was happening, but he seems to have shut it out and buried himself in work. At the time he was trying to extract adrenaline from sympathetic nerves. In a preliminary paper on that subject published in 1938, Loewi had added a footnote indicating that the work had been interrupted. What had happened was that in March 1938, only a few days after the German army arrived, storm troopers broke into Loewi's home at three o'clock in the morning, awakening him from a deep sleep. As Loewi later recalled:

> A dozen young storm troopers, armed with guns, broke into my bedroom, took me downstairs, and pushed me without any explanation into a waiting prison van that took me and others to the city jail. Later that night I was joined there by our two youngest sons, Victor and Guido. Our two elder children, Hans and Anna, were not in Austria at the time, and, of course, did not return.[2]

Loewi and two of his sons were put in jail in "protective custody," along with many other Jews.

The news of Loewi's arrest was in the newspapers, and it spread through the scientific community. Loewi's son Hans wrote to Walter Cannon and Henry Dale to ask if they could intervene in some way. After making some inquiries, Dale decided that a public or even an official protest might actually endanger Loewi. He let it be known that if Loewi could get out of Austria, he would make arrangements in England for him to be able to continue his work, and he started to explore avenues of possible support for Loewi. Walter Cannon and Loewi had previously met on several occasions, and they had a warm relationship. They had also corresponded about their mutual interests in humoral transmission and had an opportunity to talk at the 1935 International Congress of Physiology in Leningrad. Cannon wrote a letter requesting help from the U.S. State Department.[3] The letter was also signed by several eminent professors at Harvard. The reply written by Assistant Secretary of State George Messersmith was not encouraging:

> It is our primary duty to offer full and effective protection to American citizens and American interests, and hence any deviation from this policy could not be other than prejudicial. Accordingly, while fully understanding your solicitude, it is felt that we cannot take the step you recommend on behalf of Professor Loewi.[4]

Loewi was released after two months in jail. The Austrians had adopted a policy of "enforced emigration." The policy, which Adolf Eichmann administered during this period, allowed, even encouraged, Jews to leave the country, providing they surrendered almost all their assets. Loewi was permitted to leave Austria, but he had to transfer his assets to the German government, including the Nobel Prize money that had been left in a Swedish bank.[5]

Loewi left Austria for England in September 1938. He had cabled Dale, who invited him to stay at his home. Loewi's wife, Guida, was not permitted to leave until she arranged to surrender some property in Italy that her family owned for more than one hundred years. It was difficult leaving his wife behind, but the situation demanded it. In the autobiographical sketch of his life, Loewi said of Guida, "She was my faithful companion, devoting her life to my well-being for over fifty years." It was almost three years before Guida could leave Austria, and during this time she lived on a small subsistence allowance given to her after she signed a "pauper's oath." She finally gained permission to leave in 1941 and was able to join her husband, who was by then in New York. Fortunately, this was arranged just months before war was declared between the United States and Germany.

When Loewi arrived in England, he lived with Dale for several weeks. He had obtained temporary positions at the Franqui Foundation in Brussels and also at the Nuffield Institute in Oxford. He was in Belgium in 1939, but fortunately left just before the war started and the German army marched into that country. Zénon Bacq, who met Loewi at that time, wrote that: "He arrived alone, without any money, a real refugee."[6] After some negotiations, and with the help of Walter Cannon, Loewi, who was sixty-seven years old at the time, was offered a position as research professor in pharmacology at the Medical School of New York University.[7] Loewi accepted the offer and arrived alone in New York in June 1940. Loewi's account of his arrival is one of those stories that are amusing only in retrospect.

> Upon my arrival in New York harbor, a clerk prepared my papers for the immigration officer. While he was busy doing this, I glanced over the doctor's certificate—and almost fainted. I read: "Senility, not able to earn his living." I saw myself sent to Ellis Island and shipped back to Mr. Hitler. The immigration officer fortunately disregarded the certificate and welcomed me to this country.[8]

Despite living a life very different from what they were used to, Otto Loewi and Guida adjusted rapidly to living in the United States. Instead of the large house they had owned in Graz, they lived in an apartment on the west side of New York City. Loewi was happy in New York and considered it fortunate that he had had to leave Austria, because if he had remained, he would have been forced to retire as an active professor. This would have meant, he once remarked, hanging on as a permanent guest in a corner of the department "as welcome to his host, as a mother-in-law to a young married couple." As soon as he met the residence requirements, in 1946, Loewi became a U.S. citizen.

Loewi did not start any new lines of research, occupying himself in the laboratory by tying up some loose ends on problems he had previously worked on. He spent most of his time writing articles and giving lectures. The lectures often were historical accounts of the events leading up to the discovery of neurohumoral secretions, but he also liked to talk to students about their research and the traits that facilitate scientific discoveries.

A graduate student at the University of Kansas in 1952, this author remembers hearing Loewi give two lectures entitled *From the Workshop of Discoveries*. The first lecture reviewed the history of the discovery of chemical neurotransmitters, while the second was on the psychology of creativity in science. Loewi described instances where the idea behind a discovery had seemingly popped into a scientist's head while daydreaming or during an actual dream. He commented that one should not be too quick to reject such intuitive ideas, noting that if he had thought very much about the experiment he had dreamt about he would have found many reasons why it could not possibly work. He also spoke about the importance of having a wide range of experiences, not restricted to scientific topics, as they often prepare the mind for grasping the significance of unexpected observations. Loewi quoted Louis Pasteur's statement that, "In the field of observation chance favors only the prepared mind." He described how Walter Cannon's chance observation that digestive activity ceases during emotional states started him thinking about how emotions affect all bodily functions. Loewi ended his lecture by recommending to his audience, when judging the scientific promise of a man, "Do not use as the yardstick for his abilities the number or the weight of the publications, as you usually do, but the number of sleepless nights when he was struggling with problems."[9]

Loewi and Guida spent seventeen consecutive summers at the Woods Hole Laboratory on Cape Cod, where they had made many friends. Loewi enjoyed learning about the research of others. He was naturally gregarious and enjoyed chatting with students at New York University, where his colleagues called him "Uncle Otto." He loved to tell stories, which were often constructed to entertain as well as to convey some lesson. After attending a piano recital, for example, Loewi told some students, "You know, I heard Vladimir Horowitz the other day at Carnegie Hall. I must confess that when I watched the elegant fingers of the pianist

Figure 9.1 Otto Loewi talking to students at Woods Hole. It was said of him that he was "One of the most loved and respected members of the Woods Hole community" (H. B. Steinbach) and that "The fire that he needed was the enthusiasm of his young admirers" (H. Kaunitz). Cited in F. Lembeck and W. Giere, Otto Loewi: Ein Lebensbild in Documenten (Berlin: Springer, 1968), opposite p. 92.

flitting across the keyboard with a speed that my eyes were unable to follow, I began to develop grave doubts in my own theory of the chemical transmission of nerve impulses."[10]

Loewi received numerous awards and honors following the Nobel Prize, including honorary degrees from the Universities of Graz and Frankfurt. No honor pleased him more than being elected a foreign member of the Royal Society. Despite poor health, Loewi remained intellectually curious until his death. Zénon Bacq described meeting Loewi, who was in his mid-eighties, in New York in 1957:

> He invited my wife and myself to dinner in a sumptuous restaurant. His health was precarious. Mrs. Loewi looked anxious after his fits of breathlessness which interrupted his conversation, but he was extremely happy to meet an old comrade in arms of pre-war years. "Bacq," he said, "I should have been dead long ago. But everything interests me. I do enjoy living on overtime." He gave the impression of indomitable energy, like a force of nature, a kind of volcano throwing out its last fires before becoming extinct.[11]

The following year Loewi had a serious fall that broke his pelvis. Friends rallied around him and transported him to Woods Hole the next summer. Dale last saw Loewi during a visit to New York in October 1959. By then Loewi was pretty much confined to his New York apartment, having to push himself around with a walker. Guida had died the previous summer. However, Dale reported that his mind was as alert as ever.

Loewi died in his apartment in 1961 in a way he would have approved. A friend from Woods Hole had sent him a lobster and he consumed it with much pleasure, sharing a bottle of fine wine with his nurse-housekeeper. On the following morning, while engaged in a lively discussion with a friend, Loewi suddenly went silent. He had died, at the age of eighty-nine, on Christmas Day. Six months later many of Loewi's friends gathered at the Woods Hole Laboratory to commemorate his death and to bury his ashes.

Loewi had an artistic temperament, and he probably took some liberties in adding drama to his accounts of the dream that led to the discovery of neurohumoral secretions. When questioned about some of these discrepancies, van der Kloot, Loewi's colleague at New York University, said of him, "He certainly loved a good story, and they certainly showed some normal variation. He was also one of the most upright and decent men I have ever known."[12]

John Gaddum, who had been a major contributor in Henry Dale's laboratory from 1927 through 1933, wrote in an obituary for Otto Loewi that

"Every one who met him loved him as a man and respected him as a great man."[13]

Loewi had retained his interest in the arts throughout his life. The *Saturday Review* stated, "The years of Dr. Loewi have so overflowed with devotion to art, literature, music, mountain climbing, human fellowship, and the science of biology that only a book could tell his adventures."[14] Loewi rarely missed an opportunity to visit an art museum or attend a concert, and it is especially appropriate that his last publication was entitled: "A Scientist's Tribute to Art."[15]

In 1973, to celebrate Otto Loewi's one-hundredth birthday, the University of Graz commissioned a bronze bust, which was placed in the lobby outside the large assembly hall. The bust was stolen in 1985.

HENRY DALE

Henry Dale was sixty-one in 1936. As mentioned above, Dale had been heavily burdened with administrative responsibilities prior to Feldberg's arrival in his laboratory in 1933. The successes made possible by the leech assay technique that Feldberg brought to the laboratory had played a major role in rekindling Dale's enthusiasm for laboratory research. After the awarding of the Nobel Prize in 1936 and Feldberg's departure for Australia that year, however, Dale began to withdraw from the laboratory, and after 1938 he no longer did any experimental work. He still closely followed the research in all the areas in which he had worked, and he continued to attend professional meetings until he was ninety. Dale was frequently asked to summarize the presentations at scientific meetings, and his comments were often described as being more insightful about the implications of new data than those offered by the presenters themselves.

After he resigned from his position as director of the National Institute of Medical Research in 1942, Dale became director of the Davy-Faraday Laboratory at the Royal Institute of Great Britain. He had been appointed a trustee of the Wellcome Trust Fund in 1936, and as chairman of the fund in 1946 and afterward, Dale devoted considerable time to awarding research grants and scholarships to young scientists. He also served as president of the Royal Society from 1940 to 1945 and in that capacity was a member of the advisory committee to the War Cabinet during World War II. He received the Order of Merit in 1944 for his service.

Dale enjoyed participating in discussions at meetings and refused to accept honorary membership in the Physiological Society when he reached seventy years of age, as the title was generally used to honor retired physiologists. He finally consented to the honorary membership status when he was seventy-five. He was eventually forced to recognize, as most older scientists must, that he was having difficulty keeping up with all of the new technical developments. Dale once remarked, after visiting a laboratory filled with new electronic equipment, that he had begun to feel like an extinct dinosaur. He said that physiologists would soon be saying to each other when he passed: "Look, there's old Dale. He once used a [smoked drum] kymograph."[16]

Still, he kept up with the important ideas. In 1956, when Dale was past eighty, he wrote to John Eccles that he had been speculating about what is now called axoplasmic flow, the movement of neurotransmitters from the cell body of neurons to their axon terminals.

> I suppose that, like the cytoplasmic process of any part of the cell, the axon and its ending must be dependent on the nucleus of the cell body for the maintenance of their integrity and special constituents. Certainly if the axon is cut, one of the earliest effects is the disappearance of the transmitter from the axon ending, and the failure of synaptic transmission, even before the conduction of the impulse as far as the ending has been noticeably affected. One must suppose, I think, that there is a constant drift of transmitter, with its associated enzymes, from the cell body down the axon, so that there is an accumulation at the ending, and ready replacement there of what is discharged by the impulse. I have even toyed with the idea that the passage of the impulses down the axon might accelerate such a drift, but without having any notion as to how it might do so.[17]

Dale remained a prolific writer, publishing twenty-six articles between 1937 and 1942 alone. He kept up a steady pace almost until his death, writing thirteen articles after his eighty-fifth birthday. The later articles were mostly on historical aspects of science, tributes to scientists, comments on education, and obituaries of colleagues.[18] He outlived Loewi and Cannon and wrote lengthy obituaries of them that appeared in the series *Obituaries and Memoirs of the Royal Society*. Dale's obituary of his close friend Thomas Elliott was also published in the same series.

Dale had a great facility with words and expressed his ideas clearly and carefully, often with a wry sense of humor and irony. He was also kind and generous in crediting others, and while he did not hide his disagreements, he tended to express these subtly, often as reservations about whether the ideas he was questioning would have lasting value. It is unfortunate con-

Figure 9.2 Henry Hallett Dale on his eighty-fifth birthday.

sidering Dale's skill with words that he never wrote a book or taught in a classroom. Dale never held a teaching position, nor did he want one. He did, however, have a lasting impact on the many scientists who worked in his laboratory over the years.

Dale had often used his influence to help scientists who were dismissed from their positions in Nazi Germany. As mentioned several times throughout this book, he found positions for many in his own institute and helped others find temporary positions in different institutions throughout Great Britain. In 1903 Dale spent four months working in Paul Ehrlich's laboratory in Frankfurt-am-Main. After Ehrlich received the Nobel Prize in 1908, a street in Frankfurt was named Paul Ehrlich Strasse. Ehrlich died in 1915. In August 1938, the Nazis, who had been deleting references to all eminent Jews, removed his name from the street. After the war, Dale helped persuade the authorities in Frankfurt to change the street name back to Paul Ehrlich Strasse.

Dale had always been sensitive to injustices, wherever they occurred, but rather than joining political movements, he often took a personal stand. He refused to attend the 1935 Physiological Congress in Leningrad because of the crimes of the Soviet government, and he objected strongly to any governmental intrusion into science. In his 1941 Presidential Address to the Royal Society, Dale stated, "I see danger if the name of science, or the very cause of its freedom, should become involved as a battle cry in a campaign on behalf of any political system, whether its opponents would describe it as revolutionary or reactionary."

Dale was clearly referring to the Soviet Union and to Nazi Germany. In 1949 he resigned from the Academy of Science of the USSR after it dismissed the geneticists who opposed Lysenko's theories. He commented that, if this policy persisted, it would mean that "the whole fabric of exact knowledge still growing at the hands of those who have followed Mendel, Bateson, and Morgan is to be repudiated and denounced."[19]

Referring to his role on the scientific advisory committee to the War Cabinet, Dale described his opinion on the development of the atomic bomb:

> Of course, I wanted to do my best, so far as I had anything to do with it at all, to help the enterprise along, and there was the fear at the back of everybody's mind that the Germans might get the bomb first, but subject to that I had always the hope that the bomb would be demonstrated to be possible, but that it would not be used.[20]

After World War II, Dale wrote articles on the aftermath of the bomb exploded over Nagasaki, and he expressed his conviction that there must be international control over atomic bombs. Although Dale recognized the difficulty in getting countries to surrender some of their sovereignty, he made it clear, especially after the Soviet Union had the atom bomb, that he thought it essential that the United Nation's Atomic Energy Commission negotiate some way of gaining international control over any future use of atomic weapons.[21] He also wrote against the development of the hydrogen bomb:

> We begin to talk glibly, almost with a wry humour, of the possibility of a hydrogen bomb. And of this we only know that its range and power of devastation would be some yet incalculable multiple of those of any nuclear-fission bomb, so that it could serve no visible or

conceivable purpose of peace, but only that of such an immeasurable destruction or the threat of it.[22]

Dale was especially concerned about the secrecy imposed on scientific activity during the "cold war" with the Soviet Union. He often spoke about the need for science to be free and open. In 1946, at the beginning of the cold war, Dale was invited to give the Pilgrim Trust Lecture before the U.S. National Academy of Sciences. He chose as his title "The Freedom of Science," and he told his audience "that secrecy is the main infection from which science sickens today."

Dale died on July 23, 1968, in a nursing home in Cambridge, England, where he had spent the last few years of his life. He was ninety-three. His former colleagues visited him regularly, and they were often rewarded by his stimulating comments and his enthusiasm for any of the data or ideas they shared with him. Dale's many fundamental scientific contributions were reviewed in obituaries published in professional journals and newspapers around the world. One that appeared in the British journal *Lancet* stressed Dale's personal traits, which endeared him to all who knew him:

> Sir Henry Dale was not only the greatest figure on the scientific staff of the Medical Research Council, but also a pillar of its administration and a constant advisor of its officers on many matters. To these tasks he brought the same qualities that made him so outstanding in research—clarity of mind, breadth of vision, balanced judgment, unremitting care, and shrewd practicality. His help was always sought; and it was unstintingly given, with that warm-hearted consideration for others that was so characteristic.[23]

WALTER CANNON

Although he was the eldest of the three under consideration here, in his later years Walter Cannon remained the most active in refining and extending his scientific theories. Cannon had turned sixty-five in 1936, the compulsory retirement age in most universities in the United States at that time. The year before he reached this milestone, Cannon wrote to James Conant, Harvard's president, offering to resign. A Rockefeller Foundation committee had just voted to support Cannon's research for an additional five years. This was the first grant the foundation had given in support of a man rather than a specific research project. Conant responded to Cannon's letter by assuring him "that the President and Fellows have no desire to see you give up your valuable work in the Harvard Medical School."

Despite failing health, Cannon remained enormously productive professionally up to his death, and he somehow continued to find the time and energy to get involved in the social and political causes he cared about. He kept abreast of political events in part through correspondence with his many foreign students and collaborators after they had returned to their native countries, and he did not hesitate to be active in controversial organizations, even though he knew that he would probably be criticized for doing so. During the cold war, he was president of the American-Soviet Medical Society and before World War II started, he was an active member of the Joint Anti-Fascist Refugee Committee and a strong supporter of the Republicans during the Spanish Civil War. Until it was disbanded in 1939, Cannon had served as national chairman of the Medical Bureau to Aid Spanish Democracy, an organization that raised money for medical and surgical supplies and also for food and clothing. The bureau also established mobile hospital units that provided treatment for soldiers wounded during the civil war in Spain.

Cannon wrote letters and gave several speeches in support of the Republicans in Spain. In a speech at a huge rally held in the New York Hippodrome, he described what the Republican government in Spain was trying to do in resisting the forces led by General Franco. He also supported the Harvard undergraduates who collected money to purchase an ambulance to send to Spain. Before it was sent to Spain, the ambulance was driven in a peace parade organized by the American League Against War and Fascism. When these activities became known, Cannon was called a Bolshevik and accused of being anti-Catholic. (The Catholic Church was, for the most part, allied with the monarchy and with General Franco, who was leading the civil war against the elected Republican government.) Two Harvard undergraduates wrote a letter to the *Harvard Crimson* opposing Cannon's activities and the inclusion of the ambulance in a parade that they claimed was aiding Moscow and the communists. Cannon's reply made it clear that the ambulance was to be shipped to Spain with permission of the U.S. State Department, and he wrote, "The legitimate Government in Spain is fighting not only a group of Spanish fascists under Franco but also invading fascists from Germany and Italy. . . . Where the libelous letter dissolves into sound and fury is that the equipment was not used 'to aid Moscow,' but as part of a parade against War and Fascism."

Cannon had foreseen the outbreak of war in Europe and regarded the civil war in Spain as its opening salvo. He was not surprised, therefore, when war actually broke out in 1939, with England and France allied against Germany and Italy. He described his reaction to the outbreak of war in a letter to his former colleague Kálmán Lissák, who had returned to his native Hungary: "I feel at times as if the world was shaking at my feet. . . . No one knows what the outcome of this ominous war is going to be. The only faith which I cling to is that human society must persist and that there are forces at work which must establish a sort of social homeostasis in order that human society may persist."[24]

In August 1935, Cannon attended the XIVth International Congress of Physiology held in Leningrad. His experiences preceding and during the congress helped form his ideas about the Soviet Union. Pavlov and Cannon had become good friends over the years, and Pavlov had honored Cannon by inviting him to give the opening address at the congress. Their friendship had started earlier with a correspondence around their mutual interest in digestion. They had met twice before the Leningrad Congress, first when Pavlov visited in Boston and the second time when Pavlov was Cannon's guest in Cambridge.[25]

During the four months preceding the Leningrad congress, Cannon was in China. He had accepted an invitation to lecture and to demonstrate some techniques. While there, Cannon also initiated some experiments with Chinese colleagues.[26] After the stay in China, Cannon made a brief visit to Japan to visit laboratories of physiologists he knew from the literature. Cannon then traveled with his wife, Cornelia, from Vladivostok to Leningrad on the trans-Siberian railroad. The trip across Siberia was unpleasant as the train was hot and dirty, and the flies uncontrollable. The worst part, however, was seeing the many "convict camps" with barbed wire fences and guard towers and the freight trains loaded with men and women being sent east to the camps. While still on the train, Cornelia wrote to friends: "Walter is feeling the queer atmosphere of oppression and throttling that seems to underlie the simplest doings. He said later, 'Don't you feel as if you'd like to get out of this country?' There is a queer kind of horror that seizes you now and then, in these outlying regions where the veneer is thinner than I imagine it is in Moscow. The bear shows his claws in Siberia."

In Leningrad, Cannon's lecture was introduced by Pavlov. His lecture was entitled "Some Implications for Chemical Transmission of Nerve Im-

Figure 9.3   Ivan Pavlov and Walter Cannon in Boston.

pulses." After a warm tribute to Pavlov and before beginning the substantive part of his talk, Cannon expressed his feelings about the perilous state of the world and the impact that nationalism was having on science:

> Creative investigators of high international repute have been degraded and subject to privations. Some universities have been closed; others have been deprived of their ideal social function of providing a sanctuary for scholars where search for truth is free and untrammeled and where novel ideas are welcomed and evaluated. . . . The attention of the investigator must not be confined to immediately useful projects, nor to a so-called "nationalistic science," nor to a group of political ideas. . . . Every nation needs the advantages which flow from a free interchange of students and ideas.

Although Cannon did not mention Nazi Germany by name, it was clear to all that he was referring to it. The German delegates registered a protest and considered leaving the congress, but in the end they decided not to do anything that was likely to draw more attention to Cannon's remarks.

Figure 9.4   Part of a group picture taken in Professor Bykov's laboratory on the occasion of the XIVth International Congress of Physiology held in Leningrad in August 1935. From left to right: Wilhelm Feldberg, Walter Cannon, Otto Loewi, and Zénon Bacq.

Cannon did receive many compliments for his comments and, according to the *New York Times*, V. M. Molotov, the second most important person in the Soviet government at the time, personally congratulated him on his speech.[27]

When Cannon returned to Harvard following the congress in Leningrad, he was asked by many groups to talk about his impressions of the Soviet Union. His comments were balanced. He described the emphasis the country had placed on education, on activities for the young, in reducing the working hours at hard jobs, the parks, outdoor entertainment, and what seemed to be signs of great activity in building homes and factories. But he also spoke about the totalitarianism of the state and the cruelties and injustices imposed on many without any legal process, and he questioned why the present state was repeating the injustices that had led to the revolution against the old regime.

Cannon was asked to add his name to a manifesto issued by the International Committee of the Friends of the Soviet Union, but he refused to sign it, replying that "as a man of science devoted to the truth, I may be excused for insisting on the whole truth being respected." Cannon did consider accepting an offer to be president of the American Russian Institute in order to establish closer relations with Soviet scientists, but he turned down the offer when he became aware of more reports of the repressive hand of the Soviet government. Cannon refused for the same reasons to participate in a celebration of the twentieth anniversary of the Bolshevik revolution.

Cannon's health had been giving him trouble for a number of years. Early in his research career he had been warned about the possible danger of overexposure to X-rays, and he took some precautions, but little was known about the extent of the danger at the time and he apparently did not protect himself adequately. He developed painful X-ray burns on parts of his hands, and required several skin grafts. Only a few of his closest friends knew that he suffered from intense itching and dermatitis along the inside of his thighs, where he had often held X-ray tubes. A few years before his death in 1945, he had to undergo surgery to remove some skin from his lips.[28]

Despite declining health and being heavily involved in social and political causes, Cannon remained remarkably active in science. Until around 1939 he continued to do experiments on various aspects of chemical mediation of sympathetic effects, and he wrote articles and books extending his earlier ideas on different subjects. In the nine years from 1937 until his death in 1945, Cannon wrote more than forty articles. He also wrote two new books, revised an earlier book, and wrote a 230-page monograph with Arturo Rosenblueth that included some new experimental evidence they believed supported their two-sympathin theory.[29]

Cannon, who often searched for the broadest implications of his experimental results, had already written six books by 1936.[30] In 1932 Cannon summarized his ideas about homeostasis in *The Wisdom of the Body*, a book describing the different physiological mechanisms that maintain automatic control over the internal environment. The book was praised as a model of clear and lucid scientific writing; it went through several printings and was revised in 1939. The concept of homeostasis, a term Cannon had introduced, was widely adopted in many fields, from psychology to cybernetics.[31] With the encouragement of Cornelia, Cannon included an epilogue chapter entitled "Relations of Biological and Social Homeostasis."[32]

This theme was expanded in the 1939 revision of the book and also in 1942, when Cannon gave the American Association for the Advancement of Science (AAAS) Presidential Address entitled *The Body as a Guide to Politics.*[33] In the lecture, he proposed that mechanisms comparable to those that maintain the stability of the body's internal environment are needed to maintain the stability of social organizations: "Are there not general principles of stabilization. . . . Might it not be useful to examine other forms of organization—industrial, domestic, or social—in the light of the organization of the body?" He proposed several ways to correct radical economic swings, such as collecting reserves in good times that could be used to provide emergency employment and subsistence wages in bad times. This was, he felt, analogous to the body storing glucose for use in emergencies.[34]

Cannon published a new book, *The Way of an Investigator: A Scientist's Experience in Medical Research,* just months before he died in October 1945.[35] It was written for a popular audience and was meant to share what he had learned during his life in science. A review in the *Scientific Monthly* considered it "must reading for all medical students, if not all students in science in college and high school."[36]

From 1921 until his death, Cannon served on the National Research Committee (NRC) for Research on the Study of Sex, which was headed by his close friend Robert Yerkes. The NRC is the operational branch of the National Academy of Science. Cannon had long advocated funding research on sex-related topics, arguing that sexual behavior and the biology of reproduction are "fundamentally important for the welfare of the individual, the family, the community, the race. . . . Under circumstances where we should have knowledge and intelligence, we are ignorant. To a large extent, our ignorance is due to enshrouding of sex relations in a fog of mystery, reticence, and shame."[37] Although Cannon recognized the importance of studying sexual behavior in animals, perhaps because of his religious upbringing he did not believe such studies should be done on men and women. He once commented that he personally had no interest in doing sex research: "Sex is a parasympathetic function. My interest is in the sympathetic nervous system."[38] Later, the NRC, with Rockefeller Foundation money, supported Alfred Kinsey's studies of the sexual behavior of men and women.[39]

The United States declared war against Japan on December 7, 1941, after the attack on Pearl Harbor, and Germany and Italy declared war on the United States immediately afterward. Cannon had firsthand knowledge of the horrors of war from his experiences in World War I, and he described those events vividly in *The Way of an Investigator*. When he was asked to chair a committee on shock and transfusions for the National Research Council he felt he had to accept.

Although Cannon took on new research interests throughout his life, he retained earlier interests and often returned to them with a new perspective. Earlier he had written extensively, for example, on how emotional states, working through the sympathetic nervous system and the adrenal gland, could cause illness. In 1942 he wrote a paper on "voodoo death" that was published in *American Anthropologist*.[40] He had collected numerous descriptions of the death of healthy individuals who believed that a spell had been cast on them. Cannon argued that "voodoo death" is a well documented occurrence and not a myth. He proposed an explanation based on physiological changes induced by the sympathetic nervous system and adrenaline.[41] Cannon's paper encouraged others to pursue this phenomenon, and in 1957 Curt Richter, a distinguished psychologist at Johns Hopkins University, acknowledged Cannon's work when he described how "sudden death" could be induced experimentally in animals through stress.[42]

In 1939 Cannon spoke about a "law of denervation" in a lecture on denervation supersensitivity, and afterward he began work on a book on the subject.[43] It will be recalled that the sensitivity of the heart to adrenaline is greatly increased following denervation and that Cannon had earlier used such a preparation to detect the presence of adrenaline. He was also aware of many other examples of increased sensitivity following denervation, and he proposed that such supersensitivity is a general phenomenon. Cannon was ahead of his time, as we now know that following a period of deficient stimulation due to denervation, disuse, or chemical blockade, receptors increase in number and sensitivity. The reverse is also true following a period of excessive stimulation of receptors. However, understanding of receptors and the method for estimating their numbers had not advanced to the point where Cannon could offer an explanation of changes in sensitivity based on what is now called the up- and down-regulation of receptors. He did, however, propose several testable hypotheses that might explain denervation supersensitivity, and he also proposed that the phenomenon explained some of the recovery of function that takes place after brain damage.

Cannon officially retired from his position at Harvard in 1942 at the age of seventy-one. Before he retired, he tried unsuccessfully to have Rosenblueth appointed to succeed him as chairman of physiology, but there were too many people opposed to the idea. When administrators requested opinions about Rosenblueth's suitability for the position, most of the replies noted his brilliance but also included damning remarks about his personality. Philip Bard, for example, wrote, "You can be sure that if his position is such that he has to deal with a variety of individuals there will sooner or later be trouble." Bard added that Rosenblueth was devoted to Cannon and, "I have seen evidence that he is willing to put himself in the background on several occasions for the sake of Cannon's feelings. On the other hand, I feel quite sure that on many occasions he has done Dr. Cannon no good. And I, for one, rather deplore that scientific partnership."[44]

When Rosenblueth was not selected to succeed Cannon, he accepted, in 1942, a position as chair of physiology at the Instituto Nacional de Cardiología in Mexico City. In 1945 Cannon accepted an invitation from Rosenblueth to spend ten weeks in Mexico doing some experiments on denervation supersensitivity. Cannon, who was seventy-four, had been working on a book on this subject for over a year. He became quite ill during the visit and asked Rosenblueth to finish the book if his health prevented him from completing it. Cannon died in October of that year, shortly after returning from Mexico, and Rosenblueth completed the book. Because the book was not published until 1949, almost four years after Cannon died, and Rosenblueth was listed as a coauthor, it was widely assumed that Rosenblueth had written most of it. Actually, as Rosenblueth makes clear in the preface, Cannon wrote most of the book and the majority of the chapters were published just as he had written them.[45]

In one of the many obituaries written about Cannon, Henry Dale first reviewed his enormous accomplishments in many different areas of physiology and then wrote that with Cannon's passing, "the United States and the world lost one who, for a whole generation, had been recognized as a great leader in his own chosen scientific field of physiology."

Through his studies on sympathetic and adrenal adjustments of bodily function, Cannon was led to his main series of researches, dealing with the transmission of nervous effects by chemical agents. Observations of transmission by the blood, to a denervated heart, of effects produced by sympathetic nerves, would seem to have led him to the verge of discovery which [Otto] Loewi was to make, just at this juncture, by a simpler and direct method.

No mere account of Cannon's varied and uninterrupted contribution to the growth of physiological knowledge, over all those years,

Figure 9.5 Arturo Rosenblueth and Walter Cannon in Mexico City. The picture was taken only a few months before Cannon's death.

can give any adequate idea of the man, or his stimulating influence on scientific research in his own country and widely beyond it. His character was drawn on large and simple lines; he was capable of deep and loyal friendships, and readily moved to sympathy and indignation by suffering and injustice. He was a man of sensitive conscience, full of the traditions and the ideals of his native land.[46]

Zénon Bacq, the Belgian pharmacologist who with Cannon had first introduced the concept of sympathin, described his former mentor in the following words,

[Cannon] possessed all the fine qualities of the American people; he was unostentatious and had a natural and charming modesty which endeared him to his European and Chinese friends. One found in him nothing but unselfishness, generosity, kindness, but also a great strength of mind. These qualities, coupled with an astonishingly young personality, were the reason why he was surrounded, without any effort on his part, by co-workers and friends who had the deepest affection for him. Unavoidably, such a personality could not fail to fight vigorously for his ideas, but never without humor. Nothing was more characteristic of him than the important part he played in the years 1936–1939 in supporting the cause of the Spanish republicans in the States. In 1935 at the International Congress of Physiology in Moscow and Leningrad—where he met his dear friend I. P. Pavlov for the last time—he understood that fascism and nazism led straight to war. In the plenary session, he delivered a splendid eulogy to liberty, which greatly offended the delegates from some countries. Until the very last days of his life, the necessity of maintaining friendly contact between American and Soviet scientists was one of his main preoccupations.[47]

Dale, Loewi, and Cannon began their careers in a similar way. All three had decided after completing their medical training not to practice medicine. Dale and Loewi rejected a career in medicine, preferring, for different reasons, a life in research. Cannon actually enjoyed the little clinical experience he had as a medical student, and only the offer of an academic position when he graduated led to his pursuing a life in research. Probably more than the other two, Cannon was constantly considering how his research might apply to clinical problems and to everyday life experiences.

The three also differed in their willingness to be involved in social and political events. Ironically, Loewi (and also Wilhelm Feldberg), who was Jewish and most affected by political events, had the least tendency, or perhaps opportunity, to try to influence them. Loewi and Feldberg buried themselves in their research as long as possible. The Nobel Laureate Rita Levi-Montalcini, who was persecuted by the fascists in Italy, remarked that she had often wondered in later years how she could have been so enthusiastic about research during the times when the very survival of Western civilization was threatened: "The answer lies in the desperate and partially unconscious desire of human beings to ignore what is happening in situations where full awareness might lead one to self-destruction."[48]

Dale, who was not personally threatened by this kind of political persecution, was keenly aware of political events, and he held strong opinions.

He was sensitive to injustices and often invested much time and energy helping those in need. In later life he expressed his opinions by resigning from organizations or, on several occasions, by writing articles on political matters. Cannon, who was at least equally, if not more, concerned about social and political events, was much more willing to lend his name and, on occasion, to be a spokesman for organizations whose causes he believed in.

Dale, Loewi, and Cannon were all excellent mentors of the younger scientists who worked in their laboratories. Loewi once remarked that to all who worked in Dale's laboratory, "Dale became and has remained a scientific father and personal friend. . . . Small wonder that every single one of them cherishes a deep, unfaltering devotion to him."[49]

The three differed, however in their desire to teach students in the classroom. Dale acknowledged that teaching young students helped some people develop and clarify their own ideas, but he felt that this was not the case for all, and not for him. Loewi was a popular teacher of pharmacology for advanced students, and in later life, when he was no longer an active researcher, he enjoyed spending time encouraging and advising graduate students and young scientists. Cannon had a strong interest in the education of the young at Harvard, advising and supervising undergraduates, spending considerable time preparing for the classes he taught, and thinking about ways to improve education. Cannon also gave many popular lectures, and some of the books he wrote were aimed at a general audience or to young people contemplating a career in research.

The work Loewi, Dale, and Cannon started has grown in complexity beyond anything they could have anticipated. The final substantive chapter of this book describes how the work they initiated has grown and, particularly, how the concept of chemical transmission was extended to the brain, becoming what is today one of the most promising and most challenging areas of research on the nervous system.

CHAPTER 10

Brain Neurotransmitters

A New Continent to Explore

It made no difference to me initially whether GABA was a neurotransmitter. My goal was the elucidation of its function in the nervous system, whatever it might be. However, I did sense that the "transmitter question" seemed to agitate a number of physiologists.

—Eugene Roberts (1992)[1]

It would be difficult to discuss anything about the brain today without referring to chemical neurotransmitters. According to some estimates there may be as many as one hundred different chemical substances secreted by brain neurons, yet in 1950 even the possibility that neurons in the brain communicated chemically was rarely mentioned. Acetylcholine and norepinephrine (noradrenaline) had finally been accepted as neurotransmitters, but they were thought to act only at the autonomic nervous system synapses controlling the slow-responding visceral organs. How all of this changed is the subject of this chapter.

Henry Dale had briefly raised the possibility of the existence of brain neurotransmitters in his 1936 Nobel Lecture when he mentioned that acetylcholine had been found in the brain. He was alluding to work in his own laboratory by John Gaddum, who had recently reported finding it there.[2] Dale remarked that acetylcholine must be there for some purpose, and he referred to Wilhelm Feldberg's preliminary observations of physiological responses evoked by placing drugs that affect acetylcholine in the brain. However, Dale was characteristically cautious and concluded these remarks by saying, "We need a much larger array of well-authenticated facts, before we begin to theorize."

Zénon Bacq recalled that Dale did express an interest in investigating whether there are neurotransmitters in the brain, but did not know how to proceed:

I remember that Dale often showed his interest in this huge problem during informal discussions. He was inhibited because of the anatomy of the CNS, it was impossible to utilize the methods used with the heart, smooth and striped muscles, viz. isolation of an organ, complete control of its blood vessels and nerves and bioassay or chemical analysis of perfusion liquids. He asked his friends, experts in the anatomy of the CNS, but the unique suggestion has been to use a rare species of marsupial in which the isolation of a well defined part of the brain could have been possible. One had to wait for the rapid postwar development of the biochemistry of the CNS; biochemically minded pharmacologists could build progressively a convincing picture which is now universally accepted. New sophisticated techniques were introduced.[3]

In general, little consideration was given in the early 1950s to any role neurotransmitters might play in the brain despite several reports of finding acetylcholine and later noradrenaline (norepinephrine) in animal brains. In 1948 Marthe Vogt and Wilhelm Feldberg, for example, had described the concentration of acetylcholine in different brain regions.[4] By 1950 there were even reports that acetylcholine concentrations in some brain regions were increased following electrical stimulation, but this was not accepted as evidence that brain neurons normally secrete this substance.[5]

In 1954 Marthe Vogt, who had by that time left Dale's laboratory at the National Institute of Medical Research for a position at the University of Edinburgh, described the concentration of noradrenaline and adrenaline in different brain regions of cats and dogs.[6] Using Cannon's term "sympathin" to refer collectively to both noradrenaline and adrenaline, she reported that "sympathin" is most highly concentrated in the hypothalamus and other brain areas from which it had been shown that electrical stimulation can evoke sympathetic visceral responses. Nevertheless, she did not conclude that "sympathin" is a neurotransmitter in the brain, and she ended her paper with the comment that the available evidence "did not answer the question of the function of sympathin in the central nervous system."[7] It is clear from the literature of the time that even those who had played major roles in proving that noradrenaline and acetylcholine are neurotransmitters in the peripheral nervous system hesitated to conclude that they played that role in the brain.

Wilhelm Feldberg had started to investigate what acetylcholine was doing in the brain, but when his Rockefeller Foundation support ended in 1936 he had to leave Henry Dale's laboratory for a position in Australia, where he worked on other problems. Feldberg was able to return to England in 1938, but when World War II began the following year the investigation of possible neurotransmitters in the brain was not a high-priority area of research. After the war Feldberg resumed his investigations, using a cannula system that he developed for injecting drugs into the ventricles of the brain in awake and unrestrained animals.

In his 1963 book, *A Pharmacological Approach to the Brain*, Feldberg listed more than fifty different physiological and behavioral responses he had observed following injection of various drugs into the ventricles of the brain.[8] Some of these drugs, like physostigmine, were known to affect acetylcholine. Among other physiological effects, he observed changes in blood pressure, pupillary size, salivation, and respiration. He also reported behavioral responses such as increased playfulness, rage, fear, and excessive eating (hyperphagia) when he injected the anesthetic (hypnotic) chloralose into the ventricles of the brain. The hyperphagia was similar to what Walter Hess, a 1949 recipient of the Nobel Prize, had observed in cats following electrical stimulation of the hypothalamus.[9] Feldberg, however, did not mention neurotransmitters in his book, and he acknowledged that injecting drugs into the brain ventricles left open many questions about where and how the drugs were working: "To locate the site where a drug acts when penetrating the brain from the cerebral ventricles . . . is the main task anyone is faced with when applying drugs to the inner or outer surface of the brain."[10]

Even as late as 1963 Feldberg seems to have taken pains to avoid using the word "neurotransmitter" to explain any of the effects that followed the administration of drugs into the brain. Writing that same year, John Gaddum commented that although such monoamines as adrenaline, noradrenaline, serotonin, and dopamine had attracted much attention because they were found in the brain, "there is no direct evidence that any of these amines is released by nerves in the central nervous system; but there is much indirect evidence from experiments on tissue extracts suggesting that they play an important part in controlling the brain."[11]

Feldberg, who had been elected to the Royal Society in 1947, attracted a number of young scientists to work with him, and they helped to stimulate interest in developing techniques for injecting drugs into specific brain regions rather than into the ventricles of the brain. Although these younger associates were sometimes intimidated by the long hours Feldberg spent in the laboratory, they had the highest respect and affection for him.[12] One young scientist—the only psychologist working in the laboratory at the time—related how a casual remark by Feldberg forever changed the way he did research. He had proudly shown Feldberg some experimental results that had borderline statistical significance. Feldberg replied with his "distinctive German accent and a kind smile," "What are all these statistics? A good scientist doesn't rely on statistics, he gets control of his variable. You should get control over your variables."[13]

This is similar to the advice Dale gave Feldberg about fifty years earlier when he told him to keep perfecting his experiment "until your method is working to perfection." Feldberg continued to be active in the laboratory, publishing his last experimental paper in 1987 at the age of eighty-seven. He retired in 1990 following an accusation by an antivivisection group that he was mistreating animals by not properly anesthetizing them during surgery.[14] An investigation found that the accusation was exaggerated, but it may have had some merit. The affair was troublesome and painful for Feldberg, who died, not long after, in 1993.

By 1960, serotonin and GABA (γ-aminobutyric acid) had also been found in the brain, although they too were not accepted as neurotransmitters. Eugene Roberts, who would play a leading role in establishing GABA as a major inhibitory neurotransmitter, was well aware, as the quotation that heads this chapter indicates, that in the 1950s most neurophysiologists remained adamant in rejecting theories of chemical transmission in the brain. They correctly pointed out that finding a chemical substance in the brain, or even proving that it can produce various physiological and behavioral effects when injected into the brain, does not prove that it is secreted by neurons during normal synaptic transmission. Similar observations could be made, they argued, if the chemical substances were important for some metabolic function within neurons.

During the 1950s so little was known about brain chemistry, let alone brain neurotransmitters, that it was difficult even to propose any hypotheses that might explain how chlorpromazine and the other early psychotropic drugs worked. These drugs, which were being marketed successfully in Europe and North America during the second half of the 1950s, had all been discovered accidentally.[15] It could not have happened any other way, as there was no basis for predicting the effects of these drugs. Brain neurotransmitters were not implicated in any speculation about how the drugs worked. Nor were any abnormalities in neurotransmitter activity hypothesized to be the cause of mental illness as is commonly done today. The successful marketing of the initial psychotropic drugs did, however, provide a strong incentive and a source of support for investigating how these drugs worked, and this eventually led to speculation about their action on brain neurotransmitters.

One hypothesis proposed to explain psychosis, while in the end not correct, did serve to stimulate ideas about possible biochemical causes of mental illness. This hypothesis involved both serotonin and the hallucinogen lysergic acid diethylamide (LSD). In 1953 John Gaddum (soon to be Sir John), who had been a major contributor to Henry Dale's research program from 1927 to 1933, was professor of pharmacology at the University of Edinburgh. Gaddum had been studying the effects of LSD. He knew, of course, that this drug was extremely potent in producing hallucinations and delusions, major symptoms of schizophrenia. He had actually tried LSD on himself—once in the laboratory with his colleague Marthe Vogt monitoring him and another time at home while being observed by his wife and daughter. In both cases, the reaction was so severe that it was feared he was going mad.[16]

When it was learned that the chemical structure of LSD and serotonin were quite similar, Gaddum began investigating how the two interacted. At the time, serotonin was known only as a substance extracted from the intestinal tract that made blood vessels and other smooth muscles contract. Because of this action on blood vessels, it was first called vasotonin, but later the name serotonin was adopted. Gaddum found that LSD blocked serotonin's effect on smooth muscles. Shortly afterward, serotonin was found in the brain and Gaddum hypothesized that LSD might produce insanity because it blocks the action of serotonin. He proposed that serotonin might "play an essential part in keeping us sane."[17] This was in 1954, and that same year others also speculated that a serotonin deficiency might cause schizophrenia.[18] However, serotonin was not considered a neurotransmitter; rather, it was just another interesting chemical substance found in the brain. Nevertheless, the hypothesis had raised the possibility that mental illness might be caused by an abnormality in brain chemistry. This idea was the origin of the expression attributed to the neurophysiologist Ralph Gerard that: "There is no twisted thought without a twisted molecule."

In 1955 Bernard "Steve" Brodie and his collaborators at the National Institute of Health in Bethesda used a specific fluorometric assay to study serotonin in the brain. He found that LSD suppresses the action of brain serotonin while reserpine, a drug that was being used to treat schizophrenia, released serotonin from a bound state. These results led Brodie to consider that at least some kinds of mental illness might be caused by a serotonin deficiency in the brain, but he hesitated to call serotonin a neurotransmitter. Brodie concluded only that he had provided evidence that "serotonin has a role in brain function."[19] Although today serotonin is considered by many to be implicated in various mental illnesses, at the time the hypothesis of serotonin deficiency as the cause of schizophrenia was weakened when it was found that other drugs that produced hallucinations and delusions do not block serotonin.

Although a spate of possible biochemical explanations of mental illness began to be proposed, none of these early proposals mentioned brain neurotransmitters. In 1959, for example, the psychiatrist Seymour Kety, director of the Laboratory of Clinical Science at the National Institute of Mental Health, wrote a major review of the extant biochemical hypotheses of schizophrenia. The review was published in two parts as lead articles in successive issues of the influential journal *Science*.[20] All the theories Kety reviewed were based on the idea that some abnormal metabolic processes produced a toxic substance that acted on the brain. Kety concluded that all these theories of schizophrenia had serious shortcomings, but what is particularly revealing is that neurotransmitters were not referred to in any way.[21]

In a second article published in 1967, Kety and Joseph Schildkraut proposed that depression is caused by a norepinephrine deficiency. Even at this late date, however, Kety and Schildkraut hesitated to call norepineprine a neurotransmitter in the brain, although they implied that acetylcholine might be:

> While norepinephrine functions as a chemical transmitter substance at the terminals of the peripheral sympathetic nervous system, the role of the other amines [epinephrine, serotonin, and dopamine] in the central nervous system is far from clear. . . . *None of the biogenic amines has yet been definitely established as a chemical neuro-transmitter in the brain, however, and some investigators have suggested that one or more of these amines may act instead as modulators or regulators of synaptic transmission mediated by some other chemical transmitter—for example, acetylcholine.*[22]

Arvid Carlsson, who later shared the 2001 Nobel Prize in Physiology or Medicine for his work on the neurotransmitter dopamine, has described the resistance in the 1950s to the idea of brain neurotransmitters:

Our experiments in the late 1950's provided the first direct evidence for a role of an endogenous agonist [here, meaning neurotransmitter], present in brain tissue, in animal behavior, thus foreshadowing the paradigm shift from electrical to chemical signaling between nerve cells in the brain. As might be expected, these observations and interpretations at first met with considerable skepticism by some of the most prominent representatives of this field.[23]

Carlsson noted that this skepticism existed even among some of the very investigators, such as Henry Dale and his collaborators, who had done so much to prove the existence of neurotransmitters in the peripheral nervous system. This skepticism was apparent at the Ciba Foundation Symposium on Adrenergic Mechanisms held in London in March 1960.[24] Carlsson noted:

At this meeting practically all prominent workers and pioneers in the catecholamine field were present. It was dominated by the strong group of British pharmacologists, headed by Sir Henry Dale [who was eighty-five at the time]. I was impressed to see how the British pharmacologists, as well as many other former Dale associates, behaved toward Sir Henry like schoolchildren to their teacher, although some of them had indeed reached a mature age.[25]

Carlsson presented evidence that dopamine and other catecholamines acted like neurotransmitters in the brain. In his paper, which was entitled "On the Biochemistry and Possible Functions of Dopamine and Noradrenaline in the Brain," and in his Nobel Prize address in 2000, Carlsson noted that the evidence he presented in 1960 was met with "a profound and nearly unanimous skepticism." This was partly because, Carlsson remarked, dopamine had not yet been accepted as anything more than a precursor of norepinephrine (noradrenaline) and also because brain dopamine seemed to have little effect on muscles. In his Nobel Lecture, Carlsson recalled that:

Dale expressed the view that L-DOPA is a poison. . . . Martha Vogt concluded that the views expressed . . . regarding a function of serotonin and catecholamines, respectively, in the brain would not have a long life. W. D. M. Paton referred to some unpublished experiments indicating that the catecholamines are located in glia. In his concluding remarks John Gaddum stated that at this meeting nobody had ventured to speculate on the relation between catecholamines and the function of the brain. But this was what I insisted upon throughout the meeting, so the clear message to me was that I was nobody.[26]

If even pharmacologists were still skeptical about the existence of brain neurotransmitters in 1960, it is not surprising that there was strenuous resistance to the idea among neurophysiologists. Hugh McLennan, for example, wrote in his authoritative 1963 monograph *Synaptic Transmission* that: "In the vertebrate central nervous system there is only one synapse identified whose operation can, with assurance, be ascribed to acetylcholine." He was referring to Eccles' 1952 experiment demonstrating that acetylcholine is used as a neurotransmitter by spinal motor neurons. McLennan concluded that he had not found a single instance that he was willing to accept as evidence of a chemical neurotransmitter at a brain synapse.[27]

Throughout much of the 1960s it was still not clear that the various chemical substances found in the brain were neurotransmitters. This uncertainty was reflected in the common usage of the expression "putative neurotransmitter." In 1965 the psychologist Neal Miller reviewed research demonstrating that satiated animals could be made to eat or drink by inserting adrenergic or cholinergic drugs, respectively, into specific brain regions. Miller noted that because acetylcholine and norepineprine are considered neurotransmitters in the peripheral nervous system, we had a right to ask, "Does similar coding of transmission occur in the brain, and if it does, is it related to specific forms of behavior? Biochemists have shown that the hypothalamus is especially rich in both acetylcholine and norepinephrine. What are they doing there?"[28]

In retrospect, this reluctance to speak of neurotransmitters in the brain is even more surprising considering that during the 1950s and 1960s a number of new techniques were actually providing anatomical, chemical, and physiological evidence that brain neurons secrete neurotransmitters. Gas chromatography and mass spectrometry had made it possible to detect minute quantities of chemical substances released when nerves were stimulated electrically and even when they were normally active. Thus, for example, it was found that the acetylcholine concentration in the visual cortex was seven times greater when the eyes were open than when closed.[29]

Another important line of evidence for brain neurotransmitters came from the development of a new staining technique that made it possible to visualize biogenic amines within neurons. The technique was developed in the early 1960s by Swedish histochemists headed by Nils-Åke Hillarp in Götteborg and Bengt Falck in Lund. With appropriate pretreatment, norepinephrine, epinephrine, serotonin, and later dopamine could be made to fluoresce different colors. When appropriately treated, norepinephrine, for example, became a bright green, while serotonin took on a yellow fluorescence. It became possible to see norepinephrine, epinephrine, serotonin, and dopamine in the cell body (soma) of neurons and throughout their axons and terminals. By 1965 the neuronal circuits in the brain that carried epinephrine, norepinephrine, and serotonin had been mapped, and by 1971 dopamine pathways were also mapped. Because the dopamine pathway differed from the norepinephrine pathway, this supported the suspicion that dopamine was a separate neurotransmitter, not simply, as previously thought, a precursor of norepinephrine.[30] Although it was not sufficient evidence, the fact that these substances were located in specific neural pathways and not found in all neurons added to the argument that they might be acting as neurotransmitters in these brain regions.

Other techniques provided additional evidence for the existence of brain neurotransmitters. One involved the use of a device consisting of several fine glass tubes (cannulae) glued to a recording electrode. After this device was inserted into the brain, it was possible to use electrophoresis to move minute quantities of the ions of a "putative" neurotransmitter across a small group of neurons.[31] The electrode recorded changes in the firing rate of the neurons exposed to the neurotransmitter. It thus became possible to demonstrate that neurons exposed to the "putative" neurotransmitters responded by increasing or decreasing their spontaneous firing rates. One of the first results with this technique was reported by Rosamond Eccles (John Eccles' daughter) and D. R. Curtis, who demonstrated in 1958 that the Renshaw cells in the spinal cord were excited by acetylcholine.[32]

In 1961 John Gaddum described a "push-pull cannula" that could be inserted into different brain regions.[33] This device consisted of cannulae bound together and connected to precision pumps. The pumps made it possible to "push" a saline solution across a group of neurons and then withdraw ("pull") the fluid out. The withdrawn fluid could then be analyzed to determine if any "putative" neurotransmitter had been secreted by the neurons.[34] In describing the advantages of this apparatus, Gaddum wrote that,

(1) It detects the substances liberated under fairly normal conditions.

(2) It gives an indication of the turnover rather than the store, and can be used to study the effect of factors such as nervous stimulation.

(3) It should be possible to localize the site of liberation of substances in such tissues as the central nervous system more precisely than by other means.[35]

Thus, evidence began to accumulate that at least some of the chemical substances found in the brain are located in specific neuronal pathways, that they are released when the neurons are active, and that the release of these chemicals increased or decreased the firing rate of neurons exposed to them in ways that duplicated the effects of electrical stimulation.

Probably the most convincing evidence for the existence of brain neurotransmitters came from the electron microscope. This instrument was first developed by Siemens in Germany and RCA in the United States in the late 1930s, but because of World War II its potential was not realized until after the war. By the 1950s the optics of the electron microscope had been improved and techniques for fixing, staining, embedding, and sectioning neuronal tissue for examination in this microscope had been developed. Its great power of magnification made it possible to see the "ultrastructure" of neurons. In 1954 George Palade and Sanford Palay, working at the Rockefeller Institute for Medical Research, reported seeing small vesicles in neurons that they estimated to be between 300–500 angstroms in size. It was also possible to finally see the gap at the synapse, which was estimated to be approximately 200 angstroms.[36]

Two years later Sanford Palay, then in the department of anatomy of Yale University, used the electron microscope to describe more fully the anatomy of brain synapses in the rat. He noted that the clear gap, which could be seen with the electron microscope, "is impressive confirmation of the neuron doctrine enunciated and defended by Ramón y Cajal during the early part of this [the nineteenth] century." Palay commented that the dense patches seen on postsynaptic membranes where the two neurons were opposed may be "considered as the ultimate points for transmission of the nervous impulse." He also described the synaptic vesicles as being present only on the presynaptic side of the synapse, making the synapse, as Cajal had concluded "clearly polarized." Palay speculated that the vesicles may have a "direct role in the transmission of nervous impulses across the synapse."

If an analogy may be drawn between these synaptic terminals [those in the brain] and the myoneural junction which has essentially the same structure, the small vesicles may be considered as containing small units of a chemical transmitter, like acetylcholine, or precursor of this transmitter, which are discharged into the intrasynaptic space.[37]

In 1959 V. P. Whittaker in Cambridge, England, commented that "chemical transmission at nerve endings is now being increasingly discussed in terms of the 'synaptic vesicle' theory." After using a technique that made it possible to separate different components of the presynaptic terminals based on their relative density, Whittaker reported that the component containing the synaptic vesicles had the greatest concentration of acetylcholine or serotonin.[38] The conclusion that the synaptic vesicles held the chemical neurotransmitter until it was released was confirmed in 1968 when Tomas Hökfelt in Sweden combined the electron microscope and the fluorescent staining technique to demonstrate that the hypothesized neurotransmitters were indeed located in the synaptic vesicles of presynaptic terminals.[39]

John Eccles, who had led the opposition to chemical transmission, had converted completely by 1965 to the view that virtually all synapses, in the brain as well as elsewhere in the nervous system, are chemical. In his Charles Sherrington Lecture Eccles stated that both excitation and inhibition were transmitted chemically.[40] By this time several chemical substances had gained acceptance as neurotransmitters, and Eccles stated that glycine appeared to be the major inhibiting neurotransmitter in the spinal cord, while GABA probably played that role in the brain. Eccles' statement was based in part on the work of Aprison and Werman in Indianapolis, who had found that glycine is concentrated in the region of the spinal cord where inhibitory interneurons were known to be located.[41] GABA had been located in the brain, and its role as a major neurotransmitter inhibitor was supported by the finding that drugs such as bicuculine and picrotoxin, which block GABA, produce a generalized excitement of neural activity and, depending on dose, a convulsion. Strychnine, which blocks glycine inhibition at spinal motor neurons, does not block GABA-induced inhibition.

GABA and glycine are amino acids and represent a different chemical class of neurotransmitters than either acetylcholine or the biogenic amines norepinephrine, dopamine, and serotonin. In time it was learned that the amino acids, which are small molecules, are the most common neurotransmitters. We now know that in addition to the inhibitory neurotransmitters, there are many excitatory amino acids, such as glutamate and aspartate, that act as neurotransmitters in the brain. Glutamate is now considered to be the most common mammalian excitatory neurotransmitter.[42]

Several novel peptides were also shown to act as neurotransmitters. The discovery of one of these peptides, Substance P, has an interesting history. It was originally given its name by John Gaddum, who did much of the early work on this substance with Ulf von Euler when they were both working in Dale's laboratory. Gaddum suggested the name Substance P because it was a fine powder when extracted from the intestines. It was initially thought to act like acetylcholine because it dilated blood vessels and produced a contraction of the smooth muscles of the intestines. However, unlike acetylcholine, its action is not blocked by atropine. It was later found that Substance P is a peptide and that it is a major neurotransmitter for conveying sensory information about pain.[43] So by happenstance, the name Substance P became especially apt in that it could stand equally for a powder, peptide, and pain.

Still another substance found to be a neurotransmitter in the brain is histamine. It will be recalled that Henry Dale had studied histamine even before it was identified as such, but this research was done in the context of its action as a substance found in certain tissue that can produce a profound vasodilation and lowering of blood pressure. Histamine is a biogenic amine synthesized from amino acid precursors. In the 1980s histamine was found in neurons, and the distribution of this substance and its receptors in the brain have subsequently been described. The histamine neurotransmitter has been shown to have a wide range of behavioral and physiological effects involving sleep and arousal, eating and drinking, motor activity, learning, sexual behavior, aggression, and pain perception.[44]

During the 1950s and 1960s a great amount was learned about the way neurotransmitters are synthesized, stored, and released, and also how they induce changes in postsynaptic neurons. In 1970 Ulf von Euler, Julius Axelrod, and Bernard Katz shared the Nobel Prize in Physiology or Medicine. In the presentation of the award it was noted that Henry Dale and Otto Loewi and their collaborators had "showed that impulse transmission takes place by chemical means":

As with all fundamental discoveries, the discovery of a chemical mediator in nervous transmission led to revolutionary new thinking. Neurochemistry and neuropharmacology developed into rapidly expanding branches of science. A host of new questions arose. How were such highly active transmitter substances synthesized, stored, and released? How could they appear, produce their effects, and disappear within a fraction of a second, which must happen if chemical mediation was to explain the very fast chain of events taking place in nervous processes? What kind of substances were involved? Each of today's prize-winners had made his own special contribution toward solving problems in this field.

Von Euler's work on the synthesis of norepinephrine was described above. Julius Axelrod also worked primarily on norepinephrine, and he demonstrated that the action of this neurotransmitter is terminated primarily by being taken back up into the presynaptic neuron from which it had originally been released. Axelrod was a member of the Laboratory of Clinical Sciences at NIMH, which was headed by Seymour Kety. It will be recalled that Kety was interested in the possibility that schizophrenia was caused by an error in the metabolism of norepinephrine. There had been reports that adrenochrome, an abnormal oxidation product of epinephrine, was present in schizophrenic patients. It had been shown that adrenochrome could be formed from norepinephrine, but this work was not done on animals or patients. Kety was able to obtain some radioactive norepinephrine containing tritiated hydrogen (^{3}H), which had just then become available, and he used it to trace the fate of norepinephrine in animals. This work was discontinued when it did not confirm the "adrenochrome hypothesis." As there was some radioactive norepinephrine left over, Axelrod was able to use it for his more basic research. He administered it to rats and was able to show that it was accumulated in sympathetic nerves in the periphery.

It was then found that drugs like cocaine, which potentiate sympathetic effects, inhibit the uptake of norepinephrine by sympathetic nerves. This work was pursued by Axelrod and others, who eventually proved that the reuptake mechanism is the main way that norepinephrine effects are terminated in the brain as well as the periphery.[45] The reuptake mechanism was also shown to be the major way of terminating the effects of dopamine, serotonin, and GABA as well as norepinephrine. Prior to Axelrod's work, termination of all neurotransmitter action was thought to follow the model of acetylcholine, whose action is terminated by the action of an enzyme.

Bernard Katz, who shared the Nobel Prize with Axelrod and von Euler, studied acetylcholine. He was able to show that this neurotransmitter is released in small packets, or quanta, each of which produces a small electrical response at the junction between nerves and skeletal muscles. When a nerve fires more rapidly, more of these quanta are released. Katz also showed that even in a resting state neurons continually release a small number of quanta of acetylcholine; this, while insufficient to provoke a response from a muscle, is necessary to maintain the synaptic connection. Bernard Katz was still another of the many established or promising German scientists who were able to get out of Nazi Germany.[46]

By the end of the 1960s, more and more chemical substances were being considered possible neurotransmitters in the brain. Scientists recognized that it was necessary to establish some criteria that had to be met before a chemical substance could be accepted as a neurotransmitter, for its mere presence in the brain was not sufficient qualification, even if it could be shown to have some physiological or behavioral effects. At the very least, the candidate neurotransmitter had to be shown to be secreted by neurons at the synapse and to be capable of producing the same effect on postsynaptic neurons as stimulation of the presynaptic neuron. The candidate neurotransmitter had to be demonstrably present in the synaptic vesicles. There had to be evidence that the enzymes necessary for the synthesis of the candidate neurotransmitter were present in neurons and that there existed some mechanism to terminate its action, such as enzyme degradation or reuptake.

When it became possible to detect receptors for different neurotransmitters, the demonstration of their presence became an additional criterion. The concept of receptors has had a long history. As described in chapter 2, before 1900 John Langley hypothesized the existence of "receptor substances" in order to explain how a drug could produce opposite effects, either excitation or inhibition, at different sites. Over time, Langley's "receptor substance" was shortened to "receptor."[47] Although the concept of receptors was criticized as vague, the need for such a concept persisted. In the 1920s, several investigators, but particularly the pharmacologist Alfred Clark, described the properties of receptors more fully. Clark determined that drugs are generally effective only when applied to the cell surface (membrane) and not when placed inside the cell itself. He estimated the size of different drug molecules and the length of the cell membrane, and he also determined the minimum amount of different drugs that could produce a maximum response. From these data, Clark was able to argue that drugs must act on discrete receptor units on the membrane that do not occupy its entire surface.

Pharmacologists needed the concept of receptors not only to explain why a drug has different effects at different sites, but also why normally only one of the mirror images of the same molecule—the so-called left and right isomers—is effective. This was explained by analogy of the "key and lock," where the key is the neurotransmitter and the lock is the receptor.[48] Even though the arguments for the existence of receptors may now seem persuasive, pharmacologists of the time were reluctant to accept the concept. Dale, for example, remained critical of the concept for quite a while, mainly because there was no way of detecting its physical presence.[49] Before 1970 the concept of receptors was rarely discussed even at pharmacology meetings, while most neurophysiologists regarded the concept as overly imaginative theorizing and the multiplication of hypothetical receptors as an unnecessary complication.

The first receptor to be identified was the nicotinic–acetylcholine receptor. This work was initiated by Carlos Chagos in Rio de Janeiro, who studied the role of acetylcholine in activating the electric organ of the electric eel (*Torpedo marmorata*). Later studies used alpha-bungarotoxin, an active component of cobra toxin, which was found to combine with the nicotinic receptor with a high degree of specificity.[50] This receptor had been shown to control the electric organ of the *Torpedo*, but what made it possible to use bungarotoxin to identify the nicotinic receptors was the exceptionally high concentration of these receptors in the electric organ. It is estimated that the nicotinic cholinergic receptors may constitute as much as 20 percent of the weight of the electric organ in the *Torpedo* fish.

Identifying receptors for neurotransmitters in the brains of mammals, where the concentration of receptors is much lower than in the electric organ, proved to be considerably more difficult. Solomon Snyder, who was a major contributor to this work, has described how this was accomplished.[51] The identification and location of receptors for the enkephalins in the brain, which is described below, served as the model for identifying the receptors for most neurotransmitters. Basically, the technique involved "labeling" receptors by attaching radioactive substances to chemicals that bind to them. The radioactive substance might be the neurotransmitter itself or some ligand—an agonist or antagonist—that binds to the same receptor with a high degree of specificity.[52] Techniques for recovering and measuring the bound radioactive ligand were developed, and from this it was possible to determine the relative concentration of receptors in different brain areas and the relative strength of the binding affinity of different drugs competing for the same receptor.[53] During the 1970s, in addition to the enkephalin receptor, acetylcholine, norepinephrine, glycine, GABA, and histamine receptors were found in the brain.

Although most neurotransmitters were first discovered in the brain and accepted after they were shown to act on the nervous system and to have met the other established criteria, this was not always the way it happened. In the case of the peptides now known as enkephalins, for example, their existence was first inferred from the action of opium and related opioid drugs. Opium is obtained from the poppy plant, and morphine and codeine are extracts of opium. By the mid-twentieth century several synthetic opioids were also available. It was inferred that all these different opioids must be acting on a specialized brain receptor because analgesia and euphoria were obtained only from drugs with very similar chemical structures.

The search for the opiate receptor in the brain began around 1970. Although there were some suggestive earlier results, it is commonly acknowledged that Solomon Snyder and Candace Pert, then a Ph.D. student working in Snyder's laboratory at Johns Hopkins University, were the first to obtain convincing evidence of opiate receptors in the brain. This was in 1973.[54] The race was close—researchers at New York University and in Uppsala, Sweden, identified opiate receptors in brain tissue later that same year.[55]

The discovery of an opiate receptor in the brain triggered a feverish search for an endogenous opioid in the brain on the rationale that, as one investigator put it, why else "would God have made opiate receptors unless he had also made an endogenous morphine-like substance."[56] It did not require much imagination to appreciate the potential importance of finding a natural substance in the brain that could induce pleasure and alleviate pain. Several teams of researchers entered the race, but the first to succeed were Hans Kosterlitz and his young associate John Hughes, who reported in 1975 that they had extracted an endogenous opiate-like substance from brain tissue.

Hans Kosterlitz left Nazi Germany in 1934 while still relatively young. He was able to continue his education at the University of Aberdeen, where he earned a Ph.D. and a medical degree. Kosterlitz eventually became professor and head of the department of pharmacology in Aberdeen.

In 1973, when Kosterlitz had to give up his academic appointment at the compulsory retirement age of seventy, he established the Unit for Research on Addictive Drugs and continued his research on opiates. When one of his graduate students found that the mouse *vas deferens* is extremely sensitive to morphine and related opioids, Kosterlitz and Hughes used this discovery to develop a bioassay useful in the search for an opioid substance that might be in the brain.[57] They knew they would need a lot of fresh brains for this search, and decided that the pig brain might be most readily obtained from the local slaughterhouse.

John Hughes, because he was much younger than Hans Kosterlitz, had the task of going to the slaughterhouse early in the morning to collect pig brains. He later reported that a bottle of Scotch whiskey was much more helpful in getting the abattoir workers to cooperate than any story he could tell them about the importance of the science.[58] Hughes brought back carts filled with pig brains, from which were extracted various substances to test on the *vas deferens*. In 1975 Hughes and Kosterlitz reported that they had successfully isolated an opioid substance in the pig brain.[59] The substance produced the appropriate response from the mouse *vas deferens,* and the response was blocked by the specific opioid antagonist naloxone.[60] Initially, this brain extract was called an "endorphin," a contraction of endogenous morphine-like substance, but Kosterlitz and Hughes proposed the name enkephalin (from the Greek meaning "in the brain"). The more neutral term "enkephalin" was preferred because it did not restrict its effects to the known euphoric and analgesic effects of opiates.

In a second paper published the same year, Kosterlitz and Hughes reported that they had identified and synthesized two enkephalins, met- and leu-enkephalin, in the brain. The two were chemically identical, except that the former contained methionine and the latter, leucine. They wrote, "The discovery in the brain of two endogenous pentapeptides with potent opiate activity raises a number of questions that cannot be adequately dealt with in this paper. It will now be possible, however, to test the hypothesis that these peptides act as neurotransmitters or neuromodulators at synaptic junctions."[61] Among the questions Kosterlitz and Hughes mentioned as needing to be investigated was whether tolerance to and dependence on endogenous enkephalins could develop, as they did with synthetic opioids and natural opium derivatives. Many wondered whether a natural "high" could be obtained from these brain enkephalins.

Only a few weeks after Hughes and Kosterlitz published their report, Solomon Snyder, Gavril Pasternak, and Rabi Simantov reported finding the same two enkephalins in rat brains. The Snyder group also mapped the distribution of the met-enkephalin receptor in different brain areas. The high concentrations of enkephalins in the *locus coeruleus* and the *amygdala* seemed to explain the euphoric and analgesic effects of opioid drugs, as these structures are thought to play a significant role in mediating emotional responses. Similarly, the high concentration of enkephalins in the medial hypothalamus and certain regions of the spinal cord was consistent with what was known about the regulation of the affective component of pain and the transmission of pain signals.[62]

The new radiolabeling techniques eventually made it possible to identify receptors for virtually all neurotransmitters. In the 1970s both Solomon Snyder at Johns Hopkins and Philip Seeman in Toronto reported labeling two different dopamine-receptor families, now called D_1 and D_2. It was subsequently found that most neurotransmitters have more than one receptor type. In some instances numerous receptors respond to the same neurotransmitter.

In 1957, even before dopamine receptors had been identified, Arvid Carlsson and Margit Lindquist in Göteborg hypothesized about the consequences of blocking what were at the time hypothetical dopamine receptors. They were trying to explain the antipsychotic action of reserpine and drugs like chlorpromazine by determining what they had in common. Starting with evidence suggesting that both blocked dopamine transmission, Carlsson and Lindquist found that while reserpine reduced dopamine in the brain, other drugs with antipsychotic properties did not. They were especially surprised to find that drugs like chlorpromazine actually increased the amount of dopamine metabolites in the brain. Carlsson and Lindquist then theorized that the chlorpromazine-like drugs might block dopamine receptors and that this might activate a feedback loop calling for more dopamine to be released. Such a mechanism could explain the increase in dopamine metabolites despite there being no decrease in brain dopamine levels. This brilliant insight led to the discovery of the first of many mechanisms that compensate for changes induced in neurotransmitter systems. The feedback loop that Carlsson and Lindquist hypothesized was later identified, and it was shown to act on dopamine autoreceptors located on dopamine neurons. Activation of these dopamine autoreceptors regulates the amount of dopamine released. Later, the ability to tag receptors with radioactive substances made it possible to show that the number and sensitivity of receptors increase or decrease (up and down regulation) in response to blockade or excessive stimulation, respectively. Many other mechanisms that compensate for changes induced in one part of a neurotransmitter system have since been found.

It will be recalled that many neurophysiologists had rejected chemical transmission because it was assumed to be too slow to mediate the fast responses observed at many synapses. This position could no longer be defended after Paul Greengard, who shared the year 2000 Nobel Prize in Physiology or Medicine, began to describe the molecular mechanisms involved in chemical transmission of the nerve impulse. Greengard identified two different mechanisms, which he categorized as either fast or slow chemical transmission.[63] The fast chemical transmission (both excitatory and inhibitory) takes place in less than one millisecond, while even the slow transmission takes only several milliseconds—fast enough to do everything the brain is known to do.[64] Walter Cannon was certainly correct when he wrote in 1939 that not enough was known about chemical transmission to allow many of its critics to assume that it was too slow a process to do the job required.

Just when it began to appear that all synapses were mediated by chemical transmission, a few electrical synapses—estimated to be about one percent of the total—were found. Use of the electron microscope showed that the synaptic gap in what proved to be purely electrical synapses is so narrow that it is virtually obliterated. For that reason, they are called "tight junction," or "gap junction," synapses. There are no presynaptic vesicles or any other evidence of chemical transmission at electrical synapses. Transmission across this extremely narrow gap is faster than at chemical synapses, and researchers believe it can be bidirectional as well as unidirectional.

Perhaps because transmission across electrical synapses is so fast, they are found in neural circuits controlling escape behavior in invertebrates such as the squid and the crayfish. However, electrical synapses cannot be considered esoteric and unimportant, because they have been found in retinal cells of the eye, the *locus coeruleus*, the *corpus striatum*, and other areas of the mammalian brain. In the hippocampus, a region important for the formation of memories, GABA-secreting cells are linked by gap junctions in a way that facilitates a synchronous firing that maximizes their influence.[65] It is also interesting that in some animals with relatively simple nervous systems, such as the round worm (*C. elegans*), which has only 302 neurons and an estimated 10,000 synapses, half the synapses are chemical and the other half are electrical gap junction synapses.

It is ironic that although chemical transmission was once considered the exception, now it is the electrical synapse that is the exception. Moreover, our ideas about chemical transmission have broadened considerably over the years. Initially, chemical transmission was viewed as essentially doing what electrical transmission was thought to do, but in another way. That is, the action of neurotransmitters was thought to be restricted to exciting or inhibiting a postsynaptic cell. However, we now realize that neurotransmitters produce different types of changes. Rapid changes in the excitability of a postsynaptic cell occur when a neurotransmitter induces changes in the concentration of sodium, calcium, or potassium ions within that cell. Neurotransmitters acting on membrane receptors produce these changes by causing specific ion channels ("transmitter-gated ion channels") in the membrane to open or close. These changes can be completed in milliseconds.

Neurotransmitters (and hormones), called "first messengers," can also induce more enduring changes by activating so-called second messengers located in postsynaptic neurons. Many of these changes in neuronal function take place over hours, days, and weeks, not milliseconds. Through a cascading series of molecular steps, the "second messengers" can induce changes in the expression of specific genes. The changes in gene expression can alter virtually everything about a postsynaptic cell, including, among many others changes, the number of receptors, the production of neurotransmitters, and the amount of protein carrier used in the reuptake of neurotransmitters or in the production of the enzymes used to terminate the action of some neurotransmitters. Moreover, anatomical changes can be induced in, for example, the branching of dendrites and the number of spines they contain. (Synapses on dendritic spines are thought to be primarily excitatory.) These more enduring changes are thought to underlie the neural plasticity that supports learning and memory, while the faster changes are more likely to support the transfer of information through neural pathways. A single neurotransmitter may induce both the fast and the slower, more enduring changes depending on the receptor activated.

It is also known now that in addition to producing more enduring changes, chemical transmission can exert its influence at greater distances than electrical stimulation could. Whereas electrical stimulation could influence neurons located directly across the synaptic gap, some neurotransmitters travel in the blood or cerebral spinal fluid to reach receptors located a considerable distance from where they are released. In this respect, neurotransmitters are similar to hormones and are reminiscent of the common origin of sympathetic neurons and the cells of the adrenal medulla that release the hormone adrenaline into the bloodstream.

Although the distinction can be blurred, sometimes neuromodulators are distinguished from neurotransmitters. The primary function of neuromodulators is not thought to be the transmission of nerve impulses across the synapse, but rather the modification of the effectiveness of postsynaptic neurons. This may be accomplished, for example, by altering the synthesis, release, reuptake, or metabolism of neurotransmitters. Neuromodulators generally have a slower onset and longer duration of action than neurotransmitters.

More recently, the definition of neurotransmitters has had to be broadened to include some gasses, such as nitric oxide and carbon monoxide. Study of these gasses, which are called "novel neurotransmitters," has necessitated a rethinking of what constitutes a neurotransmitter. The physiological role of nitric oxide (NO) was discovered in the 1980s when Robert Furchgott found that acetylcholine does not directly cause blood vessels to relax, but does so by causing nitric oxide to be released from the endothelium cells that line blood vessels. Nitric oxide also causes the blood vessel relaxation that supports penile erection. It has been found that nitric oxide can be released in the body by many different agents. Nitroglycerin, for example, which relieves ischemic cardiac (angina) pain, causes nitric oxide to be released. In 1998 Robert Furthgott shared the Nobel Prize with Louis Ignarro and Ferid Murad for their work demonstrating that a gas can act as a "messenger molecule."

It has subsequently been shown that nitric oxide is present in central nervous system neurons and is released following stimulation of brain cells.[66] Among the brain areas most extensively studied in this regard are the cerebellum, striatum, cerebral cortex, hippocampus, hypothalamus, and the ascending reticular formation. Nitric oxide is not stored. Its synthesis is triggered on demand by an influx of calcium into the neuron. Because it is not stored, the location of nitric oxide is facilitated by tracing the enzyme nitric oxide synthase (NOS).

Studies of the possible function of brain nitric oxide are underway. There is evidence that it may play a role in the long-term facilitation of the synaptic transmission that underlies learning and memory. Eric Kandel of Columbia University and his colleagues have demonstrated that the long-lasting facilitation of synaptic transmission (long-term potentiation) may be dependent on the enzyme NOS.[67] It has been suggested that nitric oxide acts as a "retrograde messenger," potentiating synaptic transmission by either increasing the postsynaptic response or increasing the amount of presynaptic neurotransmitter that is released. There is also some evidence, obtained from "knockout mice," that nitric oxide is involved in regulating sexual and aggressive behavior.[68]

Nitric oxide seems to break every rule of nerve transmission. It is not synthesized in advance and stored in protective synaptic vesicles. Instead, it is synthesized on demand and then immediately released, normally together with other neurotransmitters. Nitric oxide does not bind to receptors on the membrane of postsynaptic cells; rather, it diffuses right through the postsynaptic membrane into the cell, where it binds to certain enzymes. It has a short half-life, measured in seconds, so there is no need for its action to be terminated by enzymatic degradation or by reuptake.

In consideration of all these differences, Solomon Snyder, who has studied the central nervous system properties of nitric oxide and other gasses, has suggested that we need a more "liberal conceptualization" of neurotransmitters. Snyder has proposed defining neurotransmitters as simply any "molecule, released by neurons or glia, that physiologically influences the electrochemical state of adjacent cells."[69] From the outset of his research he has been more interested in molecules that convey instructional messages by any means and less concerned with the criteria for calling a substance a neurotransmitter.[70] In any case, it may not be surprising that when there was a great amount of skepticism about brain neurotransmitters, a long list of criteria had to be established before a candidate neurotransmitter would be accepted. Now that chemical transmission is considered the rule, a more liberal definition of neurotransmitters seems reasonable.

We have come a long way in the past one hundred years. Around 1900 the neuron doctrine had just been accepted, although there still remained a few who continued to argue that nerve cells fibers were connected, with no gap between them. It was, however, generally accepted that there was at least a functional, if not a physical, gap that separated neurons and neurons from the muscles and glands they innervate. The acceptance of the concept of chemical transmission across that gap proceeded in stages with transitional periods between them marked by elevated levels of controversy. The first hint that a chemical process mediated neural transmission came from the observation that acetylcholine and adrenaline appeared to mimic the effects of stimulating the parasympathetic and sympathetic nervous system, respectively. The idea that these substances might be secreted by nerves was not seriously entertained, however, until Loewi's seminal experiment in 1921, which triggered both interest and controversy.

Despite the awarding of the Nobel Prize to Otto Loewi and Henry Dale in 1936, most considered chemical transmission to be a special case, applicable only to the relatively slow–responding smooth muscles of visceral organs. Any suggestion of chemical transmission at central nervous synapses was vigorously disputed. For a number of years, the dispute centered on the question of whether the innervation of skeletal muscles by spinal nerves is chemical or electrical. A breakthrough occurred when John Eccles, a major opponent of chemical transmission, became convinced by his own experimental evidence that transmission between spinal motor nerves and skeletal muscles is chemical. However, the possibility that transmission at brain synapses is chemical was still not so much disputed as ignored.

The last stage, that of proving chemical transmission at brain synapses, can be characterized as a period of persistent skepticism leading up to the time when the techniques became available that provided evidence that could no longer be ignored. Today it is accepted that transmission at over 99 percent of all synapses is chemical, mediated by one or another of an estimated 50–100 different chemical neurotransmitters that act on an even greater number of receptors.[71]

The universal acceptance of the existence of neurotransmitters stimulated some to think about their origin. Did neurotransmitters originate with the nervous system, or are their roots to be found in chemical communication that existed before the nervous system evolved? How are responses integrated in organisms that have no nervous system? Herbert Spencer Jennings, a close friend of Walter Cannon from their student days at Harvard, became a leading authority on the behavior of the one-celled animals (protozoa) that have no nervous systems. Jennings described the large range of adaptive behaviors that protozoa display, and he even argued that their behavior is modified by experience in ways that reflect a capacity for memory and learning.[72] All of this is accomplished by chemical communication without a nervous system.

A principal adaptive advantage of the early nervous system was that it sped up chemical communication. As multicellular animals became larger and more differentiated, the nervous system made it possible for distal parts of such animals to respond more rapidly and in a more integrated pattern than was possible by means of chemical diffusion alone.[73] Evolution tends to conserve what has been found useful, however, and it is now known that the primitive nervous system secretes some of the same chemicals that serve as "messenger molecules" in animals without a nervous system. These "messenger molecules" have apparently had a long evolutionary history, as receptors for many of the neurotransmitters found in higher animals have been discovered in bacteria and yeast as well as in protozoa.[74]

If this evolutionary history had been known earlier there probably would have been less resistance to the idea of chemical neurotransmitters, and the "war of the soups and sparks" might have been only a brief skirmish between disciplines defending their own turf. However, none of this was known at the time, for interest in the evolution of neurotransmitters arose only after the concept of chemical transmission became more acceptable.

In the brief epilogue that follows I offer some personal reflections on the value of history in general and this one in particular.

CHAPTER 11

Epilogue
Some Final Reflections

Those who cannot remember the past are condemned to repeat it.
—George Santayana

Never forget the importance of history. To know nothing of what happened before you took your place on earth, is to remain a child for ever.
—Source unknown

The only thing we learn from history is that we learn nothing from history.
—Friedrich Hegel

History is like a lantern on the stern, which shines only on the waves behind us.
—Samuel Taylor Coleridge

It is obvious that there are differences of opinion about what can be learned from history. While I personally believe there is much that can be gained from a study of history, I doubt that it can provide all the answers to present problems or is even much help in avoiding past mistakes. I know that it is often said that those who ignore history are destined to repeat it, but it has been my observation that those who know history are just as likely to repeat its mistakes. It is just too easy to focus on the similarities or differences between the past and the present in order to justify some action already decided on. Citing history to justify a course of action is like citing scripture. It is usually self-serving and always selective.

I did not write this historical account because I thought there were some specific lessons to be learned from it. I wrote it because I found the story fascinating and thought others would as well. I was also encouraged by discovering that despite the importance of neurotransmitters for just about everything the brain does, most people know little, if anything, about how they were discovered, the dispute over their existence, the lives of the remarkable scientists involved, or the political events that affected their lives and work.

Despite my reservations about the usefulness of history for providing specific answers to today's problems, I do believe there is much that it can offer. There is, of course, the enrichment that comes with learning how any body of knowledge has evolved. I recognize that this is a personal reaction, undoubtedly more meaningful to some than to others. However, a knowledge of history helps us appreciate that our current knowledge and convictions are only a moment on a continuum of change. This realization can make us more open to new ideas and less dogmatically certain about what we believe to be true and unchallengeable. Jonathan Cohen, a neuroscientist at Princeton, was recently asked by a reporter why he would want to participate in a symposium on Buddhism and the biology of attention. He replied that:

> Neuroscientists want to preserve both the substance and the image of rigor in their approach, so one doesn't want to be seen as whisking out into the la-la land of studying consciousness. On the other hand, my personal belief is that the history of science has humbled us about the hubris of thinking we know everything.[1]

Perhaps foremost, a knowledge of history can stimulate us to think about various aspects of any subject in ways that make these reflections richer and more nuanced. Because I have come to believe that this may be the greatest value of history, as distinguished from any pleasure to be derived from an increase in knowledge, I decided to end this book by providing some examples of the thoughts that this history brought to mind for me.

One topic, for example, to which I often found myself returning was how differences in personalities influenced the way this history evolved. Many historians regard individuals as mere "straws in the wind of history," maintaining that sociopolitical factors shape events and that scientific discoveries are made only after the intellectual ground has been prepared. However, individuals can be significant determinants of when and how events occur and sometimes of whether they even occur at all.

The three central figures in this history, Henry Dale, Otto Loewi, and Walter Cannon, differed greatly in their willingness to speculate and theorize beyond the observable facts. Dale was at one end of the continuum.

Although he discovered many of the fundamental facts necessary for proving that neurons secrete neurotransmitters, he was hesitant to theorize about the significance of those facts. He took great care in planning and executing his experiments and in checking his conclusions. His emphasis was on producing replicable results, and I believe it is true that no one ever questioned the reliability of any data he reported. At the same time, however, Dale was a "conservative empiricist," hesitant to speculate, at least in print, beyond what he believed the evidence definitely proved. Despite the merits of such caution, there is the danger that remaining too close to data may cause the forest to be lost for the trees. Dale was tantalizingly close to discovering chemical neurotransmitters in 1914 when he described the properties that acetylcholine shared with parasympathetic nerve stimulation and how this paralleled what was known about adrenaline and the sympathetic nervous system. Had he been willing to speculate about the significance of his own data, he might well have been encouraged to search for and find that acetylcholine is a natural substance in the body. When Dale (with Dudley) discovered acetylcholine in mammals fifteen years later, he used a methodology that had been available to him as early as 1914. By his own admission, he decided to look for acetylcholine in the body only after Otto Loewi's speculation and subsequent research had made the possibility of chemical transmission a subject worth pursuing. Seeing may be believing, but being able to see often depends first on believing.

Otto Loewi was much more willing to speculate about chemical transmission, even though his initial evidence was far from convincing. Having speculated about neurohumoral transmission and then been challenged by others, he was motivated to obtain much more compelling evidence in support of his speculation. His willingness to speculate had its own limits, though, and for a number of years he resisted extending his ideas on neurohumoral secretions beyond the nerves that regulate heart rate.

Cannon was at the other end of this continuum. He considered the uncovering of broad integrating concepts to be an essential part of his role as a scientist. When it occurred to him that the different physiological responses evoked by emotional states all prepare animals for "fight or flight," he seized on this integrating concept and expanded its application. He proposed that chronic exposure to stress could strain and exhaust organs and induce pathological states. Not every fact confirmed his theories. He was wrong on several occasions, for instance when he speculated that the adrenal medulla became exhausted by prolonged stress and lost its capacity to secrete adrenaline. On several occasions he became bogged down in controversy, or at least diverted by it. Cannon may also have missed his opportunity to discover chemical neurotransmitters because he was trying to defend his theory about the importance of adrenaline in coping with stress. However, Cannon's broad integrating theories, like his ideas about "homeostasis," stimulated much research and were enormously influential in diverse fields.

While considering the implications of differences in scientific styles, I began to wonder whether there might be an optimal breadth for theories in science. If a theory is too broad it might be stretched so thin that it explains nothing or might need too many *ad hoc* explanations to support it. This certainly is a possible outcome, but I rejected this thought when I considered Charles Darwin's theory of evolution. No theory about the natural world is broader in scope than the one Darwin proposed. Although Darwin had collected an enormous amount of supporting evidence for his theory, when it was published he knew virtually nothing about the mechanism of inheritance, and he was wrong in accepting Lamark's conclusion that acquired characteristics are inherited.[2] Even Darwin's core concept, that evolution works primarily through the selection and survival of the fittest, had to be modified after William Hamilton proposed that "inclusive fitness" shifts the emphasis from survival of individuals to survival of adaptive genes.[3] Yet despite the fact that Darwin's theory was wrong in some respects and incomplete in others, no other theory has had a greater influence on our thinking about the natural world or stimulated more research and discussion in diverse fields. There may be gaps in knowledge and even opposing facts, but proposing the theory often creates the incentive to fill the gaps and determine if troublesome facts really do contradict it.

It is obvious that speculation and theorizing in science involve potential risks as well as gains. It is usually not possible to know in advance which of these will predominate. Trying to support and defend a theory that eventually is proven wrong can involve a substantial loss of time, energy, resources, and, under some circumstances, reputation. Cannon's and Rosenblueth's persistence in trying to defend the "two-sympathin theory" is probably an illustrative example of the last. There is also the danger that attempting to support a theory, particularly one that involves some ego investment, can in many subtle ways decrease objectivity in designing experiments and in evaluating the results obtained.

On the positive side, however, mustering support for a theory and defending it against criticism can be exciting and energizing. It can also be a spur to collect the evidence needed to resolve the issue. While Loewi had been a productive and respected pharmacologist before 1921, his proposal that year that neurohumors regulate heart rate had a catalytic effect on his research activity. Following his speculation, Loewi's research almost immediately acquired a focus and a mobilization of resources that had not been evident previously.

The different scientific styles of Dale, Loewi, and Cannon likely reflect differences in temperament and personality. There is probably some genetic predisposition underlying these differences in temperament, but differences in scientific style are also influenced by mentors and life experiences. Evidence of both influences can be found in this history. The fact that all three made major contributions illustrates that there is no one way to make a significant contribution in science. While this may be obvious, in my experience there are many practicing scientists, and especially mentors of young scientists, who need to be reminded of it.

Another topic this history stimulated me to think more about concerns the dispute over the existence of chemical neurotransmitters and, by extension, the opposition to new ideas in general. There were undoubtedly valid scientific reasons at the time for neurophysiologists to question chemical transmission of the nerve impulse. However, these objections were to a great extent fueled by other than scientific reasons. Some neurophysiologists, for example, resented the intrusion of pharmacologists into what they considered to be their domain. For them, the controversy seemed to be a territorial dispute, and the willingness to accept evidence was often influenced primarily by which side of the argument it helped.

In a number of instances, the arguments in opposition to chemical transmission seemed to reflect a failure to think more imaginatively about what was possible. It will be recalled that many neurophysiologists opposed the concept of chemical transmission because they were convinced that it was too slow a process to evoke the fast responses that occurred at many synapses. However, as Cannon pointed out, little was actually known about how neurotransmitters act on their receptor targets and therefore there was no way of estimating the speed of chemical transmission. Only later was it learned that chemical transmission of the nerve impulse takes only a few milliseconds, fast enough for any task known.

I began to think of other ideas that were rejected prematurely because they were believed to contradict some well-established "fact," and I recalled a striking example. Once, while I was serving on a National Institute of Mental Health grant committee in the 1960s, we received a proposal from an applicant wanting to pursue some preliminary evidence that brain tissue from a donor animal might facilitate recovery of brain damage in a recipient animal. The applicant thought that the implanted tissue might establish functional connections, and he hypothesized that the adult brain seemed to have more capacity for plasticity and growth than was thought possible at the time. It is embarrassing to recall that the proposal was unanimously rejected with more than a few facetious remarks made about the naiveté of an applicant who did not know that brain neurons are incapable of compensating for damage by new growth. Today, of course, we know otherwise, and a number of laboratories and neurosurgical units are implanting brain tissue and exploring other ways to stimulate nerve growth in experimental animals and human patients. While being aware of such instances should not encourage anyone to abandon their critical acumen, it is possible that such awareness may make some scientists more open to novel ideas and less dogmatic about the reasons for rejecting them.

This history also stimulated me to reflect on the implications of synaptic transmission being chemical rather than electrical. It struck me as ironic that had the "sparks" been right our task of understanding the brain would have been easier. At least there might have been fewer variables to consider if synaptic transmission was purely electrical. While it is true that chemical transmission has opened up many possibilities for influencing brain activity with drugs, this is not as simple a task as often portrayed. According to current estimates, there may be as many as one hundred different chemical substances that affect communication between neurons. Furthermore, the rate of production and release of these different neurotransmitters is constantly changing.[4] Moreover, it is now known that most neurons secrete several different neurotransmitters ("co-transmission"), not only one as Dale had concluded. Several different peptide neurotransmitters may even occupy the same synaptic vesicle. Even if the same combination of transmitters is present at all terminals of a given neuron, differences in the receptors at the terminals will produce different results. The number of different receptors known even for a single neurotransmitter is now well beyond anything Dale could have anticipated. At this time, for example, we have discerned at least four different glutamate receptors and six dopamine receptors, and serotonin has at least fifteen different receptors, if all the subtypes are included. Furthermore, there is no assurance that future research will not discover still more receptors.

Neurotransmitter systems are in a constant state of flux. The number and sensitivity of the receptors, for example, is constantly changing, and the effect of neurotransmitters on these receptors depends on the availability of degrading enzymes or the protein carriers responsible for the reuptake of the neurotransmitters. All of these factors vary with different demands on the system. It may appear paradoxical, but as our knowledge increases and we learn more about how all these variables interact, it becomes increasingly difficult to relate any behavioral or physiological effect to a specific neurotransmitter or receptor system. Any intervention, such as the administration of an agonistic or antagonistic drug, no matter how selective its initial target, will inevitably produce a cascading series of physiological changes. Moreover, these changes will affect other neurotransmitter systems in addition to the one initially targeted. All of this knowledge has created exciting possibilities and seemingly endless realms to explore, but also an enormous and sometimes daunting challenge to fully understand or even to predict the outcome of any experimental or clinical intervention.

Still another topic I began to think about while writing this history involves the factors that influence career choices, and I started to reflect about Dale's deliberation over whether he should accept the position offered by a pharmaceutical company. Over the years I have observed many young scientists who struggled with the same decision. It will be recalled that when Dale was offered a research position in the Wellcome Laboratories his friends advised him not to sell his "scientific birthright for commercial pottage." The position turned out, however, to be a boon for Dale's career. Dale was, in part, attracted by the opportunity to obtain a "marriageable income," but the position also brought with it more support for his research than he would have had in any other position available to him at that point in his career. This support enabled Dale to make such great progress that by the end of the ten years he worked at the Wellcome Laboratory, he was recognized as one of Great Britain's foremost pharmacologists and was elected to the Royal Society. Of course, Dale had great talent, but he also had the freedom to pursue his scientific interests at the Wellcome Laboratories. In the end, however, he left the company laboratory not only because of other opportunities, but also, in part, because the commercial interests of the company were increasingly interfering with his own scientific interests.

There are today many opportunities for young scientists to work for the pharmaceutical industry, and these positions can be attractive on a number of grounds, most notably the higher salaries and the greater support available for research. However, much has changed since 1904, when Dale started to work at the Wellcome Laboratories. At that time Henry Wellcome was essentially the sole owner of the company and was therefore in a position to back up his inclination to give scientists the freedom to pursue the problems that interested them. This policy attracted a number of outstanding researchers and, in addition to Dale, three other Burroughs Wellcome scientists became Nobel Prize recipients.[5]

The situation is quite different today. Pharmaceutical companies are now huge international corporations, and some of them have an annual income that exceeds the gross national product of a number of countries. These corporations are run by CEOs and boards of directors whose decisions about the research to be pursued are greatly, when not exclusively, influenced by market considerations. Whether this influence of the marketplace turns out to be positive or negative depends primarily on the extent to which the profit motive coincides with worthwhile social and scientific goals.

Over the years, I have followed the careers of young scientists who have gone to work for pharmaceutical companies. It is probably different for established senior scientists, who may be in a position to make some demands before accepting a position with a company. For young scientists starting their careers, however, the decision to go to work for a commercial company carries a much greater risk. It has seemed to me that it has usually worked out best for those young scientists who were not strongly committed to a particular scientific problem and who were quite willing to apply their skills and knowledge to research problems selected by others. Of course, if the problem they are required to work on turns out to interest them and if it can contribute to the development of a useful drug or to scientific knowledge that can be rewarding by itself. It may be recalled that when Dale went to work for the Burroughs Wellcome Company, he was not committed to working on any particular research problem. He started to work on the ergot fungus because Henry Welcome suggested the project, surely in part because of what was thought to be its commercial potential. It was fortunate that ergot contained so many chemical substances that had properties which, in the hands of someone with Dale's ability, shed light on important physiological processes.

For young scientists strongly committed to a particular line of research the decision to work for a pharmaceutical company can involve a much greater risk. If they are lucky or astute enough to predict the future so that the company and their interest continue to coincide they can benefit from the many advantages already mentioned. Moreover, they will not have to devote the inordinate amount of time academic researchers must invest in order to get funding for their research. However, I have seen many instances where market considerations have led pharmaceutical companies to completely abandon research areas to which they once were committed. This placed young scientists committed to the now abandoned area of research in a difficult position. They often had to chose between giving up their primary research interest or seeking another position, not always an easy task at that juncture in their careers.

Career decisions always involve risks so I suppose the best one can do is to try to make informed guesses about the future. However, young scientists strongly committed to doing research in a specific area need to recognize that there are few "free lunches" in this world. A number of scientists today are electing to join the smaller biotechnology companies as these companies are generally much more committed to pursuing a particular line of research. However, because these companies tend to have "all (or most) of their eggs in one basket," the risks as well as the potential gains are generally greater. For that reason, the biotech companies tend to attract young scientists who are less "risk aversive."

These are only a few examples of the many topics that this history caused me to think about. Readers will probably have their own list. There is one additional topic that I want to mention briefly. As I became more and more involved with the lives and work of the scientists involved in this history, I began to think about how fortunate I have been to have spent most of my life in research. For me, this story illustrated much of what is best about such a life. There is, of course, the opportunity to pursue interesting problems and even though they may never be completely solved, it is rewarding to have made a contribution. As this history amply illustrates, life in research can bring you in contact with colleagues around the world. As they traveled, Dale, Loewi, and Cannon often discovered that they actually had friends with common interests who they had never met before. Most researchers have had similar experiences. Although there can be disagreements that are sometimes not pursued in the most constructive manner, this is generally the exception. In general, even disagreements about shared interests often form the basis of lifelong friendships that can enrich one's life and enhance one's work. I am not embarrassed to acknowledge that after learning from this history about scientists who unselfishly helped colleagues and who took a stand against some injustice in the world, I was moved to think about my own behavior and how I might have done more.

Finally, although I do not believe that history provides answers to any of today's problems, I am convinced that it can broaden the way we think about them. Like any good teacher, the value of history is not in the specific lessons it teaches, but in the way it influences how we think about problems.

NOTES

CHAPTER 1

1. C. S. Sherrington, *The Integrative Action of the Nervous System* (New Haven: Yale University Press, 1906), pp. 16–18.

2. There are some excellent works that describe different aspects of this history. See, for example: G. M. Shepherd, *Foundations of the Neuron Doctrine* (New York: Oxford University Press, 1991); M. Bennett, *The History of the Synapse* (Australia: Harwood Academic Publishers, 2001); M. W. Cowan and E. Kandell, "A Brief History of Synapses and Synaptic Transmission," in M. W. Cowan et al., eds., *Synapses* (Baltimore: Johns Hopkins University Press, 2001), pp. 1–87; J.-C. Dupont, *Histoire de la neurotransmission* (Paris: Presses universitaires de France, 1991); J. D. Robinson, *Mechanisms of Synaptic Transmission: Bridging the Gaps (1890–1990)* (New York: Oxford University Press, 2001).

3. Theodor Schwann's book was published in German in 1839 and translated into English in 1847. In German, it was entitled *Mikroskopische Untersuchungen über die Uebereinstimmung in der Structur und dem Wachstum der Thiere und Pflanzen* (Berlin: G. E. Reimer, 1839). See also M. J. Schleiden, "Beiträg zur Phytogenesis," *Archiv für Anatomie, Physiologie und wissenschlaftliche Medicin* 5 (1838): 137–176.

4. Experimentation with silver stains was, in part, the result of use of silver in early photography.

5. Wilhelm von Waldeyer introduced the word "neuron" and gave support and name to what became the "neuron doctrine" in his article "Über einige neuer Forschungen im Gebeite der Anatomie des Centralnerven Systems," *Deutsche medizinische Wochenschrift* 17 (1891): 1213–1218. Those opposed to the neuron doctrine considered the nervous system to be a "syncytium," a multinucleated body of protoplasm that is not differentiated into separate cells.

6. Golgi and the members of this side of the dispute were referred to as the reticularists, who argued that the nervous system was a continuous nerve net, or syncitium. The other side argued that neurons were contiguous, but not continuous.

7. S. Ramón y Cajal, *Histologie du système nerveux de l'homme et des vertébrés*, 2 vols. (Paris: Maloine, 1909–1911).

8. Much of the fundamental observation on the course of nerve degeneration after injury was done earlier by Bernard von Gudden (1824–1886) and later by August-

Henri Forel (1848–1931). As was common during their time, both men were psychiatrists as well as neuroanatomists.

9. For more about Golgi, see P. Mazzarello, *The Hidden Structure: A Scientific Biography of Camillo Golgi* (Oxford: Oxford University Press, 1999), originally published in Italian in 1996; the 1999 edition was edited and translated by Henry A. Buchtel and Aldo Badiani.

10. Golgi presented questionable evidence from studies he had conducted on the cerebellum of the brain, claiming that it showed that there are nerve fibers that are not connected to cell bodies.

11. Hans Held of Leipzig wrote a paper in support of the reticular theory as late as 1929: "Die Lehre von den Neuronen und vom Neurencytium und ihr heutiger Stand," *Fortschritte der Naturwissenschaftlichen Forschung* 8 (1929): 117–126.

12. During Cajal's visit to England in 1894 he was a guest in Sherrington's home. It has been reported that Sherrington helped prepare many of the slides Cajal used in his Croonian Lecture.

13. A more complete discussion of Sherrington's adoption of the word "synapse" is given in Bennett, *History of the Synapse,* pp. 22–24.

14. Moreover, Sherrington felt that the word "synapse" lent itself more readily to being expressed as an adjective (e.g., synaptic junction) than did some of the other alternatives under consideration.

15. Curare had originally been brought from South America, where it was used as a poison for hunting animals. Sir Walter Raleigh and other early explorers were interested in the drug, and by the latter part of the sixteenth century native preparations of the drug had been brought to Europe.

16. A good description of Claude Bernard's studies of curare is given in Bennett, *History of the Synapse,* pp. 44–45.

17. E. Du Bois–Reymond, *Gesammelte Abhandlungen zur allgemeinen Muskel- und Nervenphysik*, vol. 2 (Leipzig: Veit, 1877), p. 700.

18. Du Bois–Reymond succeeded Johannes Müller as head of physiology at the University of Berlin.

19. See M. Grundfest, "Excitation at Synapses," *Journal of Neurophysiology* 20 (1957): 316–324. When Du Bois–Reymond compared the effectiveness of electric current with mechanical, thermal, and chemical stimuli (including strychnine) in exciting nerves, he concluded that electric current ranks first, "judging from its convenience, strength, and duration" (quoted in E. Clarke and C. D. O'Malley, *The Human Brain and Spinal Cord: A Historical Study Illustrated bv Writings from Antiquity to the Twentieth Century* [Berkeley: University of California Press, 1968], p. 199). In 1860 Du Bois–Reymond wrote (*Untersuchungen über thierische Elektricität* [Berlin: Reiner]) that skeletal muscles were normally alkaline, not acid as commonly thought, but became acid after repeated contractions. The concluding point was that the "physiological reaction of muscle is accompanied by chemical changes." This is not the same as proposing that neural transmission across a synapse is chemical.

20. Striated muscles have visible striations in the tissue. A major exception to the characterization of visceral organ tissue as smooth is heart muscle, which is striated. Glands are also considered visceral organs, but they have specialized secretory cells that are neither smooth nor striated.

CHAPTER 2

1. J. N. Langley, *The Autonomic Nervous System* (Cambridge: W. Heffer and Sons, 1921), p. 44.

2. W. Langdon-Brown, *The Sympathetic Nervous System in Disease* (London: Henry Frowde, 1923), p. 1.

3. E. H. Starling, editor's preface, in W. H. Gaskell, *The Involuntary Nervous System* (London: Longmans, Green, 1916).

4. "John Newport Langley, 1852–1925," *Proceedings of the Royal Society of London*, series B, part 1 (1925): xxxiii.

5. J. N. Langley, "Walter Holbrook Gaskell, 1847–1914," *The Smithsonian Institution, Annual Report* (1915): 531.

6. W. H. Gaskell, *The Origin of Vertebrates* (London: Longmans, Green, 1908).

7. The term "sympathetic" had earlier been used to refer to what is now known as the entire autonomic nervous system. The word "sympathetic" is derived from the Latin word for "consensus." In its earlier usage the term referenced the belief that there was a certain interrelationship among the parts of the system that made them respond to each other—this is not unlike the present-day use of the expression "sympathetic vibrations." Gaskell and Langley used the term "sympathetic" to name the part of the autonomic nervous system that had its origin in the thoracic and lumbar region of the spinal cord. See the chapter "The Vegetative Nervous System" in E. Clarke and L. S. Jacyna, *Nineteenth-Century Origins of Neuroscientific Concepts* (Berkeley: University of California Press, 1987).

8. Gaskell, *Involuntary Nervous System*.

9. J. N. Langley, "On the Union of the Cranial Autonomic (Visceral) Fibres with the Nerve Cells of the Superior Cervical Ganglion," *Journal of Physiology, London* 23 (1898): 240–270 (quote from p. 251).

10. J. N. Langley and W. L. Dickinson, "On the Local Paralysis of Peripheral Ganglia and the Connection of Different Classes of Nerve Fibers with Them," *Proceedings of the Royal Society of London,* series B, 46 (1889): 423–431.

11. Nicotine was isolated from tobacco leaves. Muscarine, which is obtained from the toadstool *Amanita muscaria,* was found to have many of the same effects on visceral organs as stimulating the parasympathetic system. Pilocarpine, whose action is similar to muscarine, was obtained from the leaflets of a shrub of the genus *Pilocarpus.* Atropine was obtained from the shrub deadly nightshade, *Atropa belladonna.* Curare was obtained from South American Indians who used it on their arrows when hunting. The recipe for making curare must have differed somewhat among different tribes, and early on curare was identified by the source or by the container in which it was packaged, such as tubo-curare. The preparation of curare was a carefully kept secret known only to a few shamans of the tribe.

12. Oliver and Schäfer had been students together working under Professor William Sharpley. See E. P. Sparrow and S. Finger, "Edward Albert Schäfer (Sharpley-Schäfer) and His Contributions to Neuroscience: Commemorating the 150th Anniversary of His Birth," *Journal of the History of the Neurosciences* 10 (2001): 41–57.

13. G. Oliver and E. A. Schäfer, "On the Physiological Action of Extract of the Suprarenal Capsule," *Journal of Physiology, London* 16 (1894): 1–IV; G. Oliver and E. A.

Schäfer, "On the Physiological Action of Extract of the Suprarenal Capsule," *Journal of Physiology, London* 18 (1895): 230–276.

14. Oliver and Schäfer obtained adrenal extracts from a variety of animals, but mostly from calves, sheep, and guinea pigs.

15. F. Stolz, "Über Adrenalin und Alkylaminoacetobrenzcatechin," *Bericht deutsche Chemie Ges.* 37 (1904): 4149–4154; H. D. Dakin, "The Synthesis of a Substance Allied to Noradrenalin," *Proceedings of the Royal Society of London,* series B, 76 (1905): 491–497.

16. For more information on the history of adrenal extracts, see H. Davenport, "Epinephrin(e)," *Physiologist* 25 (1982): 76–82; H. Davenport, "Early History of the Concept of Chemical Transmission of the Nerve Impulse," *Physiologist* 34 (1991): 178–190 (esp. p. 181).

17. M. H. Lewandowsky, "Über die Wirkung des Nebennierenextractes auf die glatten Muskeln im Besonderen des Auges," *Archiv für Physiologie: Abteilung des Archives für Anatomie und Physiologie* (1899): 360–366. Because the same response was also obtained after excision of the nerve that innervates the muscles of the eye, Lewandowsky concluded that the adrenal extract acted directly on the smooth muscles of the eye.

18. J. N. Langley, "Observations on the Physiological Action of Extract of Supra Renal Bodies," *Journal of Physiology, London* 27 (1901): 237–256.

19. See J. D. Robinson, *Mechanisms of Synaptic Transmission: Bridging the Gaps (1890–1990)* (New York: Oxford University Press, 2001), p. 81, note 29.

20. J. N. Langley, "Observations on the Physiological Action of Extracts of the Supra-Renal Bodies," *Journal of Physiology, London* 27 (1901): 237–256 (quote from p. 247).

21. Langley acknowledged the overlap between his ideas about receptor substances and Paul Erhlich's theory that "side chains" of a molecule determined the affinity between a chemical substance and a particular cell.

22. Today, receptors are not thought of as endogenous chemicals, but as macromolecules located either in cells or, more commonly, on their outer membranes. These molecular configurations selectively bind other molecules, called ligands, that reach the cell in the form of neurotransmitters, neuromodulators, hormones, and the like from nerves or from the bloodstream. This binding is known to initiate a cascade of biochemical and biophysical reactions that produces changes in the structure and function of the cell.

23. J. N. Langley, "On Nerve Endings and on Special Excitable Substances in Cells (Croonian Lectures)," *Proceedings of the Royal Society of London,* series B, 78 (1906): 183.

24. For a discussion of the opposition to the Langley's concept of a "receptor substance," see A.-H. Maehle, C.-R. Prüll, and R. F. Halliwell, "The Emergence of the Drug Receptor Theory," *Nature Reviews* 1 (2002): 637–641.

25. Alfred Clark used a microinjection technique that allowed him to determine that many drugs were active only if injected onto the cell membrane, not into the cell body. He then determined the minimum effective dose and the dose beyond which further increase did not result in any greater intensity of the response. By calculating

the size of different drug molecules and the area of the cell surface, Clark was able to conclude that drugs do not act on the entire cell membrane, but only on a finite number of discrete receptors embedded in the membrane. Thus, he wrote, "measurements of the quantities of drugs that suffice to produce an action on cells, prove that in the case of powerful drugs the amount fixed is only sufficient to cover a small fraction of the cell surface. . . . The simplest probable conception of drug action is that potent drugs occupy certain specific receptors on the cell surface, and that these specific receptors only comprise a small fraction of the total cell surface" (A. J. Clark, *The Mode of Action of Drugs on Cells* [Baltimore: Williams and Wilkins, 1933], p. 87).

CHAPTER 3

1. There is some disagreement about whether more credit for these ideas should be given to Thomas Elliott or to his mentor, John Langley. See M. Bennett, *History of the Synapse* (Australia: Harwood Academic Publishers, 2001), pp. 107–108; H. W. Davenport, "Gastrointestinal Physiology, 1895–1975: Motility," in S. G. Schultz, J. D.Wood, and B. B. Rauner, eds., *Handbook of Physiology*, sec. 6, vol. 1, part 1 (Bethesda, Md.: American Physiological Society, 1989), pp. 1–101. In most of the literature, Thomas Elliott is given the credit, but this may be due to the precedent established by the writings of Henry Dale, a close personal friend of his.

2. T. R. Elliott, "On the Action of Adrenalin," *Journal of Physiology, London* 31 (1904): xx–xxi (quote from p. xxi); T. R. Elliott, "On the Action of Adrenalin," *Journal of Physiology, London* 32 (1905): 401–465.

3. Elliott, "Action of Adrenalin" (1905), p. 460.

4. Elliott, "Action of Adrenalin" (1905), pp. 466–467.

5. H. H. Dale, "Thomas Renton Elliott, 1871–1961," *Biographical Memoirs of Fellows of the Royal Society* (1961), p. 58.

6. J. N. Langley, *The Autonomic Nervous System* (Cambridge: W. Heffer and Sons, 1921).

7. Langley, *Autonomic Nervous System*, pp. 39 and 45.

8. Elliot completed his medical degree with gold medals in clinical medicine and surgery and a prize for his thesis on the innervation of the bladder.

9. T. R. Elliott, "The Adrenal Glands (Sidney Ringer Memorial Lecture, 1914)," *British Medical Journal*, June 27, 1914, 1395.

10. T. R. Elliot, "The Innervation of the Adrenal Glands," *Journal of Physiology, London* 46 (1913): 285.

11. Elliott, "Adrenal Glands (Lecture)," p. 1395.

12. For example, the pupils still dilated in response to nerve stimulation even more than a week after adrenalectomy.

13. Elliott, "Adrenal Glands (Lecture)," p. 1395. Much later, Jacques de Champlain proved that adrenaline secreted by the adrenal medulla is picked up and stored in sympathetic nerves.

14. Langley, *Autonomic Nervous System*, p. 43.

15. H. H. Dale, *Adventures in Physiology* (London: Pergamon, 1953).

16. For further information on Thomas Elliott, see "Obituary: Thomas Renton Elliott," *Lancet*, March 11, 1961, 567–568; H. P. Himsworth, "Obituary: Prof. T. R. Elliott," *Nature* 190 (May 6, 1961): 486–487; H. H. Dale, "Thomas Renton Elliott, 1877–1961," *Royal Society Obituaries and Memoirs, 1830–1998* 7 (1961): 53–74; "Obituary: T. R. Elliott," *British Medical Journal*, part 1, March 11, 1961, 752–754.

17. Cited p. 6 of J. A. Gunn, "Obituary: Walter Ernest Dixon, 1871–1931," *Journal of Pharmocology and Experimental Therapeutics* 44 (1932): 3–21.

18. Atropine was obtained from the plant *Atropa belladonna*, the deadly nightshade. The name belladonna was derived from the plant's ability in a very dilute solution to make women more attractive by dilating their pupils. In larger doses atropine is a poison and was used as such in the Middle Ages.

19. W. E. Dixon, "On the Mode of Action of Drugs," *Medical Magazine* 16 (1907): 454–457 (quote from pp. 456–457). According to Walter Cannon, Dixon later said that he had submitted the article to a relatively obscure journal because he feared the criticism of his contemporaries: W. B. Cannon, *The Way of an Investigator: A Scientist's Experiences in Medical Research* (New York: Hafner, 1945), 99.

20. Dixon did not mention what animal was used as the donor, but Henry Dale reported after Dixon's death that it had been a dog (*British Medical Journal*, May 12, 1934, 835). It is likely that Dixon purposely avoided mentioning his use of a dog so that he wouldn't arouse the ire of the antivivisectionists, who were quite militant in Great Britain at the time.

21. W. E. Dixon, "Vagus Inhibition," *British Medical Journal*, 1906, 1807.

22. H. M. McLennan, *Synaptic Transmission*, 2d ed. (Philadelphia: W. B. Saunder, 1970), p. 24.

23. Dale published data given to him much earlier by W. E. Dixon. The data only illustrated a single example of inhibition in a frog heart exposed to an extract from a dog's heart obtained after stimulation of the vagus nerve. Without appropriate controls, it is not possible to evaluate the significance of this record. Dale, *Adventures in Physiology*, p. 532.

24. Cited in *In Memory of Sir Henry Dale*, Sir Paul Girolami, Lady Helena Taborikova, and Guiseppi Nitisco, eds. (Accademia Roma di Scienze Mediche e Biologiche).

25. H. H. Dale, "Chemical Transmission of the Effects of Nerve Impulses," *British Medical Journal*, May 12, 1934, 835–841 (quote from p. 836).

26. G. Barger and H. H. Dale, "Chemical Structure and Sympathetic Action of Amines," *Journal of Physiology, London* 41 (1910): 19–59 (quote from p. 54).

27. Reid Hunt was on leave from a position in pharmacology at Johns Hopkins Medical School. In 1904 he joined the U.S. Public Health Service, and in 1913 Hunt was appointed professor of pharmacology at Harvard University. Choline had been found to be present in pig bile in 1849. In 1899 Reid Hunt began to use the acetic ester of choline, acetylcholine, in experiments, but it was not known to be a natural substance until Dale and Dudley found it in animals in 1929.

28. R. Hunt and R. M. Taveau, "On the Physiological Action of Certain Cholin Derivatives and New Methods for Detecting Cholin," *British Medical Journal*, part 2, December 22, 1906, 1788–1791. Although acetylcholine had been synthesized in 1867, little was known about its physiological action prior to Hunt's work.

CHAPTER 4

1. H. H. Dale, "Pharmacology During the Past Sixty Years," *Annual Review of Pharmacology* 3 (1963): 1–8 (quote from p. 1).

2. W. Feldberg, "Henry Hallett Dale, 1875–1968," *Biographical Memoirs of Fellows of the Royal Society* 13 (1968): 77–174 (esp. p. 87).

3. Created by Sir William Crookes during the late nineteenth century, this sealed glass tube was originally used to determine how gases would change when exposed to an electrical discharge. Wilhelm Roentgen determined it to be the source of what he called X-rays after he discovered patches of fluorescent light on various objects in the room that had been treated with the salts used on photographic plates. In trying to determine what the X-rays could not penetrate, Roentgen inadvertently discovered that the bones of his hand, which were holding a lead pipe, were displayed on a photographic plate. The story of Roentgen and his discovery is described in M. Friedman and G. W. Friedland, *Medicine's Ten Greatest Discoveries* (New Haven: Yale University Press, 1998).

4. Feldberg, "Henry Hallett Dale," p. 93.

5. Feldberg, "Henry Hallett Dale," pp. 89–91.

6. A. M. Silverstein, *Paul Ehrlich's Receptor Immunology: The Magnificent Obsession* (San Diego: Academic Press, 2002).

7. Ehrlich's ideas about receptors developed first out of his effort to understand why lead and other toxins had an affinity more for certain structures than others. He later revised his ideas about receptors when he started to do his seminal work in immunology, for which he shared the Nobel Prize in 1908. See M. Bennet, *History of the Synapse* (Australia: Harwood Academic Publishers, 2001), pp. 53–54 for a discussion of Langley and Ehrlich's theory of receptors.

8. Quoted in E. Bäumler, *Paul Ehrlich: Scientist for Life* (New York: Holmes and Meier, 1984), p. 81.

9. H. Thoms, "John Stearns and Pulvis Parturiens," *American Journal of Obstetrics and Gynecology* 22 (1931): 418–423. John Stearns, "Account of the Pulvis Parturiens, a Remedy for Quickening Childbirth," was published in the *Medical Repository of New York* (1818).

10. H. H. Dale, *Adventures in Physiology* (London: Pergamon, 1953), p. xi.

11. The substance found to produce uterine contractions was later identified as oxytocin, a hormone now known to be produced in the body. Dale described many of the properties of various ergot extracts in H. H. Dale, "On Some Physiological Actions of Ergot," *Journal of Physiology, London* 3 (1906): 163–206.

12. H. H. Dale and P. P. Laidlaw, "Histamine Shock," *Journal of Physiology, London* 52 (1919): 355. Convincing proof that histamine was present in the body was not actually provided until 1927, when Dale, with Best, Dudley, and Thorpe, was able to extract it from fresh samples of animal liver and lungs. See "The Nature of the Vaso-Dilator Constituents of Certain Tissue Extracts," *Journal of Physiology, London* 62 (1927): 397. In the same year Lewis found that histamine was liberated from injured skin and other tissue. Prior to this time, histamine had been called by its chemical name, but when it was found in animal tissue, the name "histamine," from the Greek *histos,* meaning "tissue," was introduced. Histamine is now known to be released from mast cells in the skin and

elsewhere in response to injury and to foreign protein. Histamine is also found in the brain and in peripheral nerve trunks. In addition to anaphylactic shock, histamine is known to cause pain, itching, and headaches, among other symptoms, and also to be involved in allergic reactions. In 1919 Dale gave the Croonian Lecture on the subject. See H. H. Dale, "The Biological Significance of Anaphylaxis (Croonian Lecture)," *Proceedings of the Royal Society of London,* series B, 91 (1919–1920): 126.

13. Quoted in W. Feldberg, "Henry Hallett Dale, 1875–1968," *British Journal of Pharmacology* 35 (1969): 1–9 (quote from p. 8).

14. In 1946 Ulf von Euler of Sweden, who shared the 1970 Nobel Prize with Julius Axelrod and Bernard Katz, extracted norepinephrine from the sympathetic nerves of mammals, thereby proving that it was a natural neurotransmitter.

15. H. H. Dale, "The Beginnings and the Prospects of Neurohumoral Transmission," *Pharmacological Reviews* 6 (1954): 7–14 (quote from p. 9).

16. R. Hunt and R. M.Taveau, "On the Physiological Action of Certain Cholin Derivatives and New Methods for Detecting Cholin," *British Medical Journal,* part 2, December 22, 1906, 1788–1791.

17. Letter, Dale to Elliott, December 11, 1913, Contemporary Medical Archives Center, Wellcome Institute, GC/42. Quoted in E. M. Tansey, "Chemical Transmission in the Autonomic Nervous System: Sir Henry Dale and Acetylcholine," *Clinical Autonomic Research* 1 (1991): 63–72.

18. H. H. Dale, "The Occurrence in Ergot and Action of Acetylcholine," *Journal of Physiology, London* 48 (1914): iii–iv.

19. H. H. Dale, "The Action of Certain Esters and Ethers of Choline and Their Relation to Muscarine," *Journal of Pharmacology and Experimental Therapeutics* 6 (1914): 147–190 (quote from p. 188).

20. At high doses nicotine paralyzes the nerve, allowing no further responses to be evoked.

21. In general, muscarinic sites are located at the junction of parasympathetic nerves with smooth muscles, while nicotinic sites are found primarily at the autonomic ganglia synapses and at the junction of spinal nerves with skeletal muscles. There are, however, some ganglionic synapses that are muscarinic. It is now known that there are both muscarinic and nicotinic receptors in the brain.

22. G. Barger and H. H. Dale, "Chemical Structure and Sympathomimetic Action of Amines," *Journal of Physiology, London* 41 (1910): 19–59 (quote from p. 21).

23. Dale, "Action of Certain Esters," p. 188.

24. Dale, "Action of Certain Esters," p. 189.

25. H. H. Dale, "Chemical Transmission of the Effects of Nerve Impulses," *British Medical Journal,* May 12, 1934, 835–841 (quote from p. 836).

26. Dale, *Adventures in Physiology,* p. 98.

27. E. D. Lord Adrian, "Sir Henry Dale's Contribution to Physiology," *British Medical Journal,* June 4, 1955, vol. 1, p. 1356.

28. Feldberg, "Henry Hallett Dale," p. 118.

29. Kymographic drums were cylinders attached to a motor that moved them at a controlled rate. A paper smoked over a kerosene lamp was wrapped around the cylinder. A stylus attached to an electromagnetic device left a mark on the smoked paper

when a response occurred. In order to preserve the record, the smoked paper was carefully removed from the drum, soaked in varnish, and hung to dry.

CHAPTER 5

1. O. Loewi, "An Autobiographical Sketch," *Perspectives in Biology and Medicine* 4 (1960): 3–25 (quote from p. 3).

2. O. Loewi, "Introduction," *Pharmacological Reviews* 6 (1954): 3–6 (quote from p. 3).

3. Otto Loewi's early life has been described in F. Lembeck and W. Giere, *Otto Loewi: Ein Lebensbild in Dokumenten, Biographische Dokumentation und Bibliographie* (Berlin: Springer, 1968).

4. Walther Straub was a professor of pharmacology in Marburg for part of the time that Loewi was there.

5. The point where spinal motor nerves join skeletal (voluntary, or striated) muscles is now called the "motor endplate."

6. Hans Meyer was considered one of the leading pharmacologists in Europe. In 1910, after he became professor of pharmacology in Vienna, Meyer published his famous textbook *Experimental Pharmacology as the Basis of Drug Therapy,* a book that divorced pharmacology from *materia medica* and grouped drugs on the basis of the physiological reactions they induced.

7. Loewi's colleague, a biochemist named Kutscher, had demonstrated that a pancreas could be digested until no protein was left. Loewi's speculation that animals might be able to make their required proteins from the degraded by-products was shown to be right.

8. Loewi described how he came to do this experiment in O. Loewi, *From the Workshop of Discoveries* (Porter Lecture Series 19) (Lawrence: University of Kansas Press, 1953), pp. 28–29.

9. Loewi, "Autobiographical Sketch," p. 10.

10. H. H. Dale, "Otto Loewi, 1873–1961," *Royal Society Obituaries and Memoirs, 1930–1998* 8 (1962): 67–89 (quote from p. 71).

11. H. H. Dale, "The Beginning and the Prospects of Neurohumoral Transmission," *Pharmacological Reviews* 6 (1954): 7–14 (quote from p. 9).

12. Fletcher was spending the years 1902–1905 working with Loewi in Marburg on the effect of different diuretics on the kidney.

13. O. Schmiedeberg and R. Koppe, *Das Muscarin: Das giftige Alkaloid des Fliegenpilzes* (Leipzig: Vogel, 1869), pp. 27–29.

14. Eserine is an alkaloid obtained from the Calabar bean. It had been used in the region along the delta of the Niger in trials by ordeal. It was known as early as 1905 that it increased the responses to parasympathetic stimulation. Reid Hunt showed in 1918 that it increased the effects of acetylcholine. That same year, Fühner found that eserine increased the action of acetylcholine on the leech muscle about a million times, but did not increase the potency of either choline or pilocarpine.

15. Loewi, "Autobiographical Sketch," p. 17.

16. Dale, "Otto Loewi," p. 76.

17. Loewi, *Workshop of Discoveries*, p. 33.

18. Horace Davenport may have been the first to discover the discrepancy in the dates as reported by Loewi. See H. Davenport, "Early History of the Concept of Chemical Transmission of the Nerve Impulse," *Physiologist* 34 (1991): 178–190. See also Zénon Bacq's account in his chapter "Chemical Transmission of Nerve Impulses," in M. J. Parnham and J. Bruinvels, eds., *Psycho- and Neuro-Pharmacology*, vol. 1 (New York: Elsevier, 1983), pp. 50–103.

19. Loewi, "Autobiographical Sketch," p. 17.

20. A description of the experiment was first published in a four-page article: O. Loewi, "Über humorale Übertragbarkeit der Herznervenwirkung, I," *Pflügers Archiv für die gesamte Physiologie des Menschen und der Tiere* 193 (1921): 239–242.

21. Later Loewi found that if he blocked the effect of the vagus nerve by administering atropine, it was easier to obtain acceleration when he stimulated the vagal-sympathetic nerve trunk. It was also easier to produce acceleration in the toad heart than in the frog heart.

22. See, for example, W. A. Bain, "A Method of Demonstrating Humoral Transmission of the Effects of Cardiac Vagus Stimulation in the Frog," *Quarterly Journal of Experimental Physiology and Cognate Medical Sciences* 22 (1932): 269–274.

23. During February and early March, the time Loewi did his initial experiment, frogs are just emerging from hibernation. The largest and most active frogs are the first to emerge, and it has been suggested that these were the one's selected by Loewi. See A. H. Fiedman, "Circumstances Influencing Otto Loewi's Discovery of Chemical Transmission in the Nervous System," *Pflügers Archiv* 325 (1971): 85–86. William Van der Kloot has reported that the hearts of the so-called Hungarian frogs, which are the ones Loewi used, contain a relatively low amount of cholinesterase.

24. Zénon Bacq also has written that the details of Loewi's first experiment differed from his later description. Bacq, who knew Loewi and examined his first publication closely, concluded that in most instances Loewi used only one heart. See Parnham and Bruinvels, eds., *Psycho-and Neuro-Pharmacology*, pp. 49–103 (especially p. 54).

25. Loewi, "Autobiographical Sketch," p. 18.

26. For a good review of some of the inconsistencies in Loewi's description of his experiment, see Davenport, "Early History," pp. 178–190.

27. A good description of the flaws in the initial data Loewi obtained is presented in J. Robinson, *Mechanisms of Synaptic Transmission: Bridging the Gaps (1890–1990)* (New York: Oxford University Press, 2001), pp. 63–68.

28. In German, the series Loewi published was titled "Über humorale Übertragbarkeit der Herznervenwirkung."

29. Loewi, "Autobiographical Sketch," p. 18.

30. Loewi's remark to William Van der Kloot is reported in G. L. Geison, "Loewi, O.," in C. C. Gillispie, ed., *Dictionary of Scientific Biography*, vol. 8 (New York: Scribner, 1980), pp. 451–457.

31. The importance of adding physostigmine is made clear in O. Loewi and E. Navratil, "Über humorale Übertragbarkeit der Herznervenwirkung, XI: Über den Mechanismus der Vaguswirkung von Physostigmin und Ergotamin," *Pflügers Archiv für die gesamte Physiologie des Menschen und der Tiere* 214 (1926): 689–696.

32. Ulf von Euler described the influence that Otto Loewi had on his career in U. von Euler, "Discoveries of Neurotransmitter Agents and Modulators of Neuronal Function," in F. G. Worden, J. P. Swazey, and G. Adelman, eds., *The Neurosciences: Paths of Discovery I* (Boston: Birkhaüser, 1992), pp. 180–187.

33. In 1970 von Euler shared the Nobel Prize with Bernard Katz and Julius Axelrod.

34. R. H. Kahn, "Über humorale Übertragbarkeit der Herznervenwirkung," *Pflügers Archiv für die gesamte Physiologie des Menschen und der Tiere* 214 (1926): 482–498.

35. Bain, "Method of Demonstrating," pp. 270–271.

36. Quoted in B. Minz, *The Role of Humoral Agents in Nervous Activity* (Springfield, Ill.: Charles Thomas, 1955), pp. 53–57. According to Zénon Bacq, before Loewi's first experiment on vagal slowing, Asher had used a similar technique, but he did not find any indication of a neurohumoral secretion. See Z. Bacq, *Chemical Transmission of Nerve Impulses: A Historical Sketch* (New York: Pergamon, 1975).

37. Cited in E. Clarke and C. D. O'Malley, *The Human Brain and Spinal Cord* (Berkeley: University of California Press, 1968), p. 252.

38. See Dale, "Otto Loewi" for a complete list of Loewi's publications.

39. O. Loewi and E. Navratil, "Über humorale Übertragbarkeit der Herznervenwirkung, VI: Der Angriffspunkt des Atropins," *Pflügers Archiv für die gesamte Physiologie des Menschen und der Tiere* 206 (1924): 123–134; "Über humorale Übertragbarkeit der Herznervenwirkung, VII," *Pflügers Archiv für die gesamte Physiologie des Menschen und der Tiere* 206 (1924): 135–140; "Über das Schicksal des Vagusstoffs und des Acetylcholins im Herzen," *Klinische Wochenschrift* 5 (1926): 894; "Über den Mechanismus der Vaguswirkung von Physistogmin und Ergotomin," *Klinische Wochenschrift* 5 (1926): 894–895.

40. E. Engelhardt and O. Loewi, "Termentative Azetylcholinspaltung im blut und ihre Hemmung durch Physostigmin," *Naunyn-Schmeidebergs Archiv für experimentelle Pathologie und Pharmakologie* 150 (1930): 1–13.

41. The active ingredient now known as eserine or physostigmine was originally extracted from the West African Calabar bean. It was also called "ordeal bean" because it was used to test criminals. If a suspected criminal survived swallowing the calabar powder, he was said to have passed the "ordeal" test and proven his innocence. The bean was first brought to England in 1840. Around 1863 two sets of investigators obtained the pure alkaloid active ingredient from the bean; one group named it eserine, and the other called it physostigmine. Eserine/physostigmine is now known to inactivate cholinesterase and to prolong and potentiate the effects of acetylcholine. Physostigmine belongs to a group of compounds known as reversible cholinesterase agents. Slightly before and during World War II, a group of highly toxic anticholinesterase compounds known as organophosphates was developed in Germany by I. G. Farben. These compounds were used as insecticides and for nerve gases such as sarin. Some anticholinesterase compounds are used to treat *myasthenia gravis,* a neurological disorder involving a muscle weakness due to the lack of responsiveness of the acetylcholine receptors stimulated by spinal motor secretions.

42. Cited in Clarke and O'Malley, *Human Brain,* p. 253.

43. In 1906 Barger, Carr, and Dale had isolated a pharmacologically active sub-

stance they called ergotoxine from the ergot fungus. It later was shown to be made up of three separate alkaloid substances. In 1920 Stoll obtained a purified extract of one of these, which he called ergotamine.

44. In 1922 Loewi had found that an ergot extract blocked the action of Acceleransstoff. This was based on Dale's earlier report that this extract blocked the action of adrenaline. However, this test was not considered to have ruled out other possibilities. In 1926 Loewi demonstrated that ultraviolet light inactivated Acceleransstoff just as it did adrenaline. Finally, in 1936, Loewi demonstrated that Acceleransstoff displays the green fluorescence characteristic of adrenaline. See O. Loewi, "Über humorale Übertragbarkeit der Herznervenwirkung, XVI: Quantitativ und qualitativ Untersuchungen über den Sympathicusstoff," *Pflügers Archiv für die gesamte Physiologie des Menschen und der Tiere* 237 (1936): 504–514.

45. O. Loewi, "The Humoral Transmission of Nervous Impulse," *Harvey Lectures* (1934), pp. 218–233.

46. Dale, "Beginnings," p. 11.

47. O. Loewi, "The Ferrier Lecture: On Problems Connected with the Principle of Humoral Transmission of Nervous Impulses," *Proceedings of the Royal Society of London, series B* 18 (1935): 299–316.

CHAPTER 6

1. H. H. Dale, *Adventures in Physiology* (London: Pergamon, 1953), p. xv.

2. C. H. Best, H. H. Dale, H. W. Dudley, and W. Z. Thorpe, "The Nature of the Vasodilator Constituents of Certain Tissue Extract," *Journal of Physiology, London* 62 (1927): 397–417.

3. H. H. Dale, "Croonian Lectures to the Royal College of Physicians," *Lancet,* 1929, part 1, 1179, 1233, 1285.

4. Dale, *Adventures,* p. xv.

5. Letter Dale to Richards, March 22, 1929, Archives of the National Institute of Medical Research, file 647; quoted in E. M. Tansey, "Chemical Neurotransmission in the Autonomic Nervous System: Sir Henry Dale and Acetylcholine," *Clinical Autonomic Research* 1 (1991): 63–72.

6. H. H. Dale and H. W. Dudley, "The Presence of Histamine and Acetylcholine in the Spleen of the Ox and the Horse," *Journal of Physiology, London* 68 (1929): 97–123.

7. S. Finger, *Minds Behind the Brain: A History of the Pioneers and Their Discoveries* (New York: Oxford University Press, 2000), pp. 272–273.

8. Dale and Dudley, "Presence of Histamine," p. 122.

9. Cited in O. Loewi, "The Chemical Transmission of Nerve Action," in *Nobel Lectures, Including Presentation Speeches and Laureates' Biographies—Physiology and Medicine,* vol. 2 (Amsterdam: Elsevier, 1960), pp. 416–429.

10. To get this effect, eserine had to be administered and the blood supply to the second side had to be intact: B. P. Babkin, A. Alley, and G. W. Stavraky, "Humoral Transmission of Chorda Tympani Effect," *Transactions of the Royal Society of Canada,* sec. V: Biological Sciences 26 (1932): 89–107.

11. This has been called the "Vulpian" response for Alfred Vulpian, who first observed it in 1863.

12. H. H. Dale and J. Gaddum, "Reactions of Denervated Voluntary Muscle, and Their Bearing on the Mode of Action of Parasympathetic and Related Nerves," *Journal of Physiology, London* 70 (1930): 109–144.

13. The American neurophysiologist Herbert Gasser, who later shared the 1944 Nobel Prize with his collaborator, Joseph Erlanger, suggested an alternative electrical explanation of Dale and Gaddum's results.

14. B. Minz, "Pharmakologische Untersuchungen am Blutegelpräparat, zugleich eine Methode zum biologischen Nachweis von Acetylcholin bei Anwesenheit anderer pharmakolisch wirksamer körpereigener Stoffe," *Naunyn-Schmeidebergs Archiv für experimentelle Pathologie und Pharmakologie* 168 (1932): 292–304; see also B. Minz, *The Role of Humoral Agents in Nervous Activity* (Springfield, Ill.: Charles Thomas, 1955), pp. 53–57; W. Feldberg and B. Minz, "Das Auftreten eines acetylcholinartigen Stoffes im Nebenierenvenenblut bei Reizung der Nervi splanchni," *Pflügers Archiv für die gesamte Physiologie des Menschen und der Tiere* 233 (1933): 657–682.

15. W. Feldberg and O. Krayer, "Nachweis einer bei Vagusreiz freiwerdenden azetylcholinähnlichen Substanz am Warmblüterherzen," *Naunyn-Schmeidebergs Archiv für experimentelle Pathologie und Pharmakologie* 172 (1933): 170–193.

16. W. Feldberg, *Fifty Years On: Looking Back on Some Developments in Neurohumoral Physiology* (Sherrington Lectures 16) (Liverpool: Liverpool University Press, 1982), p. 5.

17. W. Feldberg and E. Schilf, *Histamin: Seine Pharmakologie und Bedeutung für die Humoralphysiologie* (Berlin: J. Springer, 1930).

18. W. Feldberg, "The Early History of Synaptic and Neuromuscular Transmission by Acetylcholine: Reminiscences of an Eye Witness," in A. L. Hodgkin et al., eds., *The Pursuit of Nature: Informal Essays on the History of Physiology* (London: Cambridge University Press, 1977), pp. 65–83 (quote from p. 69).

19. Feldberg, *Fifty Years On*, p. 6.

20. The Rockefeller Foundation insisted on some reasonable assurance that a more permanent position would be offered after the end of the three-year period.

21. Beginning in 1933 and extending until the end of the war in 1945, the Rockefeller Foundation spent $1.5 million in identifying and assisting a total of 303 scientists and scholars from German and other countries as the war extended. Their nationalities were as follows: 191 German; 36 French; 30 Austrian; 12 Italian; 1 Polish; 6 Hungarian; 5 Czechoslovak; 5 Spanish; 2 Danish; 2 Belgian; 2 Dutch; and 1 Finnish. Of the total, 113 were in the social sciences; 73 in the natural sciences; 59 in the humanities; and 58 in the medical sciences. Seven were already Nobel prize winners, two were significant contributors to the making of the atom bomb, and many became heads of prestigious departments in U.S. universities. Ironically, the brilliant mathematics department at the University of Göttingen had been built up earlier with the aid of Rockefeller Foundation funds. When the faculty was dispersed in the thirties many of them ended up at U.S. universities. It has been said that if Hitler had intended to build up mathematics in the United States, a field that had been neglected here, he could hardly have been more successful than what his ruthlessness accomplished. For additional information see R. B. Fosdick, *The Story of the Rockefeller Foundation* (New York: Harper, 1952).

22. A copy of this letter was made available by the Rockefeller Foundation Archives.

23. For the first six months Feldberg was in London he was able to get some money out of Germany, but after that all his assets were frozen by the Nazi government.

24. Minz, *Role of Humoral Agents*. For additional information about Bruno Minz see E. Wolfe, A. C. Barger, and S. Benison, *Walter B. Cannon: Science and Society* (Cambridge, Mass.: Harvard University Press, 2000), pp. 382–383.

25. Some of this story has been described in J. S. Medawar, D. Pyke, and M. Perutz, *Hitler's Gift: The True Story of the Scientists Expelled by the Nazi Regime* (UK: General, 2001).

26. For example, in the period from 1990 to 2000, scientists in the United States received 44 Nobel prizes, while German scientists received only 5.

27. H. H. Dale, W. Feldberg, and M. Vogt, "Release of Acetylcholine at Voluntary Nerve Endings," *Journal of Physiology, London* 86 (1936): 353–379.

28. M. Vogt, "The Concentration of Sympathin in Different Parts of the Central Nervous System Under Normal Conditions and After the Administration of Drugs," *Journal of Physiology, London* 123 (1954): 451–481.

29. The test was adapted from Fühner (see below), who used the technique as a method for detecting the presence of eserine. Minz and Feldberg had turned the method around and used it to detect the presence of acetylcholine.

30. The leech that was used had originally been brought from Hungary. It was important to use the right leech, as it was found that the Hungarian leech (*Hirudo officinalis*) muscle is much more sensitive to acetylcholine than the English, German, or Israeli leech. The leech preparation was developed by Bruno Minz, a colleague of Wilhelm Feldberg. Minz published the first paper in 1932 describing how the technique could be used to detect acetylcholine. The next year, Feldberg and Minz described the use of the technique to detect acetylcholine in experiments on the cat. The original idea was based on a observation by Fühner, a German pharmacologist, who had reported in 1918 that the leech muscle's sensitivity to acetylcholine was increased a millionfold by the addition of eserine. Actually, Fühner had been using acetylcholine to detect eserine and Minz and Feldberg reversed the goal of the preparation.

31. J. C. Szerb, "The Estimation of Acetylcholine, Using Leech Muscle in a Microbath," *Journal of Physiology, London* 158 (1961): 8P.

32. Besides Wilhelm Feldberg, Dale's group included, among others: Harold Dudley, John Gaddum, Marthe Vogt, George L. Brown, A. Vartiaimen, and Z. M. Bacq. Brown, Dudley, Feldberg, Gaddum, and Vogt all were elected later to the Royal Society. Brown and Gaddum were also knighted in recognition of their scientific achievements.

33. W. Feldberg and B. Minz, "Das Auftreten eines acetylcholinartigen Stoffes," p. 657.

34. In 1936 Feldberg left England. He had been offered Rockefeller Foundation support to work with Charles Kellaway at the Walter Eliza Hall Institute in Melboune. As his Rockefeller Foundation support to work in Dale's laboratory would soon end, he accepted the opportunity. Kellaway was studying the histamine response to snake venom. This was not a new topic for Feldberg—he had published a book on histamine

with E. Schilf in 1930 and had also done some research on snake and bee venom in Berlin.

Feldberg did not remain long in Australia, for in 1938 he received an offer to be a reader in physiology at Cambridge. Feldberg was pleased to have an opportunity to return to England, and he soon gained the reputation of being an excellent teacher. He continued his research on acetylcholine, its biosynthesis, and its distribution in the gut. He also became involved in studying the electric fish *Torpedo marmorata*. The experiments were conducted at Arcachon in the south of France in collaboration with the neurophysiologist Albert Fessard. (Fessard was a neurophysiologist who had for a number of years been quite outspoken in opposing chemical transmission theories.) The electric organ of the Torpedo can be regarded as a collection of muscle end plates, so they suspected that acetylcholine might be involved in triggering the electric shock. The latency in the response when the innervating nerve was stimulated seemed to support the idea that chemical transmission was involved. Feldberg and Fessard applied cholinesterase to the electric organ, and they could then trigger the electrical discharge by applying acetylcholine to the organ. They had also begun to find evidence of the presence of acetylcholine following stimulation of the innervating nerve using the eserine-leech preparation. With these preliminary results in hand and with the help of Dale, Feldberg was able to obtain a Rockefeller Foundation grant to fly to Paris once a month to continue the research on the Torpedo with Fessard. It was 1939 and he was eager to start, but before he could do so the war broke out and the Germans were overrunning France.

Several years after the war, Feldberg was appointed head of the division of physiology and pharmacology at the National Institute of Medical Research (N.I.M.R.). It was 1949 and the institute was in the process of moving from Hampstead to Mill Hill. The research laboratory that Feldberg headed was the same one that Dale had directed for many years. In Mill Hill, Feldberg extended his work on acetylcholine in the brain. Acetylcholine had already been found in the brain, and Feldberg started to collect information about the responses that could be evoked by infusing acetylcholine and eserine into different brain sites.

35. H. H. Dale and W. Feldberg, "The Chemical Transmitter of Effects of the Gastric Vagus," *Journal of Physiology, London* 80 (1934): 16P–17P; H. H. Dale, "Nomenclature of Fibres in the Autonomic System and Their Effects," *Journal of Physiology, London* 80 (1934): 10P–11P.

36. H. H. Dale and W. Feldberg, "The Chemical Transmission of Secretory Impulses to the Sweat Glands of the Cat," *Journal of Physiology, London* 81 (1934): 40P–41P.

37. Dale, "Nomenclature of Fibres," p. 11P.

38. W. Feldberg and J. Gaddum, "The Chemical Transmitter at Synapses in a Sympathetic Ganglion," *Journal of Physiology, London* 81 (1934): 305–319.

39. A. W. Kibjakow, "Über humorale Übertragung der Erregung von einem Neuron auf das andere," *Pflügers Archiv für die gesamte Physiologie des Menschen und der Tiere* 232 (1933): 432–443. A diagram of Kibjakow's technique is reproduced in L. H. Marshall and H. W. Magoun, *Discoveries in the Human Brain: Neuroscience Prehistory, Brain Structure, and Function* (Totowa, N.J.: Humana Press, 1998), p. 168.

40. Eccles and other neurophysiologists argued, for example, that a careful analysis demonstrated that the short latency response of the heart and especially the skeletal muscles could not be induced chemically. Eccles also reported that even large doses of eserine had no appreciable effect on the fast initial response produced by a single preganglionic volley.

41. H. H. Dale, "The Beginnings and the Prospects of Neurohumoral Transmission," *Pharmacological Reviews* 6 (1954): 11.

42. One notable exception to the dissenters in 1933 was the neurophysiologist Adrian, who expressed the thought that it was not unreasonable to expect the same principle of chemical transmission to have evolved at all synapses, although not necessarily the same chemicals.

43. Dale, Feldberg, and Vogt, "Release of Acetylcholine."

44. G. L. Brown, H. H. Dale, and W. Feldberg, "Chemical Transmission of Excitation from Motor Nerve to Voluntary Muscle," *Journal of Physiology, London* 87 (1936): 42P–43P. See also Dale, Feldberg, and Vogt, "Release of Acetylcholine."

45. O. Loewi, "Salute to Henry Hallett Dale," *British Medical Journal*, 1955, 1356.

46. M. B. Walker, "Treatment of Myasthenia Gravis with Physostigmine," *Lancet*, 1934, 1200–1201. Better results were later found by treating *myasthenia gravis* with neostigmine and pyridostigmine rather than physostigmine.

47. The author was provided with copies of the nominations for the 1936 Nobel Prize for Physiology or Medicine from the Nobel Archives.

48. In 1938, the year Otto Loewi was arrested by the Nazis after the Germans had annexed Austria, Ernst von Brücke was dismissed from his professorship at Innsbruck. Although Brücke was Protestant, the Nazis charged that both his mother and wife, despite being Protestants, were of Jewish descent. Brücke was eventually brought to Harvard, but he died suddenly in June 1941.

49. All of the letters nominating Otto Loewi and Henry Dale for the Nobel Prize in 1936 were made available to the author through the kindness of the Nobel Archive Committee.

50. O. Loewi, "Problems Connected with the Principle of Humoral Transmission of Nervous Impulses," *Proceedings of the Royal Society of London,* series B, 118 (1935): 300.

CHAPTER 7

1. W. B. Cannon, *The Way of an Investigator* (New York: Hafner, 1965), pp. 59–60.

2. Z. M. Bacq, "Chemical Transmission of Nerve Impulses," in M. J. Parnham and J. Bruinvels, eds., *Psycho- and Neuro-Pharmacology*, vol. 1 (New York: Elsevier, 1983), p. 90.

3. One of those who has written that Cannon should have received the Nobel Prize is neuroscientist Ralph Gerard (R. W. Gerard, "Is the Age of Heroes Ended?," in C. M. Brooks, K. Koizumi, and J. O. Pinkston, eds., *The Life and Contributions of Walter Bradford Cannon, 1871–1945* [Brooklyn: State University of New York Press, 1975], pp. 197–208, esp. p. 199).

4. An excellent resource for information about Walter Cannon's life and work is the two-part biography: S. Benison, A. C. Barger, and E. L. Wolfe, *Walter Cannon:*

The Life and Times of a Young Scientist (Cambridge, Mass.: Harvard University Press, 1987) and E. L. Wolfe, A. C. Barger, and S. Benison, *Walter B. Cannon: Science and Society* (Cambridge, Mass.: Harvard University Press, 2000). Also useful are Cannon, *Way of an Investigator* and Brooks, Koizumi, and Pinkston, *Life and Contributions.*

5. Cannon was also described as being "better than a good sculptor" and once received a blue ribbon for a sculpture he had made of his daughter's head.

6. Walter Cannon's former students could later be found in laboratories in thirteen different countries.

7. Cannon described his ancestry and the personality of his parents in the first chapter of *Way of an Investigator.*

8. Cited in Benison, Barger, and Wolfe, *Life and Times.*

9. Cannon's diary entry, July 27, 1892; cited in Benison, Barger, and Wolfe, *Life and Times,* pp. 19–20.

10. Cited in Benison, Barger, and Wolfe, *Life and Times.*

11. Cited in Benison, Barger, and Wolfe, *Life and Times,* p. 25.

12. H. S. Jennings, *Behavior of Lower Organisms* (Bloomington: Indiana University Press, 1962).

13. Cited in Benison, Barger, and Wolfe, *Life and Times.*

14. Cited in Benison, Barger, and Wolfe, *Life and Times.*

15. For a description of Charles Eliot's presidency at Harvard, see J. M. Burns, *Transforming Leadership: A New Pursuit of Happiness* (New York: Atlantic Monthly Press, 2003), pp. 67–71.

16. Bowditch had also studied with Carl Ludwig in Leipzig.

17. A brief description of their observations appeared in the journal *Science* in 1897. Cannon and Moser were among the first to experiment with using a radio-opaque substance in food in order to follow its movement through the alimentary canal and gastrointestinal tract. Cannon also used barium sulphate. He regretted never reporting the use of the barium sulphate, for it later proved to be superior for use in clinical medicine.

18. W. B. Cannon, "The Movement of the Stomach Studied by Means of the Röntgen Rays," *American Journal of Physiology* 1 (1899): 359–382.

19. Cited in Benison, Barger, and Wolfe, *Life and Times,* p. 59.

20. W. B. Cannon, "The Case Method of Teaching Systematic Medicine," *Boston Medical and Surgical Journal* 142 (1900): 31–36.

21. One of Cannon's projects at the time involved using an experimental technique developed by Jacques Loeb to investigate whether some of the impairment from head trauma was the result of oxygen deprivation caused by increased pressure constricting blood vessels. The work on head trauma, which once again combined basic physiology with a clinical problem, attracted some attention when Cannon presented it at a meeting of the Massachusetts Medical Society.

22. Cornelia Cannon seems to have had much of her mother's energy, self-reliance, and independent spirit. Cornelia's mother, for example, gave lectures on anarchism to women's clubs. Cornelia Cannon was particularly active in the birth control movement. She wrote stories for children and several novels, one of which, *Red Rust,* was quite popular and a financial success.

23. An account of Cornelia and Walter Cannon's hazardous climb was published in the National Park magazine in 1955. Cannon described the adventure in the second chapter of his book *Way of an Investigator.*

24. H. H. Dale, "Walter Bradford Cannon, 1871–1945," *Obituary Notices of Fellows of the Royal Society, 1945–1948,* vol. 5, pp. 407–423 (quote from p. 413).

25. The committee assuaged Porter by recommending that he be appointed to a newly created chair of comparative physiology.

26. See James' letter to Sarah Cleghorn dated 1903 in F. J. D. Scott, ed., *William James: Selected Unpublished Correspondence, 1885–1910* (Columbus: Ohio State University Press, 1996), pp. 318–319.

27. Cannon's interest in the physiology of emotion had been stimulated much earlier by an undergraduate class taught by William James. James had proposed that an emotion is experienced when the brain becomes aware of the unique set of peripheral, visceral, and somatic sensations that identify each emotion. For James, emotions were the perceptions of the bodily changes provoked by different stimuli, and he noted that this was reflected in such expressions as "trembling with rage," "a sinking feeling in the stomach," and "hair standing on end." Because this theory had also been proposed by the Danish neurologist Carl Lange it is often called the "James-Lange theory of emotions."

Cannon had begun to think that William James was wrong. Several lines of evidence suggested that the brain played the primary role in the experience of emotion and that peripheral responses did not differ sufficiently to distinguish between the different emotional states. In 1913 Cannon gave an invited address at the American Psychological Association's convention in which he emphasized the primary role of the brain in the experience of emotions, but he avoided criticizing his former teacher's theory. Encouraged by friends, however, when he wrote up the talk for publication in the *American Journal of Psychology* he included a critique of James' theory: "We do not feel sorry because we cry, as James contended, but we cry because, when we are sorry or overjoyed or violently angry or full of tender affection—when any of these diverse emotional states is present—there are nervous discharges by sympathetic channels to various viscera, including the lachrymal glands" (W. B. Cannon, "The Interrelations of Emotions as Suggested by Recent Physiological Researches," *American Journal of Psychology* 25 [1914]: 252–282).

Cannon also referred to his study of emotions in sympathectomized cats. These are cats in which the chain of sympathetic ganglia have been removed. This operation blocks most of the visceral changes that occur during emotional states and also prevents the brain from receiving information about the state of the visceral organs. This was a difficult surgical procedure that Cannon had mastered. Cannon concluded that sympathectomized cats gave every indication that they were capable of experiencing normal emotions. Cannon also noted that James' theory could not explain how paraplegics paralyzed from the neck down, whose brain could not receive any sensations from the body, continued to report experiencing different emotions. One such case had recently been reported by C. L. Dana, a New York neurologist.

Much later, long after James' death, Cannon wrote a stronger critique of the James-Lange theory: W. B. Cannon, "The James-Lange Theory of Emotions: A Critical Examination and an Alternative Theory," *American Journal of Psychology* 39 (1927):

106–124. Cannon's belief that the brain played the primary role in emotions had evolved into a specific theory following the completion of the doctoral dissertation of his student Philip Bard. In 1924 Cannon and Sidney Britton had confirmed earlier reports that cats without their cerebral cortex, so-called decerebrate cats, showed intense rage, often triggered by innocuous stimuli. The following year Cannon suggested to Bard that he might pursue this observation by trying to determine what areas below the cerebral cortex were responsible for this rage. Bard transected the brains of decerebrate cats at different levels and concluded from the results that a region at the border of the thalamus and posterior hypothalamus was responsible for integrating the rage response. These observations led directly to what came to be called the Cannon-Bard "thalamic theory of emotions." The "thalamic theory" seemed to be supported by several reports from clinical neurologists that following damage to the thalamus, humans often display inappropriate emotions, sometimes crying or laughing for no apparent reason.

28. One of the publications resulting from Cannon's collaboration with surgeons who performed partial gastrectomies is W. B. Cannon and F. T. Murphy, "The Movement of Stomach and Intestines in Some Surgical Conditions," *Annals of Surgery* 43 (1906): 513–537.

29. Cited in C. K. Drinker, "Obituary: Walter Bradford Cannon, 1871–1945," *Science* 102 (1945): 470–472.

30. Reported in D. C. Schechter, "Background of Clinical Cardiac Electrostimulation," *New York State Journal of Medicine* 72 (1972): 609–610.

31. Roy Hoskins would later edit the journal *Endocrinology* for a number of years.

32. Cannon regarded most of his own work on the thyroid gland as fruitless. He later wrote that if all his effort "to obtain control of the workings of the thyroid gland could be added to the end of my years, my span of life would be prolonged, I feel sure, by some years."

33. After some unsuccessful attempts, Cannon improved his methodology by inserting a lubricated catheter in the dissected femoral vein of cats and easing it back into the heart.

34. W. B. Cannon and D. de la Paz, "The Emotional Stimulation of Adrenal Secretion," *American Journal of Physiology* 28 (1911): 64.

35. This technique was developed by Rudolph von Magnus in Utrecht.

36. Although germs of the idea had sprung up earlier, strong interest in psychosomatic medicine is thought to have started only with the publication of Helen Dunbar's book *Emotions and Bodily Changes: A Survey of the Literature on Psychosomatic Interrelationships, 1910–1933* (New York: Columbia University Press, 1935).

37. W. B. Cannon and D. de la Paz, "The Stimulation of Adrenal Secretion," *Journal of the American Medical Association* 56 (1911): 742.

38. Cannon had been interested in the thyroid gland because he thought its activity was affected by emotional states and he began to think about possible clinical implications. It had occurred to him that some abnormal thyroid conditions produce symptoms similar to those produced by excessive sympathetic nervous system activity. He had in mind hyperthyroidism, where the symptoms include high heart rate, increased metabolism, and weight loss. Cannon wondered if he might be able to demonstrate a neural influence on the thyroid gland if he stimulated the sympathetic system repeatedly.

To overstimulate the thyroid and adrenal glands, Cannon considered the possibility of connecting the phrenic nerve, which innervates the diaphragm, to nerves connected to these glands. The phrenic nerve is activated during each respiration cycle, and, he reasoned, if he could connect the phrenic nerve to the nerves innervating the thyroid and adrenal medulla, these glands would be stimulated with every breath. As Cannon had no experience suturing nerves, he asked his friend, neurosurgeon Harvey Cushing, for help. During the course of one afternoon, Cushing taught Cannon how to connected the phrenic nerve to the nerves innervating the thyroid gland and the adrenal medulla.

During the month that followed the suturing of the nerves, Cannon wasn't certain that the connections were functional. One afternoon, the animal caretaker reported that one of the cats was losing weight despite eating ravenously. When he examined this cat and others with similarly sutured nerves he found that they all had lost weight and their heart rates were accelerated. A few days later Cannon noticed that with every breath one of a cat's eyes became exophthalmic, another condition seen in hyperthyroidism. The basal metabolism of the cats was also raised, and on autopsy he found that the weight of the thyroid was three times normal. Cannon became convinced that he was able to produce the main symptoms of hyperthyroidism by excessive stimulation of the thyroid gland. He wanted to extend the use of the same technique to other endocrine conditions and decided to join the phrenic nerve to the branch of the splanchnic that innervates the pancreas to determine if he could produce diabetes by exhausting the capacity of the Langerhans cells to produce insulin.

At a presentation at the Johns Hopkins Medical Society, Cannon not only reported on his experimental technique for producing thyroid pathology, but he also added that the thyroid gland, like the adrenal gland, may also play a role in meeting emergencies: "It is not unreasonable to suppose that the thyroid gland likewise has an emergency function evoked in critical times, which would serve to increase the speed of metabolism when the rapidity of bodily processes might be of the utmost importance, and, besides that, augmenting the efficiency of the adrenin [adrenaline] which would be secreted simultaneously."

An abstract of Cannon's presentation, entitled "Some Recent Investigations on Ductless Glands," was published in 1916 (*Johns Hopkins Hospital Bulletin* 27 [1916]: 247–248).

39. Around this time, Robert Lovett, professor of orthopedic surgery at Harvard, talked with Cannon about how to measure muscle strength of polio victims, who often suffered from muscle weakness. Lovett was looking for a way to measure muscle strength in order to evaluate the changes that might occur during different exercise and massage regimens.

40. Cannon's book was well received, and the revised edition published in 1929 was translated into Russian and Chinese.

41. Cannon had become aware of Elliott's work on adrenaline and the sympathetic nervous system around 1913, and he cited it in an article published in 1914 (*American Journal of Physiology* 33 [1914]: 372, n. 1). According to Zénon Bacq, Cannon had forgotten about Elliott's 1904 paper, but one day in 1930 he arrived in the laboratory happily excited after rediscovering it with its statement about adrenaline being liberated every time a sympathetic impulse arrives at its terminal.

42. Cannon also argued that animals without the adrenal medulla might appear normal under protected laboratory conditions, but would not be able to cope with the stresses of a real-life situation. It is now known that even without the adrenal medulla animals seem to cope adequately under more natural conditions.

43. Cannon regarded the introduction of the use of the denervated heart as the most valuable result of the controversy with the Cleveland group. Cannon pointed out that the denervated heart would beat faster when adrenaline was increased by as little as one part in 1,400,000,000 parts of blood (Cannon, *Way of an Investigator*, p. 101).

44. A brief note by Cannon and Uridil entitled "Some Novel Effects Produced by Stimulating the Nerves of the Liver" was published in *Endocrinology* 5 (1921): 729–730. For a more complete account see W. B. Cannon and J. E. Uridil, "Studies on the Conditions of Activity in Endocrine Glands, VIII: Some Effects on the Denervated Heart of Stimulating the Nerves of the Liver," *American Journal of Physiology* 58 (1921): 353–364. Cannon became more convinced that the liver was the source of a sympathomimetic substance when he found that while fasted animals did not release any of this substance into the blood, in an animal that had eaten meat, a large increase in heart rate and blood pressure was seen following splanchnic nerve stimulation. See also W. B. Cannon and F. R. Griffith, "Studies on the Conditions of Activity in Endocrine Glands, X: The Cardio-Accelerator Substance Produced by Hepatic Stimulation," *American Journal of Physiology* 60 (1922): 544–559.

45. Even as late as 1929 Cannon continued to discuss these experiments as merely a response to Stewart and Rogoff's criticism, thereby overlooking the evidence pointing to the chemical mediation of sympathetic nerve impulses. See, for example, W. B. Cannon, "The Autonomic Nervous System: An Interpretation," *Lancet* 1, 1930, 1109–1115.

46. In some experiments, the adrenal glands were removed shortly before the critical test began.

47. W. B. Cannon and Z. Bacq, "Studies on the Conditions of Activity in Endocrine Glands, XXVI: A Hormone Produced by Sympathetic Action on Smooth Muscle," *American Journal of Physiology* 96 (1931): 410–411. See also W. B. Cannon and Z. Bacq, "The Mystery of Emotional Acceleration of the Denervated Heart After Exclusion of Known Humoral Accelerators," *American Journal of Physiology* 96 (1931): 403; H. F. Newton, R. L. Zwemer, and W. B. Cannon, "Studies on the Conditions of Activity in Endocrine Glands, XXV: The Mystery of Emotional Acceleration of the Denervated Heart After Exclusion of Known Humoral Accelerators," *American Journal of Physiology* 96 (1931): 377–391. An account of this work is also given in Z. M. Bacq, *Chemical Transmission of Nerve Impulses: A Historical Sketch* (New York: Pergamon, 1975), pp. 38–39.

48. Cannon and Bacq, "Studies, XXVI," p. 408.

49. The anatomical arrangement of the sympathetic nervous system, with its chain of ganglia that permits vertical communication between different nerves, facilitates the triggering of the system as a whole (see fig. 1 above).

50. The preganglionic parasympathetic neurons tend to be long, and they synapse on postganglionic neurons close to or within the organ that is innervated. This anatomical arrangement does not have the same potential for integrating a group of responses that exists in the sympathetic nervous system (see fig. 1).

51. Cannon and Bacq, "Studies, XXVI."

52. George Parker did much of this work in his seventies. At the age of seventy-six, Parker was awarded a prize by the American Philosophical Society for his studies of the control of adaptive color changes in fish.

53. In a speech given in 1931 Cannon contrasted Parker's ideas about the origin of the humoral substances with his own and what he mistakenly believed were those of Thomas Elliott. Cannon's speech was delivered in 1931 to the Association for the Study of Internal Secretions.

54. George Parker published the book *Humoral Agents in Nervous Activity, with Special Reference to Chromatophores* in 1932 (Cambridge University Press).

55. W. B. Cannon, "Chemical Mediators of Autonomic Nerve Impulses," *Science* 78 (1933): 43–48.

56. After Zénon Bacq worked with Cannon, he had an opportunity to work with G. L. Brown in Henry Dale's department at the National Institute for Medical Research in Hampstead.

57. H. W. Davenport, "Signs of Anxiety, Rage, or Distress," *Physiologist* 24 (1981): 3.

58. Davenport, "Signs," p. 3.

59. N. Wiener, *Cybernetics; or, Control and Communication in the Animal and the Machine* (Cambridge, Mass.: M.I.T. Press, 1948). In this book Norbert Wiener mentions regularly attending Rosenblueth's roundtable discussion group, and he acknowledged that the book "represent[ed] the outcome, after more than a decade, of a program of work undertaken jointly with Dr. Arturo Rosenblueth . . . a colleague and collaborator of the late Dr. Walter B. Cannon." In 1945 Norbert Wiener and Rosenblueth spent ten weeks together in Mexico City, after the latter had returned to Mexico.

60. A. Rosenblueth, "The Sensitivization by Cocaine of Gastric and Uterine Smooth Muscle to the Inhibitory Action of Adrenin," *American Journal of Physiology* 98 (1931): 186–193. Otto Loewi had earlier demonstrated in a different way that prior treatment with cocaine potentiated the response to adrenaline.

61. W. B. Cannon and A. Rosenblueth, "Studies on Conditions of Activity of Endocrine Glands, XXVIII: Some Effects of Sympathin on the Nictitating Membrane," *American Journal of Physiology* 99 (1931–1932): 398–407.

62. Cannon, "Chemical Mediators," pp. 45–46.

63. W. B. Cannon and A. Rosenblueth, "Studies on Conditions of Activity in Endocrine Glands, XXIX: Sympathin E and Sympathin I," *American Journal of Physiology* 104 (1933): 557–574. See also Cannon, "Chemical Mediators."

64. Ergotoxine is one of a number of different alkaloid substances derived from the ergot fungus. They all have marked effects on the uterus and cardiovascular system, in part because of their ability to block peripheral adrenergic activity. See chap. 4 above for an extended discussion.

65. It was well known that adrenaline makes the nictitating membrane contract and the uterus relax. When they tested sympathin obtained by stimulating the hepatic nerve, it had the same effect as adrenaline on the nictitating membrane, but not on the uterus. However, when they obtained the sympathin by stimulating the nerve that innervates the duodenal artery, both contraction of the nictitating membrane and relaxation of the uterus occurred (see Cannon, "Chemical Mediators," p. 46).

66. Cannon described these ideas about the two sympathins in a review article en-

titled "The Story of the Development of Our Ideas of Chemical Mediation of Nerve Impulses" (*American Journal of Medical Science* 188 [1934]: 145–159).

67. Z. Bacq, "Walter B. Cannon's Contribution to the Theory of Chemical Mediation of the Nerve Impulse," in Brooks, Koizumi, and Pinkston, eds., *Life and Contributions*, pp. 73–74.

68. See, for example, H. H. Dale, "The Beginnings and the Prospects of Neurohumoral Transmission," *Pharmacological Reviews* 6 (1954): 7–14, esp. p. 7.

69. W. B. Cannon and A. Rosenblueth, *Autonomic Neuro-Effector Systems* (Experimental Biology Monographs) (New York: Macmillan, 1937).

70. See, for example, Bacq, "Cannon's Contribution."

71. A copy of Göran Liljestrand's review in the Nobel Archives was provided to the author. It should also be noted that Liljestrand was the teacher of Ulf von Euler, who won the Nobel Prize in part for determining that norepinephrine (noradrenaline) is the main neurotransmitter at sympathetic nerve terminals.

72. See B. Cannon, "Walter B. Cannon: Personal Reminiscences," in Brooks, Koizumi, and Pinkston, eds., *Life and Contributions*, p. 167.

73. W. B. Cannon, "The Nobel Prize in Physiology and Medicine," *Scientific Monthly* 44 (1937): 195–198 (quote from p. 198).

74. Letter from Henry Dale to Walter Cannon, November 15, 1936.

75. Lissák and some collaborators also measured the duration of the diminution of acetylcholine in a nerve after it had been severed and demonstrated that the disappearance of this substance was correlated with the loss of effectiveness of the severed nerve to excite skeletal muscles. Cited in W. B. Cannon, "The Argument for Chemical Mediation of Nerve Impulses," in M. B. Visscher, ed., *Chemistry and Medicine: Papers Presented at the Fiftieth Anniversary of the Founding of the Medical School of the University of Minnesota* (Minneapolis: University of Minnesota Press, 1940), pp. 276–291.

76. W. B. Cannon and K. Lissák, "Evidence for Adrenaline in Adrenergic Neurons," *American Journal of Physiology* 125 (1939): 765–785.

77. Cannon and Lissák, "Evidence," p. 774.

78. R. P. Ahlquist, "A Study of the Adrenergic Receptors," *American Journal of Physiology* 153 (1948): 586–600.

79. Isoproterenol, for example, was effective only at beta sites. Noradrenaline (norepinephrine) primarily acted on alpha receptors, while adrenaline (epinephrine) acted mainly on beta receptors.

80. It was several years before the dual–adrenergic receptor theory was widely accepted. Ahlquist, a pharmacologist, had initially submitted his manuscript to the *Journal of Pharmacology and Experimental Therapeutics*, but the editor rejected the manuscript mainly because of what was judged to be its audacious and irreverent tone. See G. O. Carrier, "Evolution of the Dual Adrenergic Receptor Concept, Key to Past Mysteries and Modern Therapy," in M. J. Parnham and J. Bruinvels, eds., *Discoveries in Pharmacology*, vol. 3: *Pharmacological Methods, Receptors, and Chemotherapy* (New York: Elsevier, 1986), pp. 217–218.

81. U. S. von Euler, "A Specific Sympathomimetic Ergone in Adrenergic Nerve Fibers (Sympathin) and Its Relations to Adrenaline and Nor-adrenaline," *Acta physiologica scandinavica* 12 (1946): 73–97. For historical reflections see U. S. von Euler, "Dis-

coveries of Neurotransmitter Agents and Modulators of Neuronal Functions," in E. G. Worden, J. P. Swazey, and G. Adelman, eds., *The Neurosciences: Paths of Discovery I* (Boston: Birkhäuser, 1992), pp. 181–187. In von Euler's 1946 paper he noted that it had recently been shown that noradrenaline can be formed in the body by demethylation of adrenaline.

82. A. Rosenblueth, *The Transmission of Nerve Impulses at Neuroeffector Junctions and Peripheral Synapses* (New York: Wiley, 1950).

83. Dale, "Beginnings," p. 8.

84. H. H. Dale, "Chemical Transmission of the Effects of Nerve Impulses," *British Medical Journal*, May 12, 1934, 835–841 (quote from p. 837.)

85. H. H. Dale, *Adventures in Physiology* (London: Pergamon, 1953), p. 98.

86. Bacq, *Chemical Transmission.*

87. U. von Euler, *Noradrenaline: Chemistry, Physiology, Pharmacology, and Clinical Aspects* (American Lecture Series 261) (Springfield, Ill.: Charles Thomas, 1965), p. 4.

CHAPTER 8

1. J. C. Eccles, "Synaptic and Neuromuscular Transmission," *Ergebnisse der Physiologie, biologischen Chemie und experimentellen Pharmakologie* 38 (1936): 339–444 (quote from p. 397).

2. Z. M. Bacq, *Chemical Transmission of Nerve Impulses: A Historical Sketch* (New York: Pergamon, 1975), p. 51.

3. Z. M. Bacq, "Chemical Transmission of Nerve Impulses," in M. J. Parnham and J. Bruinvels, eds., *Psycho- and Neuro-Pharmacology* (New York: Elsevier, 1983), pp. 92–93.

4. In a later account, after he had already changed his mind, John Eccles noted the opposition to chemical transmission expressed by his fellow neurophysiologists at a 1939 symposium on the synapse. He wrote that at the meeting "Lorente de Nó, [Herbert] Gasser and [Joseph] Erlanger were strongly in favor of electrical transmission" (J. Eccles, *The Physiology of Synapses* [New York: Academic Press, 1964]).

5. J. Erlanger, "The Initiation of Impulses in Axons," *Journal of Neurophysiology* 2 (1939): 370–379 (quote from p. 371).

6. D. Bronk, "Synaptic Mechanism in Sympathetic Ganglia," *Journal of Neurophysiology* 2 (1939): 380–401 (quote from p. 382).

7. J. S. Cook, " 'Spark' vs. 'Soup': A Scoop for Soup," *News in Physiological Sciences* 1 (1986): 206–208.

8. Both letters are cited in *In Memory of Sir Henry Dale,* Sir Paul Girolami, Lady Helena Taborikova, and Guiseppi Nitisco, eds. (Accademeia Roma di Scienze Mediche e Biologiche).

9. Cited in *In Memory of Sir Henry Dale.*

10. Cited in *In Memory of Sir Henry Dale*

11. See Eccles' letter dated July 28, 1943, and Dale's reply dated August 22 in *In Memory of Sir Henry Dale.*

12. In 1966 John Eccles came to the United States, taking a position at the American Medical Association's Institute for Medical Research in Chicago. In 1968 he moved

to the State University of New York at Buffalo, where he remained until retiring in 1975 at the age of seventy-two. Eccles then moved to Switzerland, where he died at the age of ninety-four. During the latter years, including the postretirement years, Eccles wrote books on the relation of the brain and mind, the emergence of consciousness, and the concept of the self as well as a work on the evolution of the brain and a biography of Charles Sherrington.

13. W. Feldberg, "The Early History of Synaptic and Neuromuscular Transmission by Acetylcholine: Reminiscences of an Eye Witness," in A. L. Hodgkin et al., eds., *The Pursuit of Nature: Informal Essays on the History of Physiology* (Cambridge: Cambridge University Press, 1977), p. 72.

14. J. C. Eccles, "Under the Spell of the Synapse," in F. G. Worden, J. P. Swazey, and G. Adelman, eds., *The Neurosciences: Paths of Discovery I* (Boston: Birkhäuser, 1992), p. 161.

15. B. Katz, "Bernard Katz.," in L. Squires, ed., *The History of Neuroscience in Autobiography*, vol. 1 (Washington, D.C.: Society for Neuroscience, 1966), p. 373.

16. A. G. Karczmar, "Sir John Eccles, 1903–1907, Part 1: On the Demonstration of the Chemical Nature of Transmission in the CNS," *Perspectives in Biology and Medicine* 44 (2001): 81.

17. Feldberg, "Early History," p. 67.

18. J. C. Eccles, "Synaptic and Neuro-Muscular Transmission," *Physiological Reviews* 17 (1937): 538–555.

19. Eccles, "Synaptic and Neuro-Muscular Transmission," p. 551.

20. W. B. Cannon, "The Argument for Chemical Mediation of Nerve Impulses," *Science* 90 (1939): 521–527.

21. Quoted from a similar article published the following year: W. B. Cannon, "The Argument for Chemical Mediation of Nerve Impulses," in M. B. Visscher, ed., *Papers Presented at the Fiftieth Anniversary of the Founding of the Medical School of the University of Minnesota* (Minneapolis: University of Minnesota Press, 1940), p. 281.

22. Eccles, "Synaptic and Neuro-Muscular Transmission"; J. F. Fulton, *Physiology of the Nervous System* (New York: Oxford University Press, 1938).

23. Cannon, "Argument for Chemical Mediation," p. 527.

24. Eccles, "Under the Spell," p. 161.

25. Birdsey Renshaw had discovered the role of small "internuncial cells" in creating recurrent inhibition of spinal motor neurons: B. Renshaw, "Influence of Discharge on Motor Neurons Upon Excitation of Neighboring Motor Neurons," *Journal of Neurophysiology* 4 (1941): 167–183

26. J. C. Eccles, "From Electrical to Chemical Transmission in the Central Nervous System," *Notes and Records of the Royal Society, London* 30 (1976): 219–230 (quote from p. 225).

27. L. G. Brook, J. S. Coombs, and J. C. Eccles, "The Recording of Potentials from Motorneurons with an Intracellular Electrode," *Journal of Physiology* 117 (1952): 431–460 (quote from p. 455).

28. In addition to sodium, Hodgkin, Huxley, and Bernard Katz later explained that the resting potential of a neuron also depended on the movement of potassium

and chloride ions; see A. L. Hodgkin and A. F. Huxley, "A Quantitative Description of Membrane Current and Its Application to Conduction and Excitation in Nerves," *Journal of Physiology* 117 (1952): 500.

29. Eccles and his colleagues were correct in principle, if not in all the details. It is now known that a collateral branch from the spinal motor nerve axon secretes acetylcholine to stimulate the small Renshaw cells, which in turn secrete an inhibitory neurotransmitter.

30. From Eccles' letter to Dale dated September 17, 1953, cited in *In Memory of Sir Henry Dale*.

31. Eccles, "From Electrical to Chemical Transmission," p. 226.

32. H. H. Dale, "The Beginnings and the Prospects of Neurohumoral Transmission," *Pharmacological Reviews* 6 (1954): 11.

33. Dale, "Beginnings," p. 10. Dale was repeating an earlier comment attributed to von Bruecke.

34. H. H. Dale to J. C. Eccles, August 22, 1943, cited in Eccles, "From Electrical to Chemical Transmission," p. 224.

35. J. F. Fulton, *Physiology of the Nervous System,* 3d ed. (New York: Oxford University Press, 1949), p. 73.

36. Eccles, "From Electrical to Chemical Transmission," p. 223; A. Fessard, "Transmissions, synaptiques ganglionnaire et centrale: Discussion," *Acta internat. physiol.* 59 (1951): 605–618.

37. C. T. Morgan and E. Stellar, *Physiological Psychology* (New York: McGraw-Hill, 1950).

38. R. W. Gerard, "Discussion (Symposium on Neurohumoral Transmission, Physiological Society of Philadelphia, Sept. 11–12, 1953," *Pharmacological Reviews* 6 (1954): 123–131 (quote from p. 123; see also p. 126).

39. A. K. McIntyre, "Central and Sensory Transmission (Symposium on Neurohumoral Transmission, Physiological Society of Philadelphia, Sept. 11–12, 1953)," *Pharmacological Reviews* 6 (1954): 103–104 (quote from p. 103).

40. W. S. Feldberg, "Transmission in the Central Nervous System and Sensory Transmission (Symposium on Neurohumoral Transmission, Physiological Society of Philadelphia, Sept. 11–12, 1953)," *Pharmacological Reviews* 6 (1954): 85–93 (quotes from p. 85).

CHAPTER 9

1. "Austrian Receives Award. Clerical Press Plays Down Nobel Award to Jewish Scientist," *New York Times*, November 2, 1936, p. 12.

2. From "An Autobiographical Sketch," reprinted in F. Lembeck and W. Giere, *Otto Loewi: Ein Lebensbild in Dokumenten* (Berlin: Springer, 1968), pp. 185–186.

3. The concerted effort made by Henry Dale and Walter Cannon to help Otto Loewi is well documented in E. L. Wolfe, A. C. Barger, and S. Benison, *Walter B. Cannon: Science and Society* (Cambridge, Mass.: Harvard University Press, 2000), pp. 386–392.

4. As cited in Wolfe, Barger, and Benison, *Science and Society,* p. 387.

5. Loewi's Nobel Prize money had to be transferred to a German bank. According to the Nobel records, in 1939 Loewi was awarded a grant of 5,000 Swedish crowns for his research in consideration of the money he had had to surrender to the Germans.

6. Z. M. Bacq, *Chemical Transmission of Nerve Impulses: A Historical Sketch* (New York: Pergamon, 1975), p. 99.

7. See discussion in Wolfe, Barger, and Benison, *Science and Society,* pp 388–392.

8. O. Loewi, "An Autobiographical Sketch," *Perspectives in Biology and Medicine* 4 (1960): 23.

9. O. Loewi, *From the Workshop of Discoveries* (Porter Lecture Series 19) (Lawrence: University of Kansas Press, 1953).

10. Reported in D. Lehr, "The Life and Work of Otto Loewi," *Medical Circle* 9 (1962): 126–131.

11. Bacq, *Chemical Transmission,* p. 100.

12. William van der Kloot (1990), as cited in a personal communication by H. Davenport in "Early History of the Concept of Chemical Transmission of the Nerve Impulse," *Physiologist* 34 (1991): 185–186. William van der Kloot was Loewi's colleague at New York University.

13. J. H. Gaddum, "Prof. Otto Loewi," *Nature* 193 (1962): 525–526.

14. Quoted in the *New York Times* (Dec. 27, 1961, p. 21, col. 1) obituary for Otto Loewi.

15. O. Loewi, "A Scientist's Tribute to Art," in *Essays in Honour of Hans Tietze, 1880–1954* (New York: Gazette de Beaux Arts, 1958), pp. 389–392.

16. Cited in W. Feldberg, "Henry Hallett Dale, 1875–1968," *Royal Society Obituaries and Memoirs, 1830–1998* 16 (1970): 77–174 (quote from p. 159).

17. Letter from Dale to Eccles, August 25–26, 1954, cited in J. C. Eccles, "From Electrical to Chemical Transmission in the Central Nervous System," *Notes and Records of the Royal Society, London* 30 (1976): 219–230 (letter cited p. 228).

18. A complete list of Dale's publications is provided in Feldberg, "Henry Hallett Dale, 1875–1968."

19. Cited in Feldberg, "Henry Hallett Dale, 1875–1968," pp. 152–153.

20. Cited in "Obituary: Henry Hallett Dale," *Lancet,* August 3, 1968, 288–290.

21. H. Dale, "The Atomic Problem Now," *Spectator,* September 30, 1949, 409.

22. H. Dale, "What Nagasaki Meant," *Spectator,* July 13, 1951, 54–55.

23. "Obituary: Henry Hallett Dale," *Lancet,* August 3, 1968, 288–290.

24. Quoted from a letter dated January 22, 1940, cited in Wolfe, Barger, and Benison, *Science and Society,* p. 447.

25. Cannon had also intervened in mobilizing support for Pavlov, when, following the Russian Revolution in 1917, there were somewhat exaggerated (as it turned out) reports that Pavlov was starving in St. Petersburg. They first met when Pavlov visited the United States in 1922, and they met again at the International Congress of Physiology held in Boston in 1929. The two had a warm reunion, and Cannon took Pavlov to his summer house in Franklin, New Hampshire, where he met with much of the Cannon family. The third meeting between Cannon and Pavlov was at the congress in Leningrad in August 1935.

26. When Cannon arrived in Shanghai with his family, they were greeted by Dr. J. H. Liu, who had been Cannon's student at Harvard in 1915. Cannon, who always prepared diligently for talks, gave lectures at the National Medical College in Shanghai and at the Army Medical College in Nanking. In addition, he gave several public lectures on the physiology of emotions and the importance of a prepared mind in scientific research. Cannon received enthusiastic applause after all his lectures. His wife was also invited to lecture about her work with the Birth Control League of Massachusetts, and wherever they traveled in China she made contact with people involved in the birth control movement. In Peking (now Beijing) they were greeted at the railroad station by the entire physiology department of the Peking Union Medical College. Cannon gave three lectures a week at P.U.M.C., talking about the chemical mediation of nerve impulses, the physiology of emotions, sham rage, and homeostasis. The lectures were so well attended that there was only standing room and many could not even get into the room. He also worked in the laboratory, starting experiments on denervation sensitivity with some of the staff and students and demonstrating his technique for performing a total sympathectomy.

Cannon was kept so busy in China that Cornelia wrote in a letter home: "He might as well be in Cambridge for, except for the rickshaw ride to the P.U.M.C. every day and weekends in the country, he sees only rheostats and smoked drums and the insides of cats." His hosts gave him a Chinese name that captured what they regarded as his main traits—friendliness and diligence. Later, after returning to Harvard, Cannon accepted a Chinese student from the P.U.M.C. to work in his laboratory.

The Cannons also visited the Japanese University of Mukden, where Cannon enjoyed his time with Professor Kuno, a gentle Japanese scholar whose friendliness to the Chinese was a sharp contrast to the swaggering Japanese military Cannon had seen on brief visits to Manchuria and also Korea.

27. V. M. Molotov's official position at the time was chairman of the council of the people's commissars.

28. Some details of Cannon's medical condition and the eventual cause of his death are described in R. Spillman, "Editorial: Walter Bradford Cannon, 1871–1945," *American Journal of Roentgenology, Radium Therapy, and Nuclear Medicine* 55 (1946): 94–96. According to Spillman's account, Cannon's severe itching was caused by leukemia cutis (*mycosis fungoides*), and his eventual death was due to malignant lymphoma.

29. W. B. Cannon and A. Rosenblueth, *Autonomic Neuro-Effector Systems* (Experimental Biology Monographs) (New York: Macmillan, 1937).

30. W. B. Cannon, *The Mechanical Factors of Digestion* (London: Arnold, 1911); W. B. Cannon, *Bodily Changes in Pain, Hunger, Fear, and Rage: An Account of Recent Researches Into the Function of Emotional Excitement* (New York: Appleton, 1915); W. B. Cannon, *A Laboratory Course in Physiology* (Cambridge, Mass.: Harvard University Press, 1923): W. B. Cannon, *Traumatic Shock* (New York: Appleton, 1923): W. B. Cannon, *The Wisdom of the Body* (New York: Norton, 1932); W. B. Cannon, *Digestion and Health* (New York: Norton, 1936).

31. Cannon first used the term "homeostasis" in a 1925 speech honoring Charles Richet. A paper based on the speech, entitled "Physiological Regulation of Normal

States: Some Tentative Postulates Concerning Biological Homeostatics," was published the following year (*Transactions of the Congress of American Physicians and Surgeons* 12 [1926]: 91). He next delivered a lecture to psychiatrists and neurologists in New York entitled "The Sympathetic Division of the Autonomic Nervous System in Relation to Homeostasis." Shortly afterward, Cannon gave a lecture in Chicago entitled "Functions of the Sympathetic Nervous System in Maintaining the Stability of the Organism." This idea was more fully described and illustrated in a review article entitled "Organization for Physiological Homeostasis" (*Physiological Reviews* 9 [1929]: 399–431) and in two articles published in 1930, the Linacre Lecture in Cambridge and one at the Sorbonne.

In his book *Cybernetics; or, Control and Communication in the Animal and the Machine* (New York: M.I.T. Press 1961), p. 115 and elsewhere, Norbert Wiener acknowledges his indebtedness to Cannon and the concept of homeostasis.

32. Cannon was writing during the 1930s, when there was a worldwide depression and war was imminent in several parts of the world. He wrote that it was time to consider whether the wars and famines, the inequalities in distribution, and the economic swings that lead to periods of "boom and bust" and depressions might be avoided or ameliorated if some protective, stabilizing mechanisms were in place.

33. W. B. Cannon, *The Body as a Guide to Politics* (Thinker's Forum 15) (London: Watts, 1942).

34. Cannon argued that stabilizing mechanisms in the social realm would enhance, not curtail, freedom, just as the automatic regulation of the internal environment of the body frees an organism to engage in other activities. People in dire economic straits, Cannon wrote, are not truly free men, and he quoted the lord chancellor of England who had declared that "Necessitous men are not, truly speaking, free men." Cannon elaborated on the many different types of social upheavals that can impose severe hardships on individuals. He noted that: "A new machine may be invented which, because it can do the work of thousands of laborers, throws thousands of laborers out of their jobs and in such disasters individual members of the social organization are not responsible for the ills which circumstances forces them to endure."

Only thoughtful advance planning, he argued, could ameliorate these social and economic problems: "Various schemes for avoidance of economic calamities have been put forth not only by dreamers of Utopias but also by sociologists, economists, statesmen, labor leaders, and experienced managers of affairs. . . . Communists have offered their solution of the problem and are trying out their ideas on a large scale in the Soviet Union. The socialists have other plans for the mitigation of the economic ills of mankind. And in the United States where neither communism or socialism has been influential, various suggestions have been offered for stabilizing the conditions of industry and commerce. . . . The multiplicity of these schemes is itself proof that no satisfactory social scheme has been suggested by anybody. The projection of the schemes, however, is clear evidence that in the minds of thoughtful and responsible men a belief exists that intelligence applied to social instability can lessen the hardships which result from technological advances, unlimited competition and the relatively free play of selfish interests."

Today, when Republicans seem to favor an unfettered, free market economy, it will surprise many that Cannon, who favored economic restraints and planning and advocated "New Deal" policies before there was a "New Deal," always voted for Republican presidents. Cannon never voted for Franklin Roosevelt despite the fact that his son-in-law, Arthur M. Schlesinger Jr., an advisor to Roosevelt, tried unsuccessfully to persuade him that he and the president held similar views.

35. W. B. Cannon, *The Way of an Investigator: A Scientist's Experiences in Medical Research* (New York: Norton,1945).

36. A. J. Carlson, review of *The Way of an Investigator*, in *Scientific Monthly*, October 1945, p. 117.

37. The quote is from the report Cannon wrote as chairman of the subcommittee responsible for summarizing the mission of the NRC Committee for Research on the Study of Sex. The full report is available in the Archives of the National Academy of Sciences, U.S.A.

38. B. Cannon, "Walter B. Cannon: Personal Reminiscences," in C. M. Brooks, K. Koizumi, and J. O. Pinkston, eds., *The Life and Contributions of Walter Bradford Cannon, 1871–1945* (New York: State University of New York, Downstate Medical Center, 1975), p. 162.

39. When Kinsey's books were attacked by some congressmen, the Rockefeller Foundation withdrew its support in 1954.

40. W. B. Cannon, " 'Voodoo' Death," *American Anthropologist* 44 (1942): 169. Reprinted in *Psychosomatic Medicine* 19 (1957): 182–190.

41. Today the emphasis would be on the role of stress in triggering responses from a circuit including the hypothalamus and pituitary and adrenal cortical hormones. See E. M. Sternberg, "Walter B. Cannon and 'Voodoo Death': A Perspective from Sixty Years On," *American Journal of Public Health* 92 (2002): 1564–1565.

42. C. P. Richter, "The Phenomenon of Sudden Death in Animals and Man," *Psychosomatic Medicine* 19 (1957): 191–198. In one series of experiments, Richter demonstrated that severe stress shortened the survival time of animals placed in extreme conditions.

43. W. B. Cannon, "Law of Denervation (Hughlings Jackson Memorial Lecture)," *American Journal of Medical Science* 198 (1939): 737–750.

44. Philip Bard's letter is dated December 15, 1939. It is quoted as cited in Wolfe, Barger, and Benison, *Science and Society*, p. 475.

45. The book is W. B. Cannon and A. Rosenblueth, *The Supersensitivity of Denervated Structures: A Law of Denervation* (New York: Macmillan, 1949). Of its twenty-two chapters, eleven were completely finished by Cannon, and these appeared in the final version essentially as he had written them. Rosenblueth finished five chapters that Cannon had not completed and wrote six additional chapters.

46. H. H. Dale, "Prof. W. B. Cannon," *Nature* 158 (July 20, 1946): 87–88.

47. Bacq, *Chemical Transmission*, pp. 100–101.

48. R. Levi-Montalcini, *In Praise of Imperfection: My Life and Work* (New York: Basic Books, 1988).

49. O. Loewi, "Salute to Henry Hallett Dale," *British Medical Journal*, 1955, p. 1357.

CHAPTER 10

1. E. Roberts, "GABA and Inhibition: Command Control in Nervous System Function," in F. Samson and G. Adelman, eds., *The Neurosciences: Paths of Discovery II* (Boston: Birkhäuser, 1992), p. 92.

2. Chang and Gaddum had reported finding acetylcholine in the brain, as well as other organs, in six different species. See H. C. Chang and J. Gaddum, "Choline Esters in Tissue Extracts," *Journal of Physiology, London* 79 (1933): 255–285.

3. Z. M. Bacq, "Chemical Transmission of Nerve Impulses," in M. J. Parnham and J. Bruinvels, eds., *Psycho- and Neuro-Pharmacology* (New York: Elsevier, 1983), p. 88.

4. W. Feldberg and M. Vogt, "Acetylcholine Synthesis in Different Regions of the Central Nervous System," *Journal of Physiology, London* 107 (1948): 372–381.

5. J. D. Robinson, *Mechanisms of Synaptic Transmission: Bridging the Gaps (1890–1990)* (New York: Oxford University Press, 2001), pp. 75–76.

6. M. Vogt, "The Concentration of Sympathin in Different Parts of the Central Nervous System Under Normal Conditions and After the Administration of Drugs," *Journal of Physiology, London* 123 (1954): 451–481.

7. Vogt did note that the distribution of sympathin might suggest that it has a neurotransmitter role, but she then listed a number of reasons to be cautious about drawing that conclusion. For example, it was known that adrenaline and noradrenaline were present in high concentrations in gliomas, where they could not serve as neurotransmitters. She also noted that it was possible that adrenaline and noradrenaline were in the brain to modify transmission of other substances, noting in this context that adrenaline had been reported to modify the response to acetylcholine.

8. W. Feldberg, *A Pharmacological Approach to the Brain: From Its Inner and Outer Surfaces* (Baltimore: Williams and Wilkin, 1963). See table on p. 42. This book was based on a 1961 lecture presented at Washington University in St. Louis, Missouri.

9. Feldberg explained his results by assuming that the anesthetic had paralyzed the hypothalamic region that B. K. Anand and John Brobeck had called the "satiety center" because its destruction produced animals that exhibited almost insatiable eating. B. K. Anand and J. R. Brobeck, "Hypothalamic Control of Food Intake in Rats and Cats," *Yale Journal of Biology and Medicine* 24 (1951): 123–140. Feldberg, *Pharmacological Approach,* p. 55.

10. Feldberg, *Pharmacological Approach,* pp. 106–107.

11. J. H. Gaddum, "Chemical Transmission in the Central Nervous System," *Nature* 197 (1963): 742.

12. The affection and respect that the young scientists working with Feldberg had for him is reported in Bacq, "Chemical Transmission," pp. 100–101. A similar account was given by Robert Myers (personal communication, August 30, 2001), a young American scientist who had the opportunity to work with Wilhelm Feldberg.

13. Personal communication from David Longley, November 19, 2003.

14. Animal rights activists had gained access to his laboratory under the pretext of making a film of his research. Parts of the film were submitted to the Home Office with the accusation that his experimental animals were inadequately anesthetized. Feldberg's Home Office license was revoked. The report by the Medical Research

Council's enquiry published in 1991 was less critical of Feldberg, although it recognized that some breach of regulations had occurred. This report recommended a "tightening of controls," but by this time Feldberg had made the decision to retire from laboratory work. A summary of Wilhelm Feldberg's life and scientific work can be found in *Biographical Memoirs of the Royal Society* 43 (1997): 145–170.

15. The story of the accidental discoveries of the initial psychotropic drugs is described in E. S. Valenstein, *Blaming the Brain: The Truth About Drugs and Mental Health* (New York: Free Press, 1998).

16. John Gaddum's experience with LSD is recounted in D. Healy, *The Creation of Psychopharmacology* (Cambridge, Mass.: Harvard University Press, 2002), p. 204.

17. J. H. Gaddum, "Drugs Antagonistic to 5-Hydroxytriptamine," in G. E. W. Wolstenholme and M. P. Cameron, eds., *Ciba Foundation Symposium on Hypertension: Humoral and Neurogenic Factors* (Boston: Little, Brown, 1954), p. 77. The first report of finding serotonin in the brain appeared in B. M. Twarog and J. H. Page, "Serotonin Content of Some Mammalian Tissues and Urine, and a Method for Its Determination," *American Journal of Physiology* 175 (1954): 157–161. It is telling that this paper was originally submitted to another journal in 1952 but was rejected by the editor as unimportant. This incident is reported in B. M. Twarog, "Serotonin: History of a Discovery," *Comparative Biochemistry and Physiology* 91 (1988): 21–24. For a brief discussion of the discovery of LSD and its connection to serotonin see Valenstein, *Blaming the Brain,* pp. 12–15.

18. D. W. Woolley and E. Shaw, "A Biochemical and Pharmacological Suggestion About Certain Mental Disorders," *Science* 119 (1954): 587–588. Woolley and Shaw wrote: "pharmacological findings indicate that serotonin has an important role to play in mental process and that the suppression of its action results in a mental disorder. In other words, it is the lack of serotonin which is the cause of the disorder. If now a deficiency of serotonin in the central nervous system were to result from metabolic rather than from pharmacologically induced disturbances, these same mental aberrations would be expected to become manifest. Perhaps such a deficiency is responsible for the natural occurrence of the diseases."

19. B. B. Brodie, A. Pletscher, and P. S. Shore, "Evidence That Serotonin Has a Role in Brain Function," *Science* 122 (1955): 968.

20. S. S. Kety, "Biochemical Theories of Schizophrenia," *Science* 129 (1959): 1528–1532; 1590–1596.

21. Kety, "Biochemical Theories."

22. J. Schildkraut and S. S. Kety, "Biogenic Amines and Emotion: Pharmacological Studies Suggest a Relationship Between Brain Biogenic Amines and Affective State," *Science* 156 (1967): 21.

23. A. Carlsson, "A Paradigm Shift in Brain Research," *Science* 294 (2001): 1022.

24. Published in J. R. Vane, G. E. W. Wolstenholme, and E. O'Connor, eds., *Ciba Foundation Symposium on Adrenergic Mechanisms* (Boston: Little, Brown, 1960).

25. Carlsson, "Paradigm Shift," p. 1024.

26. A. Carlsson, "A Half-Century of Neurotransmitter Research: Impact on Neurology and Psychiatry," *Les Prix Nobel 2000,* pp. 242–243. See also A. Carlsson, "Arvid Carlsson," in L. Squire, ed., *The History of Neuroscience in Autobiography,* vol. 2 (San Diego: Academic Press, 1998), pp. 30–66.

27. H. McLennan, *Synaptic Transmission* (Philadelphia: Saunders, 1963).

28. N. E. Miller, "Chemical Coding of Behavior in the Brain," *Science* 148 (1965): 330.

29. B. Colier and J. E. Mitchell, "The Central Release of Acetylcholine During Stimulation of the Visual Pathway," *Journal of Physiology, London* 184 (1966): 239–254. Also see B. Colier and J. E. Mitchell, "The Central Release of Acetylcholine During Consciousness and After Brain Lesions," *Journal of Physiology, London* 188 (1967): 83–98.

30. The development of the fluorescent stains and their use have been described in A. Carlsson, "Perspectives on the Discovery of Central Monoaminergic Neurotransmission," *Annual Review of Neuroscience* 10 (1987): 19–40.

31. Electrophoresis entails the use of electricity to move molecules in a controlled direction.

32. D. R. Curtis and R. Eccles, "The Excitation of Renshaw Cells by Pharmacological Agents Applied Electrophoretically," *Journal of Physiology, London* 141 (1958): 435–445.

33. J. Gaddum, "Push-Pull Cannulae," *Journal of Physiology, London* 155 (1961): 1–2.

34. In some of the initial experiments with push-pull cannulae, the acetylcholine secreted by central nervous system neurons was detected by the leech muscle technique.

35. J. H. Gaddum, "Push-Pull Cannulae," *Journal of Physiology, London* 153 (1960): 1P–2P.

36. An angstrom (Å) is equal to one ten-millionth of a millimeter. There are approximately 25 millimeters in an inch. George Palade and Sanford Palay first presented these results at the 1954 meeting of the American Association of Anatomists in Galveston, Texas. G. Palade and S. Palay, "Electron Microscope Observations of Interneuronal and Neuromuscular Synapses," *Anatomical Record* 118 (1954): 335–336.

37. S. Palay, "Synapses in the Central Nervous System," *Journal of Biophysical and Biochemical Cytology* 2.4, suppl. (1956): 193–201.

38. V. P. Whittaker, "The Isolation and Characterization of Acetylcholine-Containing Particles from the Brain," *Biochemical Journal* 72 (1959): 694–702.

39. T. Hökfelt, "Electron Microscope Studies on Peripheral and Central Monoamine Neurons" (thesis, Karolinska Inst., Stockholm, 1968). For more details of the early investigations with the electron microscope see Robinson, *Mechanisms,* pp. 106–112. What was in the synaptic vesicles was later determined by a chemical analysis of the synaptosomes, the terminals that contain the vesicles. The synaptosomes were separated from the rest of the cell by using a centrifuge. Through several steps, it was possible to concentrate the synaptosomes based on differences in weight from other parts of the neuron. A chemical analysis then could determine if the candidate substance was concentrated in the region of the neuron containing the synaptic vesicles.

40. J. C. Eccles, *The Inhibitory Pathway of the Central Nervous System* (Sherrington Lectures 9) (Springfield, Ill.: Charles Thomas, 1969), p. 112.

41. M. H. Aprison and R. Werman, "The Distribution of Glycine in Cat Spinal Cord and Roots," *Life Sciences,* 1965, 2075–2083.

42. Glutamate was first discovered in crayfish in the 1950s, and in the 1960s electrophoretic studies demonstrated that it was effective on mammalian brain neurons.

43. Susan E. Leeman is credited with isolating and characterizing the peptide Substance P: M. M. Chang and S. E. Leeman, "Isolation of a Sialogogic Peptide from Bovine Hypothalamus Tissue and Its Characterization as Substance P," *Journal of Biological Chemistry* 245 (1970): 4784–4790.

44. For a good review of the study of histamine in the brain, see R. S. Feldman, J. S. Meyer, and L. F. Quenzer, *Principles of Neuropsychopharmacology* (Sunderland, Mass: Sinauer, 1997), pp. 445–454.

45. For more of the history of the discovery of the reuptake mechanism, see S. Snyder, "Forty Years of Neurotransmitters," *Archives of General Psychiatry* 59 (2002): 983–994.

46. Bernard Katz was born in Leipzig, Germany, in 1911. He was the son of a Russian-Jewish fur trader. In 1935, after experiencing a number of anti-Semitic incidents, he left Germany. He had read a paper by the British physiologist Archibald V. Hill, whose own interest in how muscles are innervated was stimulated by John Langley. Hill, a 1922 Nobel Laureate, was a strong critic of Hitler. When Katz showed up on his doorstep as a refugee from Hitler, Hill decided to take a chance on him. Katz worked with Hill in London and later with John Eccles in Sydney. He returned to England (University College, London) with a Royal Society fellowship arranged by Hill. Bernard Katz was knighted in 1966. He died in 2003 at the age of ninety-two. See obituary, *New York Times*, April 25, 2003, p. A31.

47. Henry Dale and George Barger may actually have been the first to use the word "receptor," but, ironically, they criticized the concept as unnecessary, offering other explanations of how a drug might have different effects. See G. Barger and H. H. Dale, "Chemical Structure and Sympathomimetic Action of Amines," *Journal of Physiology, London* 41 (1910): 21.

48. In 1894 Emil Fischer, who was awarded a Nobel Prize in 1902 for his work on the cleavage of sugars by enzymes, introduced the analogy of a "lock and key" as a way of thinking about the specificity of chemical reactions. E. Fischer, "Einfluss der Configuration auf die Wirkung der Enzyme," *Berichte der deutschen chemischen Gesellschaft* 27 (1894): 2985–2993.

49. Dale's reservations about the concept of receptors are noted in the reminiscences of W. Patton, in "On Becoming and Being a Pharmacologist," *Annual Review of Pharmacology* 26 (1986): 10.

50. R. Miledi, P. Molinoff, and L. T. Potter, "Isolation of the Cholinergic Receptor Protein of the *Torpedo* Electric Organ," *Nature* 229 (1971): 554–557.

51. Snyder, "Forty Years"; S. H. Snyder and G. V. Pasternak, "Historical Review: Opioid Receptors," *Trends in Pharmacological Sciences* 24 (2003): 198–205.

52. Receptor binding studies can be done either by combining the radioactive ligand with a homogenate of brain tissue or by placing it on a thin slice of brain mounted on a glass slide. The latter technique has the advantage of preserving some of the anatomical relationships and of using different radioactive ligands on adjacent brain sections.

53. A good description of radioligand binding is presented in Feldman, Meyer, and Quenzer, *Principles of Neuropharmacology*, pp. 43–51.

54. C. Pert and S. Snyder, "Opiate Receptor Demonstration in Nervous Tissue," *Science* 179 (1973): 1011–1014.

55. L. Terenius, "Stereospecific Interaction Between Narcotic Analgesics and a Synaptic Plasma Membrane Fraction of Rat Cerebral Cortex," *Acta pharmacologica et toxicologica* 32 (1973): 317–320; E. J. Simon, J. M. Hiller, and I. Edelman, "Stereospecific Binding of the Potent Narcotic Analgesic (^{3}H) Etorphine to Rat Brain Homogenate," *Proceedings of the National Academy of Sciences of the United States of America* 70 (1973): 1947–1949.

56. This statement was made in the early 1970s by Avram Goldstein at Stanford University.

57. Before they adopted the mouse *vas deferens* test for detecting the presence of morphine-like substances, Hughes and Kosterlitz had been using the guinea pig *ileum* as a bioassay.

58. An account of John Hughes' trips to the slaughterhouse is given in J. Goldberg, *Anatomy of a Scientific Discovery* (New York: Bantam Books, 1988), chap. 1.

59. J. Hughes, "Isolation of an Endogenous Compound from the Brain with Pharmacological Properties Similar to Morphine," *Life Sciences* 88 (1975): 155–160; J. Hughes, T. Smith, H. Kosterlitz, L. Fothergill, B. Morgan, and H. Morris, "Identification of Two Related Pentapeptides from the Brain with Potent Opioid Agonist Activity," *Nature* 258 (1975): 577–579.

60. The test involved stimulating the sympathetic nerve, which caused the *vas deferens* to contract. Morphine and other opioids blocked the sympathetic response. It was later learned that the opioids blocked the sympathetic release of noradrenaline.

61. H. Kosterlitz and J. Hughes, "Some Thoughts on the Significance of Enkephalin, the Endogenous Ligand," *Life Sciences* 17 (1975): 91–96 (quote from p. 95).

62. For additional references see R. Simantov and S. Snyder, "Morphine-Like Peptides in Mammalian Brain: Isolation, Structure Elucidation, and Interactions with the Opiate Receptor," *Proceedings of the National Academy of Sciences of the United States of America* 73 (1975): 2515–2519; R. Simantov, A. M. Snowman, and S. Snyder, "A Morphine-Like Factor 'Enkephalin' in Rat Brain: Subcellular Localization," *Brain Research* 107 (1976): 650–657. To locate the met-enkephalin receptor Snyder and his collaborators used the antibody-immunohistological technique.

63. P. Greengard, "The Neurobiology of Slow Synaptic Transmission," *Science* 294 (2001): 1024–1030.

64. The excitatory fast transmission normally uses glutamate as the transmitter; the fast inhibitory transmission generally uses GABA as the neurotransmitter.

65. T. Fukada and T. Kosaka, "Gap Junctions Linking the Dendritic Network of GABAergic Neurons in the Hippocampus," *Journal of Neuroscience* 20 (2000): 1519–1528.

66. S. K. Kulkarni and A. C. Sharma, "Nitric Oxide: A New Generation of Neurotransmitter," *Indian Journal of Pharmacology* 25 (1993): 14–17; S. H. Snyder and C. D. Ferris, "Novel Neurotransmitters and Their Neuropsychiatric Relevance," *American Journal of Psychiatry* 157 (2000): 1738–1751.

67. H. Son, R. D. Hawkins, M. K. Kiebler, P. L. Huang, M. C. Fishman, and E. R. Kandel, "Long-Term Potentiation Is Reduced in Mice That Are Doubly Mutant in Endothelial and Neuronal Nitric Oxide Synthase," *Cell* 87 (1996): 1015–1023.

68. The evidence has been reviewed in Snyder and Ferris, "Novel Neurotransmitters."

69. Snyder and Ferris, "Novel Neurotransmitters," p. 1748.

70. Personal communication to the author, January 19, 2004.

71. The estimate of 50–100 neurotransmitters includes acetylcholine, the biogenic amines (epinephrine, norepinephrine, serotonin, dopamine, histamine), amino acids, peptides, gasses such as nitric acid and carbon monoxide, and also neuromodulators. See Snyder and Ferris, "Novel Neurotransmitters." The number of receptors change so often that any number mentioned is likely to be outdated before this book sees the light of day. At the time of writing it is generally agreed that there are six different dopamine receptors and five different classes of serotonin receptors. Moreover, some of these receptors have several subtypes; counting subtypes, then, there are as many as fifteen different serotonin receptors. In addition to the receptors that exist on postsynaptic neurons, there are so-called autoreceptors on presynaptic neurons that modify the rate of release of the neurotransmitter.

72. Paramecia, for example, have an oral groove that serves as a mouth and cilia, hairlike protoplasmic processes that cover most of their bodies. The cilia propel paramecia toward food and then sweep the food into the oral groove. The cilia also sweep harmful substances away, while also propelling the animals out of danger. Jennings observed that the responses of animals without a nervous system are not machinelike and invariable. He noted that the responses of these animals change with their state, such as when they last received food or whether previous experience with a given stimulus was beneficial or harmful. Because a paramecium changes its response to stimuli depending on experience, Jennings concluded that these animals have some capacity for memory and learning. Moreover, because he noted that a given chemical stimulus does not evoke the same response from all parts of a protozoan's body, he concluded that the responses must be determined by the "release of certain forces already in the organisms." This is similar to Langley's proposing the concept of "receptor substances" in order to explain why adrenaline has different effects on different organs in mammals.

Jennings also argued that protozoa respond to all the different classes of stimuli that regulate the behavior of animals with nervous systems. For example, paramecia respond to chemical, temperature, mechanical, light, water pressure, gravity, and electrical stimuli. The following is an often-cited statement by Jennings:

"The writer is absolutely convinced, after a long study of this organism, that if Amoeba were a large animal, so as to come within the everyday experience of human beings, its behavior would at once call forth the attribution to it of stages of pleasure and pain, of hunger, desire, and the like, on precisely the same basis as we attribute these things to a dog. . . . In conducting objective investigations we train ourselves to suppress this impression, but thorough investigation tends to restore it stronger than at first" (H. S. Jennings, *Behavior of the Lower Organisms* [Bloomington: University of Indiana Press, 1962], p. 336).

It is of some interest that early in his research career, Henry Dale also studied the behavior of paramecia (H. H. Dale, "Galvanotaxis and Chemotaxis of Ciliate Infusoria," *Journal of Physiology* 26 [1901]: 291–361.)

73. The development of a vascular system made it possible for chemicals to circulate more efficiently around the body. With the development of specialized secretory

cells that coalesced into glands, hormones secreted into the vascular system provided an additional route for chemical communication. However, although the vascular system increased the efficiency of chemical diffusion, nerve conduction combined with neurotransmitter secretion is many times faster.

74. For example, alpha and beta adrenergic receptors and also receptors for dopamine, serotonin, acetylcholine, glutamate, GABA, and many peptides have been found in bacteria, yeast, and protozoa. It is thought that many of the major neurotransmitter receptor classes were present even before there were animals.

CHAPTER 11

1. Cited in the *New York Times*, September 14, 2003, sec. 6, p. 46.

2. Darwin's theory of natural selection of inherited differences was critical for understanding how species evolved. He had no basis for understanding how inherited differences are transmitted, however, and he never abandoned the Lamarkian theory that acquired characteristics are inherited.

3. W. D. Hamilton, "The Genetic Evolution of Social Behavior," *Journal of Theoretical Biology* 7 (1964): 1–52. The concept of inclusive fitness makes evolutionary sense of altruistic behavior and various social behaviors that can favor risking survival to support close relatives (kin), with whom many genes are shared.

4. For example, Paul Greengard has described one of the many factors that influence the amount of neurotransmitter released. Greengard described two different pools of synaptic vesicles: a pool located close to or against the axon terminal, and a reserve pool located at a distance from the terminal. The neurotransmitter is released only from the pool close to the terminal membrane, but the ratio of reserve to releasable pools constantly changes in response to physiological demands on the cell. P. Greengard, F. Valtorta, A. J. Czernik, and F. Benfenati, "Synaptic Vesicles, Phosphoproteins, and Regulation of Synaptic Function," *Science* 259 (1993): 780–781.

5. The three other Nobel Laureates who did a substantial amount of their work in Burroughs Wellcome Laboratories were George H. Hitchings, Gertrude Belle Elion, and John R. Vane.

A HEART TURNED
EAST

A HEART TURNED EAST

Among the Muslims of Europe and America

ADAM LEBOR

LITTLE, BROWN AND COMPANY

A *Little, Brown* Book

First published in Great Britain in 1997
by Little, Brown and Company
Copyright © Adam LeBor 1997

A CIP catalogue record for this book
is available from the British Library.

ISBN: 0 316 87803 0

Set in Times by M Rules
Printed and bound in Great Britain by
Clays Ltd, St Ives plc

Little, Brown and Company (UK)
Brettenham House
Lancaster Place
London WC2E 7EN

For my parents, for everything

'God has made you brethren one to another, so be not divided. An Arab has no preference over a non-Arab, nor a non-Arab over an Arab; nor is a white one to be preferred to a dark one, nor a dark one to a white.'

Muhammad speaking at Arafat, shortly before his death

'The western countries have the material goods and we have spirituality. But you cannot live only with material things and you cannot live only with spirituality. So we are like a couple, a man and a woman, and to have normal happiness we have to share them together.'

Mustapha Tougui, lecturer in Islamic theology,
the Paris Mosque

CONTENTS

ACKNOWLEDGEMENTS

Without the help of dozens of people – Muslim, Christian and Jew – family, friends and colleagues, spread across a dozen countries, this book would never have been written. They all deserve a mention, but my thanks must go first to my agent Jennifer Kavanagh, in whose office the idea for *A Heart Turned East* was conceived. She had the faith to take on an untried first-time author, encourage me – and cajole when necessary – to carry out this project, from initial idea to publication. All that holds too for my editor Alan Samson, whose unflagging enthusiasm was always a source of inspiration, and whose advice has kept me motivated and focused. I am also indebted to Professor Akbar Ahmed, of Selwyn College, Cambridge, for his support and help, especially for his valuable guidance on points of Islamic theology.

Then, the others: their numbers are legion. Some took to the road with me, others I met on the way, but they all helped by being there, especially in the war-zones. In Bosnia and the former Yugoslavia: Peter Maass (best slalom driver in the Lasva Valley), Patrick Bishop, Cathy Jenkins, Michael Montgomery, Tim Judah, Yigal Chazan, Adrian Brown, Charles Lane, Isabelle Laserre, Laura Pitter, Joel Brand, Matt Frei, Paul Lowe, Amra Abadžić, Miloš Vasić, Dessa Trevisan, Philip Sherwell, Richard Carruthers, Tony Gallagher, Martin Nangle, Julian Borger, Mirijana Miličević, Ian Traynor, Tamara Levak, the reception staff at the Hotel Esplanade in Zagreb, everyone at the BBC Kiseljak office in the winter of 1992/1993, and Kurt Schork, who once drove me down Sniper's Alley to Sarajevo airport in a soft-skinned VW Golf. Thanks, Kurt.

In Hungary and the Balkans: Chris Condon and Simon

Evans at the Budapest International Press Centre. Chris for his rigorous reading of this manuscript and thoughtful editorial advice; Simon, together with Tom Popper, heroic transcribers of many interview tapes. My much-loved friends: Nora Milótay, Vesna Kojić, Vlada Vlaškalić, John Nadler, Julius Strauss, Steve Boggan, Meriel Beattie, Dóra Czuk, Miles Graham, Desmond McGrath, Colin Woodard, Rick Bruner and my brother Jason. Thanks also to Anna Szőke, my muse and more while I wrote most of this book. Also, everyone at *Budapest Week*, my workplace in the summer of 1995, particularly Krisztina Krizsán, Carl Kovac and Stephen Loy.

In Turkey: Andrew Finkel, for good-humoured expert advice. In Albania: Justin Leighton, best friend and road-trip companion for almost a decade and hopefully many more. In Germany: Frederick Schulenberg, Heidi Modro and Henning Angerer. In France: Emmanuelle Richard, Elizabeth Johnston, Amy Oberdorfer, Sacha Smith, John Randall, Fred Guilledoux, Frederick Tristram, Muriel Bouquet and Peter Sinclair. In the USA: Sam Loewenberg, for his unfailing lengthy hospitality, Daniel Felber and Joshua Freeman. In Britain: Dr Ahmad Khalidi, Rabbi David J. Goldberg, and Tony Earnshaw at the *Yorkshire Post*.

Among my colleagues at *The Times*: former foreign editors Richard Owen and Martin Ivens, for first sending me to Bosnia and printing my despatches as I wrote them, which is as much as any journalist can ask for. Also at *The Times*: Anne McElvoy, Michael Binyon, Tom Rhodes, Michael Evans and especially Julia Llewellyn-Smith, a reassuring voice at the end of the telephone during some dangerous days. But my biggest thanks there go to Roger Boyes, a fine friend and mentor, ever-ready with advice and novel ideas.

I am indebted also to Steve Crawshaw and Godfrey Hodgson, who first sent me abroad to Budapest as a correspondent for the *Independent*. Thanks also to my other editors who have shown consistent faith in me over the years: Susan Ward Davies at *Elle*, David Meilton at the *European*, Andrew McLeod at the *Scotsman*, Joe Millis at the *Jewish Chronicle*, Sandra Harris at *Business Life* magazine, Maggie Pringle and the late Les Daly.

There are several others too, in various countries, who agreed to be interviewed, and pointed me in the right directions. They have asked that their names not be published. Doubtless there are some individuals whom I have forgotten to include. To all, I say thank you.

I would also like to acknowledge here the many works of Professor Bernard Lewis, historian of Islam, which have inspired and instructed me during the writing of this book, and in addition, H.T. Norris' treasure-trove of a work: *Islam in the Balkans*. Any errors are mine.

Finally I must thank the several organisations that provided me with goods and services. Dr János Muth at Compuserve Hungary, who provided me with a complimentary account and invaluable access to Compuserve's databases, while I researched the lives of Muslims in the West. British Rail International, for providing me with several complimentary Interrail 26+ tickets with which I travelled across Europe and the staff at Phipps PR for helping arrange them. Finally, thanks to Mr Bashir of Regent's Park Mosque, who gave me a beautiful copy of the Quran. *Shukran*!

INTRODUCTION

'He who leaveth home in search of knowledge, walketh in the path of God,' said Muhammad. He also instructed his followers to 'seek knowledge even unto China'. I didn't go to China, but I did visit a dozen countries across Europe as well as the United States while researching this book. My aim was simple: to travel to where Muslims live in the West, speak to them, report on their lives and aspirations, and give them a voice. And from that, analyse the implications of this fresh stage in history, when for the first time, both Europe and the United States are home to a sizeable part of the *ummah*, the brotherhood of the world's Muslims.

Like every traveller before me I carried my own baggage: of stereotypes and preconceptions about Islam and Muslims, which gradually shrank and withered as I roamed the continent and entered, however briefly, their lives, and they, mine. I found an extraordinary national and cultural diversity that enriches life in Europe and the United States. Western Muslims are Bosnian soldiers in Sarajevo, Turkish musicians in Berlin, Algerian artists in Marseilles, Pakistani lawyers in Bradford, exiled Arab dissidents, modernising Muslim scholars in London and Islamic political activists in Washington DC. Each of their lives was shaped by Islam, but each was also moulded by their new homes. From these something new and unique is emerging: a modern, western Islam.

Just as Islam evolves in the environment in which it exists, for life in medieval Muslim Spain was radically different to present-day Saudi society, each of the West's Muslim minorities is grappling with issues specific to the new lands among which they now live.

This book opens in Sarajevo, where for over three years Muslims struggled just to stay alive as the world's first ethnocide to be broadcast live on prime-time television unfolded, literally, on their bloodied doorsteps. In France, Muslims must grapple with Islamophobia, a problem certainly not helped by the spate of bombs on the Paris metro last summer that were attributed to Algerian Muslim terrorists. Even so French Muslims have never suffered the murderous attacks of German neo-Nazis on Muslim immigrants. Most French Muslims are in fact French citizens. Only a tiny minority of Turkish immigrants in Germany have German citizenship, a result of that nation's obsession with nineteenth-century ideas of pure bloodlines, and a citizenship law dating back to 1913, the Imperial Naturalisation Act, specifically designed to keep the *Volk* pure from foreign blood, then Jewish.

For French Muslims the issue is dignity and respect. For German ones, it is the security that citizenship would provide. For British Muslims, the overwhelming majority of whom hail from the Indian sub-continent, it is a combination of both, combating an imperialist legacy that imported them from former bits of empire such as India and Pakistan to do the dirty work in Bradford factories that had been rejected by native Britons, but at the same time denying Muslims the legal protection of the Race Relations Act. In Britain, though, just as in the United States, Muslims are slowly building their own political lobbying organisations to try to make their voices heard in the West's corridors of power. Many, such as the American Muslim Council, or the Bradford Council of Mosques, draw directly on the lessons of the Jewish immigrant experience, and it is heartening that the Jewish Board of Deputies is actively advising Muslim groups on issues such as securing the provision of halal meat in British schools.

Modern Muslim life also demands compromises from Muslims, or at least an understanding of how politics work in the West. When, for example, Bradford Muslims incinerate a copy of *The Satanic Verses*, nailed to a stick like a literary crucifixion, tongues of fire charring the pages as it burns in front of a crowd of angry Muslim demonstrators, that image rebounds around the world, confirming the atavistic fears of many in the West that Islam is set on an inevitable collision course with life in a liberal western democracy.

As Ayad Abu Chakra, a senior editor at the London-based Saudi-funded newspaper *Ash-Sharq Al-Awsat* (Middle East) told me, there must also be a two-way process and Muslims must give as well as take, if they wish to be accepted. 'What is required is understanding by both sides. The Muslims might not like this, but I think I have to be realistic, you don't expect people to accept you in their culture if you are not ready to concede something. I for one have no right to impose my values on the British. I am gaining from them, I can give to them if they want to get something from me, but I don't impose myself on them. But if they want to know more about my culture I am ready to give them, to give them the best I can. We need concessions from both sides. A civilised society has to be open-minded.'

Which also means being open to the contribution that Muslims want to give to the West. An Islamic perspective can bring much to life in a modern industrial society, says Dr Dalil Boubakeur, rector of the Paris Mosque. The alternative, of continued prejudice and marginalisation, will only help spawn an angry Islamic underclass that will turn increasingly to a radical, confrontational Islam and polarise further the divisions between the West's Muslims and their host countries. Better by far, then, to let Muslims in the West introduce a new approach – or rather a much older one – founded in spiritual values, rather than material ones, says Dr Boubakeur.

'Muslims who are living in another country than their original one have a new way to discuss life, religion, nationality, philosophy, which are not only about the problems of life in their own countries. Islamic life can be a part of communal life in France, and Muslim values can help provide answers to problems of modern life, to make an *Ijtihad* [intellectual renewal]. For instance on artificial insemination, that brings a lot of theological problems in Islam, on abortion, on genetic engineering, problems of malformed fetuses.'

The right to eat *halal* meat in Bradford, or to wear a headscarf to school in Bordeaux, this is the stuff of Muslims' daily struggle to live an Islamic life. But from these debates spring other, more fundamental long-term questions that also demand answers. Modern Muslim life in Islam's new lands – where events in Algeria rebound almost instantly in Amiens – means a reconsideration of the nature of Europe, its ideal and reality, as the slowly-uniting continent struggles to forge its structure for the twenty-first century. It demands a new look at the very nature of nationalism and the cultural diversity of the modern nation-state; about citizenship laws and ethnic identity; and how Muslims can combine their loyalty to a single nation-state with a supra-national religious ideal.

About living a life in the West, with a heart turned East.

Running to avoid sniper fire, Sarajevo

Chapter 1

SARAJEVO: THE
SHOOTING GALLERY

*'From the ruins of a whole people, imperial ambitions
were already springing, which might one day threaten
all southern Europe.'*
John Reed, Serbia, 1915

'We must fight. The Muslims have the Islamic virus.'
Vojislav Savić, Bosnian Serb soldier, Zvornik, 1995

It was the smelliest and worst-fitting garment I had ever worn.
Designed apparently for someone five feet tall and four feet
wide, the flak-jacket hung stiffly on my shoulders, its rigid con-
tours wide and unyielding as I tried to bend down. Its armpits
stank of stale sweat and it lacked both a chest-plate to stop a
sniper's bullet and a slide-down groin flap, the latter especially
coveted by male journalists in Bosnia after a photographer with
Associated Press lost a testicle to mortar shrapnel.

But its large pockets were roomy enough to stash the foreign
correspondent's essential war-zone kit of notebooks, torch, Swiss
Army knife, cigarettes and hip-flask. And even this unlikely life-
saver was much more protection than most Sarajevans ever had.
Later that day, gratefully encased in the shrapnel-proof Kevlar,
I would nervously ask the foreign press, sitting at the bar in the
wrecked Sarajevo Holiday Inn, if I should sleep in it as well,
triggering guffaws of laughter. The black monster later disappeared, handed down the chain of reporters from *The Times*
who covered Bosnia. By the autumn I was kitted out in over
£1200-worth of body armour and helmet, an outfit that weighed
so much that one day in Mostar an old Muslim man even

offered me his seat, so hot and breathless was I in my armadillo-style gear. But that was several months away and my priority in July 1992 was getting into Sarajevo.

At five in the morning the Zagreb sky was slowly lightening as I picked up the flak-jacket, together with my luggage, and tramped downstairs to the lobby of the $158-a-night Hotel Esplanade. Its haute cuisine, chilled white Croatian wine and urbane staff made it a favourite haunt of the foreign press corps covering the wars in former Yugoslavia. Like the Commodore in Beirut and the Holiday Inn in Sarajevo, the Esplanade, wartime headquarters of the Gestapo, haunt of arms dealers, mercenaries and prostitutes in the Yugoslav wars, has entered journalistic folklore. Its ornate lobby and ballroom drip history and Balkan intrigue but I had more mundane concerns. The night manager summoned the bill as I checked out, again, and began another day trying to get into Sarajevo. There, under the gaze of the world's television crews, the unique culture of Bosnian Muslims, with its triple Turkish, Slavic and Muslim heritages, was being targeted for annihilation.

For a city under siege, blasted for months, and to be blasted for years to come by Bosnian Serb artillery, the guns a gift from Serbian president Slobodan Milošević's military henchmen, getting into Sarajevo was generally straightforward. The United Nations Protection Force, UNPROFOR, precursor of NATO's I-FOR, ran regular flights in and out of the city, open to aid agency staff, relief workers, journalists, UN personnel, foreign diplomats, almost anyone with an interest in the Bosnian war, except of course the people trapped inside the city. But the airlift often stopped and I had already spent three futile days sitting on the runway watching stationary C130s. Even the hotel staff raised an eyebrow, remarking dryly as I checked in again for the fourth time in as many days, 'Back again, Mr LeBor?' In fact I wasn't even sure that I wanted to get to the Bosnian capital, then one of the most dangerous places in the world, but this was my trial for a new job with *The Times*. If I filed enough stories, and came home in one piece – instead of the wooden boxes or air ambulances that too many of my colleagues left Sarajevo in – I would get the post.

No wonder the airlift was named 'Maybe Airlines' by the UN. The soldiers at the airport claimed that the name meant

'Maybe you will get shot down and maybe you won't,' but it seemed to me it meant 'Maybe you will get on a flight and maybe you won't.' The wasted days triggered a curious new emotion: the feeling of being desperate to get to a place I was secretly terrified of arriving at.

But just three hours later that morning we were roaring down the runway at Zagreb airport, strapped into a German C160 transporter plane. The plane's nose lifted and we sailed up into the clear Croatian sky, heading into Europe's latest battlefield, where Bosnian Serbs saw themselves as the new Crusaders against a wholly imagined Islamic threat. On board UN flight 267 were just over ten tons of food, several journalists cracking jokes and trying to look nonchalant about the possibility of being killed or injured over the next few days, and the aircrew, for whom this was just another day's work. It wasn't even nine o'clock yet, but we passed the brandy round and the plane banked, vibrating with the noise of the propeller engines, as we turned out to sea, out of range of the anti-aircraft guns of the Serbian army in occupied Croatia and flew over the Adriatic.

Standing up next to the piles of cooking oil and corned beef, the bland diet of Sarajevo's besieged citizens, I looked out of the window to see the Croatian coastline stretched out seventeen thousand feet beneath us, a beautiful panorama of jagged rocks with distant archipelagos dotting the azure sea. In the distance the mountains of central Bosnia that would, a few months later, see the start of the war between Bosnian Croats and Muslims, shimmered in the morning heat. The sea and the hills looked like a shot from a tourist-board video, from the C160's window, but Bosnia's welcome that summer was always violent and sometimes fatal. We flew down the coast over Zadar, the ancient Roman port of Split, and then descended to land at Sarajevo airport. It took just forty minutes to leave tranquil early-morning Zagreb and land in hell on earth.

I had seen the wreckage of war before, in the Croatian war, and heard the boom of shells fired in anger. The vista of destruction in the debris of former Yugoslavia was almost universal: a house outside Osijek hit by an air-to-ground missile, collapsed into rubble with its red roof tiles scattered outside; the almost delicate symmetry of shrapnel spray marks that a mortar left when it exploded across the ancient paving stones of

Dubrovnik; the eerie ghost town that was Vukovar, with its empty shells of houses, where an immaculate multi-volume biography of Tito lay among the glass, symbolism almost too obvious to be notable. In Slavonski Brod I had sat silently in a hospital basement with a doctor while we listened to a whistle. It was clear and sharp, subtly altering its tone until it ended in a not very distant explosion. It was the tune a shell played as it soared overhead and all conversation in the room stopped when it cut through the autumn afternoon air. Life itself seemed to be suspended, hanging on a thread of a Serb gunner's trajectory, while we waited to hear where the shell landed. It missed the hospital and landed in an apartment block. A few minutes later the first casualties were brought in on trolleys, rushed into a makeshift operating theatre in the hospital's gymnasium. The floor was awash with blood, crimson rivulets flowing back and forth, and pools of red sticky under the bright lights.

But nothing, not even the reports of the Sarajevo press corps that I had read and watched, prepared me for the horrible reality of the Bosnian capital that would be made bloody manifest later that morning, although the airport was a good introduction. Its wrecked buildings were carpeted with shards of glass, glinting in the hot morning sun, the remaining windows shot through. Spent cartridges littered the inside of the terminal, a reminder of the battle between government troops and the Bosnian Serbs at the start of the war. The tense UN soldiers stood guard behind sandbagged positions made it clear that we must leave the area immediately, for standing on the Tarmac we were in easy range of both sides.

It was the noise, a racket that sounded in the city for years, that was the most unnerving. It resounded all day and most of the night, echoing around the mosques and churches, the narrow Ottoman streets, the maze of the smashed shops and stalls of Baščaršija, the old market section, and across the former stadium for the 1984 Winter Olympics, where Torvill and Dean had won their gold medal, now a giant graveyard. The Bosnian Serbian artillery provided the back beat, with a regular boom and rumble of the 155mm shells that the soldiers, safely hidden in their bunkers in the surrounding hillside, liked to lob into streets and markets crowded with civilians and children. The mortars added a lighter note, and made a sharper bang when they exploded downtown or in the suburbs. When things got too slow, there was the rattle of machine-gun fire and the whoosh of rocket-propelled grenades to up the tempo of death, punctuated by the sharp snap of sniper fire, like a dry branch cracking underfoot. Surrounded by hills, spread out down the banks of the river Miljacka, Sarajevo was a giant shooting gallery.

How could this happen in 1990s Europe, asked the world, as it watched the destruction of this once-beautiful, cosmopolitan Ottoman town with its medieval mosques and craftsmen's shops and narrow Turkish streets? When Iraq invaded Kuwait the world's armies amassed to expel Saddam Hussein's forces. But the west would never launch 'Operation Balkan Storm', said Sarajevans. For there was no oil in Bosnia, and anyway, for western politicians and diplomats, Muslim life is cheap.

From the airport I scrounged a lift from a television crew and we raced down Sniper Alley as fast as possible, the driver swerving and weaving like a racing professional as we bumped down side streets and back alleys to avoid any sniper fire. I looked around the almost empty streets and scribbled random words into my notebook, barely legible scrawls about destruction, artillery and incoming fire. We hurtled past the UN Checkpoint that controlled access into the airport, past the shell holes punched into the apartment buildings, the road signs riddled with bullet holes, the sand-bags and tank-traps, the mangled cars, the anti-sniping barricades and, bizarrely, several pedestrians, including a middle-aged woman wearing a green top picking a careful path to somewhere. But mostly the streets were empty, for the Bosnian Serb gunners had weaved a web of dread and terror through the city, one so thick it was almost tangible. It was a feeling of dread I had never experienced before. The very air itself seemed tensed and ready for violence, swollen with anticipation for the next impact that would surely tear it open with high-explosive and shrapnel, hurling bricks and mortar, limbs and blood across the pavements and along the road. Something coiled tightly, twisting in my stomach, as I shrank into my flak jacket, a fear raw and visceral such as I had never known before. It was the fear for my life.

My destination was the Holiday Inn, a hotel whose amazing powers of endurance have now ensured it a place in history. The

11

Holiday Inn was located in one of the most dangerous areas of the city, a couple of hundred yards away from the front-line. Its glass entrance had been shot to pieces and its rooms were in easy reach of Serb sniper, rocket and shellfire. Guests were requested to keep their curtains drawn, so as not to draw incoming fire, and a notice proclaimed that they would be asked to leave if they failed to do so. Many of the rooms had been wrecked and there were gory urban-myth stories about a supposed corpse that still lay somewhere on the tenth floor. I had tried to book a room before leaving home in Budapest for Sarajevo and had eventually got through to reception. The staff were not encouraging to would-be guests. 'Don't come. It's very dangerous,' I was told.

It was. The front entrance was covered by a Serb sniper and so guests and staff used the back door. The building shook regularly as it took yet more hits from the Serbs' high-calibre machine guns. When a staff member had to repair something in front of the building the manager stood at the door with a submachine gun, ready to give him covering fire. The receptionist kept an AK47 under the counter. With a curfew in force there was little to do in the evening, except drink and watch the battles erupting outside. The view from the hotel's upper floors of the sudden flash of the explosions and arcs of tracer fire back and forth through the darkness had a terrible beauty, if that is the word. Newcomers were barraged with survival tips, and that was just for staying alive inside the building. Use this staircase, don't use that entrance. Stay away from that side of the hotel, get a room on this side.

It seemed worse at night, when the silence of a city under curfew was broken in the early hours of the morning by the whoosh of rocket-propelled grenades past my bedroom window. One evening we were drinking in the foyer bar when the alcohol reached saturation point and I needed to visit the toilet. I couldn't find the way and was scared to blunder about in the darkness in a semi-drunken haze. A Bosnian soldier escorted me to the gents', his Kalashnikov at the ready in case of incoming fire. It was the first time I had been accompanied to the toilet for many years. The rules of survival inside the Holiday Inn were complicated, but you learnt fast. And even though the hotel was directly in the line of fire it survived, just, thanks to a complicated arrangement of black market deals that kept the Bosnian Serbs supplied with a good proportion of the dollars and Deutschmarks that poured in from the foreign press corps, and which were passed across the nearby front-lines to former friends and neighbours.

Somehow, even among the carnage, the niceties of luxury hotel etiquette continued. The toilets were cleaned daily, and a fresh paper wrapper appeared each evening on the seat, proclaiming it hygienically disinfected. A pianist played each afternoon on the mezzanine floor, giving a spirited rendition of some Viennese waltzes, followed by Chopin's *Revolutionary Etude*. Waiters dressed in tuxedos and bow ties gliding back and forth with trays of iced drinks only added to the air of surreality. One evening as we sat around mulling over the day's events, discussing whose turn it was to buy the drinks, a shell exploded loudly somewhere nearby. 'Whose round is it anyway?' someone asked. 'Bosnian army, I believe,' came the reply. The staff too kept a morbid sense of humour. One morning the press came down for breakfast to find the receptionist carefully sorting through a pile of publicity leaflets, emblazoned 'Holiday Inn, Sarajevo, Yugoslavia'. Black marker in hand, he carefully inked out the word 'Yugoslavia' on each flyer before neatly piling them up again. 'What are you doing?' someone asked. 'Dead country,' he replied, shaking his head.

But standing in the foyer that first morning, hot, confused, and scared, I watched in dismay as a blood-stained and disorientated CNN reporter suddenly wandered in, repeating over and over how frightened he was to anyone who would listen. Everyone rushed over to help him. Red splotches were spattered across his shoulder and he was bleeding from glass splinters as he walked across the foyer. He was, by Sarajevo standards, comparatively lucky, although his colleague travelling in the same vehicle was not.

The CNN van had taken a direct hit on Sniper Alley, just a short while after we had sped down the long straight road. The camerawoman travelling with him had been hit in the face by a Serb anti-aircraft bullet. It blew off her jaw and she was medivaced out by the UN. She lived, but she would never look the same again. For weeks afterwards the white CNN van stood alone in the corner of the Holiday Inn's underground car park, for when death can come in an instant and luck means all, the van gave off bad vibrations and nobody would park near it.

That was the morning's casualty toll, at least among the press. In the city outside the Bosnian Serb gunners and snipers continued their work, maiming and killing, and the government troops, outgunned and outnumbered, fought back. Later that evening as we chatted to some Canadian UN soldiers, a photographer slowly limped in. He had been shot through the leg that afternoon as he took pictures of soldiers in action on the front-line. He had even seen the soldier taking aim at him, he explained indignantly, but his cries not to shoot went unheeded.

As the graffitto on Sniper Alley said: 'Welcome to Hell.'

Hell had come to the heart of Europe, but Bosnia's neighbours did little to put out its fires. The grand notion of a common European home did not apply to Muslims, even European ones. Diplomats and envoys such as Lords Owen and Carrington, the best drones in suits the West could offer, flew in and out of former Yugoslavia, trying to stop the war. They apportioned equal blame to all parties and tut-tutted at the reluctance of the multi-national Bosnian government – nearly always reduced to the description of the 'Muslims' – to roll over and give up, all in the best tradition of Foreign Office appeasement. The international community soon excused itself from intervention to stop the Bosnian Serbs and save Bosnia's Muslims, by claiming the Bosnian conflict was a 'civil' war, yet as a conscience-salver simultaneously imposed sanctions on Yugoslavia for Belgrade's role in fomenting the conflict. One of the more illuminating examples of the role of hypocrisy in international relations.

In Bosnia words as well as people are casualties of war. Orwell's Ministry of Truth is alive and flourishing: not in its supposed natural home, the now-vanished Communist bloc, but in the luxurious corridors of the UN in New York and Geneva. There a 'safe haven' was one of the most dangerous places in the world and a 'peace plan' is code for surrendering to Bosnian Serb ethnocide. One of the more revolting sights of the Bosnian war was watching the sleek UN diplomats pretend to anguish over the carnage they could stop with one phone call to NATO, before they waddled self-satisfiedly off to the next Security Council session.

Lords Carrington and Owen failed to stop the war. All the UN's peace plans and proposals failed until they recognised the single salient fact of the war in Bosnia. The conflict was spawned and directed in Belgrade by Serbian president Slobodan Milošević, implemented on the ground by his henchmen, former psychiatrist and Bosnian Serb leader Radovan Karadžić – the clearest case yet of 'physician heal thyself' – and General Ratko Mladić, the Bosnian Serb military leader and the man with the bloodiest hands in Europe. It was planned years in advance, which is how the Bosnian Serbs, with their Yugoslav allies, were able to seize over half the country in just a few months. The arms were smuggled in, openly imported, or just handed over to the Bosnian Serbs by the Yugoslav National Army (JNA), the Bosnian Serb militias formed, the sites selected for the camps and the methodology of ethnic cleansing honed to maximum efficiency.

Europe was about to witness its first ethnocide since the end of the Second World War – the destruction of Muslim Bosnia, its buildings, its culture, its people and even its history. Many months later I encountered Radovan Karadžić in Geneva at one of the seemingly endless series of peace conferences organised under UN auspices to try and make the Bosnians roll over and give up. He stood in front of the microphone, flanked by his bodyguards, bullish and simultaneously aggrieved at the failure of the world to understand the terrible Islamic menace to Europe that only the Serbs could stop. 'You, who wades in blood, how can you sleep at night?' I wanted to yell at him. Instead I asked his view of some arcane point of the latest maps to slice up Bosnia.

Whitehall and Washington, Bonn and Brussels, all know this which is why Milošević, Karadžić and Mladić have been named as potential war criminals, but Milošević especially enjoyed the attentions of the world's statesmen and had a ready welcome in the international corridors of power when he chose to talk peace. One day he faces possible arrest, the next he is an honoured guest in the company of world statesmen, drinking slivovitz with American diplomat Richard Holbrooke in Dayton, Ohio as they carve up the corpse of Bosnia together on the diplomatic table. How they must have laughed at night for years in Belgrade and Pale, capital of the self-proclaimed quasi-fascist

Bosnian Serb Republic, as they danced rings around the hope-less emissaries from the self-serving international bureaucracies that are supposed to represent world security, and the barrage of futile acronyms in whose name they speak: the UN, NATO, EU, WEU, OSCE and NAA with their useless maps, and plans for new zones, republics, confederations, even minute details of fly-overs and motorways.

Practitioners of violence such as Karadžić and Mladić under-stand one word: force. The summer 1995 massacres at Srebrenica when thousands of Bosnian Muslim men were slaughtered en masse, in a Nazi-style killing field, were too much even for the West to stomach, and waves of airstrikes followed that finally brought the Bosnian Serbs' puppet master, Slobodan Milošević, to the negotiating table. It was the worst atrocity in Europe since the Second World War. 'The Bosnian Serbs made a fatal error. They overplayed their hand when they went into Srebrenica,' said US Defense Secretary William Perry. Which begs the question as to what exactly, in the West's eyes, would have been a well-judged hand for General Mladić? But if the West could stop the war then, how much easier it would have been to hit Mladić's forces in 1992, how many more people would still be alive, how many homes still standing.

That said, the common view of the Bosnian war as the Serb, and Bosnian Serb, people, solidly united against all Muslims is both wrong and simplistic. For a start, it neglects the role of Croatian President Franjo Tuđjman in the death of multi-national Bosnia. This is the ultra-nationalist leader who once felt it necessary to apologise to Jewish organisations for his anti-semitism and the man in charge of the nearest thing to a quasi-fascist regime in Europe, where unwelcome local election results are annulled and troublesome Croat journalists are arrested. Croatia has the dubious honour of initially being refused entry to the Council of Europe, thanks to its human-rights record. Or lack of one. For Tuđjman's government organised, armed and financed the Bosnian Croat army, admittedly an occasional ally of Bosnian government troops, but far better known for its siege of Muslim Mostar, a mini-war that far out-classed the siege of Sarajevo in its brutality and destruction. Tuđjman was as keen as Milošević to slice up Bosnia – the only difference was he tried to present his regime as an oasis of western-style democracy in a swamp of Balkan nationalism. He failed.

At the same time the old Yugoslav ideals of multi-nationalism still lived, and live on, among many Serbs. A largely unreported footnote to the Bosnian war was the continued presence within Serbia proper of a large Muslim community in the Sandjak region, some of whom left to fight for the Bosnian government army. Serbia's tragedy was that a whole section of society that could form a liberal opposition to the war – students, writers, artists and much of the intelligentsia – simply uprooted and left the country in their tens of thousands, leaving the field clear for the nationalists. Several came to Budapest, and became close friends of mine. They too have scars, psychological rather than physical, for their country, the old Yugoslavia, no longer exists. At the same time, it was a peculiarity of the Bosnian war that some Muslim refugees fled to Belgrade, of all places. At first glance this appears a strange choice, but the city, while home to plenty of war profiteers and gangsters, kept enough of its cosmopolitan spirit alive to take them in.

Many Bosnians now admit they should have seen what was coming back in June 1991, when war erupted in Croatia and Slovenia. Then the JNA metamorphosed into the protector and armer of the rebel Serbs in Croatia, handing its extensive arsenal – for this was a force armed and trained to withstand possible attack from both NATO and the Warsaw Pact – to General Mladić's rebel Serb soldiers fighting newly independent Croatia. At the same time some Bosnian Muslim soldiers in the JNA fought under the Yugoslav flag, in effect aiding the Serbs during the Croatian war. They failed to realise that the lessons Serb fighters learnt in Vukovar and Pakrac, of shelling civilians, terrorising and killing anyone in their path, the ABC of ethnic cleansing, would be soon put to good use in Višegrad and Zvornik. That some Bosnian Muslims fought with the JNA would also be well remembered by the Bosnian Croats a couple of years later.

After Slovenia and Croatia declared their independence in June 1991, war in Bosnia was almost inevitable. The state in the making had several choices: to remain in rump Yugoslavia, in effect being absorbed into Greater Serbia; join Croatia in a federation; or go for independence. Bosnian Serb leaders repeatedly warned that the Serbs would not accept an independent Bosnia, which they claimed would be dominated by Muslims. Karadžić and his henchmen launched the most effective propaganda drive against the idea of a multi-national Bosnia since Goebbels' work for the Nazi cause. 'The Muslims are trying to dominate Bosnia. They want to create an Islamic state here, but we Serbs are not going to let them. You cannot force Christians to live in a Muslim state. It's just like Lebanon. We do not want to live under Oriental despotism,' said Karadžić.

The propaganda was directed at an international audience as well as those living inside Serb-controlled Bosnia. Bosnian Serb officials handed reporters covering the conflict a slickly produced 92-page booklet entitled *Bosnia-Herzegovina: Chronicle of an Announced Death*, which was an apt title as they had first pronounced the capital sentence on the nascent state. Its pages were filled with supposed captured documents, produced by Bosnian Muslims, claiming for example that: 'For one dead Muslim one hundred Serbs will be liquidated and for one wounded Muslim (depending on the severity of the wound) 10–15 Serbs.' A picture of a Muslim resting at the Zagreb mosque was captioned: 'A rest at the assembly centre in the Zagreb mosque before setting out to the holy war in Bosnia-Herzegovina.' The Serb cause also went online, and those interested could download more of the same from Serbnet, on the Internet. The various newsgroups soon filled up with furious arguments between rival supporters of each cause as the Bosnian war spread into cyberspace.

It was that fantastic vision of some reborn Ottoman empire run along Iranian lines that Karadžić used to rally the Bosnian Serbs to his bloody standard. It was a perfect example of the 'Big Lie' technique, for the more nonsensical the claims of the supposed Islamic dictatorship that was on its way, the more many Bosnian Serbs believed them. Reports that Bosnian Serb women would be forced to wear the veil and their men press-ganged into the mosque, in an independent Bosnia, were readily believed by the often unsophisticated rural Bosnian Serbs, even though they had lived among, and sometimes married, their Muslim neighbours for decades.

One of the grimmer ironies of the Bosnian war was that for

all the Serbs' supposed horror of the Bosnian Muslims' Turkish heritage, Milošević's Serbia far more resembled the declining Ottoman empire than the present day Europe it claimed to be protecting from the Islamic menace. From the country's rotten and corrupt bureaucracy and the paranoid fear of strangers in border areas, to Belgrade's use of assassination and murder as a tool of political control, modern day Serbia is a grandchild of Ottoman Turkey. Even the sawing oriental rhythms of Serbian nationalist 'turbo-folk' music draw on Turkey's musical heritage.

Ancient folk memories of the Turkish occupation, more recent ones of Croat and Muslim atrocities in the Second World War, the lack of any developed democratic culture or strong independent national media in Communist Yugoslavia, all meant that Bosnian Serb Television found a ready audience for the stream of lies it pumped out every night on its equipment, stolen from the BBC. In fact the roots of this maniacal nationalism extended back over a century. Here are the thoughts of Petar Petrović-Njegoš, poet, Montenegrin prince and monk, published in 1847, on the future of Islam and the 'Id Muslim festival in the southern Slav lands (Miloš Oblilić, a Serbian knight, supposedly killed Sultan Murad I at the battle of Kosovo):

> So tear down the minarets and mosques,
> and kindle the Serbian yule logs,
> and let us paint our Easter eggs.
> Observe the two fasts honestly,
> and as for the remainder, do as you like.
> I swear to you by the creed of Miloš Oblilić
> and by the trusty weapons that I carry,
> our faiths will be submerged in blood.
> The better of the two will rise redeemed.
> The 'Id can never live in peace
> with Christmas Day.

Almost 150 years later his instructions would be comprehensively carried out.

Under Karadžić's instructions many Bosnian Serbs boycotted the March 1992 nationwide referendum on independence, claiming that the plebiscite violated the principle of tri-national constitutional parity. There was here a legitimate political point.

Nobody could force the Bosnian Serbs to live in an independent Bosnia if they didn't want to. But, had there been the political will among the Bosnian Serb leadership, a solution could have been found to the conundrum of Bosnia's three nationalities. Joint Bosnian-Yugoslav citizenship; federation with Yugoslavia for a Serb-dominated region; confederation; cultural autonomy, political autonomy, territorial autonomy for a Bosnian Serb entity, open borders, any or all of these could have provided an answer. But the Bosnian Serb leaders were never interested in a peaceful solution. They had a more final version in mind.

In response they held their own referendum, which produced a result overwhelmingly against independence. Fighting erupted at the end of March and the eastern Bosnian city of Bijeljina was the first town to fall to the Bosnian Serbs. Its capture produced one of the most famous images of the war: a Bosnian Serb soldier, cigarette in hand, lazily kicking a pile of Muslim corpses. That shot handed the Bosnian government an early propaganda coup and has been used in pro-Bosnian posters across the world. The fighter was a member of the Tigers militia, commanded by Željko Ražnjatović, a.k.a. Arkan, and another on a list of alleged war criminals. Rumour had it that the soldier was subsequently severely, possibly fatally, disciplined, after the picture was published, but if so his only crime, by Bosnian Serb standards, was to be immortalised on film in the act of atrocity.

Whether they were religious or secular, Bosnian Muslims dismissed as ridiculous the Bosnian Serbs' barrages of propaganda, that claimed the government of President Alija Izetbegović aimed to set up an Iranian-style Islamic state. Drinkers of slivovitz, strong plum brandy, eaters of pork, for many Bosnian Muslims their only connections with Islam until the war were that they had names like Amra and Emir and left their shoes outside their houses. Bosnian Muslims were largely secular and those that were religious emphasised that they were 'European Muslims', something quite different to the Ayatollahs of Iran and the Islamic clergy of Saudi Arabia. All were proud of their multi-ethnic heritage and traditions of co-existence with both Orthodox and Catholic Christianity and Judaism.

As even one prominent Bosnian Serb leader explained to the

Belgrade magazine *NIN* in June 1990: 'In Serbia, the media are wrongly speaking of the dangers of [Islamic] fundamentalism in Bosnia . . . Our Bosnian Muslims are Slavs, of our same blood and language, who have chosen the European life along with their Muslim faith . . . We Serbs are closer to our Muslims than we are to that Europe.' But back then Radovan Karadžić was presenting a different face to the world. Even now though many Bosnian Serbs would agree with Karadžić's old analysis. Many have stayed loyal to the idea of a multi-ethnic Bosnia, and fight in the Bosnian army, such as General Jovan Divjac, or just try and survive in government-controlled towns such as Sarajevo and Tuzla. Loyal Bosnian Serbs who fight under the government's flag are usually singled for especially harsh treatment if they are captured by their religious kin on the other side of the front-lines, who regard them with loathing as traitors.

Until the outbreak of war many Bosnian Muslims were largely uninfected by the virus of nationalism that lurks just beneath the surface of so many southern Slavs, and were quite content in Yugoslavia. Bosnia's Muslims did not have their own independent state under Tito, but then neither did anyone shell their towns and cities. Unusually for the former Yugoslav republics, Bosnian public buildings often still had pictures of Tito hanging in offices. Even the barber in the dreary central Bosnian city of Zenica, where I once had my hair cut, still worked under Tito's benevolent gaze.

Right up to the outbreak of war, many identified themselves as Yugoslavs, rather than Serbs, Muslims or Croats, especially those of mixed marriages. This cosmopolitan tradition was highest in Sarajevo, where, according to the Bosnian government's Institute for Statistics, the capital reported the highest intermarriage rate for the whole country, with 34.1 per cent in 1991. But many cynics understandably wonder how, if the city was truly as tolerant and multi-cultural as everyone claimed, it degenerated so quickly into Europe's longest running urban-battleground since the Second World War. After the Dayton Peace accords were signed most Bosnian Serbs on the Serb side of the former front-lines left en masse for Serbia proper. The few who remained suffered repeated harassment by angry Muslims.

In Sarajevo, even as the JNA dug trenches around the city to prepare for the Bosnian Serbs' bombardment, few believed that a war was coming. When the war erupted in April 1992 the defence of the city was led by criminals who knew how to use guns and where to get them, such as the notorious 'Juka', who ran a black-market and extortion operation in the wartime city. Juka was repeatedly wounded and seemed to be held together by a collection of nuts and bolts but carried on fighting bravely and, it must be said, profitably. He later turned against the government and joined forces with the Bosnian Croats and was found shot to death near a motorway in Aachen, Germany.

A few prescient Sarajevans formed the 'Patriotic League' and secretly reconnoitered the JNA trenches being dug around the town, but their warnings went almost ignored. 'Nobody believed that war was coming. I didn't believe it,' said Arman, a former Sarajevo businessman turned fighter in the early days of the war. 'I remember that when someone asked me to join the Patriotic League early in 1992 I just laughed at him and said I was a pacifist. A few months later, after the war started, we met downtown. He was in a uniform, carrying a pistol. He shouted over to me: "Hey, Mr Pacifist! How are you now?"'

Rifles and pistols were passed from hand to hand, as during the Jewish Warsaw Ghetto revolt in 1943, and Bosnian soldiers ran up and down the trenches firing from different positions pretending to be more numerous than they were. By May 1992, one month into the war, there was still no war cabinet and no proper defence of the city, notes Misha Glenny in his book *The Fall of Yugoslavia*. 'When I ask Defence Minister Jerko Đoko what his strategy for the defence of the city is he replies: "Actually I come from Mostar and I've only been here a year so I don't know the city very well."' The Bosnian soldiers died in their hundreds, partly as a result of President Izetbegović's poor planning for war. 'Citizens of Sarajevo, sleep peacefully, there will be no war,' he famously proclaimed in a broadcast on 4 April 1992. Probably only he knows what kind of assurances he was given by Europe and America, that the West would ensure his nascent state's survival when he decided that Bosnia-Herzegovina should be an independent country without the consent of the Serb minority. Izetbegović's initial strategy, it seemed to many, was to let as many civilians be killed as necessary, until the West was galvanised into action to save Bosnia. It failed.

He also foolishly believed the repeated promises of the Yugoslav army that its soldiers were neutral and would stand between the two sides as a buffer. Muslim fighters were so nervous that Izetbegović would disarm their Green Beret militia that they met in secret. A day after the attack on Bijeljina the Green Berets stopped a JNA convoy on the road to Sarajevo. Hidden among the bananas were rockets and ammunition. President Izetbegović ordered that the convoy be allowed to continue on its way. The supposed great international Islamic plot to establish an Iranian-style Islamic state at the gates of Vienna did not seem very effective, at least militarily.

Many Serbs and Croats saw their Muslim compatriots as rather sleepy and disorganised. Just as British people tell 'Irish' jokes, Yugoslavs told Bosnian Muslim jokes, about a pair of peasants called 'Muljo and Suljo', mocking their supposed rural parochialism, although Sarajevo was at least as cosmopolitan as Belgrade or Zagreb. The latest joke about them is quite short: Muljo and Suljo are dead.

So were Senad Sačirović, Mirijana Milanović, Branko Pletikoša and Leo Sternberg. Just four of the Sarajevans laid to rest in one morning at the Lions cemetery overlooking the Olympic complex. The graveyard was divided into three sections: Muslim plots were marked with a star and crescent; Christian ones with a crucifix, and atheists with a star, symbol of the old Yugoslavia. Leo Sternberg was buried with the atheists, as Sarajevo's Jewish cemetery is now a front-line position.

Grief hung heavy in the air like a fog of human pain among the forest of cheap wooden grave markers. They stretched for rows up the hillside, each ending in the same year: 1992, the year of the start of the war. Bereaved relatives stood hunched over the fresh mounds, uniformed soldiers and policemen trying to keep their features composed while mothers, daughters and sisters sobbed out their pain. A woman's scream, sharp and agonised, pierced through the sounds of mourning as Branko Pletikoša's coffin was lowered into the ground. Space for corpses was so precious that old bones had been churned up in the ground. As I walked nearer to the funeral of Senad Sačirović, where his crying relatives kneeled down intoning the Muslim prayers for the dead, fragments of long-buried arms and legs poked through the earth. It was a truly dreadful place and I found myself biting my lip to keep my composure and stop myself from crying. A Muslim, a Serb, a mixed Serbo-Croat and a Jew. Bosnians could no longer live together in peace, but in death at least they were united, their funeral eulogy the rattle of machine-gun fire nearby.

But even in the midst of carnage, daily life, of a sort, sputtered on. Some cafés still functioned, where patrons nervously sipped black coffee among the wreckage, their faces tense as they puffed hard on cigarettes, their pistols and sub-machine guns lying ready on the table. An underground theatre was staging a play called *Die Like a Man*, in fact a comedy about two men and a girl. Death could, and did, come at any moment, but as Sarajevans said, nobody wants to stay indoors all day. Men and women still went to work in the morning, crunching through the broken glass, the women with make-up freshly applied as a personal gesture of defiance, the men clutching their briefcases. At each intersection downtown they stopped, checked for snipers and then ran across.

Even the traffic police carried on stopping cars for speeding and breaking the rules of the road. As one policeman said: 'Just because there's a war on, everyone thinks they can drive through red lights.' Police regularly pulled over our car, a beaten-up old Lada purchased in Split, driven by photographer Paul Lowe, because it lacked number plates. The plates, emblazoned with the Croatian red and white chequerboard, had sensibly been thrown away on the drive into the city before reaching the Bosnian Serb lines. But repeatedly sitting in the car in the middle of an often empty road, prone and exposed, while a Bosnian policeman demanded to see our documents and shellfire echoed around us was an unnerving experience. Eventually we tried to register the vehicle with Bosnian plates, and after a morning spent traipsing from office to office, we finally found the right one. Unfortunately though, the car could not be registered in Sarajevo for the moment, the clerk explained, laughing. Nobody had ever registered a foreign car in Bosnia before and so the procedure didn't exist.

But Sarajevo's veneer of normality was thin. One that could be instantly smashed by the curl of a Serb finger on the trigger a kilometre away in the hills, as we soon discovered one morning.

There was a set routine for getting out of the Holiday Inn, demanding a series of racing turns as fast as possible to avoid any sniper fire. As the passenger it was my job to check for other vehicles, for Paul had no time to look. 'Clear right?' he snapped, as we bumped out of the Holiday Inn's underground car park, racing up the concrete path onto ground level, whizzing past the danger spot where a Serb sniper often took pot-shots at passing cars. 'Clear left?' he demanded, and then we were off, heading downtown. The pavements were comparatively crowded at nine that morning as we weaved along the road and, the destruction aside, the city looked almost normal.

Until a bullet smacked into a building a few yards away and we leapt out of the car. Fear filled the air as the street erupted into pandemonium. Terrified pedestrians started screaming and running for cover. We ran too, rushing towards a doorway where, it seemed, twenty people were crammed into space for three or four, all scrabbling to get off the exposed pavement. An old lady sat in the middle of the road wailing out her terror, unable to move fast enough or get up alone. I watched scared and ashamed that I didn't go to help her, although someone braver did.

Somewhere in the hills a Bosnian Serb was smiling.

His commander General Mladić would have laughed as well, had he witnessed the parade one Saturday afternoon at the Bosnian Army headquarters downtown. If this was the crack force of fanatical Bosnian Mujahideen ready to bring Saudi Arabian-style Islam to Sarajevo, that so terrified the Serbs, they really had very little to worry about.

The soldiers, in name at least, lined up as the officer called them to attention. They looked as though they had just walked in off the street, which most of them had. Dressed in shirts and jeans, tracksuits and training shoes, the recruits, some with earrings and ponytails, looked around nervously. One gangly youth displayed all the awkwardness of adolescence, smiling eagerly with darting glances from side to side as he tried to keep in step. He was barely out of childhood and should have been out playing football or chasing girls, but instead he was going to war. Artists and accountants, musicians and mechanical engineers, anyone who could point and shoot a gun was being drafted.

The scene was a modern-day version of George Orwell's opening to his book *Homage to Catalonia*, when he first arrives at the barracks to join the rag-tag International Brigade. And just like the Republican forces of the Spanish Civil War, the Muslim-led army of independent Bosnia said it too was fighting against fascism.

The Old Town First Brigade was composed of grinning teenagers, tattooed bruisers and a sprinkling of middle-aged men with grimmer faces. They rose at seven for three hours of exercises, with breakfast at ten, followed by a course in shooting and destroying tanks. After two or three days' instruction they were despatched to the front-line. 'We teach them how to fight and how to go to war. We fight like partisans, because we have no weapons for any kind of fight. Their education is on the front-line. We have no choice,' said instructor Nijaz Sabgica. The Bosnian army was massively outgunned, but morale was high at the *ad hoc* military academy. Soldiers just back from the trenches, or about to be despatched, milled about, while in the corner a fighter in his twenties filled his pockets with hand-grenades. The strains of music from the Bosnian Army band drifted down the corridors. Couriers ran to and fro with messages and documents to be signed and orders to be ratified.

A few fighters had weapons, ancient bolt-action rifles and the occasional Kalashnikov. Adison Nezirović had a screwdriver. He pulled it out, exclaiming scornfully: 'What we have we use. I don't even have a knife.' Adison, son of a Muslim mother and Croat father, in former life an importer of china and cutlery, had seen action. An articulate young man in his twenties, he looked sadder and wearier than his new brothers-in-arms in the courtyard.

His was a familiar Sarajevo tale of betrayal, but one no less poignant for its telling. 'I've seen people I know fighting on the other side. Five of us were in school together, in the same class, and we saw our friend Cile fighting with the Serbs. We arrested him and told him to go home. We told our chief that we know him and we don't want him to fight. He went back to the hills. Don't ask me if I would kill him. I don't want to talk about it, because he was a very good friend of mine. Every Serb thinks the same now and I don't know how it can happen. I couldn't believe it when I saw him. He slept in my house a hundred times and now he is fighting against me and my friends.'

Adison had been wounded three months earlier, with shrapnel in his back, but he claimed that he was not scared before combat. 'If you don't go to the war, if they come here, they will kill you. I felt very bad the first time I killed someone with a hand-grenade. They were coming out of a car, I saw it and threw the grenade. I had never killed an ant before, but I had to do it because I was afraid for my people. At that time I wasn't sure what was happening but I thought now I have killed in my life. But in war you had to do it. Everyone said. I have to protect my mother and sister, but I don't know what I will do when the war is over, maybe go to Canada. I don't want to be here.'

Here was Sarajevo, capital of newly independent Bosnia-Herzegovina, but even many Bosnian Muslims were not quite sure what that meant. One day they were Yugoslavs, strolling by the Miljacka river, the next they were sheltering in their basements because their Bosnian Serb neighbours were shelling them on account of their Muslim religion.

So who, exactly, are the Bosnian Muslims? They are probably the only nation in the world to be defined by a religious criteria, in their case Islam, but at first glance that raises more questions than it answers. How can a people be described as Muslim, when most of them were secular unbelievers? And where does that leave Orthodox Bosnian Serbs and Catholic Bosnian Croats who want to be Bosnians first and Serbs and Croats second, not to mention Bosnian Jews and Gypsies?

For Serbian and Croatian nationalists the question is irrelevant. They believe Bosnian Muslims are really Orthodox Serbs or Catholic Croats who took a theological wrong turn and converted to Islam when Bosnia was part of the Ottoman empire and should now return to the fold. No Bosnians means no Bosnia. The logical conclusion of this argument were the secret accords reported to be signed by Presidents Tuđjman and Milošević carving up Bosnia between them, even as hostilities erupted between Serbs and Croats.

Certainly there is an argument that there is no such thing as a Bosnian *per se*, just as one can argue that there is no such thing as, for example, a Slovak. Like Serbs and Croats, Bosnian

Muslims are southern Slavs (Yugoslavs, in Serbo-Croat). They speak the same language, Serbo-Croat, and have no specific racial features to distinguish them from Serbs and Croats. Bosnian Muslims are dark-haired and brown-eyed or blond and blue-eyed, just like Serbs and Croats. All three peoples originated in the great migrations of Slavic tribes from the Volga basin over a thousand years ago that spilled across eastern Europe through Bulgaria and former Yugoslavia to the Czech Republic and Slovakia. But by these criteria Serbs are really Croats and vice-versa. Or even Poles, or Slovaks. What differentiates Bosnian Muslims from Serbs and Croats is their religion, Islam; their hybrid Turkic-Slavic culture and of course their history of continuous occupation on the same territory, adequate criteria in 1990s Europe for statehood.

The collapse of the Soviet Union and Communism has triggered an avalanche of new nation states which barely match Bosnia's claim to nationhood, yet which have been welcomed into the international community. Until 1 January 1993 there had never been an independent Slovakia, for example, although a Nazi puppet state had existed in the Second World War. There was not even a referendum on independence in Slovakia, but its claim to nationhood has gone largely unchallenged by most of the world. In Bosnia there was, and a majority voted for independence, even with the Serb boycott. So why should Bosnia be any different and its status as a independent state be continually marginalised and prevented from buying the arms to defend itself? The answer, many Bosnian and foreign Muslims believe, is because of the country's Islamic heritage. Fear of Islam has prompted the world to stand by while the country was destroyed by marauding Serbs.

But Bosnia-Herzegovina has at least as good a claim to statehood as Croatia and Serbia, no matter whether its citizens pray in a church, mosque or synagogue, if that claim is based on a desire of a majority of the population for self-determination and a history as a national entity. Bosnia was politically independent by 1100, but it was not until the late twelfth century that an independent country emerged, led by the 'Great Ban' Kulin. Medieval Bosnia had its own script and its royal family's symbol was the fleur-de-lys, now readopted by the current Bosnian government as the country's symbol, to the chagrin of many devout Muslims there who would have preferred an Islamic crescent.

Bosnia also had its own church, a key element of national identity in the Balkans. The Bosnian Church appears to have been a hybrid of Catholic and Orthodox, a fitting mix for a region that straddled the theological divide. The key event in Bosnian history was the disappearance of this Church and the replacement of Rome by Mecca as Bosnians steadily deserted Christianity for Islam after the arrival of the Turks in the fifteenth century. Many writers have sought to explain this switch by claiming that Bosnian Christians were in fact Bogomils. Bogomil(ism), meaning 'love of God' in Slavic, was a heresy modelled on Manichean dualism, a Babylonian doctrine based on the eternal conflict between light and dark. As Bogomils were perpetually at the mercy of either Rome or Byzantium, the argument goes, with no external protector of their own, they were more susceptible to the lure of Islam which gave them full citizenship in the Ottoman empire. Other historians, such as Noel Malcolm, completely reject the Bogomil hypothesis. He points out that Bosnian Christians used the sign of the cross, held masses, drank wine and ate meat, all forbidden for Bogomils.

Either way, probably by the late sixteenth or early seventeenth century Muslims were a majority in Bosnia. Just like the Bedouin, the Turks, the Gulf Arabs and North Africans, the Bosnians simply chose Islam. Many Bosnians' links to Christianity were tenuous, and converting to Islam brought immediate benefits under Ottoman rule. The first was the privileged status of Muslims: non-Muslims were discriminated against by the *Kanun-i-rayah*, which governed the status of subject peoples. Conversion also brought enhanced career prospects. Anyone hoping for a position as an Ottoman functionary had to be a Muslim.

Ottoman Islam also offered a civilised life-style for its subjects, at least by sixteenth- and seventeenth-century standards, as long as they observed the law and accepted Ottoman rule. Religious minorities, including the Serbian Orthodox Church, were allowed their own courts and given the powers to settle their own internal affairs. But wrongdoers, real or imagined, who threatened the stability of the Sublime Porte, were subjected to revolting punishments such as impaling. The prisoner would be held down and a stake fed into his rectum, passing through his body out of a cut in his shoulder, before he was raised up on the pole like a human kebab, to flop helplessly in agony. This was a calculated brutality, designed to control often restless subject peoples. The Ottoman executioners were specially trained in their job, easing the stake through so it bypassed vital organs. The wretched prisoner would often survive for several days, dangling helplessly as a grisly warning to passers-by of the price of law-breaking under the Turks.

That was the downside of Ottoman rule. So was the hated practice of *devshirme*, the state-organised kidnapping en masse of young males to Istanbul where they would be converted to Islam and trained in the army or Ottoman civil service. The benefits could be considerable for towns such as Travnik and Banja Luka, which quickly became important local centres. The Turks erected an urban infrastructure; not only mosques, but also markets, bath-houses, schools and bridges. Ghazi Husrev Beg, who lived in the sixteenth century, is immortalised in this Sarajevo folksong praising his endowments:

> *I built the medresa* [school] *and imaret* [public kitchen]
> *I built the clock tower by it a mosque*
> *I built Taslihan and the cloth market*
> *I built three bridges in Sarajevo*
> *I turned a village into the town of Sarajevo*

Serbian towns too flowered under Ottoman rule. By the eighteenth century Užice, for example, had eighteen mosques, as well as the usual municipal buildings, and became a noted centre of Islamic learning. Under Ottoman rule Bosnians evolved their own Turkic-Slavic hybrid culture, just as the Bosnian Church adopted some but not all of the trappings of Catholicism. Bosnian Muslims wore Turkish dress but few learnt Turkish and married women were only half-veiled: 'Go to Bosnia if you want to see your wife,' said one Turkish saying.

Until the collapse of the great Austro-Hungarian and Ottoman empires in the First World War, the question of Bosnian independence was largely irrelevant. Bosnia was annexed by Austria in 1908 and later became part of the

Kingdom of Serbs, Slovenes and Croats, set up after the First World War. But during the Second World War, as victory drew nearer for Tito's Communist partisans, the question of Bosnian nationhood became a vitally important issue. At that time Bosnia had been absorbed into the Independent State of Croatia (NDH), a Nazi puppet-state led by Croatian fascist Anté Pavelić.

In 1942 the Muslim leadership sent a memo to Hitler, boasting of their supposed Gothic origins, complaining of killings of Muslims by Pavelić's Ustaša troops and asking for Bosnian autonomy. That was denied but the memo led to the formation of the first Bosnian SS division composed of Muslims, the 13th Handar, named after the Turkish knife, a move actively encouraged by Haj Amin El-Husseini, the Grand Mufti of Jerusalem, fervent anti-Semite and ally of Hitler. Spurred on by Bosnian Imams, at one stage 21,000 men joined up. Fifty years later the Bosnian Serbs would well remember the SS Handar division. But relations between the NDH and the Muslims soon drastically deteriorated after religious leaders across the country issued protests about the NDH's brutal treatment of Jews and Serbs, and by the end of 1944 the SS unit was disbanded as Muslims flocked to the partisans.

Post-war Yugoslavia though failed to deliver on its promises of religious freedom. Its first twenty years saw Islamic courts suppressed, veils outlawed together with teaching of children in mosques, and only one state-controlled Islamic association allowed. It was not until 1974 that nationhood for Muslims of both Bosnia and the Sandak, Muslim-inhabited parts of Serbia and Montenegro, was enshrined in the Yugoslav constitution, the first time in post-war Europe that a nation had been created on religious criteria. As Yugoslavia began to implode in the late 1980s and early 1990s Bosnia was pulled in two directions, between ethnic nationalists who saw themselves as Croats, Serbs or Muslims first, and Bosnian nationalists who supported the idea of a multi-ethnic state. The Muslims were not immune to the idea of ethnically based nationhood: the first nationalist party to be formed in Bosnia was their Party of Democratic Action, the SDA, in 1990.

The choice of Alija Izetbegović as leader of the SDA and subsequent president of Bosnia-Herzegovina gave Bosnian

Serbs their greatest propaganda coup. Born in Bosanski Samac in August 1925, he was sentenced to three years in prison in 1946 as Tito's secret police, beginning their crackdown against Muslim activists, charged him with 'pan-Islamic' activities. Tito's security services watched his development with interest over the next few decades and monitored his works such as *Islam Between East and West* and *Problems of the Islamic Renaissance*. In 1983, together with other Muslim nationalists, he was put on trial for 'counter-revolutionary acts derived from Muslim nationalism'. He received a fourteen-year sentence, of which he served six years. The prosecution's case rested on the *Islamic Declaration* that he had written in 1970, translated copies of which are easily obtained at the Serbian press centre in Belgrade, and are continuously quoted by Serb spokesmen.

The declaration appears to call for the establishment of some kind of Islamic order, based on the experience of Pakistan, another country where Islam defined a nation as well as a religious allegiance, although nowhere does it mention Bosnia by name. But the work, a rambling and contradictory treatise, is so hedged with qualifications that its alleged Islamic radicalism was best seen in the eye of the beholder. The declaration's repeated emphasis on the individual's conscience and moral behaviour places it firmly in the modern reformist tradition of Islam, and far from the Iranian or Saudi models, according to analysts such as Yahya Sadowski at the Brookings Institution.

Such important distinctions were lost on the Yugoslav prosecutors. At his trial Izetbegović defended the declaration as offering 'a vision of a democratic and humanistic social order', and said an Islamic order could only be established with a majority Muslim population. Izetbegović's defenders say the *Islamic Declaration* must be taken in context as a dissident document, written under a Communist regime, that used Islam as a tool to express opposition to a repressive Marxist dictatorship. If other Yugoslav leaders were judged on what they stood for twenty years ago, Serbian president Slobodan Milošević would remain a vote-rigging despot and Croatia's Franjo Tuđjman an authoritarian ultra-nationalist. Both of which are, in fact, still fairly accurate descriptions.

Defenders also point to the sentiments of Izetbegović's later book, *Islam Between East and West,* published in 1980. Writing in the *New Republic*, Professor Fouad Ajami of Johns Hopkins University described this confused tome, jammed with references to everyone from Albert Camus and Frederick Engels to Bertrand Russell, as an amateurish attempt to marry Bosnian Muslim traditions with those of Anglo-Saxon Europe. Hardly the views of an Islamic fundamentalist. This attempt to weld Islamic and modern European culture and norms was also evident in the interview Izetbegović gave to the first issue of the magazine *Ogledalo* [mirror], published in 1990, a few months before the first outbreak of war in the former Yugoslavia, where he refers to Jean-Paul Sartre. 'For the Bosnians, the fact that the people aren't the same, but different is totally natural. Hell is not otherness (remember Jean-Paul Sartre) but something you accept and understand. From the beginning we have been getting used to people that belong to other religions and nations. This situation is an extra experience and an extra wealth.'

But for Bosnia, that wealth, of four religions – Islam, Catholicism, Serbian Orthodoxy and Judaism – and as many cultures, now exacted a high price. A short walk through Sarajevo showed the true cost. Hurtling around town by car and living in the Holiday Inn, journalists were to some extent insulated from the lives of average Sarajevans. It was the trip on foot to Amra Abadžić's house in the old Turkish part of the city – a frightening combination of walking, sprinting and hiding in doorways – that brought home the grim reality of daily existence in the Bosnian capital.

Black-haired, slim and fine-boned, Amra was a law student and Communist party member before the war. She carried a Makarov pistol in her handbag and looked like a model, which may have been an added incentive to pay her family a call. I met Amra through Paul Lowe and they have since married, giving at least one happy ending to a story out of Bosnia. We walked out of the Bosnian army headquarters where her father Fuad was based, tramping downtown through the wreckage of war: shops with their fronts blown out, twisted remains of cars splayed across the road and everywhere, building façades pitted with shells and bullet holes. In the background the Sarajevo symphony played on, with its orchestra of rifles, mortars and artillery. For half a mile or so we ambled down the city's main street, named after Marshal Tito. Amra and her sister Elma walked along quite casually, but my stomach was clenched with tension. We were exposed; the only people on the street.

At every intersection, favourite targets for snipers, we stopped, checked and ran across. Then we came to a small precinct, empty apart from smashed cars and twisted lamp-posts. 'Here we must run,' they said as we stopped in a doorway. 'Go across the square and over to that corner.' I ran as fast as I ever had, fear spurring me on. So fast in fact that I almost barged Amra and Elma out of the way as we reached the other side. From there, hearts pounding, we trekked up a steep hill into the warren of the old city. We tramped across gardens, up dusty alleys until finally we were home. As we went further into the relative safety of old Sarajevo, where children played in the streets, the tension began to drain away. But it never went.

Like many Bosnian Muslim families, the Abadžićs had always thought of themselves as Yugoslavs. Now though, they had discovered they had a new nationality as Bosnian Muslims – even Amra, the former communist. 'I loved Tito very much. When we were young they taught us about him. We sang for him and danced for him. I loved him for twenty years. I don't blame him for anything, he is part of my life. It was a good time under Tito. There was no war, we were together and we believed in something. Maybe it was stupid but at least we were together,' she said. As we spoke her mother Munira served us coffee, chocolate and biscuits. Lines of tension were etched onto her face and she looked tired and haggard.

Her husband Fuad had already been wounded three times, once as he stood outside the front door. In civilian life Fuad made belts, but now he was a Bosnian army officer. 'Before the war I could never have imagined this situation, that I would be a soldier or commander. I never thought about war. But if we all stay at home who will be in the front-line fighting? Once this war is over I will go back to my leather job. I want to do that,' he said.

Like most families in Sarajevo the Abadžićs survived with the help of UN food aid. Bread, beer and ice-cream were the only food readily available then, although if you were prepared to take the risky journey to the Bosnian Croat barracks, eggs, meat and vegetables were on sale, for Deutschmarks. A typical aid package might contain sardines, soap, pasta and a tin of goulash. The goulash, explained Munira, could be stretched to several meals by serving it with pasta. 'Oil and sugar would be more use. You can't live on a tin of goulash every two or three months. We do appreciate the UN aid of course, but if you wait for that you will die.' The war governed every aspect of daily existence, said Amra. 'In the beginning I felt awful, but now I'm used to it. I arrange my day around the war. I spend my day finding water, meeting my friends. We have good friends, our neighbours, we sit with them and play the guitar and listen to music or talk and play cards. You feel better when you have friends. You become really close when you spend all day and night together in the shelter.'

One evening, back home in Budapest, I turned on Sky television to see the belligerent *Daily Mail* columnist Richard Littlejohn hosting his chat show. Among his guests were the *Daily Mail*'s star feature writer Ann Leslie and Carol Sarler, a columnist with the *People* newspaper, two heavy hitters from Fleet Street's battalion of media worthies. Somehow the mention of Sarajevo came up. 'Sarah who?' quipped Littlejohn. His crowing guests guffawed with laughter. After years of war, the slow death of the Bosnian capital was, for these people, worth no more than a snigger.

But down on Marshal Tito boulevard, and high on the slopes of the Lions cemetery, nobody was laughing.

Refugees from Banja Luka, Travnik

Chapter 2

BOSNIA: ISLAM REBORN

'They are doing this to us because we are Muslims. We are all Muslims enough to be killed . . . this war can continue to the day of judgment if necessary.'
Dr Mustafa Cerić, Imam of Bosnia

'Allahu Egber [God is Great].'
Written in the dust on a BBC armoured vehicle,
Vitez, central Bosnia

When the morning sunlight glinted on the river Lašva, cutting through central Bosnia, and the mist rose through the bumpy mountain tracks snaking behind the front-lines, it was hard to believe that we were crossing a war-zone. Streams poured crystal water through the tree-covered hills, dogs barked in the distance, birds sang out their early morning call and schoolchildren trudged along the road to Vitez on their way to classes, brightly coloured rucksacks on their backs. The soldiers at the checkpoints would lazily wave us through, as they lifted up the wooden pole blocking the road, their presence the only blight in this pastoral idyll. That, and the frequent funerals, as another cross or Muslim marker was added to the little forests of graves that spilled down the verdant hillsides.

But when tension rose between Bosnian government forces and their occasional allies in the Bosnian Croat army (HVO), usually a prelude to all-out fighting, the trip from the Bosnian Croat village of Kiseljak, just outside Sarajevo, to the old Turkish town of Travnik, was an unnerving journey. Then the checkpoints were transformed into combat positions and it was best to drive up very slowly. The fighters were alert and tensed, their Kalashnikovs cocked as they stopped each vehicle to check the occupants. Soldiers laid two lines of land-mines in a careful path that curved towards their position, often reinforced with a machine-gun. A British officer at the UN base in Vitez claimed that the mines were usually empty or not primed, but we weren't about to check when we gingerly drove between them. The sand-bags would come out as well, together with lines of tank-traps, giant crosses of steel rails welded together that were slowly rust-ing, but no less menacing for it.

The journey after dark was the most unnerving. One night as we returned from the government-controlled town of Zenica, speeding away after an air-raid siren sounded, we reached the Bosnian Croat checkpoint at the Vitez turn-off. Heavily armed soldiers watched us warily in the darkness. A torch flashed in the night and an HVO fighter pulled us over. 'Go slowly, slowly,' he cautioned. 'The Muslims have put up barricades down the road.' Our interpreter Mirijana suddenly became very nervous, for she was a Bosnian Croat, and if Muslim soldiers had set up new checkpoints on the way home life could get complicated very quickly. Mirijana was a delightful young woman, twenty years old, pretty and brave, who had been a student in Sarajevo. 'Oh Sarajevo, how I would love to go back to Sarajevo,' she would sigh as she looked longingly down the road that led to the front-line Serb positions at Ilidža. We would pick her up most mornings from her parents' house just outside Kiseljak before setting off in search of a story. Her father had run a small haulage business before the war, but now his trucks sat rusting in the garden. We thanked the HVO troops for the warning and slowly set off again down the badly lit road, past the Dutch UN troops at Busovača, until a torch beam again cut through the night. There were no barricades. A handful of ragged Bosnian army troops stood by the side of the road as we stopped. One peered inside the car and stared dolefully at us, saying only, 'Help us. Help Bosnia,' before letting us drive on.

In the government-controlled towns of Travnik and Zenica, Sarajevo and Tuzla, Bosnia's Muslims were at first puzzled, then bitter and incensed at the failure of the West to come to their aid. Food, medical supplies, even British toilets in prefabricated units, all these trundled up the dirt roads from the UN warehouses on the Croatian coast. Welcome as the relief aid was, it was not enough. Loyal Bosnians, of all faiths, wanted weapons, preferably artillery, tanks and fighter planes. Or as Nađija Ridić, a Bosnian army officer based in Travnik, put it: 'If someone kills your children with a hand-grenade, you don't want to be given a lunchbox.'

The Bosnian army, slowly evolving from a ragged band of fighters armed with pistols and ancient bolt-action rifles into a coherent fighting force, was a unique military phenomenon. For the first time since the collapse of the Ottoman empire soldiers in a Muslim-majority army – together with a minority of loyal Bosnian Serbs and Croats – were fighting in mainland Europe. But they were vastly outgunned by the Bosnian Serbs, who under the orders of Serbian President Slobodan Milošević had been handed the JNA's arsenal. Together with the HVO, government troops held barely a third of Bosnian territory. The Serb onslaught had been stopped, and a village here, a ridge there recaptured, but without arms deliveries it would take decades and thousands of lives to regain enough territory to make Bosnia a viable state.

But the West refused to supply weapons to Sarajevo, thanks to the UN arms embargo on all countries of the former Yugoslavia. Diplomats claimed the decision would slow down the fighting and help the cause of peace, but the reality on the front-lines of the arms embargo was to draw a moral equivalence between Muslim victims and Serb aggressors, further increasing Muslim anger. More importantly though, the arms embargo was providing an opportunity for less scrupulous, or perhaps more scrupulous, depending on your viewpoint, states to step into the military gap. If Washington would not help, then Tehran and Riyadh would. Bosnian Muslims had little sympathy for these Islamic countries' theocratic systems of government, but after months of pleading in vain with the West for guns they gladly accepted military and financial aid from Middle Eastern countries.

Fighters too came, hard-faced men from Saudi Arabia and Algeria, some veterans of the Afghan war, all ready to die in what they saw as a 'Jihad', a holy war to save their Muslim brethren. In fact there was little Islamic brotherhood in evidence between the Mujahideen and Bosnian soldiers. The foreign fighters were shocked at Bosnians' liberal ways and the presence of women fighters on the front-line, the Bosnians were alienated by the Mujahideen's Islamic orthodoxy. One UN officer gleefully related how, when a Mujahid ordered a Bosnian woman soldier off the front-line, she shot him. But ultimately a single armed Arab fighter was worth any number of European diplomats mouthing platitudes about the latest Serb atrocity.

For Bosnian Serbs the presence of the Mujahideen was the proof they had been waiting for that Bosnian Muslims were Islamic fundamentalists. At the Bosnian Serb headquarters in Banja Luka, epicentre of ethnic cleansing, Serb officers showed visiting journalists a picture, supposedly found on a dead Saudi soldier, of a Mujahid holding up the severed head of a Serb. In Brussels and Washington too, diplomats feared that Bosnian Muslims would turn into Iranian-style Islamic radicals. But in front-line towns such as Travnik, where the Mujahideen prayed at the sixteenth-century multi-coloured mosque, a more subtle process, of awakening and realignment, was unfolding.

Travnik was for centuries a vital staging post on the trade routes from the Balkans to Vienna and Trieste. It had been home to Muslims, Croats, Serbs and a now-vanished Jewish community for centuries but the domes and the minarets of the mosques, the Turkish fountains and water towers, the urban heritage of the Ottoman empire dotted through its streets and alleys marked it as, culturally, a Muslim town. The city was immortalised in *Travnik Chronicle*, by Ivo Andrić, the only Yugoslav to win the Nobel Prize for Literature. 'No one in Travnik ever thought it was a town made for simple and everyday happenings,' he wrote. 'It is like a half-open book on whose pages on one side and the other are gardens, small streets, houses, fields, graveyards and mosques, like illustrations.'

Andrić was more prescient than he knew. Travnik, the disembarkation point for thousands of ethnically cleansed refugees, was regularly shelled by the besieging Serbs, although the presence of British UN troops in their Warrior armoured vehicles on patrol sometimes seemed to reduce the intensity. Still, plenty of shrapnel scars were splashed across houses in the ancient city centre. Some were large gashes from ground-burst shells, while others were a filigree of tiny holes, from rounds that exploded in mid-air, showering schools, markets and people with a deadly steel rain.

Sometimes the streets would rapidly empty as though the

inhabitants had been suddenly sucked away, alerted by some communal sixth sense and doors would suddenly close. Or the sirens would ring out across the narrow, winding lanes, warning of impending attack and we would jump into our Lada Niva four-wheel drives and tear off as fast as possible while the inhabitants fled to their basements. Important buildings such as the headquarters of the Bosnian army and the HVO were boarded up, to protect against shrapnel and blasts. So was the multi-coloured mosque, with its minaret unusually positioned on its eastern side and clutch of hairs said to come from the beard of Muhammad. The mosque, in the centre of town, was always crowded with soldiers returning to the front, or just back from the fighting.

In Travnik at least, the mosques were still standing. In Serb-controlled towns such as the northern Bosnian city of Banja Luka, Bosnian Serbs not only forced Muslims from their homes, killing and raping as they went, but also comprehensively destroyed Muslims' history. Every one of Banja Luka's centuries-old mosques, part of a UNESCO heritage site, has now been destroyed. Even the Muslim cemetery has been bulldozed and the rubble of the mosques dispersed to prevent them ever being rebuilt. Just as the Nazis attempted to eradicate Jewish history in the Holocaust, by destroying cemeteries, synagogues and schools, the Bosnian Serb army was dedicated to wiping out Bosnia's Muslim heritage and history. One of the more shameful footnotes to the Bosnian war has been the encouragement that the Serbian Orthodox Church has given to its congregants as they set about destroying every vestige of Islam in Bosnia. The God of the Serbs, some believe, has no time for notions of brotherhood or universal spirituality. As one Orthodox priest reportedly said of the destruction of the Banja Luka mosques: 'They were in the wrong place.'

At the Blue Water hotel in Travnik, now a Bosnian army headquarters, Emir Tica was bitter at the West's refusal to arm Bosnia. Sophisticated and well-travelled, this U2 fan now had a job organising transport for the Bosnian army. Each time we met he looked increasingly fatigued, but he still carried an increasingly dog-eared travel guide to the city, hidden away in his combat fatigues, to show new foreign journalists that once Travnik had been a popular tourist destination.

A clear stream gurgled down the side of the hotel in Travnik, and thick wooden boards had been placed at its sides to protect against shrapnel and flying glass. The dining room was crowded with soldiers, men and women in fatigues, as he spoke. 'If the Islamic countries break the arms embargo it is the shame of Europe. Europe made the embargo and cut off our chance to survive but the Islamic countries try to help. But if somebody wants to help me and I have to hide that, it's wrong. Somebody wants to save my life, so why should I have to hide it from the people who officially want to help.'

Like every Bosnian Muslim, Emir emphasised his European heritage. 'I want to be part of Europe. When I come to London I am more European than people from Paris or Berlin. Whether or not Europe wants me, I belong to it. I listen to U2 and Madonna. I have friends in Budapest and London. For me Europe does not mean living in Europe, it means the European dream. European governments don't listen to the people who want to help us. But where is Europe when somebody wants to shoot me? When we worked in Germany nobody was scared of Islamisation. Now when they try to kill us in Bosnia, they are. I only started to feel like a Muslim when they put a map on the table and separated Bosnia into Croat, Muslim and Serb.'

In towns such as Travnik, nearby Zenica, Vitez and Sarajevo, right across the sliver of government-controlled Bosnia, a new Bosnian Muslim identity was emerging, sired by both the West's inaction and the arms supplied by the Islamic world. None of the Bosnian Serbs seemed to realise, or more likely care, that the harder and more destructive their onslaught against their Muslim neighbours, the more they would turn to Islam to see the answers for the destruction and death that was spreading all around them in this peculiarly vicious and murderous conflict.

Every mosque reduced to rubble made Serb propaganda of a Bosnia allied to the Muslim world instead of the West more likely to be a self-fulfilling prophecy. Part of this was a strain of 'Ottoman-nostalgia', a longing for the days when it was Istanbul, not Belgrade, that dictated events in Bosnia. Then Muslims were the élite and it was Christians and Jews that were *rayahs* or subject peoples. Another was a slow shedding of western culture in favour of a Muslim one, based on the precepts of Islam. Young

Muslim women were exchanging jeans and skirts for the veil and baggy trousers, both sexes were greeting each other with the Arabic 'Marhaba' instead of 'hi' and starting to spend their free time in the mosque instead of bars and clubs.

Other Bosnian Muslims, especially victims of ethnic cleansing, began to compare themselves to the Palestinians, another dispossessed Muslim nation. 'We are similar. We will fight until we get our land back. Even if they give the Muslims a tiny bit of land we will have a liberation organisation and it's one hundred per cent certain I will join it. I will fight until the last,' said Samir Bidić. Bidić, 25, escaped from the Serb detention camp of Keraterm and made his way to Zenica. His home was a space on the floor of a fetid sports hall, shared with dozens of other malnourished, coughing refugees. 'As long as there are Muslims I will fight. Sometimes I get so tired and sick and I think I should leave everything. Then I get angry with Europe and I want to take a knife and kill Serbs.'

Even in the yellow-painted school building in Travnik, now home to hundreds of refugees huddling in dank, crowded rooms, the niceties of Muslim village life were carefully preserved. The soldiers, just back from the nearby front, left their shoes outside the room while they brewed coffee in a small brass pot over an open flame. Outside it was so cold that the refugees' washing had frozen solid into a jumble of ice-covered sweaters and trousers. Despair hung in the air, together with the fog of cigarette smoke. But in similar shabby and grimy refugee centres across Bosnia, and across Europe, Islam was taking root among a nation dispossessed, its homeland laid waste. The very methodology of the Bosnian war was helping this radicalisation. Fighting, at least at the outbreak of war, was organised to a different set of rules to conflicts such as the Second World War. Bosnian Serbs avoided military engagements against other soldiers as much as possible. Instead they shelled their target for a few days, sent in the paramilitaries to slaughter a few families and then the buses arrived to ship out the Muslims en masse, such as the Naderević family.

Amla Naderević told me her story one freezing winter morning in the Travnik schoolhouse that was now her home. She was seventeen, with short dark hair and remnants of puppy fat. But her voice was older than her years and she had a faraway look in her eyes as she spoke in the chill of the schoolroom. Hers was a common story, simply told.

'My family comes from Brezičani, a village near Prijedor, it was burnt down. We were brought here by bus in October 1992, but we didn't have to buy a ticket because our house was burnt down. We were given a lift by the Serb Red Cross to Mt Vlašić where we had to cross the front-lines and no-man's land. We were accompanied by Serb gunfire and they tried to take our money and jewellery. My father was in a camp at Prijedor for four and a half months. These were the worst days of my life. He was taken in June and held until October.

'I can say I hate Serbs but we are not capable of doing such things. It's hard to say what the real cause is, it's somewhere in the dark side of the brain. They represent a country which is capable of doing this for its own prosperity, to move people around and kill them. I can't explain why, perhaps it's because they want their own country to be bigger, stronger, greater. I had Serb friends but they were forbidden to contact us and forced to stay away.

'I work from 8.00 am to 2.00 pm as a civilian protection officer here. I have a job trying to help people who are suffering, dealing with problems delivering food, medicine and so on. The conditions here are very miserable, it is very cold, several degrees below freezing and the food supplies are very bad. Still I feel optimistic; after all the suffering we went through we are stronger and capable of enduring anything. All the people here, whatever faith they follow, have right on their side and they will win. My home has been burnt down but I will build another one. It's hard to think about the future. My best wish is for a normal life. Schools, parties, meeting people and to be at home. I won't stay here but someday I will return to Travnik, but next time as a tourist, just for fun.'

Bosnian Muslims accepted help from the Islamic world because they had been forgotten by Europe, said Effendi Nusret Abdibegović, the Imam of Travnik. 'To be a Muslim [now] is absolutely different. If you asked me before the war I would say it's nice to be European Muslim. But now I have to say that we don't feel like we used to feel in Europe. We are part of Europe and part of that is our religion. We still feel European because we felt that Europe will protect us. In Europe democracy is a symbol of a free life.'

Shrapnel scars pitted the walls of Imam's Abdibegović's house, a minute's walk from the multi-coloured mosque. As we spoke he served coffee and biscuits, the sweets still wrapped in their Arabic wrappers. 'But Europe forgets that we are also part of it. We have accepted help from our Muslim brothers because it's their duty to help us. We accept all kinds of help. If Milošević and Karadžić say they will defend Europe from Islam we say we will defend Europe from the Serb religion. Comparing us before the war and now, I wouldn't say Islam has grown stronger. In some places it's ruined, such as Bijeljina. But Muslims are awakening to their identity. Still the Muslims countries could do more. They must not allow what happened to Palestine happen to Bosnia. They should use their oil money.'

Foreign Muslims, particularly Iranians, were using their oil money to provide guns as well as biscuits. The many Islamic relief organisations based in Zagreb were running streams of relief convoys into central Bosnia packed with food and medical aid for the refugees. The Muslim hard men congregated in Zenica, home to the all-Muslim 7th Brigade. Members of the 7th Brigade, part of the Bosnian Army Third Corps, were either practising Bosnian or foreign Muslims. Unlike most Bosnian soldiers they were often hostile to the press. In fact it was best to avoid them completely, as BBC reporter Alan Little and his crew discovered in Zenica one day. Filming in the city the crew were arrested by soldiers from the 7th brigade and held for hours. The Muslim soldiers were so suspicious they even copied down the serial number on a Jubilee line ticket. The BBC crew had a lucky escape. A few months later a British aid driver was killed by Muslim soldiers in Zenica in a military-style execution. President Izetbegović triggered controversy across government-controlled Bosnia when he attended a commemoration of the brigade's founding in January 1995. Together with fellow presidency member Eyup Ganić he defended his attendance in a statement: 'The unit in question saved people from genocide. Their religion aided them in that and was a source of their strength.'

Life in Zenica became increasingly unpleasant for the remaining Serbs and Croats. They were harassed by Muslim fighters and many fled across the lines to their Serb brethren. There were reports of ethnic cleansing by government troops, arrests and imprisonment of non Muslims. Muslim fighters were also blamed for a massacre in a Bosnian village, when over thirty corpses were discovered. However, when it came to the slaughter of the innocents, government troops were always in the third division, trailing far behind the Bosnian Serbs and Croats.

Sometimes, hidden under the boxes of flour and cooking oil that trundled into Travnik and Zenica, lay freshly greased Kalashnikovs and boxes of ammunition. Much of the government forces' arms and ammunitions had been paid for by Islamic countries, working under the cover of relief organisations. One Iranian plane packed with weapons had been turned back at Zagreb airport, reportedly after the Croatian government had been tipped off by US intelligence.

Two years into the war a second attempt was more successful and any tip-offs from the CIA were ignored by the Croats. An Iranian air force transport plane, loaded with sixty tons of explosives and material for weapons production, landed at Zagreb in May 1994, its cargo destined for the Bosnian army. The Iranians maintained an unusually large embassy in Zagreb which was a central point for arms supplies to the Bosnian government forces. Tons of arms still sat at Maribor airport, in Slovenia, after a botched attempt to fly them to the Bosnian army via Hungary and the Arab world. But despite the UN arms embargo, a trickle of mostly small arms arrived in central Bosnia, depending on the current state of relations between Zagreb and Sarajevo. Plenty of western aid organisations and the United Nations were also bringing in aid convoys, but each Islamic relief convoy emblazoned with a green crescent logo, crammed with goods covered in Arabic writing, reinforced the growing impression that only Islamic countries cared about Bosnia.

But Christians too came to help, and increasingly, to fight, said Hamed Mesanović, Travnik's deputy commander. 'We have French, British, Italian, Hungarian and Slovenian volunteers. They are all fighting against evil. Their military significance is not great but they boost our forces. Their presence is morally significant. Some of the foreign Islamic fighters might be motivated by the possibility of fighting a holy war but I only wish there were more of them. They are welcome and their presence shames Europe.'

At one stage there seemed to be so many British volunteers (they said they weren't mercenaries because the pay was virtually non-existent) in the Bosnian army, that it became an almost daily occurrence to see a soldier wearing the blue and white Bosnian fleur-de-lys speaking in a broad Glasgow or Manchester accent.

There was John from Blackpool, a computer programmer who had previously fought with the Croatian army. He would often drop by the BBC office in Kiseljak, provoking fury from Kate Adie when he brought his weapon indoors with him. John was sick, suffering from a kidney infection, not helped by a diet for weeks on end of rice and beans. Like many British soldiers in the Bosnian army he had nothing but praise for the welcome he received from the people for whom he was fighting. 'In some ways they are more cultured than we are. They have manners that we have forgotten. They have a kindness and regard for each other. When a man sees you in the street he will walk up, shake your hand and say how are you? Even if he sees you every day. It is just a politeness. If you go into their homes they will give you everything they have. They are starving in some areas but they will give you the food from their mouths. It's very embarrassing sometimes to go into someone's house and be given a meal and then you realise it's their rations for a week or more,' he explained.

There were other, gorier stories as well, of throat-slitting and slaughter. 'I've killed Chetniks [Serbs] personally and in actions, I've been in countless actions. I've killed one as close as you are to me. He was a sentry and my patrol was attacking a village where the Chetniks had a front-line and they had two sentries patrolling in the woods. I killed one with a knife, my friend killed another with a knife. They had to be dealt with, quietly.' Another afternoon at the village of Pazaric just outside Sarajevo we met a middle-aged man with a bag emblazoned with a Red Cross. He claimed he was a field medic from Derbyshire. A couple of minutes later five young Britons appeared, all former soldiers in the British army, their faces covered in camouflage cream, full of bravado as they headed to the front nearby. They all claimed that the television pictures of Serbian detention camps had triggered their decision to fight for Bosnia. But there was always the suspicion that the better opportunities for killing people in the Lašva valley than in Lancashire also helped influence their decision. At least two British volunteers would later die a horrible death.

Ted Skinner and Derek Arnold, from Liverpool and Glasgow respectively, were a common sight at the Blue Water hotel in Travnik, eating their lunch of rice and bean soup with their comrades in arms. They were well known to the Bosnian Muslim fighters, having trained many of them. Skinner, in his late thirties, had just spent Christmas staring down his Zastava rifle's night-sight near Travnik before killing a Serb sniper, he said. When he shot his enemy he felt nothing, he claimed, but the killing sometimes haunted him in the early hours of the morning. Both men worked as instructors teaching raw recruits one end of an AK47 from the other. They described themselves as 'soldiers of misfortune' and said they supplied their own uniforms and field dressings, although their guns were issued by the Bosnian army.

Like most foreign fighters they were suspicious of outsiders at first, and reluctant to talk. But after they had seen us a couple of times in the army headquarters, they slowly opened up. Skinner was an intelligent and articulate man who drew a parallel between the Bosnian and Spanish civil wars. 'That was a practice run for Germany and a testing ground for the Condor legion. It would be terrible if the same happened here. If they get away with it in Bosnia who's to say they won't in Kosovo?' But it was clear that the two men, both British army veterans, had a cause in the Bosnian army that they could never get in the civilian world. Arnold also taught field medicine to trainee medics. 'They had four days to learn. A lot of them are kids, sixteen, seventeen or eighteen and many are girls. Some of them are instinctively good fighters who would be welcome in anyone's army.'

Like many in Travnik that winter they sensed the coming battle between the HVO and the Bosnian army. They were scathing about the shaven-headed Croat fighters who swaggered around the old town's bars and cafés, shiny hand-grenades clamped to their belts. Skinner and Arnold's bodies were found a couple of months later, tied to trees near the front-line at Turbe, just outside Travnik. They had been tortured and then shot. Some blamed the HVO, others the Mujahideen. The two were reportedly abducted from their flat in Travnik. They said they had ten years of military experience but it all proved useless. In the end, they were in too deep.

So, it appeared, were we, when we awoke one morning at 6.00 am to an early morning call of repeated rifle fire right outside our bedroom window in Vitez. We had arrived the night before at the Muslim-run guest house above a petrol station a couple of miles from the British UN base. The Bosnian soldiers downstairs assured us that everything was calm, and that the only recent trouble was an attempted armed robbery, which in Bosnian terms made the hotel reasonably safe. I noticed that one fighter had an AK47 clip in his trouser pocket, but it was late, we were tired, we had just been turned back from a checkpoint because of fighting nearby and we had to sleep somewhere. In the dark we didn't notice that a network of slit trenches had been dug around the forecourt. We had checked into a battlefield and we would be lucky to check out again alive.

It was the eve of the latest phase of the Bosnian war: the fight between government forces and their supposed allies against the Bosnian Serbs, the Bosnian Croats. The HVO had issued an ultimatum to government forces to submit to its authority in areas with a heavy Bosnian Croat presence by 15 April 1993. Under the leadership of the ultra-nationalist Mate Boban, the HVO attempted to carve out its own Croat mini-republic of Herceg-Bosna, with its capital in Mostar, which he and President Tudjman planned to absorb into Croatia proper, even while the HVO was nominally allied to Sarajevo. Bosnian Croats claimed that Herceg-Bosna, a quasi-fascist statelet modelled on the Bosnian Serb Republic, was merely a temporary measure to ensure their protection while the war lasted. They were lying. One afternoon in Mostar Muslim leaders showed us a school registration form for the city, overprinted with the Bosnian Croat chequerboard. At the bottom, in tiny letters, was the printing imprint date, January 1992. The war in Bosnia started in April 1992.

Like the Bosnian Serbs, with whom they would soon be fighting against government forces, the Croats had long prepared for their annexation attempt. In Herceg-Bosna just as in Serb-controlled Bosnia, Muslims were expelled en masse, murdered or incarcerated in concentration camps. Sporadic fighting between government and HVO troops had erupted since the previous autumn, and now an all-out war was about to explode. The Bosnian army refused to sign and the ultimatum expired at midnight. Six hours later the shooting started. Outside our front door. When the crump of mortar fire was added to the rattle of machine guns we decided to get out as fast as possible. The mortars were close enough for us to hear their baseplates rattle as the shells soared upwards. But it was hard to leave as there was a battle going on downstairs just a few feet away. As we stood in the corridor deciding what to do a bullet smashed through a window, flying right by us. The staccato cracking speeded up and the motel's windows and bathroom mirrors shattered all around us as the bullets smashed into the rooms.

We instantly dropped to the ground, noses virtually scraping the carpet as we crawled along the floor and gathered our luggage. Donning our flak jackets and helmets we sat on the stairs and waited while the battle unfolded a few feet beneath. There was nowhere to run, nowhere to hide. Shouts and screams filled the air as I listened, trying to dam the rush of nightmare scenarios pushing their way into my head and considering the possibility that we might not get out alive. Not only were we caught in a real live battle, even worse, we were trapped above a garage and its great tanks of petrol while bullets flew all around us. Peter Maass, a friend and colleague from the *Washington Post*, held the keys to our Lada Niva in his hand and pointed to the pocket in his flak jacket where he was placing the keys. The implication was clear: if something happens to me, this is your ticket out. Even my other friend and travelling companion, Cathy Jenkins of the BBC, a real veteran of the Yugoslav wars, was starting to look worried, for this was way beyond any adventure. It would make a great story around the dinner table of course, if we made it out alive.

The fighting lasted about a quarter of an hour, although at the time it seemed longer. Then a Bosnian Croat fighter ran up the stairs towards us, his face covered with a ski-mask and his AK47 cocked and ready for action. Here at last was my chance to shout the phrase I had carefully memorised in Serbo-Croat for just this sort of unhappy occasion: 'Foreign journalists! Don't shoot!' But all I could remember was 'novinar' (journalist). I shouted it anyway. Several times. It must have done the trick for the fighter, who thankfully was at least wearing proper fatigues with brigade badges and wasn't a freelance bandit, stopped. He demanded to see our press cards, which we all scrabbled to produce. 'Who are you?' we asked carefully. 'Don't ask. You don't want to know,' he replied.

He and his colleagues then put on a fine display of destruction for us, jumping in and out of rooms, waving their guns around and leaping through glass doors in their search for more Muslim soldiers, even though we said the other rooms were all empty, and emptying their magazines all over the place. Here at last was Rambovic in action, literally under our noses. It was a truly terrifying but also bizarre experience, for in a way I had moved beyond the Sarajevo-style fear of being shot and shelled when caught in a war-zone. Part of my brain just seemed to switch off and I watched the soldiers in a sort of interested, but dispassionate daze, like seeing a 3-D film. It is a curious feeling to contemplate the possibility of your own death in a war. Then the HVO soldiers beckoned us downstairs.

A stench of cordite filled the air and a hand-grenade lay on the floor. The young Muslim soldiers from the previous night stood against the wall, their hands in the air, their faces fearful. A Croat soldier ushered us out, gun in the air, ready to give us covering fire if there was a counterattack from government troops. Another waved us off with a Nazi style-salute. Thankfully our Lada was a hardy beast and Peter drove off down the road, weaving to avoid any sniper fire, until we reached the British base, where we were served a full English breakfast by the doughty and unflappable British UN troops. I couldn't eat anything. In fact I think I went into some sort of shock, for it was not until lunchtime that I was able to hold a reasonably coherent conversation. I tried to read a book, *A History of the World in 10½ Chapters* by Julian Barnes, to take my mind off the shooting and shelling all around us, but the words just danced across the page.

Incredibly, even in the midst of a battle unfolding around them, the daily necessities continued. Housewives hung their washing out and old men stood on their doorsteps wondering what the day held in store. This bizarre mix of the banal and the deadly that was Bosnia, was perhaps best summed up by a

British officer as a 'Bosnian mind-fuck'. Certainly my repeated trips to the war-zone began to fuck up my mind as my mental recovery period became longer and longer each time I left. I became irritable, short-tempered and twitchy. Unexpected noises made me jump. They still do, in fact. When I began to dream about Bosnia, even at home in Budapest or London, I knew it was time to think about getting out. I still remember one dream vividly. I was out on the front-line with a group of Bosnian army soldiers who were setting up a mortar position. But the mortar was pointing straight upwards, instead of being angled towards the enemy. 'If you fire that it will come straight back down and land on our heads,' I exclaimed to the fighters. I never found out what happened in that dream, for the telephone rang then, waking me.

We spent the rest of the day in a house where the British troops had set up a press base. Their presence was comforting but our immediate surroundings only made me more nervous. A Bosnian Croat, the local postman, sat at the bottom of the garden, every now and again lazily firing his high calibre mortar. The doors shook and the windows rattled every time the postman knocked and we all moved underground into the cellar. From the windows we watched the battle unfold as soldiers dodged between houses, some of them on fire. In the evening a British army doctor showed us how to apply an intravenous drip. I paid close attention thinking it might be a useful lesson but couldn't remember much. It was the longest day of my life.

I did not return to Bosnia after that. The next morning I hitched a lift back to Split in a UN vehicle with a British army soldier. We drove through the mountains on back roads, where fear and tension hung in the air like an almost tangible, brooding presence. Compared to Vitez, Split was a paradise, a gorgeous seaside town with atmospheric Roman ruins, where we dined every night on grilled seafood and watched leggy blondes zip around town on their boyfriends' scooters. It was hard to believe the war-zone was just a couple of hours' drive away, but the holiday cabins near the beach that were once filled with revellers were crowded with Bosnian women refugees, smoke from their cigarettes drifting across the lines of washing strung between the huts.

*

Fighting between the HVO and government forces lasted almost a year. In spring 1994 Bosnia, Croatia and Herceg-Bosna signed a peace accord, brokered mainly by the USA and in part by Iran, that called for a joint Muslim–Croat federation in Bosnia, leaving an opening for the Bosnian Serbs to join. Under pressure from Washington and the threat of sanctions against Croatia the two sides made peace and Mate Boban, leader of Herceg-Bosna, was replaced. The war in central Bosnia between the HVO and government forces stopped almost overnight. The territories that Bosnian Croats and the Bosnian army currently control will be joined in a Muslim–Bosnian Croat federation, probably linked to Croatia, while the majority-Muslim part of Bosnia juggles its western and Islamic heritages.

That spreading Islamisation called for a delicate balancing act by Bosnian commanders, and other officials. One morning we drove from Kiseljak to Visoko, a stronghold of government forces also facing the front-line, shelled daily by the Bosnian Serbs. We took the only safe way in, across a winding mountain track that cut behind the front-lines. Ice crackled under the tyres as we climbed through the mountains outside Sarajevo past the Bosnian army checkpoint. The scenery was breathtaking: clumps of trees hugged the steep hills as the winter sun reflected off the snow. The boom of artillery echoing again around the valley reminded us this was a war-zone.

Visoko greeted us with the usual panorama of Bosnian destruction: buildings with roofs like collapsed soufflés, houses with massive holes punched through their walls, and streets empty apart from the frantic barking of stray dogs. A small picture of the Ayatollah Khomeini was taped to a cabin outside army headquarters and nobody was available to talk to us. Visoko's hospital had gone underground because of the constant bombardment and boxes of Iranian Red Crescent aid were piled up at the entrance. Two wounded Algerian soldiers lay on their beds in the basement. 'I came here for the idea of Islam,' said one, but as soon as he started talking the doctors quickly stopped the interview, fearful that if we reported his presence it would trigger a fresh Serb bombardment of the hospital.

Ultimately it was this deep sense of betrayal that Europe had deserted Bosnia, more than the Muslim aid convoys or the Mujahideen, that was helping create a vacuum steadily being filled by Islam. Nor was this confined to Bosnia proper. The ripples of Islamisation also reached the hundreds of thousands of Muslims forced from their homes, listlessly passing their days away in the network of refugee camps over Europe.

In places such as the Nagyatad centre in southern Hungary, just a few miles from the Croatian border, and home to dozens of mainly Muslim Bosnian refugees, young women such as Mirsada spend their days studying the Quran and praying. One room there has been converted to a mosque, its white walls covered with Arabic inscriptions and carpets on the floor. In Yugoslavia, Mirsada, an eighteen-year-old from Visegrad, could not easily live as a religious Muslim, she said, but the war has provided a curious form of liberation. 'In Visegrad I could not talk about my religion or practise it this much, but now I have learnt a lot more about it. Because of my situation, like many young Bosnians, I am much more interested in Islam. The war has made me religious.'

Or as Dr Mustafa Cerić, leader of Bosnia's Muslim community, said one afternoon in his office at Zagreb's mosque, its entrance crowded with Bosnian refugees: 'They are doing this to us because we are Muslims. We are all Muslim enough to be killed. Muslims are coming together because their suffering is the same and no one wants to take them. But Muslim blood is not worth the same as Muslim oil.' He was eloquent in his anger at the West's inaction as his people were slaughtered. 'People feel they are not wanted in Europe as Muslims. But we are not Arabs or Turks. We didn't come here. People said "never again" after the concentration camps, but it happened again, in their neighbourhood, not in Africa, not in Asia, but here.

'Now Bosnian Muslims are witnessing a national transformation. The Muslims that were communists before know what power is. They are frustrated and they feel betrayed by Europe and they cannot ask Moscow [for help]. The people who have a faith have an explanation why this is happening. Now more people are turning to Islam. They feel betrayed by the West and by Europe. If Europe wants humanism and religious tolerance we are here and it should say we will not allow these people to be harmed. What is happening to us is a sin for Europe. No one can tell me that Europe can have a clear conscience over the Bosnian Muslims.

'What is happening to Bosnian Muslims is not war, it is not happening by chance. It was planned for us. No matter what we tried, the Serbs rejected every possible solution. This is why we are disillusioned with European values. We believed too much in Europe. I don't know a single Bosnian Muslim who wanted an Islamic state run on the Islamic religion. There are rules in Islamic laws, giving rules for life, but according to Islamic law nothing from Islam can be enforced on non-Muslims. Even if a Muslim wanted to make a Christian obey Islamic law he cannot do that. We have come to the bottom of the bottom. Now life for us is something else. What is better? To live in falsehood or die in truth? This war can continue to the day of judgment if necessary. We have no choice, no place to go and nothing else to lose.

'We are not a big united people who can create an Islamic state to threaten Europe. But the western world lost its chance to help us and get applause. By this inaction and by just observing it might lead us in a very dangerous direction. Some will look for revenge and they will have nothing to lose. We still control our people but I don't know how long anyone will be able to control them. We are asking please help us while we are still sane.'

Part of the tragedy of Bosnia's Muslims was they also believed too much in the Islamic world as they looked eastward for help. Considering the wealth of the oil-rich Arab nations, the response of many Muslim governments has been unremarkable, with more sound and fury than hard cash, apart from the Iranians and, to a lesser extent, the Saudis. Certainly the regular convoys of relief aid and ammunition helped many Bosnian Muslims stay alive and fight back against the Serbs. But the lack of political action, particularly by the Arab world, which has oil production as a powerful weapon, has left a gap readily filled by Tehran. 'I am a Palestinian from Gaza and I know what it means to leave your country,' said Kamal Mosleh, a relief worker based in Zagreb. 'The reaction of the Islamic world is very weak. They could have given more and done more. All the money has come from the people except from Saudi Arabia.'

Some Islamic countries have also offered UN peace-keepers, including Iran which said it was ready to send 10,000 men, but the UN rejected the plan. Most of the Islamic relief effort has been financed by private donations and Muslim countries have taken in just a tiny fraction of the refugees that poured out of Bosnia, although this is also due to Bosnian Muslims' reluctance to leave Europe and live in the Arab world. Money allocated to relief work is often squandered, on publications like *Sarajevo*, a thick glossy full-colour magazine printed in Arabic.

Muslim states see Bosnia as a recruiting ground for their own brand of Islam and spend as much time manoeuvring for advantage as actually helping refugees, said one senior Islamic aid worker with extensive experience of working in Bosnia, who must remain nameless. 'Muslim non-governmental organisations are run by amateurs, instead of doctors or people with experience. They are sitting on millions of dollars and they don't know what to do with them, so they produce a colour magazine. You can see the level of political commitment in the number of embassies in Sarajevo [in 1993]. Only Iranians and Turks are there, and not a single Arab country. Their involvement is to pour in relief supplies, and that's because the people in their own countries are clamouring for them to do something.'

Instead of uniting, Islamic countries were using relief aid as a weapon in the struggle to establish a bridgehead in Europe between competing regimes such as the Sunni Saudi Arabians and the Shiite Iranians, the aid worker said. 'It's not even effective. The Arab countries should have tremendous clout. In 1973 in the Yom Kippur War they cut off oil supplies. They could hold the world hostage, but instead aid has become a political weapon in Muslim relief organisations because there are so many different groups such as the Saudis and the Iranians. The Iranians will not work in an area where there are no [rival] Saudis.

'Don't think the Iranians are in Sarajevo to express solidarity with their fellow Muslims. That would be naive in the extreme. Once they are in the field they have the power to use aid as a weapon: "We will give you aid, but you must follow our line." As Bosnian Muslims become more politicised the different Muslim states will move in an organised way to get a foot in Europe. The

Saudis demand that women refugees wear the veil. This is the sinister aspect of the Muslim relief effort. Make them Muslims and then we will give them aid,' he said.

Iran has skilfully exploited the Bosnian war to try and build a bridgehead in Europe. By openly breaking the arms embargo and supplying arms, fuel and weapons-manufacturing equipment it has left the West with a dilemma. Now, after the Dayton accords, the arms embargo is lifted – a move also supported by many American politicians – and Iran is seen to have set the international agenda and gains kudos, in both Bosnia and the wider Islamic world. When the arms embargo stayed, Iran gained influence in Bosnia as the country's patron, ready to supply the embattled government with what the West refuses. Both Washington and Tehran regard each other with loathing, but on the issue of Bosnia they became de facto allies. Towards the end of the war, there were repeated stories of US-organised arms drops to Tuzla, always denied, of course, by US officials.

Either way Tehran has laid firm foundations for future influence in Sarajevo and also in Zagreb. After Iranian foreign minister Ali Akbar Velayati visited Sarajevo in May 1994 to open a new embassy there, he discussed setting up a joint economic commission with Bosnia and Croatia. Meanwhile Saudi Arabia promised to send two Jumbo jets to fly Bosnian Muslim pilgrims from Sarajevo to Mecca, a move that looked feeble in comparison. Once again the Ayatollahs of Tehran had outmanoeuvred the White House and the United Nations to set the Bosnian agenda, while Washington and even the Arab world looked on helplessly.

The very words western politicians and many reporters have used during the Bosnian war have also accelerated this process of Islamisation and realignment away from the West. The mainly Muslim Bosnian government army, with its significant minority of Serb and Croat soldiers and officers, especially in and around Sarajevo, is inaccurately referred to as the 'Muslim soldiers'. The Muslim-led Bosnian government, with its Serb and Croat officials and ministers becomes 'the Muslims'. Land held by the Bosnian government, an internationally recognised sovereign state and member of the United Nations, becomes 'Muslim-controlled territory', and so on. No matter how many times Bosnian leaders such as President Izetbegović emphasised their commitment to a multi-ethnic society of Bosnian Muslims, Serbs, Croats and Jews – and the adherents of these faiths who stayed faithful to Sarajevo are testament to his sincerity – the language of diplomacy ghettoised their government.

The Bosnian war was described as a 'civil war', fought by 'warring factions', a description especially favoured by UN spokesmen. At times in Geneva during the many futile peace conferences UN officials could barely contain their impatience with the Bosnian government for refusing to surrender at least half their country to the Bosnian Serbs. Of course if they were only 'Muslims', their claim to the land was much weaker.

As the West sows, so shall it reap. As one soldier put it, referring to claims that Bosnian Muslims wanted an Islamic regime: 'Europe will bring about here the very thing it fears.' For young Muslims like Samir, a twenty-year-old soldier waiting outside the multi-coloured mosque, the West was doing nothing for Bosnia while Islamic countries were ready to help. 'Islamic countries will give us guns because there is no one else to help. We were waiting for Europe, but now there is no other way because we were betrayed by Europe.'

Both Izetbegović and Silajdzić have made visits to Libya and Iran, seeking financial and military aid for Bosnia. These links, which Bosnians defend as unavoidable given the West's failure to stop the Serbs and the continuing UN arms embargo, were being watched with concern in western capitals.

'Europe is accusing us of an Islamic fundamentalism that is a danger to Europe, but this notion of Islamic fundamentalism is very hard to defend yourself against,' said Dr Mustafa Cerić. 'It is a political category of conflict between Iran and the US. For the Serbs someone is an Islamic fundamentalist if he has an Islamic name. There are staunch communists who imprisoned Muslims and are now charged with being Islamic fundamentalists.'

In fact for Muslim activists like Melika Saliahbegbosnawi, jailed for two years under the Communists for her beliefs, Izetbegović's government was not Islamic enough. The former dissident was angry because a Bosnian soldier had reprimanded her in the streets of Sarajevo for dressing as an orthodox Muslim woman in headscarf and long sleeves.

'At the beginning of the war I was stopped on the street by a Bosnian soldier. He said, "I'm going to arrest you and Allah." He even had a Muslim name. I said you can arrest me but you cannot arrest Allah. The new government is doing the same as the communists against religious people. The Bosnian government is afraid to be seen as Islamic fundamentalist because then they won't get help from the West. The communist authorities did the same thing.'

A former party member and graduate of Zagreb university who also studied in Paris, Saliahbegbosnawi started wearing the chador at the time of the Iranian revolution in 1979. 'That's when I started to wear Islamic clothes, but that revolution had nothing in common with my revolution. I wear the chador because it keeps human morality and a law always gives a rule for society. Women are beautiful and if they adorn themselves they will attract someone from the street and they will try to get her. If a woman is uncovered she attracts men through her body as an object. If a woman covers herself she will force men to see her other possibilities and potential as a woman. Love from this kind of respect is more lasting.'

Like many Sarajevans, she had a narrow escape from death when a missile destroyed her house. 'I escaped three minutes before it came in. I just thought something was going to happen. I lost everything, my books, my manuscripts and my furniture. Now I write and try to live a normal life. I never run on the streets because I don't want them to destroy my dignity. I'm not an animal who is scared. I believe that if there is a missile to kill me it is my destiny.'

She was scathing about President Izetbegović's poor planning for war. 'Izetbegović sparked the fire of nationalism among Muslims as that was the one chance for him to get a party and then get power. He used to say that Bosnia was three nations and now he exudes nationalism. He plays the role of Gandhi to establish and defend the country through diplomacy. He cannot do it through power but you need power to get the respect of your enemies. We knew that war started in Slovenia and Croatia. He didn't want to arm the Bosnian people and prepare but he is prepared for bleeding.

'I love and respect Ayatollah Khomeini but humanity did not get a chance to learn the truth about his deeds. But during his lifetime people from the Majlis [Iranian parliament] changed the objects and aims of the revolution. Muslim ladies were treated in communist propaganda as ignorant and repressed but I chose this. I know what is going on. I am very peaceful but unhappy. I would be happy if I could call people to light – Islam. But their ears are full of false promises from the government. Izetbegović has nothing to do with Islam. Everything in the flag comes from pre-Islamic times. But that is wrong because these people are Muslim in their way of life. Islam is composed of a culture and a life and that is nothing to be ashamed of. I would like to see a civil society and I would like to establish Islamic law and for God's law to rule the world, but it's too early, you have to call people back to God. I don't want the morality that prevails in the West. That's immorality if you want to plunder people or lie to them.'

Just as in Travnik and Zenica, the old idea of a multi-national Bosnia was also fading in Sarajevo as the Islamic wing of the SDA party began to gain strength in the capital. The shadow of a one-party state fell again on Bosnia, although this time it was Muslim and authoritarian, rather than atheist and communist. Even loyal Bosnian Serbs and Croats began to feel marginalised. The city too changed, its fabled cosmopolitanism fading as more and more sophisticated urbanites left, their places taken by refugees from the countryside who hated Serbs and Croats for killing their families and destroying their houses and were not impressed with talk of Bosnia's multi-ethnic heritage. Concern spread among the political classes at the growing influence of the SDA and in February 1995 five opposition members of Bosnia's collective presidency, Serb, Croat and Muslim, published a statement in the Sarajevo daily newspaper *Oslobođenje* warning that the future of a secular, multi-cultural Bosnia was under threat. 'Each weakening of multi-culturalism diminishes the prospect for liberating the country and establishing a new democratic society,' they said.

But if the city could live through over three years of war, it could cope with SDA Islamic activists. The new general manager of Sarajevo Radio and Television banned short skirts and leggings for female workers, reported Tom Gjelten in his book *Sarajevo Daily*, but Gjelten still watched an uncut version of

Basic Instinct, complete with Sharon Stone crossing her legs while free of underwear on television one night. When Džemaludin Latić, editor of the pro-SDA weekly *Ljiljan* wrote a series of columns condemning marriage between Muslims and other faiths as a 'catastrophe' he triggered an immediate broadside from Slavko Santić, a columnist on the city's *Oslobođenje* newspaper – whose ethnically mixed staff had never missed an edition throughout the war – attacking his intolerance.

This move away from a multi-ethnic Bosnia was a direct result of the logic of the international peace plans that sought to divide the country along ethnic and religious lines. Diplomats too opened the path for a Muslim national state. Under the Dayton accords, Bosnia is to be divided into two ethnic entities: one Serb and one Muslim–Croat federation, the former with 49 per cent of territory, the latter 51 per cent. The pretence of a unitary state is to be maintained, as Sarajevo keeps control of trade, finance and international relations, but the entities control their own armed forces and police. Dayton is a diplomatic fig-leaf that justifies Bosnian Serb ethnocide and the Serb entity will not have to wait long before fusing with Serbia proper, with the Bosnian Croats perhaps following suit.

Not surprisingly, just as Bosnian Serb and Croat nationalists sought their own states, some Muslims began to follow their example. In the winter of 1993, for the first time, Bosnian leaders formed a self-appointed Muslim assembly in Sarajevo, that excluded Bosnian Serb and Croat leaders loyal to the government. As they pounded the tables and waxed lyrical about Ottoman rule, they simultaneously called for a Muslim national state in Bosnia, and blamed the West for forcing them into an ethnic ghetto in the first place. By then President Izetbegović's SDA had already voted in favour of ethnic partition of Bosnia. But even then, with the multi-ethnic ideal in its death throes, while most of the delegates called for a Muslim Bosnia, they demanded one based on the idea of nationality, rather than religion.

This future Bosnia could be a Muslim national state, but largely secular, much as Israel is a Jewish state but a secular and democratic one where religious identity has fused with nationality to spawn a new nation. Most of the delegates at the Sarajevo Muslim assembly drank alcohol and few read the

Quran or attended services at the mosque, but their religion both formed and framed their embryonic nationality. This latest stage in the (not necessarily religious) Muslimisation of Bosnia was merely a continuation of events almost twenty years ago, when Tito's Communists gave Bosnian Muslims the status of a nation. Now, in the fires of war, that nation was defining itself. But whatever the final size and type of the so far still-born Bosnia – independent Muslim majority nation-state, Muslim-Croat federation or new mini-Yugoslavia – it is clear that once the war is over, the new Bosnia will in many ways be more Muslim than its predecessor, the Yugoslav Federal Republic of Bosnia, ever was. Economically if not politically, Bosnia will be a Muslim Bantustan in the heart of Europe, propped up in part by diplomatic and financial aid from the Muslim world.

It could also be a political springboard into Europe for Iran, and some Arab states as well, all of whose aid will long be remembered with gratitude, thus adding another perpetual destabilising factor to the Balkans, an already deeply unstable region. Its angry and resentful population scattered inside its borders, or exiled in their hundreds of thousands across Europe and the world, could provide ready recruits for any future Bosnian Liberation Front, ready to use force for Sarajevo's cause, their rage fuelled by militant Islam. As Croatian professor of philosophy Žarko Puhovsky, a former advisor to President Izetbegović, told me: 'There is a simple formula for the future of Bosnia. The smaller the territory, the more the regime will be drawn to Islamic fundamentalism.'

Bosnia will also be part of an arc of Muslim-majority states stretching from central Asia, through Turkey, to Albania, another former part of the Ottoman empire where Islam is being reborn, on the shores of the Adriatic. Even as the Bosnian war broke out, Turkish officials saw it as an opportunity to expand Ankara's political fiefdom. The UN security council deployed Turkish peace-keeping troops in Bosnia, to the anger of its traditional pro-Serb enemy, Greece. The decision reversed a long-standing policy that countries with a historic involvement in a war-zone should not provide peace-keeping troops. In January 1996 Turkish general Ersim Yaltsin signed an agreement with Bosnian army commander Rasim Delić to provide military training. For the Turks it will be a homecoming of sorts to a land now wrecked and shattered, where billions of dollars will be spent on reconstructing a country that need never have been destroyed in the first place. Turkey will be aided by the United States in beefing up the Bosnian army, as part of Washington's geo-political strategy to build an arc of allies stretching from Bosnia, through Albania, Turkey, Jordan and Israel.

And where Washington goes in Bosnia, Tehran soon follows. Few countries are currently receiving so much simultaneous attention from the USA and Iran, arch-enemies in so many other ways. In July 1996 a high-level Iranian delegation, led by First Vice-President Hassan Habibi, visited Sarajevo where its members pledged $50 million to help reconstruct the country, through financing companies and repairing infrastructure. It was a shrewd move by the Iranians to help Bosnian businesses, for that will give Tehran a foot-hold in the Bosnia economy as it steadily grows. The agreement also singled out a factory in Zenica, home of the remaining foreign Mujahideen and the all-Muslim 7th Corps of the Bosnian army, for Iranian economic assistance.

Now even *The Economist* carries advertisements for banking officials to work in Bosnia to help reconstruct the economy. Every country's war has left physical destruction and wreckage in its wake. What was unique about the Bosnian war was that it was the world's inaugural prime-time ethnocide, diligently reported, televised and watched for over three years before the West called a halt to the killing. For the first time in history, a country was laid waste, its people massacred and Bosnia's Slavic-Turkic culture was destroyed, much of it live on television. Muslims the world over believe, probably rightly, that had the Bosnian Serb guns been pointed at Christians the killings would have been stopped long ago. The Croatian war of independence, after all, didn't even last a year. The Slovenian war lasted ten days.

Not only mosques, houses and people, but also libraries have been destroyed, as Bosnian Serbs turned their guns on Bosnia's very history. In just one day, 17 May 1992, Sarajevo's Institute for Oriental Studies was destroyed. It held thousands of books, manuscripts and Ottoman documents, much of Bosnia's historical records, until they were incinerated in a hail of deliberately targeted rockets.

Every house, every field, every yard of Bosnian territory that was lost to the Serbs steered Bosnia's Muslims away from the West and their hopes of European intervention and towards the East and the Muslim world. In the fields of destruction, Islam was laying its roots, nurtured by Iran and the Arab world. 'Religion is the only stronghold we have, and the support we draw from it,' said Mustafa Bećirević, chief Imam for Croatia and Slovenia. 'The war showed Bosnian Muslims that they do exist as a people and the Serbs cannot tell them what is worthwhile in their culture. Even as victims they discovered this. Muslims have now become conscious of their force; politically, militarily, religiously and culturally.'

Even now, for me, the very word Bosnia still triggers a chain of images and memories. Of being stuck outside Novi Travnik with a column of refugees and the sudden flinch of shock and fear on a young woman's face as she crouched, suddenly shrinking into herself, when a mortar exploded fifty yards away from us, kicking up a plume of earth. Of a ridge near Gornji Vakuf and a turbanned Imam standing silently among the clutch of Muslim soldiers, like a scene from a Second World War partisan film, as they slowly advanced down the hill towards us. Of the Koševo hospital in Sarajevo where Kemal Karić, a four-month-old Muslim boy lay on his back in a hospital cot, the stump of his amputated leg waving uselessly in the air as he grinned and gurgled. Of a young Bosnian Serb soldier in hospital in Belgrade and the sheet lying flat on the hospital bed where his legs should have been. Of an old lady sitting screaming in the street in Sarajevo. And the black ski mask on the face of the Croat fighter in the hotel outside Vitez as he advanced up the stairs towards us, his Kalashnikov in his hand.

The West is now weary of images like these, as Bosnia slides down the newslists at newspapers and television stations. But they are images which the Muslim world will long remember.

Hungarians in the Rudas Turkish Baths, Budapest

Chapter 3

EUROPE'S FORGOTTEN
ISLAMIC HERITAGE

'In the mosques there is no more prayer
In the fountains no more ablution
The populous places have become desolate
The Austrian has taken our beautiful Buda'
      Seventeenth-century Ottoman song lamenting
            the Austrian capture of Budapest

'I am the garden, I awake adorned in beauty: Gaze on me
well, know what I am like . . . What a delight for the
eyes! The patient man who looks here realises his
spiritual desires.'
      Verse by the fourteenth-century poet Ibn Zamrak,
            inscribed on the walls of the Alhambra,
                        Granada, Spain

The mosques and *hamams* (baths) of Bosnia have vanished
under Serb shells, or carefully placed explosive charges. But in
Budapest scattered jewels of Ottoman architecture remain.
Pashas and Imams, Sufi dervishes and Ottoman merchants, perhaps even Sultan Suleyman the Magnificent himself once
soaked in the sulphurous, slightly radioactive waters of the
Rudas Turkish baths' thermal springs, on the banks of the
Danube in Budapest. I spend every Saturday morning there,
one of a less illustrious group of bathers than our Ottoman
predecessors; foreign correspondents planning the weekend's
parties, Hungarian businessmen with clanking gold chains and
bracelets, their arms and backs covered with exotic tattoos, and
weary factory workers, soaking away a week on the shop floor.

But we all share in our enjoyment of one of Europe's finest remaining hamams. Budapest is dotted with thermal springs and four hamams still remain from the time when Hungary was ruled by the dictates of the Sublime Porte in Istanbul. Lazing in the hot water, gazing at the domed roof while sunlight cuts through the steam rising from the main pool, it's simple to reinvent yourself as an Ottoman tradesman who has just dismounted from a days-long Caravanserai trek from Istanbul up through Sofia and Belgrade. Built by Szokoli Mustafa, the sixteenth-century Pasha of Buda, who is still commemorated on a plaque by the tap dispensing sulphurous mineral water, the Rudas is an extraordinarily atmospheric place, that retains its original cupola, vaulted corridor and main octagonal pool.

Bathing in Budapest has been a tradition since the arrival of the Romans, but it was under Ottoman rule that the culture of the baths flowered. Islam demands that its followers follow a strict set of rules before praying five times a day, washing and abluting in the prescribed manner.

Every mosque must have a supply of clean running water for worshippers to cleanse themselves. Water, or more specifically, running water has always had a specific motif in Islam, partly perhaps because of the faith's origins in the deserts of Arabia, but also for its life-giving and sensual qualities. Paradise for believers, as described in the Quran, will be a garden of bliss, filled with crystal fountains and springs, fruits and sweetmeats, as verses 15 to 19 of Sura 56 describe:

> *(They will be) on couches*
> *Encrusted (with gold and precious stones)*
> *Reclining on them*
> *Facing each other*
> *Round them will (serve)*
> *Youths of perpetual freshness*
> *With goblets, (shining) beakers*
> *And cups (filled) out of*
> *Clear-flowing fountains:*
> *No after-ache will they*
> *receive therefrom, nor will they*
> *Suffer intoxication*

The previous Sura, 55, describes in detail the two gardens where the Blessed will be able to stroll in the afterlife, dark green in colour from plentiful watering, according to the Quran. Verses 66 to 67 outline the pleasures that await there:

> *In them (each) will be*
> *Two springs pouring forth water*
> *In continuous abundance*
> *Then which of the favours*
> *Of your Lord will ye deny*

There will be plenty of beautiful young virgins as well, among the fountains and ornate couches, as verses 70 to 74 describe:

> *In them will be*
> *Fair (maidens), good, beautiful*
> *Then which of the favours*
> *   of your Lord will ye deny?*
> *Maidens restrained (as to*
> *Their glances) in (goodly) pavilions*
> *Then which of the favours*
> *   of your Lord will ye deny?*
> *Whom no man or Jinn*
> *Before them has touched*
> *Then which of the favours*
> *Of your Lord will ye deny?*

None, obviously. The Islamic paradise is a far more sensual and exotic one than the Christian version, with its promises of fulfilling all of believers' sensual appetites. A hint of its spirit lives on at the surviving *hamams* but the Ottoman tanneries, mosques and *medresas* (schools) that once filled Budapest's streets have vanished, and only a few remnants of that era remain across central Europe.

Further east, Romania was once a bridge between western and Arab culture. Its provinces of Wallachia and Moldavia were by the eighteenth century important centres for the study of Arabic. The first Arabic typography appeared in Wallachia, before being transferred to Aleppo, in Syria. Romania even once boasted a centre for Sufis, Islamic mystics, on an island in the

Danube, long since submerged. In the Serbian capital Belgrade, once the former frontier city between the lands of the Sultan and those of the Hapsburg Emperor, hundreds of mosques and hamams have long since been demolished. Now only one mosque still stands, where nervous Muslims pray in a city home to many of those Serbs who strove to eradicate Bosnian Muslim culture.

Overlooking the Danube, at the top of Budapest's last surviving Ottoman street, a steep incline named for Gül Baba, is his tomb. The bustle of the capital and one of the main bridges across the Danube are just a few hundred yards away, but the tomb is one of the most peaceful places in the city. Perched on a ridge over-looking the river, the site of the tomb, the most northern place of pilgrimage for Muslims, is suffused with the sense of tran-quillity that is the hallmark of Islamic places of worship.

Gül Baba was a holy man and a high-ranking member of the Bektashi Sufi order of Islamic mystics. His name means 'father of the roses' and according to local folklore, he was the man who introduced the flower to Budapest, thus giving the name Rózsadomb, Rose Hill, to the area around Gül Baba street. Several beds of roses surround the small domed structure. In fact he died during the thanksgiving service for the capture of Budapest on 2 September 1541, just after the King Matthew church had been converted into a mosque. Inside the mau-soleum are verses inscribed by the Turkish traveller Evliya Çelebi when he visited the tomb in 1663, as well as antiquities and furnishings donated by Hungarian Muslims. Sadly, the adjacent hotel that once catered to Muslim pilgrims is no more, but Çelebi's writings on the Magyars remain. Çelebi was quite a fan, understandably preferring their company to those of the Austrians he had met:

> They are clean in their ways and in their eating and honour their guests. They do not torture their prisoners as the Austrians do. They practise sword play like the Ottomans. In short, though both of them are unbelievers without faith, the Hungarians are more honourable and cleaner infidels. They do not wash their faces with urine every morning as the Austrians do, but wash their face every morning with water as the Ottomans do.[1]

In fact the appearance of the first Muslims in Hungary pre-ceded Gül Baba and the Ottomans by over five hundred years. But apart from mentions in specialist academic journals, Hungary's Muslim history remains largely unknown even to most Magyars, a representative and telling example of Christian Europe's rewriting of history in favour of the cross rather than the Islamic crescent. According to the Hungarian Chronicle known as the *Anonymi gesta Hungarorum*, a group of Muslims arrived, probably from Bulgaria, in the tenth century. During the tenth, eleventh and twelfth centuries Hungary's Muslims were granted several freedoms and given civil and military duties, serving the border guards. Other Muslim communities were settled by the country's then frontier, marked by a wooden stockade between the Danube and Lake Balaton. When the Tatars invaded in 1241, they laid waste to their settlements, and the rest of the country. A description of early medieval Hungary (whose territories then stretched much further than its present borders) survives in the writing of Abu Hamid, a Spanish-Arab traveller and writer from Granada who arrived in the region of the Slavs and the Magyars in the middle of the twelfth century. In his book *Tuhfat al-Albab* he describes what he found in Bashgurd, as the area was known in Arabic:

> It has seventy-eight cities, each one of which is like Isfahan and Baghdad wherein is to be found (God's) bounty, favour and abundant blessing and a luxury and easiness of living that can neither be accounted for nor quantified. My eldest son, Hamid, there married two ladies from among the Muslim noblemen and he (by them) begat male offspring.[2]

Hamid's example has since been followed by many expatriates even now entranced with Hungary's luxury and easiness of living. But for all the centuries that the country was the cross-ing point between Islam and Christianity, and the cultural fertilisation that sprang from its mixtures of populations – who then included Muslims from Spain and Bulgaria, Arabs and native converts – its Islamic heritage has been in large part not only physically destroyed, but also deliberately ignored. Just as the Communists rewrote the past, according to the dictates of ideology, many Christian historians of Europe have ignored, or at best marginalised the role of the Ottoman empire and Islam in developing the continent.

This is a Europe-wide trend. The inventions and discoveries of Muslim mathematicians and scientists, doctors and astronomers have been deliberately sidelined, their contributions often ignored in favour of a historiography that claims a seamless progression from the Dark Ages to the Renaissance, under the benign patronage of Italian and French nobility. This view is nonsense. In mathematics alone for example, two of the most important numeric processes and sciences, without which the present-day computer would never have existed, algebra and algorithms, were invented by Arab mathematicians. Both are Arabic loan words, that still retain the prefix Al (the), linguistic remnants of a rich and intricate culture that flourished while most Celts and Anglo-Saxons were still lurking in damp and freezing wattle and daub huts. Suggest to most Magyars, like most Europeans, that Islam has helped developed their continent and civilisation and they will shudder with horror, having been indoctrinated with the view that the arrival of the Turks was an unmitigated disaster. It was not, for the Ottomans brought a civic infrastructure, a network of schools, kitchens and public bath-houses in their wake, as well as religious freedom for Jews and Christians. Certainly Budapest's Jews knew what lay ahead when in 1683 the Hapsburg guns pounded the hills of Buda, where their community had flourished under Ottoman rule. They fought hard on the side of the Sultan's armies, for Muslim rule was always preferable to that of the Catholic church which often brought persecution and pogroms.

In one building, in the southern Hungarian town of Pécs, the cross has even physically displaced the crescent. This pretty town of sloping streets filled with tile-roofed houses was once an important Ottoman centre. Now its squares are filled with American soldiers in I-FOR, the NATO force in Bosnia, whose border is just a couple of hours' drive away. At the top of Széchenyi square stands the tall imposing building of the Gazi Kasim Pasha mosque. Unfortunately while the building retains its Islamic interior, including Arabic calligraphy, it has been turned into a Catholic church, and the resulting mutant hybrid is an unsettling and insulting attempt to appropriate a building designed to worship Allah for the cause of Jesus and Mary. Better preserved, and unaltered, is the Jakovali Hassan Mosque, with its surviving friezes, miner (pulpit) and white walls. That said, the Ottomans converted plenty of churches into mosques.

The idea of the rapacious Turk is encapsulated in books such as *Egri Csillagok* (*Stars Over Eger*), the story of a group of young Hungarians who make a trek to Istanbul to rescue their captured friend. The book is riddled with ethnic stereotypes of the cruel, lascivious Turk, all portrayed through the writer Géza Gárdonyi's narrow prism of nineteenth-century catholicism. *Stars Over Eger* is one of the most famous books in the country, required reading for every school-child, and has shaped several generations' perception of Muslims, Turks and Islam.

August 29, 1526, when King Lajos II and fewer than 10,000 Hungarian knights were routed by 80,000 Ottoman troops at Mohács, is commemorated by many Hungarians as one of the worst disasters to befall the country. It marked the start of Turkish rule and the era of the northernmost expansion of the Ottoman empire which at one time reached the gates of Vienna. King Lajos, who drowned in a stream when he was trapped by his armour, was mourned even by Suleyman the Magnificent who led the Ottoman forces. 'May God be merciful to him and punish those who misled his inexperience. I came indeed in arms against him, but it was not my wish that he should be thus cut off while he had scarcely tasted the sweets of life and royalty.'

Most Hungarians now view the start of the Ottoman occupation as the beginning of a dark age, of desolation and oppression when Christian Hungary fell under a cruel Muslim yoke. Undoubtedly cruel practices such as *devshirme* (boy tribute) ripped apart families across the Ottoman empire. But the arrival of the Ottomans also gave young foreigners across the empire undreamed-of opportunities in the capital Constantinople. It brought new ideas, new scientific discoveries and new ways of thinking to Europe. In fact the lives of many of the Ottomans' new subjects, whether Hungarian, Serb, Romanian or Bulgarian, flourished under the Sultan's rule. Hungarians such as Ibrahim Muteferrika, a Christian convert to Islam, wrote one of the first books to be published from the first Turkish press, which he had founded, a lengthy work outlining the necessary changes in Ottoman military strategy.

History records though that others fared less well, such as Murad, a teenager who was captured by the Turks at the battle of Mohács and converted to Islam. He became a successful interpreter in the Ottoman foreign service, composing a missionary treatise in Turkish and then in Latin and even translated Cicero for Suleyman the Magnificent. Unfortunately Murad remained more Magyar than Muslim and was sacked for persistent wine drinking. Turkey also provided a refuge for Hungarian refugees after the failed eighteenth-century revolution of Ferenc Rákóczi, when the independence fighters fled the rule of Vienna for sanctuary in Turkey. His memory is still commemorated in a museum at the eastern town of Tekirdag. A second wave of Hungarian refugees fled to Turkey after the failed revolution of 1848, and helped modernise Turkey's civil service.

Contemporary chroniclers, such as John Shirly, author of *The History of the Wars of Hungary (Or, An account of the muiseries of that Kingdom)*, published in London in 1685, record how, when it came to cruelty and religious persecution, it was the Catholic church, rather than the Ottoman Muslims that set the pace, especially when dealing with Protestants:

> Anno Domini 1605. Notwithstanding the former Defeat of Botscay, other intestine Troubles arose in the Kingdom, occasioned by the Bishops in their late assembly, passing a Decree, cruel and bloody, *viz*. That all of the Reformed Religion should be Burnt or Banished: Against which, notwithstanding the Nobility, as well of one Religion or another, Protested. Yet the Reformed Churches were Seized, publick and private Worship forbidden; also the reading of the Holy Bible . . . Such a desolation was made by the discontented Persons in arms, that the Turks stood amazed to see themselves out-done in cruelty.

The Hungarian Catholic bishops' call for the Protestants to be burnt or banished would find an echo centuries later in the Nazi puppet state, set up by Croatian fascists in alliance with the Catholic church in 1941. Its policy for Orthodox Serbs was:

'Kill one third, expel one third and baptise one third.' Perhaps not surprisingly, passing through Hungary in 1790, the Ottoman envoy Azmi Effendi records that Hungarians showed extreme good-will towards both him and the Ottoman empire.

Here then is the clash of cultures that is so often written about by the media pundits on Islam: not the supposed inability of Islam and democracy to co-exist, but the refusal or more likely inability of most Europeans to realise the contribution that Islam and Muslims have made to Europe, in every way from architecture to religious freedom, of which Hungary is an excellent example. Christianity, whether Catholic, Protestant or Orthodox, has claimed all of central Europe and eastern Europe and its history for its own, reducing the Ottoman empire and Islamic culture to a series of cultural stereotypes of ravagers of maidens and despoilers of land, when, if anything, the reverse is true, for the Catholic crusaders left a far wider and deeper river of blood in their wake than the Ottoman armies.

Over three hundred years after the fall of Budapest to the Hapsburgs, a small Turkish community has once again settled in Budapest, although none can trace their lineage in Hungary back to the Ottoman empire. Turan Aktas's Török Büfe (Turkish Buffet) at the city's Józsefváros (Joseph town) market is the meeting point for the city's nascent Turkish community. There Turkish businessmen drink tea from small gold-rimmed glasses, to be slowly sipped with the first cigarette of the day. Breakfast is lentil soup and lamb kebab, eaten to the accompaniment of Turkish pop music. The smell of roasting lamb fills the morning air while Turkish businessmen chat and joke over their food.

Most of the 1,500-strong community are businessmen, importing textiles, running restaurants or travel agencies. 'It is the best of friendship between Hungary and Turkey, in business, culture we are similar to each other,' says Sedat Harputluoglu, a 32-year-old businessman from Istanbul. But even now the Magyar–Turkish connection is an ambivalent friendship, even one-sided, say many Turks, who say their compatriots' warm feelings towards Hungary are not always reciprocated. 'In Turkey everyone sees Hungary not as a friend, but as a brother.

But Hungarians have a different view because of the 160 years of occupation,' said one Turkish diplomat. That some Hungarians take an unenthusiastic view of their distant cousins annoys some Turks now living here. 'We didn't spoil the country. We have been to a lot of places and we never touched religion or the land. We were here for 160 years but show me one Hungarian who speaks Turkish,' said one Turkish businessman.

As the Ottoman empire advanced up through the Balkans and into central Europe, however, Islam's western-most land, cradle of much of the Muslim world's finest achievements in science, culture and architecture, was rapidly collapsing under a Catholic onslaught. The last Islamic outpost in Spain, Granada, fell in 1492 to the armies of Ferdinand and Isabella, who allowed safe passage out of the city for those Muslims who could not live under the rule of the cross. The legend goes that at the final ridge overlooking Granada, Boabdil, the city's last Muslim king, turned back one last time to see his lost lands. 'You should weep like a woman for the land you could not defend like a man,' his mother taunted him as they stood on the outcrop now known as *El Ultimo Suspiro del Moro*, the Moor's Last Sigh.

The victory of the Spanish Catholics was a disaster for Iberian Jews and Muslims but led to an undreamed-of prize for the Ottomans. On 31 March 1492 Ferdinand and Isabella issued the Edict of Expulsion, that all Jews must leave Spain by the last day of July. Don Isaac Abravanel, leader of the community, pleaded in vain for the order to be rescinded. 'Thrice on my knees I besought the King. Regard us, O king, use not thy subjects so cruelly. But . . . the King hardened his heart against the entreaties of his supplicants,' he wrote.

In Istanbul, Sultan Bayezit II was amazed at the stupidity of the Spanish throne. 'Do they call this Ferdinand a wise prince who impoverishes his kingdom and thereby enriches mine?' he supposedly asked when he learnt that his adversaries were expelling en masse many of Spain's most skilled professionals, from doctors and scientists to traders and teachers. If Spain's Jews were not welcome there any longer, they could make a home in the Ottoman empire. The Sublime Porte was flung open to welcome the exodus, leading to a flowering of Jewish life in places as far apart as Salonika and Sarajevo, Cairo and Cyprus. Putting down new roots in the Ottoman empire, Jews spoke of the Sultan's tolerant realm in wonder. It was: 'A broad expansive sea which our Lord has opened with the rod of his mercy. Here the gates of liberty are wide open for you that you may fully practise your Judaism,' wrote one Jewish immigrant to his friends. Even now, among the remnants of the small Jewish community in the Bosnian capital, there are Jewish families with Spanish surnames, who trace their roots to the expulsion in 1492 and who still know some Ladino, the Judeo-Hispanic argot of Iberia.

For decades after 1492 Jews, dissident Christians suffering under Catholic rule, Muslims and Moriscos – Muslim converts to Christianity – poured into the Ottoman empire bringing a whole universe of knowledge and expertise to revitalise the Sultans' realm. When Granada fell Ferdinand and Isabella had promised to honour the religious rights of the remaining Muslims, but that promise was soon broken. Ten years after the expulsion of the Jews the Muslims were given the same choice: convert or go. Most went. A hundred years later the Moriscos were thrown out as well. Certainly there were some restrictions on Jews and Christians under Ottoman rule, but compared to life under the tutelage of the Vatican, the Sultan's lands were a paradise, especially by sixteenth- and seventeenth-century European standards. Jews and Christians were *dhimmis*, that is, people of the book, whose prophets are recognised as part of the series of God's messengers, of which, Muslims believe, Muhammad was the last. They were restricted in minor matters such as the weapons they could carry and the animals they could ride. *Dhimmis* also had to pay higher taxes. But as long as they recognised the ultimate supremacy of Islam, they were free to live and run their communities. It was a supremely sensible as well as a humane policy. Jewish refugees from Spain brought sciences and knowledge virtually unknown to the Ottomans until then. Their contribution was especially important in medicine, printing, and military science, and the Turkish armed forces began to incorporate western military techniques in their warfare manuals. By expelling his Jews King Ferdinand had invigorated the scientific, intellectual and military life of the Ottoman empire.

Among the massive exodus of skilled professionals were, of course, a good number of Jewish physicians, for in fifteenth-century Spain, as now, Jewish mothers liked to boast of 'my son, the doctor'. In his book *The Muslim Discovery of Europe*, the distinguished historian of Islam Professor Bernard Lewis details how the influx of Jewish doctors revitalised Ottoman medicine:

> By the sixteenth century, Jewish physicians, most of them of Spanish, Portuguese or Italian origin, were common in the Ottoman empire. Not only the sultans, but many of their subjects had recourse to these physicians, whom they recognised as representing a higher level of medical knowledge.[3]

Christian visitors were often dismayed at the influence that Jewish doctors had on the Ottoman sultans, who became so numerous that they soon formed their own corps of court physicians, separate from their Muslim colleagues. One of the first ever works on dentistry was written by Moses Hamon, a Jew of Spanish origin who became chief physician to Suleyman the Magnificent.

Islam's western bastion was founded in 711 when the Spanish city of Toledo fell to the Moorish – North African Muslim – armies. Under the leader of their commander Tariq – whose name is commemorated in the word Gibraltar, Arabic for Jebel Tariq, or mountain of Tariq – the Moors defeated the Visigoths at the battle of Janda and their adversaries' kingdom collapsed. City after glorious city soon fell to the advancing Moorish forces: Cordoba, Granada, Seville, Valencia. Cordoba became the jewel of Europe, where learning, art and culture flowered under Muslim guidance. 'Islam was overextended in Spain; it thus made its accommodation with its habitat, ruled with a light touch,' writes Fouad Ajami:

> At its zenith in the tenth and eleventh centuries it was to fashion a society of tranquillity and brilliance. Its cities thrived. Cordoba's population approximated a quarter of a million people; it was unmatched by any European city of the time. Its only rivals were the cities of Baghdad and Constantinople.[4]

The finest monument from Spain's Islamic era is the Alhambra in Granada. Built by the Nasrids in the fourteenth century, it was not so much a fortress as a mini-metropolis. Inside its walls were gardens, palaces, fountains, a mosque, houses and shops. For Victor Hugo it was a 'palace that the genies have gilded like a dream and filled with harmony'. Even Granada's blind beggars boasted of its glories, lamenting a cruel fate that deprived them of the chance to see its marvels, as a means of eliciting sympathy and bigger donations. There was nothing like the Alhambra in the Christian world, no palace that could compare to its light and tranquil series of patios, courtyards and gardens, where the sound of water gently rippling through fountains was a permanent backdrop. Even the walls are bedecked in Arabic poetry, prayers and praise for God. Thankfully, the Alhambra never fell in battle and survives still, as a monument to a civilisation now almost vanished.

But the scientific achievements of Islamic Spain and the eastern Muslim empire live on. Medicine, mathematics, astronomy and navigation, all these arts flowered from Toledo to Baghdad, as Muslim scholars followed the prophet Muhammad's injunction to study and seek knowledge. The *hadith* (sayings) of the prophet include:

> To seek knowledge is obligatory on every Muslim.
> Seek knowledge even unto China.
> The ink of a scholar is holier than the blood of a martyr.
> Those who go out in search of knowledge are fighting in the way of Allah until they return.

The great age of Islamic scientific discovery began, according to legend, with a dream. The ninth-century Caliph al-Ma'mun, who ruled the lands of Islam from his palace in Baghdad, dreamt he met a man with a fair and ruddy complexion, a broad forehead, joined eyebrows, a bald head and a generally pleasant appearance. It was the Greek philosopher, Aristotle. A short dialogue followed.

'What is good?' asked the Caliph.

'That which is good in the mind,' replied the master of philosophy.

'And what comes next?'

'That which is good in the law.'

'Then what?'

'That which is considered good for the people.'[5]

Aristotle also advised the Caliph to guard the doctrine of *tawhid*, or oneness of God. It was all useful advice. Under al-Ma'mun's reign reason and a spirit of scientific enquiry triumphed, and the work of the world's first great international multi-lingual and multi-disciplinary scientific academy, Baghdad's House of Science, reached its zenith. Al-Ma'mun invited linguists, doctors, astronomers and scientists of every discipline from all over the world to Baghdad and provided them with lodgings, financial support and facilities to allow them to carry on their research. The results were astonishing. Works were translated from Greek, Syriac, Hebrew, Sanskrit and Persian into Arabic, triggering an explosion of ideas as the heritage of the great cultures of the Mediterranean, the Arab, Persian and Indian worlds flowed across the academy's busy corridors. For the first time, an international body of knowledge was being shaped into a single corpus, in one language. As this vast corpus of science germinated and cross-fertilised in ninth-century Baghdad Arab scientists leapt ahead of their European counterparts in fields as diverse as optics and astronomy, medicine and, especially, mathematics.

Arab mathematicians had of course never seen a laptop computer such as the Apple PowerBook on which this book was written. But the works of al-Khwarizmi and al-Kashi, in which they developed our current system of numbers, drawing on Indian mathematical systems based on units of ten and zero, laid the intellectual foundations for the mathematical calculations that would, centuries later, result in enormously powerful computer chips and the whole revolution in information technology that flows from that. It is ironic that some Muslim regimes now seek to control the flow of that information in and out of their countries and try and censor the Internet. Caliph al-Ma'mun would not have approved. But he might have welcomed the state-of-the-art technology used by the Islamic Saudi dissident Dr Muhammad al-Mass'ari as he runs his anti-government campaign from his base in north-west London. There banks of fax machines and computers send his message all over the world, the latest manifestations of intellectual advancement in mathematics that began in Baghdad a millennium ago.

Even now our system of numbers is referred to as Arabic numerals, in contrast to the ungainly Roman system, which has survived only on Roman monuments, watches and the credits at the end of film and television programmes. Like every scientist then and now, Arab mathematicians drew on what had been discovered before, and drew on Indian and Chinese numerical ideas. But while it's unclear who exactly invented the zero – the Arabic word for which, *sifr*, has become the Latin cipher – it was Muhammad ibn Musa al-Khwarizmi who was responsible for its systematic introduction into mathematics. The word Algebra comes from the title of al-Khwarizmi's great work *Al-Jabr wa Al-Muqabala,* written during the reign of Caliph al-Ma'mun. It drew on Indian, Euclidean and Babylonian methods but was unique in its systematic cataloguing and rendering of mathematical problems and the accompanying proofs and solutions. Al-Khwarizmi also developed the idea of logarithms, a corruption, in fact, of his name. Knowledge soon begat knowledge and from this small stream of mathematical discoveries a great waterfall of scientific advancement quickly burst forth.

As well as being a mathematician al-Khwarizmi was also a geographer and a philosopher. In 830 he translated Ptolemy's epic work on geography. Under Caliph al-Ma'mun's tutelage he drew up a map of the world, adding the Muslim lands, that divided the world into seven zones in accordance with climatic changes. His work was used for centuries by western geographers. Together with his colleagues he developed an astronomical counting table by which the earth's distance and width could be surveyed, with the logical conclusion that the planet was not flat, but round. Al-Khwarizmi's work was later translated into Latin by Adelard of Bath, and helped shape western European astronomy.

The Key to Arithmetic, by Jamshid ibn Masud al-Kashi, written in Samarkand in 1427, is one of the finest works of Arab mathematics. A manual for the use of everyone who needed to understand numbers, from astronomers to clerks and surveyors, it also included a comprehensive examination of the workings of decimal fractions. In fact these had already appeared in the works of al-Uqlidisi, a mathematician based in tenth-century Damascus. 'Al-Kashi's novel treatment of the subject thus anticipated similar developments in Europe by about two hundred years,' writes A.I. Sabra in *The Scientific Enterprise*, part of *The World of Islam* compendium.

But if one man embodies the spirit of Islamic scientific enquiry and discovery, it is Hakim Abu Ali Abdullah Husayn Ibn Sina, known in the West as Avicenna. Centuries before the Renaissance that began to drag Europe out of the dark ages, Avicenna, born in 980 in Bokhara, central Asia, was the ultimate Renaissance man, preceding Leonardo da Vinci by almost half a millennium. He is perhaps better described by his Arabic honorific: the Sheikh of the peak of knowledge.

Avicenna was already a Hafiz – one who has memorised the entire Quran – by the age of ten. Six years later he turned to medicine, in which field his achievements would be best remembered and which still influence western doctors today. At sixteen he cured the Samanid Sultan, Nuh ibn Mansur, of a serious ailment and was then appointed his court physician. Other doctors were frequent visitors at his rooms, discussing his new discoveries and formulae. Like Caliph al-Ma'mun he was strongly influenced by Aristotelian ideas and served as a bridge between Greek and Islamic thought. His output was prodigious: he wrote over 200 works covering everything from medicine, physics, chemistry, mathematics, music, economics and religious and moral philosophy. In his book *Kitab al-Insaf* (Book of Impartial Judgment) written when he was twenty-one, he poses and answers 28,000 questions on the nature of Divinity. But it was his two great medical works, *Kitab ash-Shifa* (the Book of Healing) and *Al-Qanun fi'l at-Tibb* (the Canon of Medicine) that remain his greatest legacy to medical science, the latter, according to the *Encyclopedia Britannica*, 'the most famous book in the history of medicine, in both East and West'. His achievement was to codify a work that would serve as the basis of medical science in European universities for centuries.

Over a million words long, the five volumes of the *Canon of Medicine* utilised every medical knowledge from every system of health and healing in the then known world, including Tibetan, Greek, Chinese, Arab, Persian and Chinese. It is the single most influential work of medicine in medical history. Avicenna also codified the scientific principles that led to the development of modern botany and pharmacy and invented the process of distillation, extracting oil from roses. His ideas of the importance of a balanced metabolism still influence practitioners of homeopathy today.

The twelfth-century Syrian writer Usama ibn Munqidh has left us a vivid description of the different approaches of European and Arab doctors, quoted in Professor Lewis' *The Muslim Discovery of Europe*. The words are those of Thabit, a Syrian Christian doctor, who was called to treat some patients by a Crusader baron. He was away for only ten days, for his diagnoses were overruled by a European doctor, with gruesome results for the patients:

They brought me two patients, a knight with an abscess on his leg and a woman afflicted with a mental disorder. I made the knight a poultice, the abscess burst and he felt better. I put the woman on a diet and kept her humour moist. Then a Frankish physician came to them and said to them: 'This man knows nothing about how to treat them!' Then he said to the knight: 'Which do you prefer, to live with one leg or to die with two?' And the knight said: 'To live with one.'

Then the physician said: 'Bring me a strong knight and a sharp axe,' and they brought them. Meanwhile I stood by. Then he put the sick man's leg on a wooden block and said to the knight: 'Strike his leg with the axe and cut it off with one blow!' Then, while I watched, he struck one blow, but the leg was not severed; then he struck a second blow, and the marrow of the leg spurted out and the man died at once.

The unlucky knight at least died quickly. Next in line for the European's brutal attentions was the woman with a mental disorder, as Thabit describes:

The physician then turned to the woman and said: 'This woman has a devil in her head who has fallen in love with her. Shave her hair off.' So they shaved her head, and she began once again to eat their usual diet with garlic and mustard and such like. Her disorder got worse and he said:

'The devil has entered her head.' Then he took a razor, incised a cross on her head and pulled the skin in the middle until the bone of the skull appeared; this he rubbed with salt, and the woman died forthwith.

Then I said to them: 'Have you any further need of me?' and they said no and so I came home, having learnt things about their medical practice which I did not know before.[6]

Jewish culture too flourished under Muslim rule, especially in Islamic Spain. Far from the academies of Baghdad, Spanish Jews developed their own western European intellectual ideas, that drew on Judaic, Muslim and Christian thought. Jewish academies opened in Granada, Toledo, Barcelona and Cordoba. It was a golden age, still referred to by that title by Jews, an era when thought flowered and Jewish life enjoyed a renaissance of freedom and civil rights that was unrivalled in history.

Even now few Jews know that the greatest intellectual of medieval Judaism, Moses Maimonides, put many of his original thoughts to paper not in Hebrew or Spanish, but in Arabic, the language of the Quran. Known as Rambam, Moses ben Maimon was born in Cordoba in March 1135. Like his Muslim philosophical predecessor Avicenna, he was a doctor as well as a religious thinker, for it was symptomatic of intellectuals under Islamic rule that they specialised in several disciplines, their knowledge advancing through the symbiosis of subjects. Like Caliph al-Ma'mun, founder of Baghdad's House of Science, Maimonides believed that reason should be a guide for life, without sacrificing the Bible's tenets. His *Guide to the Perplexed*, written in Arabic, draws on rationalist ideas and led to major disputes within the medieval Spanish Jewish community.

But as he pored over his text the shadow of the cross fell over Muslim Spain: the *Reconquista*, and the Catholic advance was proceeding. Toledo, Tudela and Saragossa had all been lost to the Muslims. Cracks appeared in the edifice of Muslim tolerance. In 1159 Maimonides' family fled to Fez in Morocco from Cordoba after suffering persecution from Islamic extremists. In Fez too they could find no peace and they finally settled in Fostat, near Cairo. There he practised medicine and eventually became court physician to Sultan Saladin, precursor of the

Spanish Jews who, centuries later, would bring their medical skills to the court of Sultan Suleyman the Magnificent.

Ghosts and customs still remain scattered across southern Spain, remnants of a unique Muslim–Judaic–Hispanic civilisation, destroyed by Catholic royal edict and burnt on the pyres of the Christian Inquisition, a civilisation whose harvest of science and culture flowers still to enhance our lives. Catholic families who for centuries lit candles on a Friday night, the start of the Jewish Sabbath; Spanish lyrics sung to Arabic rhythms and the jewels of Islamic architecture. That Hispanic–Arab connection has also been given new life by ventures such as the new Euro-Arab Management School, in Granada. Promoted by the EU and the Spanish government, the school aims to train a generation of Arab managers and teachers of management. Spain proudly promotes its Islamic heritage through its vigorous tourism industry marketing. An exhibition of Islamic art in Granada entitled *Al-Andalus*, the Arabic word for Spain, in 1992, the 500th anniversary of the expulsion of the Jews, attracted over two million visitors.

And did not the Moorish Sultan Boabdil's last sigh echo through the centuries across the streets and plazas of Madrid, when in October 1991, Arabs and Israelis met there for the slow start of the Middle East peace process? The Spanish capital was an apt venue for the talks. For if Jews and Muslims could flourish together there a millennium ago, perhaps Israelis and Arabs could once again learn to live together in peace, starting on Spanish soil.

If medieval Baghdad, Cordoba and Granada were the sites of the zenith of Muslim civilisation, then 1990s Sofia is – apart from the wrecked cities of Bosnia – its near nadir. A swastika marks the wall of the sixteenth-century Bani Bashi mosque in the downtown city centre, and on a sticky summer's afternoon the inside of the building is almost deserted, apart from a few worshippers staring contemplatively at the building's white walls. Trams trundle by outside on the hot dusty street as vendors sell beers and chocolates to passers-by, while the idling men that are a permanent feature of every Balkan city sit on a low wall nearby, barely visible beneath plumes of cigarette smoke.

In Bulgaria, as across all of eastern Europe, the collapse of Communism has triggered an upsurge in nationalism and a search for a spurious pure ethnicity, in a region that for centuries has been home to a myriad of peoples. Just as in Bosnia, Slavic orthodox nationalists have made Muslims their new *bête noire*, although this small south Balkan country is not about to erupt in a Muslim–Christian war. Bulgaria's Muslims are mainly ethnic Turks, with a small minority of Pomaks, Slavic Muslims. Both are an easy target for the country's small but violent neo-Nazi movement.

'Now we have problems with skinheads, you can see outside on the wall. In the last four or five years the situation for Muslims is worse. It's the same as in Germany. The police do nothing,' says Bilal Halil-Darcan. At twenty-five, Halil-Darcan is young to hold the post of Imam, but Islam in Bulgaria is a long way from re-establishing itself from the neglect of the Communist era, let alone enjoying any kind of renaissance. Halil-Darcan's wife is reluctant even to venture outside alone, for her covered head marks her as a Muslim and she has been abused and intimidated, he explains. 'If I go outside on my own, I don't have any problems, because nobody knows that I am a Muslim. But if I go outside with my wife, her head is covered, it is very dangerous. Once at 8.00 pm, we were coming to the mosque, some nationalists attacked us both. Now my wife is always with me. People look at me with dirty looks, although very occasionally people are nice.'

As one of just two European countries to share a border with Turkey, Bulgaria was a natural first stop on Islam's northward and eastward expansion. Ottoman cultural influences still linger. As in all of the Balkans, Bulgarian coffee is served thick and strong in small cups, Turkish-style – which itself came from the Bedouin method of preparation – and Turkish influence is evident in the country's cuisine, a mix of fish, kebabs and salads. Like Albanians and Bosnian Muslims, other former Ottoman subjects, Bulgarians are also proud of their country's history of religious tolerance and co-existence, in part a legacy of Ottoman policies. Back in the thirteenth century these lands were once the site of the adventures of the legendary warrior saint Sari Saltik, a leader of Muslim Turkic nomads, whose name is still revered across the southern Balkans and the Middle East. The subject of many epic tales of conquest and miracles, Sari Saltik was a prototype for the super-hero of today's comic books and his adventures draw on the fantastic events related by Scheherezade, narrator of the *One Thousand and One Nights*. His feats denote an intricate culture of story-telling, part of an Islamic cultural mosaic that in Bulgaria is now much reduced and dilapidated. Saltik takes to the air on a giant bird; he is anointed with salamander grease that becomes a winged horse that is both heat-proof and fire-proof; he flies over water on his prayer rug; he converts the prince of Christian Georgia to Islam and cuts the head off a seven-headed dragon in Kalagria (now known as Kilgra, near the Bulgarian city of Varna). He is even credited with leading a band of seventy disciples into Russia and Poland, eventually reaching Gdansk, where, the story goes, he killed the patriarch and set about converting the locals to Islam.

But for Bulgarian Muslims today, reality is more prosaic. Imams travel to outlying villages to give crash courses in Islam and its five pillars, for many have forgotten the principles of their ancestral faith. 'Only a very small minority come to the mosque. For the last fifty years, under time of Communism, religion was forbidden, especially Islam. But now people are starting to come, a new generation, more young people. All the teachers in the Islamic school are from other countries, there are no Bulgarian teachers. Some Bulgarians have gone abroad to study Islam in Arab countries and will be able to help. But now it is very difficult,' says Halil-Darcan.

Of all the countries that claimed to be Communist, it was the Bulgarian regime of Todor Zhivkov that showed itself most willing to use nationalism for political expediency. In 1989, in a last desperate bid to shore up the crumbling Bulgarian Communist system the Zhivkov government launched a political blitzkreig against the country's million or so ethnic Turks. They were protesting against attempts to make them adopt Bulgarian names, and a wave of bans against Turkish-language newspapers and even conversations in Turkish. Thousands of Muslims poured across the border, seeking refuge in Turkey, with tales of being given just a few hours to pack, before being expelled. Just as in Bosnia, European Muslims were being ethnically cleansed, although the term was not then in common use. And just as in Bosnia, most of western Europe reacted slowly and with great reluctance against the deportations.

Over seven years later, a key figure in the 'assimilation' process, as the anti-Turkish policy was known, Ilcho Dimitrov was once again stalking Sofia's corridors of power, this time as Minister of Education and Science in the democratically elected government. Now Bulgarian Muslims fear not a new attempt at 'assimilation', but a more subtle clampdown on their rights and national identity. 'I am worried that Turkish television will be forbidden. In many towns it is already forbidden for muezzins to call people to prayer from the mosque, although not in Sofia. They are all small steps,' says Halil-Darcan.

For Mehmed Beytulla, an MP in the ethnic-Turkish Party of Rights and Freedoms, the presence of Dimitrov in government is bad news. 'We have misgivings about him, because during the Zhivkov regime when the assimilation policy was going on he had the same position. We are afraid the government will start a policy of limiting the rights of the Turkish minority in Bulgaria. We don't believe they will start changing names again, but some things they have done seem to indicate a process of restricting the rights of the Turkish minority. I'm meeting the vice-minister of education, because they sent letters to regional inspectorates instructing them not to hire full-time Turkish language teachers.' At the same time the government has reinstalled its own place-men, who held office under Communism, as leaders of the Muslim religious community, said Beytulla.

Like most Bulgarian Muslims, Beytulla, 62, is not religious. Rather Islam is for him a badge of cultural and ethnic identity, part of his Turkish heritage. 'My father and grandfather were deeply religious although I am not and I don't go to the mosque. There were some difficulties during the "assimilation" process. We could not bury our dead according to the Muslim rights, during 1984 to 1989. My father and mother died then so I could not bury them according to Muslim rituals. It was forbidden. I am a Bulgarian citizen, but my nationality is Turkish. My home country is Bulgaria, I love Bulgaria, and will do everything to help my country. But my nationality is Turkish because I love my mother tongue, my culture, I keep my traditions and the rituals, they are very precious to me.'

Like ethnic Hungarians in Slovakia and Romania, Bulgaria's

ethnic Turks seek not political or territorial self-determination, but a form of linguistic, and cultural autonomy. Much as the Ottoman empire granted its subject peoples. As Beytulla says: 'We want the conditions to develop Turkish culture in Bulgaria, we don't have book printing, theatres or folk troupes. Also we would like the Turkish language to be studied as a mandatory subject at school for children belonging to the Turkish minority instead of as an option.'

On the other side of the Balkans, in Albania, Islam is enjoying a renaissance, but this is a very particular rebirth. This spectacularly beautiful country, the only European state to join the international Islamic Conference organisation, was once the most isolated corner of southern Europe. It was virtually sealed off for decades under the rule of the Communist dictator Enver Hoxha who in 1967 outlawed religion, closing or demolishing many mosques. Now money is pouring in from the Middle East, in an attempt to build a bridgehead in Europe for Saudi and Iranian-style Islam. Nominally at least 70 per cent of Albanians are Muslims, but while they welcome the chance to rebuild their mosques and revitalise their Islamic culture, they often have radically different ideas about how to live the life of the faithful to their Arab and Iranian co-religionists.

Organisations such as the Abu Dhabi Welfare Organisation (ADWO), based in a large house in the back streets of Tirana, have set up a nationwide network of aid projects, such as repairing houses and sponsoring orphans. Money for many Muslim aid projects is funnelled through a joint Arab–Albanian bank, but it's a touchy subject. Muslim aid workers deny that their groups are active in Albania only because it is a Muslim-majority country. 'There is no difference between us and Christian organisations. We have no Islamic aims, for us there is no difference between Muslims and non-Muslims,' said the ADWO's Osama Lila. But posters on the office wall explain how to pray according to Islamic ritual, and the Arabic writing has been covered over with instructions in Albanian. I ask Mr Lila where else his organisation is active. 'We have branches in other countries,' he explains. 'Somalia, Pakistan, Kashmir, Bosnia and Macedonia for example.'

A couple of miles away from the ADWO's office a present

from King Fahd of Saudi Arabia sits on the sideboard in Bashkime Korca's house. She has never met the Saudi Arabian monarch, but the Quran she received from him, via an Islamic relief organisation, is one of her greatest treasures. Like many Albanian Muslims, Korca and her husband Shafiq have received aid from Islamic welfare organisations based in Saudia Arabia, Kuwait and Britain, often gifts of sugar and flour when Islamic holidays occur.

And with the food and clothes comes the message that Islam, specifically Arab Sunni, or Iranian Shiite Islam, is the answer for Albania. Tirana airport is crowded daily with bearded Arabs from the Gulf states, flying in and out. Over 400 new mosques have been built in Albania since the end of the hated Communist regime, mostly financed by foreign Muslim countries. Under the old system mere possession of a Quran, or a Bible, brought a prison sentence of several years. But now almost every village has a new Muslim prayer house and travelling across Albania, it's a common sight to see a concrete shell of an almost finished mosque dotting the landscape.

'Now there is freedom in Albania I go to the mosque every Friday. When I am at home we pray five times a day. Under the old system religion was illegal, but now there is a place in our souls to know peace,' said Korca, an activist in the Tirana Women's Islamic Organisation. 'For fifty years there was no faith here, because it was taken away. Now I have entered the Islamic faith and we have to start again from the beginning. My life has changed since I became a Muslim. The faith is in our soul and I feel more at peace. I am satisfied that I see this and the change in our family. Even my daughter prays in the morning.'

Islam was established in Albania under Ottoman rule which lasted over 400 years, from the late fifteenth century to 1912, when Albania declared its independence. Albania supplied a stream of Grand Viziers to the Ottoman rulers in Istanbul and Albanians rose to prominence all over the empire. An Albanian even led the defence of Buda against the Austrians. Vizier Abdurrahman Abdi Arnaut Pasha, the last Turkish governor of Buda, was killed still fighting at the age of seventy as the advancing Hungarians and Austrians took the Castle in 1683.

Islamic officials say there is now an Islamic renaissance in Albania as more young people turn to the ancestral faith of their fathers. 'There is a renewal of Islam in Albania, because the young people are free to come to the mosque and more and more are coming. But that is natural, because over seventy per cent of Albanians are Muslims,' said Ramazan Rusheku, vice-chairman of the Muslim community.

One of these young people is Adam Nelku. Dressed in wrap-around sunglasses, black jeans and matching t-shirt and a reversed baseball cap, Nelku, 21, who rides a pink and white bicycle, does not fit the stereotypical western view of what a Muslim should look like. 'I accepted Islam two years ago after I studied the Quran and saw that it was the word of God. To be a Muslim is the best life in the world, to love people and to worship. Islam says that all Muslims are brothers. People in the West have the wrong idea about Islam, they should study it.' I met Nelku at the Kokonozit mosque, a few minutes' walk from the Muslim community offices. The Kokonozit mosque was named for a grand old Tirana family, but its glory days were long gone, its extreme dilapidation testament to the destruction and neglect of the Hoxha regime. A thick layer of dust covered the mosque's floorboards and the once intricate and colourful floral motifs on the walls were peeling off. Its plaster was cracked and pitted and its window frames bent and collapsing. The minaret had been demolished and still lay outside in a pile of rubble and the ceiling was full of holes. But it was still a mosque, and as Nelku knelt to pray a sense of serenity filled the room.

For Hafiz Shaban Saliaj, the mosque's Imam, the grimy room was full of memories. It was at a ceremony at the Kokonozit mosque in 1952 that he became a *Hafiz*, one who has memorised the entire Quran. 'We learn the Quran by heart, in case it ever gets lost. Then we can get it back in a few hours,' he explains happily. A genial man in his early sixties, Imam Saliaj is very pleased to see western visitors at the mosque that means so much to him. 'Even in these conditions Muslims carry on their faith. We waited for this day, because it says in the Quran to be patient. When religion was banned I carried on secretly, and even in the workplace I gave hints about religious questions. I was very careful in the beginning. I worked in a laboratory with eighty women and I would drop extracts from the Quran and other sayings into the conversation, and they liked it because they had never heard them. I secretly kept a copy of the Quran in my room. The director told me: "If all the workers were like you we would have Communism now!" But Allah protected me.'

I first visited Albania in March 1989, back when the 'Land of the Eagle', as Albanians refer to their homeland, was an isolated Balkan backwater in the grip of the world's most maniacal Marxist regime, apart perhaps from North Korea. The only way in was to pretend to be a football fan, visiting for the England versus Albania game, as foreign journalists, like virtually everything else, were banned and whatever was not banned was compulsory, as the old Communist joke had it. In Albania though, it was true. Religion, private cars, private tourism, western books and newspapers, all these were illegal. The Land of the Eagles seemed more of a bizarre curiosity than a proper country, where the best-known film star is Norman Wisdom, where a nod means 'No' and headshake 'Yes'. Even beards and flared trousers were outlawed at one time. Looking out from my hotel at ten o'clock in the evening I saw a cleaning lady sweeping the deserted streets outside the Edhem Bey mosque, one of the few to survive Hoxha's campaign of destruction.

Now Tirana is a city transformed into a buzzing Balkan minimetropolis. The main square is jammed with cars wandering across intersections, new hotels are springing up and bars and cafés are crowded into every corner. Downtown is jammed every night with Albanians of all ages strolling up and down and chatting to their friends, as in any Mediterranean city. Small kiosks and stalls jam the pavements, selling everything from bananas and shaving cream to beer and chewing gum. Tirana still buzzes with the heady air of freedom. Everywhere there are friends talking, gesticulating, and loudly greeting each other, sipping thick coffee from tiny cups and blowing plumes of smoke into the air as they light the next in a chain of cigarettes. The streets are filled with the sounds of hooting cars as a procession of battered old Mercedes, Audis and BMWs meanders up and down the boulevards. Tirana by night is an engaging slice of Balkan disorder.

It is hard to reconcile this relaxed chaos with the image many have in the West of life in a Muslim country. But to be a Muslim in Tirana is something quite different to being a Muslim in Tehran. The roots of Albanian Islam are Ottoman. In Albania, just as in the rest of the Ottoman empire, religious denominations enjoyed a good measure of freedom under the rule of Constantinople. Albania was also for centuries an important centre for Sufism, Islamic mysticism, and when Kemal Atatürk closed the Sufi centres in Turkey, the headquarters of the Sufi Bektashi movement moved to Albania.

The Bektashis, as part of Islam's heterodox tradition, incorporated many of the traditions of pre-Islamic central Asia in their rituals. By their wanderings and easy-going emphasis on spirituality – in contrast to Arab formalism – they played an important role in the spread of Islam through the Ottoman empire. They emphasised spiritual communion with God through prayer and meditation, rather than the importance of orthodox Islamic ritual. Women are admitted to the *tekke* (prayer house) without a veil and are recognised as having equal rights to men. A Bektashi meeting might include a meal where a sheep will be slaughtered, and washed down with wine – forbidden for Muslims – before the start of religious discussion. Under Enver Hoxha Albania's Bektashi heritage was almost wiped out. Of fifty-three *tekkes*, only six were left standing. In the mid-1940s there were about 285 Bektashi Babas and dervishes, both grades of membership in the Bektashi hierarchy. By 1993 there were five Babas and one dervish left alive. The Bektashis met their deaths in prison, or at the hands of Hoxha's executioners. Their beliefs though, live on. In March 1991 the Bektashi headquarters in Tirana, formerly converted into an old people's home, reopened. Speakers from all of Albania's four main religious traditions spoke at the opening – Bektashi Sufis, Sunni Muslims, Catholic and Orthodox. Each led the crowd in prayer, and each paid homage to Albania's multi-faith heritage.

This tradition of tolerance of different faiths and a liberal interpretation of the Quran has helped ensure good relations between Muslims and their Orthodox and Catholic neighbours for hundreds of years and still continues today. 'There is no danger of Islamic fundamentalism here. Many Muslims think that all this talk of fundamentalism comes from people who are hostile to the Quran. Albanian tolerance could be an example to other Islamic countries, but this is something that comes from the Quran. Europe is afraid of Islam because Islam preaches equality between people, and says do not obey other people, only God,' says Ali Hoxha, a teacher of Islamic studies.

But now many Albanian Muslims fear that their country's tradition of tolerance is under threat from foreign missionaries. Bahai travel teachers, Evangelical Christians, Muslim emissaries from the Gulf States, Mormons, Jehovah's Witnesses, representatives of virtually every faith, major and minor (except Judaism, which does not believe in proseletysing), are all pouring into Albania. For them the country is now a supermarket of souls, where a naive and unsophisticated population that was sealed off from the rest of the world for almost fifty years offers easy spiritual pickings.

On a sticky summer night, in the lounge of Edward Isufi's flat in a crumbling apartment block in central Tirana, a dozen or so young and middle-aged Albanians are earnestly grappling with the theological intricacies of the Bahai faith. Isufi was born a Muslim but has now adopted Bahaism. 'My life has changed since I became a Bahai. I am more optimistic about my new life. Bahais are working for a new world, for peace, for unity. In origin I am a Muslim but religion was forbidden and for a long time we had a space in our heart. My friends and colleagues have a very good impression of the Bahai faith, they told me this is the religion for the future of the world,' says Isufi, a professor of civil engineering at Tirana University.

The Bahai faith was founded in the nineteenth century by the Persian Bahaullah, a.k.a. Mirza Husayn Ali. Liberal and humanistic, Bahais, who claim believers in virtually every country, reject slavery, polygamy, religious prejudice and call for equal rights for both sexes, world peace and harmony. Krishna, Buddha, Moses, Zarathustra, Jesus, and Muhammad are all recognised as prophets, a divine chain culminating in, say Bahais, Bahaulla himself. But the Bahai claim that another prophet followed Muhammad drives Muslims into a fury, including Albanian ones. 'Bahaullah is an Indian deceiver who wants to revise the Quran. He was just a normal man. I'm not angry about him, I'm just revolted by the ignorance of people who accept his word as truth,' says Ali Hoxha. 'Bahaism is not a true religion, they are just deceivers. Because religion was forbidden here many people have minds like a *tabula rasa* and the Bahais take advantage by giving material things like gifts or scholarships.'

Isufi is hosting a 'fireside', a sort of multi-lingual informal chat where, in a babble of Albanian, Italian, Spanish, French and English, new Bahai converts are discussing their faith that preaches love, peace and universal brotherhood. Trays of bananas and peanuts are passed round together with glasses of fizzy orange juice as the foreign 'travel teachers' – missionaries by any other name – from Iran, birthplace of the faith's founder Bahaullah, Italy and the United States discuss the religion's precepts. 'This is the most common way that the Bahai faith becomes known, through personal contact,' says Iraj Sabet, who fled his homeland in Tehran for Switzerland after the Islamic regime launched a wave of horrific persecution against the country's Bahai community. Sabet is now a regular visitor to Albania, where, say Bahai officials, about 12,000 Albanians have converted in the last few years.

A mile or so from the Bahai centre, at the Haxhi Hafiz Dashi mosque, I get a more guarded welcome from some worshippers. This is where the emissaries from the Arab world, bearded young men with sunglasses, come to pray after parking their new four wheel drive cars outside. Upstairs on a balcony sits a middle-aged Egyptian Sheikh, who lives and sleeps here on a mattress on the floor. He ushers me in, and politely serves cold drinks, while beckoning me to sit down. He is neither friendly, nor unfriendly, but he won't tell his name and turns all my questions back on me, turning my attempt at an interview into a tutorial on Islam which is a complicated business, as he speaks in Arabic which is then translated into Albanian and then English.

Here I feel that same sense of distance and suspicion that radiates from many radical Islamists towards the West and westerners. The Sheikh has met my type before: 'I travel to all Muslim countries on my own expenses. This is the third time I have met journalists and they always ask when did you come and why?' He does have some thoughts on the *jihad,* or holy struggle, that is a duty of every Muslim. *Jihad* can be interpreted as covering a range of obligations, from spiritual self-purification to waging a holy war. 'During the Communist regime the *jihad* was mixed with mud, but now it can shine again and the light from this *jihad* will never fade away.'

The Sheikh's ambivalent welcome annoys many of the Albanian Muslims at the mosque, who assure me that in their country everyone is welcome to visit and ask whatever they want. In fact the Albanian liberal interpretation of Islam has led to several disputes between local Muslims and Arab emissaries, just as has happened in Bosnia, where foreign Muslim fighters were shocked to see Bosnian Muslims drinking alcohol and women fighters on the front-line. 'There are some who try and make Albanian Muslims more religious, but the desire and the facts are sometimes different,' says Bardhye Fico, a Muslim official in the Department of Religion. 'They try to make more Albanian women wear a scarf, but that is not Albanian. There is too much speculation on this.'

Many Albanian Muslim women, especially in the cities, such as Bashkime Korca, refuse to cover themselves, seeing the veil as a retrograde step. 'We have our rules that when women go to the mosque their heads and arms must be covered, so we do. But when we have finished praying we take the coverings off. We want to be good Muslims but European Muslims, not Arab ones. There is a big difference between European and Arab Muslims. We cannot be like them and we don't want to be. In the past women here were covered and followed the rules like Arab women, but then we changed. There is nothing to be scared of in Islam. There is peace in our soul but nothing more.'

Budapest, Pécs, Belgrade, Sofia and Tirana. All were once important Ottoman cities, where Islam and Muslim culture flowered in the mosques and *medreses*, *tekkes* and *hamams*. In Hungary and Serbia, those countries' Islamic heritage has been demolished and marginalised, apart from a few scattered remains. In Sofia, Bulgarian Muslims fear a new wave of oppression from the former Communists still in power. But perhaps it will be in the ramshackle mosques and dusty streets of Tirana, and other Albanian cities, that a new Balkan Islam can flourish, synthesising and drawing on everything from Ottoman tolerance, Bektashi eclecticism, liberal Sunni orthodoxy and secular feminism. Europe's least developed country could yet spawn a modern Islam that combines the heritage of the Sultans with the demands of the modern age.

Mosque, Brick Lane, London

Chapter 4

LONDON:
BEIRUT-ON-THAMES

*'Living here means you are influenced by British cul-
ture, British democracy, British traditions, that
unconsciously seep into your character. There are many
things about being here which have given me a fuller and
rounder understanding of the world and of myself as
well. As far as I am concerned London is a very good
place to be.'*
Dr Ahmad Khalidi, Palestinian author, academic
and adviser to the PLO leadership

*'London is very bound up in the affairs of the Middle
East. Look at the al-Mass'ari affair. Britain is not run on
the principles on which he wants Saudi Arabia to be run,
but he needed Britain to be the way it was in order
to carry on an Islamist cause from here.'*
Dr Rosemary Hollis, Middle East Programme Head,
The Royal Institute of International Affairs

The world's first Islamic cyber-revolutionary chose an unlikely
base for his international campaign to topple the desiccated and
increasingly fragile edifice of the Royal House of Saud, but the
beauty of cyberspace is that location matters far less than the
quality of the telephone lines. Even quiet, pedestrian London,
NW10, has no shortage of these. And should Dr Muhammad
al-Mass'ari, Secretary General and spokesman of the
Committee for the Defence of Legitimate Rights (CDLR), ever
return home in triumph to Saudi Arabia from his exile in a

north London suburb, then Willesden will finally be assured of its place in the history books.

Its quiet streets of 1930s houses, with their middle-class Indian, Afro-Caribbean and Jewish communities, have few claims to fame. But events in at least one residence were watched avidly by virtually every player in the Middle East, both in the public sphere and the shadow world, from the CIA and MI6 to the Saudi Arabian embassy and the Foreign Office. From there al-Mass'ari ran his campaign through telephone calls, faxes and e-mail – he also has a site on the Internet's World Wide Web – to remove the monarchy from power in Saudi Arabia and replace it with an Islamic democracy, a movable feast of a political concept, and one which is perhaps best described as being in the eye of the beholder. Information and intelligence on events in Saudi Arabia pours out of the fax machines every hour, sent in by sympathisers and supporters of the CDLR across the whole spectrum of Saudi society, from officials in the military and intelligence services, to businessmen and Islamic clerics. All are keen to see the moribund monarchy replaced by a government better able to respond to what they believe are the national and international needs of Saudi Arabia and its population.

Al-Mass'ari – whom the British government has repeatedly, and unsuccessfully tried to deport as a hindrance to British–Saudi relations – is proud of his technology. 'We use the telephone and fax. We use a system of safe calls, which works very well as long as the telephone is not under surveillance. They can call us and there is no way the call can be traced, we use an 0800 number in the USA, through the British Telecom operator to make collect calls, or on credit cards. We have e-mail but we don't use that so much because many people in Saudi Arabia only use computers for games or word-processing or education, and are not yet computer literate,' says the man who was formerly one of Saudi Arabia's top theoretical physicists, and eminent scholar of the *Shariah*, Islam's code of law.

Al-Mass'ari's rows of state-of-the-art computers contradict the stereotypical views that many in the West have of Middle Eastern Muslims, portraying them either as roaming camel-traders and indolent oil-rich sheiks, or dour nonagenarian Ayatollahs passing *fatwas* (judgments) on the dusty plains of

Persia. And unlike in Iran, Mass'ari's concept of Islamic democracy has no special place for the Islamic clergy as a separate and powerful political class. The Islamic revolutionaries of Saudi Arabia, in fact across the Middle East, draw on professional technological and scientific backgrounds that firmly root them in the 1990s. Although while they are happy to benefit from a western scientific education, many Islamists maintain a visceral hostility to the liberal values of the societies that trained them. But just as science flourished under the reign of Caliph al-Ma'mun in ninth-century Baghdad, under scientists who saw their experiments as part of a holistic unity that combined God's will with the expansion of knowledge, Islamists such as al-Mass'ari embrace new technology as part of Islam's mission. To see Islam as not scientific is to have the eyes of a fool, he believes.

'People who think that must be idiots, this is absolute nonsense. They think this because of misconceptions and because of the demonisation campaigns against Islam, through the centuries, that it is anti-science. It is not only not true, it is a joke and should not be discussed. Now in Saudi Arabia the most pro-regime people are the classical clergy. They think in terms of Islamic law which has been developed over the centuries, but they are unable, because of their limited scope of thinking, to go back to the original sources in the Quran which must be applied to the current situation and analyse other points of view. They don't have a critical attitude.' Now Islam is enjoying an intellectual renaissance, he says. 'We are coming out of a deep time, a very low time in Islamic culture and civilisation, with no critical thinking and analysis. Most of the scholars and clergy of Islam are quite backward in the sense they are supporting these regimes out of fear. But the ones who are not overloaded with historic rubble of old ideas, who have sharpened their analytical tools, and studied physics, mathematics, they are now accepting change and improvements.'

It is in London, rather than Cairo or Beirut, Damascus or Riyadh, that these 'changes and improvements' are being fashioned. Positioned halfway between the Middle East and the United States, with easy access to Europe, the hub of a global communications network, and with decades-old ties to Islam's lands, London has now become the *de facto* intellectual capital of the Middle East. Speech is free, no British government hand censors newspapers or television reports – although Saudi-backed London-based newspapers such as *Ash-Sharq Al-Awsat* (*Middle East*) cannot criticise their paymasters – and opposition groups enjoy the sort of freedoms here they could never exploit at home.

As long as exiles do not break Britain's laws, or advocate violence or terrorism, there are no legal restrictions on political activities for Arab opposition groups in London, whether their members arrive as tourists, students or claim political asylum. London's large community of Arab exiles makes the city a bridgehead between the Levant and the West, bringing substantial amounts of capital and investment, and adding a vibrant patch of colour to the city's multi-cultural mosaic. Britain's tradition of Arab scholarship gives the many analysts watching events in the Middle East an open window onto the Arab world. But every benefit has a cost as well, and the downside was never more evident than in July 1994 when two bombs exploded outside the Israeli embassy and the office of the Joint Israel appeal.

Two major international Arabic language newspapers *Ash-Sharq Al-Awsat* (*Middle East*) and *Al-Hayat* (*Life*), are published in London, and the city is also home to an Arabic satellite television station and, of course, the BBC Arabic service, respected across the Middle East for its impartiality and accuracy. The English–Islamic media arc extends also to Pakistan, with the Urdu-language newspaper *Jang* and the English language *The News*, both published in London and Pakistan. Many exiled Muslim dissidents take ready advantage of British freedoms that they would never allow their own citizens, should they ever take power. Radical nationalists, pan-Arabists, Islamists, democrats, liberals, Baathists, monarchists, every shade of Arab, Kurdish, Turkish and Iranian political opinion can be found in Britain's capital. The newsagents in the central London streets of Queensway and Edgware Road, tucked between the Arab cafés, restaurants and estate agents, are crowded with piles of newspapers, magazines and bulletins, all calling for everything from parliamentary democracy, to Islamic theocracy, or a revival of

Baathism, that hybrid ideology of socialism and Arab nationalism. Every major and most minor opposition groups have a presence in London.

They include: Algeria's Islamic Salvation Front; the Palestinian Hamas; the Tunisian An-Nahda (renaissance) movement, led by Raschid al-Ghannouchi; secular Iraqi opposition organisations such as the Free Iraqi Council and even Pakistan's Altaf Hussain's Mohajir Quami Movement, based in the north London district of Muswell Hill, as for some reason radical Muslim groups feel quite at home on the capital's quiet suburban streets. Most of these are banned in their mother country and the presence of so many Muslim revolutionary organisations, many of whom are in regular close contact by telephone, fax and e-mail has even prompted talk of London being the base for a new 'Fundamentalist International', a sort of Muslim Comintern, where Islamic revolutions are planned and emissaries despatched with secret orders. These organisations are tolerated for several reasons, although many in the Home Office and the Foreign Office would like to see them expelled because of the difficulties they can cause to Britain's domestic security, and the way they disrupt trade and diplomacy with the Arab world. To change the laws and outlaw political activity would be a legal quagmire, because there is, after all, still a tradition in Britain of giving asylum to foreigners seen as trouble-makers at home; they provide useful intelligence from their networks of supporters at home and also, there is always an outside chance that one or other group may come to power in what is still an unstable region, and hopefully will then remember who its friends were in the wilderness years.

As well as a flourishing Arab community – of businessmen and women, students and holiday makers, as well as Islamic activists – London is also home to a large Jewish community, among which I grew up. I was born in London, grew up and went to school there. As a child I first saw Arabs in Edgware Road and Oxford Street, shopping or chatting in one of the many Arab cafés around Marble Arch, and at Speaker's Corner at Hyde Park, where I sometimes spent Sunday afternoons watching the passionate and ferocious arguments between Jews and Palestinians. My ancestors are Jews from Lithuania and

Belarus, on one side, mingled with English, Scots and Irish on the other. My paternal grandparents left the Czar's lands at the end of the last century, heading west for a better life, where no Cossacks on horsebacks would thunder past their doors, heralding the latest pogrom. Like many Jews of that great migration, they settled first in London's East End, before moving westwards across the capital to greater comfort and prosperity.

History laid its hand lightly on us, compared to those left behind in eastern Europe. My grandmother lived in White-chapel's Sydney Street, where as a young girl she witnessed the siege, when Russian anarchists holed up in a house nearby fought a gun battle with the police, an operation directed by the young Winston Churchill, then Home Secretary. Some relatives became capitalists, running small factories making clothing or opening shops, others, communists and ardent socialists, demonstrating for Spain during the civil war and throwing bricks at Mosley's fascists as they marched in Cable Street in the East End. Britain has been good to my family, and all have prospered to some degree, living there in peace and safety. The Holocaust passed at least my immediate relatives by, as everyone had left in time. Still, I am sure that old family friends of my great-grandparents and unknown distant relations must have taken that final walk to the trenches outside town before tumbling into the mud and the bodies, or into the gas chambers of the German concentration camps. Now that same immigrants' path across London on which my grandparents travelled is filled by Muslim arrivals. The streets and squares of Whitechapel and Spitalfields that once resounded to Yiddish and Polish are filled with Urdu-speaking Pakistanis and Bangladeshis and the smell of chicken soup has been replaced by curry.

My first school was the North-West London Jewish Day School, the second the Jewish Free School, in Camden, London. Not surprisingly Islam, apart from the Golden Age of Spanish Jewry, did not figure greatly in the syllabus at either, although at both we were taught plenty about Israel and Zionism. But even among the Yiddish-speaking Ashkenazi Jews of north-west London cultural bonds remained with the Middle East and their distant estranged cousins, the Arabs. The Yiddish greeting *Sholom Aleichem* (peace be upon you), derived from Hebrew, is virtually identical to the Arabic greeting *Salaam Aleikum*. On trips to Israel I picked up a few words of Arabic, learning that many were identical or extremely similar to Hebrew, such as the word for house, *beit*, or one, *wahad* in Arabic, *achad* in Hebrew. Theologically, culturally, and linguistically, Islam and Judaism have far more in common with each other than Christianity, even though that religion was also spawned in the Middle East and was for a long time an alien implant in white Europe, which many Europeans seem to forget when they decry Islam as something strange to this continent.

Both Islam and Judaism are monotheistic and recognise the oneness of God, expressed as *tawhid* in Arabic (which takes its root from the word *wahad/achad*), and repeated assertions of God's unity in Hebrew prayers. The concept of the trinity, or praying to an individual, even a deity in human form such as Christians believe Jesus was, is anathema to both. However, Islam recognises all the major prophets of Judaism, such as Abraham and Moses, as well as Jesus and Mary as prophets of God, a chain of revelation culminating in Muhammad, the last and final messenger. Both faiths emphasise the timelessness and holy revelation of their scriptures, the Torah and the Quran respectively. And like Judaism, Islam evolved through a system of interpretation and intellectual questioning, in its case of the meaning of the Quran, and the *hadith*.

Muhammad made quite stringent efforts to convert the Jews of Arabia to Islam, although he failed, Jews then as now being a stubborn bunch, and the Jewish community of Medina was eventually killed, the first stage in the elimination of the ancient Jewish presence in Arabia, apart from Yemen. At the birth of Islam the first Muslims even prayed toward Jerusalem, not Mecca. Both faiths share strict dietary laws, and the laws of *halal* (for Muslims) and *kashrut* (for Jews) are so similar that even the extreme radical Islamist Sheikh Omar Abdul-Rahman, no friend of Jews and now imprisoned in the United States, exhorted his followers to buy kosher meat if no *halal* was available. The family too, of paramount importance for Jews, is the centrepiece of Islamic society. Parents are respected, their wishes obeyed. Aurangzeb Iqbal, a Muslim lawyer in Bradford, told me that when his young clients were arrested and came to him for help, the first thing they asked was not what will happen to them, but 'What will my Mum and Dad say?' No wonder that on my travels I felt far more at home in Muslim Bosnia than in Bavaria, shelling aside.

Growing up in London, immersed in Jewish culture, for some paradoxical reason by my late teens I wanted to know more about Arabs and Islam. A few years later I realised that this was quite a common reaction – on my bookshelves stand several books about the Middle East and Islam by Jewish authors. At school I was not exactly taught that the Arabs were my own personal enemy, but it was understood that our interests were diametrically opposed. I grew up in the 1960s and 1970s, at the time of the Six Day War in 1967 and the Yom Kippur War in 1973. The aim of the Arab armies was simple: to eradicate the state of Israel and throw the Jews into the sea (no distinction was made between Jews and Israelis). The Holocaust was still fresh in my parents' and grandparents' minds. The Arab armies were the heirs of the wartime Mufti of Jerusalem, Haj Amin El-Husseini, good friend of Hitler. They were the soldiers of Nasser, who had hired Nazi German rocket scientists to try and destroy Tel Aviv. Once again it was us against them, although this time 'them' spoke Arabic instead of German. Actually many of us didn't like the Israelis at school very much. They stuck together, spoke Hebrew all the time, and worse, always got to go out with the prettiest girls in class.

But slowly I began to realise the tangled relationship between Jews and Muslims was a bit more complicated than Nasser and Sadat *v* Golda Meir and Moshe Dayan. At school we learnt about the Golden Age of Spanish Jewry, one of the transcendent moments in Jewish history, when culture, religious life, philosophy and science flowered in medieval Spain, all under Muslim rule. And then there was Auschwitz, the very nadir of Jewish life, or rather death. We studied the Holocaust in depth, watching grisly but necessary films of stick men and women in striped pyjamas peering out from behind barbed wire, bodies lying starved on the pavements of the Warsaw Ghetto, bulldozers shoving piles of wasted bodies into trenches. 'Never again!' we cried before singing the Israeli national anthem.

But why were we singing the Israeli national anthem? We weren't Israelis, we were British. If Israel was so great, as our teachers kept telling us, then why didn't they go and live there?

Looking back, it seems almost our whole education, and our developing identities as Jews, were refracted through a prism of Israel and Zionism. Either way it seemed to me that the logic of hating all Arabs was slightly faulty. True, the Arabs wanted to annihilate Israel, and all its Jewish inhabitants, or at least they said they did, repeatedly, their bloodthirsty rhetoric a substitute for their inability to do so.

Still, several millennia of by no means perfect, but nevertheless reasonable – in comparison to Christian Europe – co-existence showed that such virulent hatred between Arab and Jew was a comparatively new phenomenon, dating more or less from the evolution of political Zionism at the end of the nineteenth century and the establishment of the State of Israel. Across the Middle East, just as in Serbia, the lack of a democratic, pluralistic culture, the absence of a free press and political repression combined to shape a population easily susceptible to hate campaigns and demonisation of a supposed enemy. For the Serbs, the devil was Bosnia's Muslims. For the Arabs it was, and for many still is, the Jews. Even now, the sales assistants at the Muslim bookshop on Charing Cross Road will happily sell customers works purporting to outline the mechanics of the mythical 'International Jewish Conspiracy'. In Belgrade and Baghdad, Damascus and Tripoli, the ruling élite well knows how to transfer their subjects' anger and alienation onto an outside enemy, instead of their own governments. For Arabs Israel was an alien implant in the Middle East, the ultimate symbol of their impotence in the face of the West, that had sliced up the Ottoman empire like a Sunday roast, drawing arbitrary borders to create states such as Jordan and Iraq to suit imperial interests, deposing and imposing monarchs and leaders on a never-ending merry-go-round of place-men and stooges.

But who had so far killed the most Jews in history? It was not Muslims or Arabs who built the death camps of the Holocaust, but Christians, Catholic and Protestant Germans, with their copies of Goethe in one hand and the gas chamber lever in the other. It would make more sense to go to war against Germany.

If Israeli is a colonialist implant, so are most Arab regimes, which is in fact the position that many radical Islamists take, denouncing everyone from Saddam Hussein to Shimon Peres as the lackeys of western imperialism. They have a point. Until

Winston Churchill invented – and I use the word advisedly – Iraq, for example, at the post-war carve-up of the Middle East, there had never been such a country. Present-day Iraq comprised three separate Ottoman provinces: Mosul in the north; Baghdad in the centre and Basra in the south. The area was under British military rule, an army which included my paternal grandfather. There is a picture of him at home, sitting proudly on a horse outside Baghdad railway station. The provinces were home to substantial Kurdish, Assyrian, Shiite and Jewish minorities but Winston Churchill ignored advice that the arbitrary unification of such a diverse area would eventually lead to disaster. The Gulf War dispute over Kuwait proved the naysayers correct. T.E. Lawrence, better known as Lawrence of Arabia, was in Cairo with Churchill, and later boasted that he was one of a small group who had designed the modern Middle East over dinner. They did a bad job. It was a faulty structure, with serious inherent flaws whose consequences are still with us now.

I also learnt that across the Arab world, in Egypt and Syria, Morocco and Iraq, Jewish communities had thrived for millennia, tracing their roots back to the Babylonian exile, all once an integral part of the Middle East's ethnic mosaic, but now vanished apart from a few remnants. Their lives then under Islam were not perfect, but the institutionalised discrimination of the Islamic world was far more benevolent than in Christian countries. It was far easier, and pleasanter, to live as a Jew in sixteenth-century Cairo or Baghdad than, for example, Venice or London. Venetian officials then forced Jews to live in an area known as the *ghetto nuovo*, a cannon foundry – the origin of today's word ghetto – surrounded by high walls, with ten watchmen paid for by the Jewish community. In England Edward I had expelled the Jews en masse in 1290, and while a few individuals remained, no sizeable community returned until the mid-1650s after the English Civil War.

It was not too fanciful that I sat next to some of the descendants of Babylonian Jewry in the sixth form common room. Some of my schoolfriends' families had left Arab countries such as Iraq or Yemen, to live in Britain. They even spoke Arabic among themselves, rather than English. It took me a while to realise that when we played poker at lunchtime the muttered conversation in Arabic behind me was my opponent's cousin telling him what my hand was. But how could you be a Jewish Arab, I wondered? One schoolfriend, whose family had left Baghdad, once confided to me that he didn't know what nationality he was; not English, because he was born in Iraq, no longer Iraqi, because he was a Jew living in London, and certainly not Israeli, because he had no desire to move to Tel Aviv. It was my first experience of the dispossession of exile, although at the age of sixteen I didn't really understand its subtleties, or its power. Fifteen years later, in the refugee camps of Bosnia, I would do.

That is my background. But if this book has any specific genesis it was a spring morning in 1980, when I was working in a wood factory on a kibbutz in northern Israel, before starting university. We stopped for breakfast and I noticed that workers split into two groups: Jewish Israelis, speaking Hebrew, and Muslim (or Christian) Israeli Arabs, speaking Arabic. There was no mutual animosity, but the division was clear and, it seemed, natural there, in the sense that this was the correct order of things. But why? This was a kibbutz after all, an outpost of radical socialism and communitarianism, almost Communism, in fact. An Arab though could not join the kibbutz, because that option was only open to Jews. The Arab workers were hired hands, wage-slaves in the Marxist ideology to which the kibbutz still professed to adhere. There was a contradiction here. There seemed only one way to try and solve the contradiction, for with whom do you make peace, if not your enemies? I decided to learn Arabic, that language that was grammatically and linguistically so similar to the Hebrew that I then spoke reasonably well, and to study Islam, as I did at Leeds University. I also took a summer course in colloquial Palestinian Arabic at the Hebrew University of Jerusalem in August 1982.

From there I took a bus to Cairo, where I first heard the call of the muezzin to prayer. I still remember the sound now, as it reverberated through my room at the Anglican Hospice, on the island of Zamelek in the cool of an August morning. The long, stretched-out consonants of classical Arabic, and how they filled the dawn silence of Cairo, soaring across the Nile, summoning the believers to the mosque. Here you are no longer in Europe, with its bland safe certainties, the muezzin said to me, here you sleep and awake in the *Dar al-Islam*, the House of

Islam, where other values reign. Something deep inside me shivered and stirred when I heard him, an ancient ancestral memory, stretching across the millennia, to a time of deserts and exile, of miracles and patriarchs, and an awesome God who shepherded his troublesome people out of Egypt across the Sinai desert to Israel.

Islam and Judaism: so similar but somehow so separate, two tongues, two cultures, two religions; entwined for centuries but now split asunder, like a once married couple who have lived together for years, know each other's every characteristic, but are now divided by a legacy of mutual acrimony and bitterness.

My half-remembered Arabic came in useful at the CDLR's headquarters in Willesden. The door was opened by a sleepy teacher of Arabic from Saudi Arabia, who lets me use the telephone to find al-Mass'ari, as our arrangements became muddled and he is not there. Eventually he arrives, a bearded, extrovert man, dressed in a long white jelabia that reaches almost to the floor. For dissidents such as al-Mass'ari, even though he has an American wife, London, home to foreign correspondents from all over the world, is the ideal centre of operations. 'For the international media I need to be in an English-speaking country, so what other English-speaking countries are there? The choice was down to Britain or America. Frankly we would not feel safe in America. We were following the case of Sheikh Omar Abdul-Rahman, which was a frame-up, and America has a very clear agenda in the Arab peninsula, to keep the puppet Saudi regime in power. My wife is American, she had a very bad experience when I was in jail, from the embassy who were very uncooperative. They neglected her. But here we have all the facilities to speak out, and we don't have to worry too much about a frame-up. Britain has been quite tolerant, maybe they have an agenda for this, but Britain is the safest place.'

It's well known across the Middle East, and among many London-based Arabs, that al-Mass'ari's enemies would like to see him silent and or dead. Criticising Arab regimes can be a perilous business. Arab diplomats here, he claims, have been involved in plans to silence him. They even asked Mossad, the Israeli intelligence service, to help, he said, but Mossad declined to get involved. 'Mossad so far has refused to cooperate with such extreme measures as assassination. They said they didn't want to get involved, and that they had their own problems. But one Arab official told them, "But you did assassinate [PLO official] Abu Jihad in Tunisia," and the Mossad said, "Yes we did indeed assassinate Abu Jihad, but we regret that we did that, it would have been better not to do it." Even the Mossad is wiser.'

The late Ayatollah Khomeini was reported as saying that 'there is no fun in Islam', which might be true under the reign of the Ayatollahs but al-Mass'ari takes a radically different view. He is articulate, learned and loquacious, entertaining even, his conversation spinning off on tangents, peppered with *hadith*, verses from the Quran, ranging from musings on Islamic history and theology to the duty of married couples to satisfy each other sexually, adopting whichever position takes their fancy. No wonder he drives the Saudi royal family into such a furious frenzy, for he runs a slick PR operation and knows what the international media need for a story: human interest, a sense of outrage at injustice and amazement at the cravenness of the British government in bowing to Saudi demands to try and expel him. But the al-Mass'ari case is significant as more than just a simple black and white deportation trial. It highlights London's role as the epicentre of events in the Middle East, it shows how decisions made in Riyadh and Jeddah to push for his deportation from Britain reverberate across England and it also illustrates how, despite many Muslims' claims of a widespread Islamic bias in Britain, even an unlikely candidate such as a Saudi Islamic dissident can become a *cause célèbre*, with support across almost the whole political spectrum, even be invited to appear on the television programme *Have I Got News For You*.

Dr Muhammad al-Mass'ari's trek from Riyadh to Neasden began at 2.30 am on 15 May 1993. His American wife Lujain al-Iman, then aged 26 and seven months pregnant, woke next to her husband in the large house they inhabited on the campus of King Saud University in Riyadh, to the sound of bumps and banging. Writing in *Vanity Fair*, Leslie and Andrew Cockburn report what happened next:

Crashing through the garage entrance to the house, a dozen or so heavily armed *Mabaheth* – secret police – had entered the servants quarters, on the ground floor. They were Bedouin tribesmen from the Saudi interior; red and white checked *ghutrahs* masked their faces so that only their eyes showed. They held a maid at gunpoint until she waved them toward the study, where Mass'ari's eldest son eighteen-year-old Anmar was sleeping. They woke him and smashed his head against the wall, shouting, 'Where is your father?'

The masked Bedouin crashed through the bedroom door and dragged al-Mass'ari away. Lujain suffered an asthma attack while the *Mabaheth* dragged away books, pulled down the curtains and slashed the carpets. For good measure they confiscated her asthma medicine and her American passport as well. Al-Mass'ari's arrest received scattered publicity in the international press, but there was no hint of his future status as international *cause célèbre*. Earlier that month, he had, together with a group of lawyers, judges and professors, launched the CDLR after many meetings over a series of months that aimed to build an embryonic opposition. In a repressive monarchy that saw no role for independent political organisations, he and his colleagues had taken a brave and public stand: they demanded a measure of accountability from the House of Saud for its actions, and criticised King Fahd, the Keeper of the Two Holy Shrines (Mecca and Medina) on Islamic grounds. The members of the CDLR were not, to be sure, western liberals or democrats. They based their criticisms of the Saudi regime on their interpretation of the Quran. Which made them even more dangerous to King Fahd, and even more of a target, for they could not be dismissed as western interferers.

Before he was arrested al-Mass'ari had told his wife to go to the American embassy for help. None came, and Lujain was cold-shouldered. Her husband meanwhile was in Al-Hayer prison, outside Riyadh. He was imprisoned for six months and repeatedly beaten on the soles of his feet. Warned after his release that the government was planning to execute him on trumped-up charges, he was brought by the Saudi underground to Britain. In

London he started his bulletin, *Monitor*, cataloguing events in Saudi Arabia, which was faxed and e-mailed back to the Kingdom. In Mecca and Medina, Riyadh and Nejd, copies circulated through the underground opposition movement passed by hand to hand, or faxed to offices and universities. *Monitor*, available on the Internet's World Wide Web, is a mixture of polemic against King Fahd and the Saudi royal family, facts and analysis. It is extremely anti-western in tone, implacably opposed to the Arab–Israeli peace process, and presents a different shade of the CDLR to the one al-Mass'ari gives to western journalists. Here are some extracts from number 19, published in March 1995, addressed to the Saudi Royal family:

YOU HAVE TURNED US INTO A NATION OF BEGGARS, OUR LAND IS A WESTERN COLONY

The whole nation has been deprived of its Islamic character – deprived of its mastery of the land. You have reduced their country into a colony of Superpowers. Their wealth and resources are daily being looted by monsters whose greed knows no bound. You have created animosity between their neighbours and weakened the ability of the military to defend their own land and security.

. . . Despite the disinformation, our people are only too well aware of your immoral relations with your genocidal Zionist friends. You are protected by the Americans and continue to sustain the illusion that they will protect you permanently. Think what happened to another American-lover, the Shah of Iran. There is no one left with any interest or conviction in your leadership!

THE LAST STRAW – EL AL TO FLY IN SAUDI AIR SPACE

The Zionist Minister of transport declared that the Saudis would soon allow the airplanes of El Al airlines to fly in Saudi airspace. This is considered an important step in the normalisation of relations between the two countries, the result of open and secret meetings. In yet another treacherous development, Bandar Ibn Sultan was accompanied

113

on one of his trips to the Kingdom by a group of Jewish business men who are being persuaded to invest in the Kingdom.

That al-Mass'ari was free to publish and distribute *Monitor* from London drove the Saudi royals into a frenzy. Riyadh quickly piled on the pressure to expel him, twisting the arms of the British weapons industry. Arms manufacturers were warned that vital arms deals between Britain and Saudi Arabia, worth hundreds of millions of pounds, were being threatened by his continued presence in London. The industrialists buckled quickly. A memo written by the Chief Executive of Vickers, Sir Colin Chandler, a former head of arms export at the Ministry of Defence, published in the *Guardian* in January 1996, cast a revealing light over the shadowy web of finance and armaments that binds Britain to Saudi Arabia. Dated 6 September 1995, it said that Britain passed secret intelligence on Saddam Hussein to Saudi Arabia to appease Riyadh's fury. More ominously, it referred to 'direct Saudi intervention' against him, warning that it would be difficult as 'he is, as you probably know, the son of a leading cleric'.

The later revelation that Andrew Green, Britain's ambassador to Saudi Arabia, had been a non-executive director of Vickers since April 1994 surprised few with any knowledge of the links between Britain's arms industry and the palaces of Riyadh.

Government ministers flapped and flailed under Saudi pressure, claiming a plethora of excuses as to why al-Mass'ari must leave, for, of all places, the Dominican Republic. Prime Minister John Major said al-Mass'ari was an illegal immigrant who had sought to create an 'unsettled relationship' with Saudi Arabia. Home Office minister Anne Widdicombe was more honest when she said: 'We have close trade relations with a friendly state which has been the subject of considerable criticism by Mr al-Mass'ari.'

By attempting to deport him the government forged an unlikely alliance between left and right, says the Saudi dissident. 'They didn't understand how much it would backfire. It united the imperialists who cannot live with the idea that the empire is gone, and the humanists who cannot live with the idea that basic freedoms are threatened. That leaves only a minority of arms manufacturers and the poor workers in the north-east, who really have a case. . . . Have you seen *Spitting Image*, the last one? It showed a cabinet meeting, and someone said and now for the al-Mass'ari case. One minister said: "Couldn't you have found a suitable lie, instead of telling the truth about balancing economic interest against human rights?" The other holds a bucket and replies, "No we couldn't. We have been in power for seven years and all our lies are gone now. Look in the bucket, it's empty!",' he exclaims, laughing uproariously.

The al-Mass'ari case has also complicated relations between Britain's Muslim community and their Christian compatriots. On one hand, everyone from crusty old peers to liberal *Guardian* readers has rallied to his Islamic standard, but at the same time, the government's attempts to throw him out because ministers fear the impact of his message are compared by many Muslims to the state protection given to Salman Rushdie. The talk now in the mosques is that the British government supports free speech to attack Islam, but not to criticise trading partners.

'Many Muslims in Britain felt very negatively about my story, compared to Salman Rushdie. They wonder what the British believe, whether freedom of expression is a British right, or a universal one. Salman Rushdie was defended on UN conventions that guarantee basic human rights, not because he is British. So tell us clearly what you mean, if these are basic human rights, then al-Mass'ari should enjoy them, if not, if they are only for British people then shut up. You have to be consistent. Rushdie insulted Islam, al-Mass'ari insulted the [Saudi] royal family, so what does freedom of expression mean? Because if you only allow what you like and prohibit what you don't like, this is not freedom of expression, this is a dictatorship. This is a very difficult argument to overcome. Salman Rushdie also cost the taxpayer money. The broken relations with Iran must have lost hundreds of jobs in cancelled contracts. But those workers have not been considered. Are they less worthy than the people working at Vickers? They are all worthy human beings.'

Which brings us to the Rushdie case. Most Muslims believe that Rushdie, in *The Satanic Verses*, set out deliberately to insult Muhammad and court controversy and publicity in what they see as one of the basest ways possible: by apparently both satirising Islam in general and insulting it in specific ways as well, such as giving a clutch of prostitutes the same names as Muhammad's wives. If that was the author's tactic it worked, not wisely but too well. Unfortunately for Rushdie events spun out of control, for he was caught up in an intra-Iranian power struggle as various factions jostled for position and influence as Ayatollah Khomeini lay on his death bed. That he was born into a Muslim family only increased his peril. Briefed by his son Ahmad, Khomeini issued a *fatwa* (judgment) that Rushdie was an apostate, one who has renounced the Muslim faith, an act punishable by death under Islamic law. Tehran radio carried a report of the ailing Ayatollah's pronouncement:

> The author of the book *The Satanic Verses*, which is contrary to Islam, the Prophet and the Quran, together with all those involved in its publication who were aware of its content, are hereby sentenced to death . . .
> I call on zealous Muslims to execute them promptly on the spot they find them, so that no one else will dare to blasphemise [*sic*] Muslim sanctities.

To underline the seriousness of the threat, the Iranian 15th Khordad Foundation offered a reward of $2.6 million to anyone who killed Rushdie. He has been under police protection ever since, although he appears with increasing frequency at London literary parties. Understandably fearful for his life, he converted to Islam, but then appeared to backtrack, further inflaming Muslim anger. So, I ask al-Mass'ari, is there an Islamic justification for the *fatwa* on Salman Rushdie? The answer, as I have grown to expect, is complicated.

'The question is whether or not he is in the Islamic domain. There you may punish him because it is prohibited to attack God and the prophets who are mentioned in the Quran. But you have to distinguish between Muslim and non-Muslims. In the case of Muslims, he will be regarded as an apostate, depending on your interpretation of the apostasy law. One school says that his family relations will be severed and so on and he will not be executed. This is a minor school, very minor, but it does exist. The other is that he should be executed. But if he is a non-Muslim, this has been a problem for Islamic thinkers all through history. For example there is the question of what to do with a Jew who says Mary, peace be upon her, was an adulteress. The first thing is that we believe there can be no compulsion in religion. So if that is his belief, for us it is a disgusting belief, but he has it. Others say no, she enjoys the protection and the dignity that the Quran guarantees for its prophets. The other school says no, he enjoys the protection of his faith and religion and we cannot force him not to say that. For me the answer is that is his point of view, you cannot punish him if he is non-Muslim, if he is in the Islamic domain.

'Outside the Islamic domain you could condemn him. I don't think Salman Rushdie was an apostate, because he was never a Muslim in the beginning, and he is entitled to his point of view. My view is that if you insult certain personalities, there may be some punishment, not necessarily up to death, to keep public discussion away from insult and degradation. You may very well believe in the immaculate conception, maybe Mary was married, maybe it was from Joseph the carpenter, or maybe from something else. If I say that it is not insulting, but if I say she was an adulteress, that is insulting. But I don't think the punishment should go to death, maybe some jail term, whips or lashes, whatever the Islamic state adopts. In his writings now Rushdie is insulting Hindus, so he seems to be a specialist in outraging other people. I can express my point of views respectfully. Consider the Pope: I have respect for the man, but I don't believe in his infallibility. I would not call the Pope a dog or a *schweinhund*, the dog which cares for pigs, the worst kind of dog. That's very insulting.'

As our discussion ends he asks if I have a faith myself. 'Yes, I am Jewish,' I reply. Not for a nano-second does he pause.

'Ah, Jewish, oh yes, I was walking down the street the other day and I saw this lovely old Jewish couple coming towards me and the man asked if I knew where the Jewish cemetery was. I said of course, which one do you want, liberal or conservative!'

But a few months after our meeting, the good Doctor was taking a less ecumenical position on Jewish–Muslim relations. Press reports quoted an article in his newsletter *Al-Huquq*, that described Jews as 'a nation of cowards' and reportedly claimed, 'Their annihilation is not difficult for Muslims'. Furious MPs, including Greville Janner, *de facto* parliamentary spokesman for the Jewish community, demanded the Crown Prosecution Service rescind its decision not to prosecute al-Mass'ari for what Janner regarded as incitement to racial hatred. The Saudi dissident then claimed in an interview with Chrissy Iley of the *Sunday Times* – who summed him up as 'an Islamic equivalent of a wide boy' – that the whole furore had erupted because of a mistranslation, and the *Al-Huquq* article referred only to those Jews living in Israel. Quite.

A few miles from the CDLR's headquarters in Holborn, central London, are the offices of the Saudi-supported *Ash-Sharq Al-Awsat,* one of the Arab world's most important newspapers, whose pages are a forum for debate on almost every issue affecting the future of the Middle East, except anything critical of the House of Saud. A picture of a smiling Queen Elizabeth greeting the Saudi King Fahd is proudly displayed in the newspaper's foyer, a pictorial remnant of Britain's long ties to the Arabian Gulf.

Britain still has a special place in the Muslim world, says Ayad Abu Chakra, a senior editor at the newspaper. That influence is based on several pillars: an imperial legacy as one of the powers who dismembered the Ottoman empire to create new states such as Jordan and the Gulf Emirates; the works of writers such as Richard Burton, translator of *One Thousand and One Nights* and military adventurers such as T.E. Lawrence, who helped lead the Arab revolt against the Turks. British schools and universities have also educated a generation of Muslim leaders and politicians, such as Muhammad Ali Jinnah, founder of Pakistan, who studied law in London and practised for several years. 'Britain enjoys a unique position in the Muslim world. As a mandatory power in Asia and Africa Britain interacted with the Muslim communities, and with the intellectual trends in the Muslim world. You cannot underestimate the influence Britain had on the Indian sub-continent. Jinnah was a lawyer in the Temple and the head of the Muslim League. In the First World War the British promoted the idea among Arabs (of Arab nationalism) that "you are Arabs, why should you obey a Turkish sultan, he is not a Muslim sultan, he is a Turkish sultan". It created a legacy of goodwill until 1948, when the state of Israel was created.'

The orientalist view of the 'noble bedouin', untouched and uncorrupted by the modern world, still widespread among certain parts of the British upper-classes, has also helped Britain's cause in the Levant. 'The British are easily charmed by the East, British travellers like Burton were fascinated by the East, they were fascinated by its sentimentality and the spiritualism. Look at the way they dealt with the beauty of their surroundings, the romanticism of Lawrence. He was sitting in a tent with an old bedouin sheikh, explaining how advanced our modern observatories are, all the stars you can see and the poor old sheikh said, "that's great, I'm very impressed, but we see someone you don't see, beyond the stars, and that is God". For someone like Lawrence, an intellectual, he thinks this man has something, he is not dumb.'

As a journalist and self-described 'semi-academic' Abu Chakra represents a different breed of Arab intellectual to al-Mass'ari, still Muslim of course, by birth, culture and belief, but tending more to modern secularism than Islamic revivalism. For him Islam, as a religion of about a billion believers spread across the world, is a mosaic of cultures, each with its own colours and textures. A Shiite Ayatollah in Iran, an African Muslim musician in Senegal, a Turkish businessman in Istanbul, an Arab journalist in London, all may share the same core religious beliefs, but their lives are a reflection of their own national cultures as well, which may not always include the idea of universal brotherhood.

'You cannot simplify Islam, just as you cannot simplify Christianity. You cannot say that Christianity is the brand that I define as Christianity. Christianity for a monk on Mount Athos in Greece is his brand, but if you go to the Bible belt in America, it is about baptism and evangelical Christianity. Islam is a world culture, it is a huge culture, a global religion. Someone in Bradford, involved in the mosque with a puritanical approach to Islam, does not behave in the same way as a secular Muslim in Istanbul, or as an Arab nationalist who thinks that Arab nationalism is based on Islam in Syria and Egypt. They are all totally different. What Islam means to X is not necessarily what it means to Y. To me, I believe that Islam is a global religion, it can encompass many trends, you cannot tighten it up to the extent that it suits your own definition. I perhaps am a very bad example, I am a very secular man myself, but I don't believe that you serve the interest of your religion if you try to make it narrow, narrow its scope and make it exclusive. Inclusivity is the source of greatness of every religion.'

As an international capital, home to a large Arab community of exiles, dissidents, businessmen, students and tourists, London can be host to a cultural symbiosis between the West and the Middle East, says Abu Chakra. Arabic language newspapers, television stations, cafés, restaurants, clubs and mosques bring a new facet to London's ethnic mosaic. 'In the West there is enough social flexibility to accept these things. I believe that multi-cultural societies thrive. Take America for example, the top scoring students in the universities are Chinese or Indian or are from outside America. It is not a source of bringing these cultures down. It is creating more diversity, enriching the culture and bringing a multi-faceted angle and face. The problem that we have in the West today is that some politicians are charlatans and they play the race card all the time.'

But this must also be a two-way process, he says, and Muslims must give as well as take, if they wish to be accepted. 'What is required is understanding by both sides. The Muslims might not like this, but I think I have to be realistic, you don't expect people to accept you in their culture if you are not ready to concede something. I for one have no right to impose my values on the British. I am gaining from them, I can give to them if they want to get something from me, but I don't impose myself on them. But if they want to know more about my culture I am ready to give them, to give them the best I can. We need concessions from both sides. A civilised society has to be open-minded.'

But not every import from the Middle East to London is welcome. The cycle of violence in the Israeli–Arab conflict spilled over onto the streets of London in July 1994, when a car bomb exploded outside the Israeli embassy in Kensington. Fourteen people were injured in the attack, which followed repeated rumours of an impending operation against Israeli targets and Jewish organisations. Four more people were hurt the next day when a bomb exploded outside the offices of the Joint Israel Appeal in north London.

For Middle East expert Hazhir Temourian, an Iranian Kurdish exile now living in London, the increasing Saudi dominance of the Arabic-language international media, and the presence of so many Arab journalists, indeed a substantial Muslim minority at all, is not healthy for Britain. 'The Sheikhs of Araby want to use London to set up newspapers to spread propaganda among the other Muslims who live in Europe, and London is a good communications centre for that. Britain allows this. It is supposed to have a tight immigration policy, but every year it allows hundreds of Arab journalists to be recruited by Saudi Arabian Sheikhs and princes who own the Arab press here, and after four or five years working here, they all become British citizens, because they don't want to go back. They like life here and I don't blame them. I have met some of them, who have taken me out to lunch and have begun to trust me. They admitted that they are Muslim revolutionaries, passing themselves off as moderates, in order to be brought to Britain by the Arab press. London is now the biggest television satellite centre for the Arab world. I believe that Britain is just storing up trouble.'

I met Temourian, who is a regular contributor to the BBC and *The Times*, for tea at the Athenaeum, one of the first division English gentlemen's clubs where crusty old bastions of the British establishment meet to bemoan the state of the world over tea and cakes, or an excellent dinner washed down with vintage claret. Pink-faced gentlemen seated in leather armchairs dozed off around us in the large lounge or perused the day's newspapers, while white-jacketed waiters brought a pot of tea. It was a quintessentially English scene, and for all his background as an Iranian Kurd, Temourian looked very much at home. For him, like many Middle Eastern exiles, Britain is still the home of democracy and democratic values. And those values, he believes, are threatened by the influx of a large Muslim minority, many of whom, he says, do not agree with western liberal values or mores and seek to impose an Islamic lifestyle.

'The Muslim intellectual, or person of Muslim background, who has a constructive manner, he is not going to be a source of trouble. But the ones who are culturally very Islamic, they could be. The Muslim who comes here, who believes that his own society is superior, he comes here to take advantage of prosperity, but keeps his militancy. In my opinion liberal Muslims are cheats who are trying to endear themselves to western society. The liberal argument about multi-culturalism is naive, it assumes that everybody is going to be as good as you are. I don't believe in a multi-cultural society, they all come to grief. I don't believe that in the name of tolerance you should tolerate intolerance.

'I've told Muslims in Britain that if you really believe that Muslim society is superior to this society, then why did you come in the first place? They become angry, they become abusive, I say really you owe it to yourselves and your children to go back, you can never turn this into a Muslim society by force, but they do have such dreams. Once I was in hospital, visiting an Iranian friend, who was here for heart treatment. Because this man had received government-subsidised currency, the local Mullah from the Iranian embassy came to see him. This Mullah was quite new in Britain and he had never seen me. He asked me what my name was and I had to give a false name in case this patient was in trouble for meeting me. The Mullah became quite complacent and he asked me how many children I have. I said two, he said, "Is that all?" I said why, what do you mean. He said, "We cannot conquer these people with tanks and troops, so we have got to overcome them by force of numbers." He was very sincere and I don't blame him. He is commanded by his God that it is better for the world if everybody turns into a Muslim.'

As an immigrant himself, Temourian says he can articulate what he believes are the secret fears of many in the liberal middle classes, who are scared to speak their mind for fear of being accused of racism. 'The answer is for people like me, especially immigrants who care about liberal values, to be more active, to write about it. It's easier for us to say this, if you wrote you would be described as a fascist. I cannot be accused of being a white supremacist, or a racist. I believe, and I think lots of people in Britain believe, although because of political correctness they won't write about it, especially if they are British, that Muslims are unlike the previous waves of immigrants. Even in their own countries, these Bengalis and Pakistanis are regarded as hillbillies. They are not even from the capitals, but from the backward countryside, they come without a word of English and the British allow them to bring their own Mullahs who are extremely hostile to the spirit of enlightenment. They discover themselves as a minority, they feel rejected and so they grow more attracted to the only thing they have in common together, Islam.'

London has also played a pivotal role in the burgeoning Arab–Israeli peace process. Some of the first meetings between the two sides took place in the British capital, and it was there that the early tentative feelers toward a rapprochement of the decades-old conflict were extended. Both Israelis and Arabs say that it is often more productive for them to meet in the British capital than in Cairo, Amman or Tel Aviv. For Palestinian intellectuals such as Dr Ahmad Samih Khalidi, born in Beirut in 1948 and a scion of an ancient Jerusalemite family, London is the place where Israeli and Arab could talk freely.

'London was the place where they met for the first time, it began in London, when Hanan Ashwari met an Israeli university professor, after it was suggested to him that he meet a senior PLO figure. Much of the debate does take place here. London and England are the venue for many other kinds of meetings from places such as St Anthony's College, where Arabs and Israelis meet and talk, to Chatham House, and studies, conferences, lectures. There is a whole environment which has been here for a long time. The British role in the Middle East was absolutely paramount, and for the Palestinians it was absolutely crucial. People come here for serious academic work on the subject, because you have all the state archives here, partly because there is a cultural affinity and respect for British education.'

The British capital is also a place where the many currents of contemporary Arab thought meet and coalesce. 'In many cases the Arab world is not really conducive to serious thought. If you think of yourself as a free thinker, and you want to practise what you preach or write what you believe it's much easier here. The atmosphere in much of the Islamic world today and in the recent past is not only repressive but pretty much closed. In my case it's not just that, I studied here in Oxford and London. One of the main factors in the last decades has been the demise of places like Beirut, especially for the Levantine Arabs. Physically Beirut is being rebuilt, but I don't think that intellectuality will return.

'From the mid 1980s, there has been an accelerated migration to the West to London and Paris, for people like myself and other categories – for want of a better word, intellectuals – the Palestinian and Lebanese business communities, which thrived in Beirut, many of them have moved to London. It is the closest major capital to the Middle East and a crossing point where you get the currents from across the Atlantic. In many senses London has become the intellectual capital of the Middle East. If you go to Queensway, and walk up and down the road, go to the newsagents and look at the very wide array of newsletters, magazine and journals, about 60 or 70 per cent will be printed in London.'

In one way it is a telling comment on the intellectually stifling atmosphere in much of the Arab world, rather than the inherent merits of London, that has led many Muslim thinkers such as Khalidi to base themselves in Britain. But personal considerations as well play a part, and a move back to Israel/Palestine would entail a massive upheaval for him, his wife and four children. But somehow, by retaining a geographical distance from the Middle East, Arab intellectuals also retain a more critical perspective on their homelands. 'In an ideal sense I suppose it is sad that the centre of intellectual life is in Europe, but in a real sense, given the situation in the Arab world, it probably isn't, because at least you have some focal point for all this cross-fertilisation for all these ideas. And they do, in one way or another, percolate back, such as in the case of al-Mass'ari, who is a very interesting man. You can see that at a distance, a relatively small outfit, with a relatively modest means, can have a major impact, possibly in a way they would not, if they were there.'

Just as events in Turkey or Algeria rebound through immigrant communities in Germany and France, the twists and turns of the Arab–Israeli struggle reverberate across London. Speaking at the Sharm el-Sheikh anti-terror summit in March 1996, that followed a wave of Hamas suicide bombings in Israel, John Major promised to clamp down on British-based groups that he said provided support for Hamas, the Islamic Resistance Movement in Jordan, Gaza and the West Bank. MI5, Britain's domestic intelligence service, had for several months been investigating support groups for radical Islamists based in London, partly at prompting from Israel that London was the 'focal point' for Hamas fundraising, with over £6 million a year being supposedly funnelled to its underground networks through charitable fronts in London. (That Israel had, not so long ago, encouraged the build-up of Hamas as a counterweight to the PLO in the Gaza Strip and occupied West Bank was conveniently forgotten.)

What followed the Sharm el-Sheikh gathering was not a wave of arrests, or the discovery of caches of arms and explosives, or even a list of Jewish or Israeli potential terrorist targets, but a single frozen bank account. It seemed a poor haul, but then the logic of politics and diplomacy is that something must be done, and preferably in public. The Charity Commissioners took over the funds of Interpal, a Palestinian charity based in north-west London, that in 1995 raised half a million pounds for humanitarian work in Palestine. Interpal's Vice-Chairman Essam Mustapha strongly denied that his organisation had any connection to Hamas, or engaged in any political activities either in Britain or abroad, and pointed out that its work on the ground was coordinated with both the Israeli government and the Palestinian National Authority. 'We do go through local authorities, whether that is the Israeli authorities or the Palestinian authorities. We have good relations with all of them and we have had no allegations from them against Interpal,' he said in an interview with the Muslim weekly newspaper *Q News*. Muslims dumped the blame for the Interpal affair into the laps of Israel and British Jews. For many Muslims across the political spectrum the Interpal affair was an excuse to lapse back into hoary old stereotypes about the supposed influence of the 'Zionist lobby', code for British Jewry, and fuel the anti-Semitic conspiracy theories that still hold considerable sway among Muslims.

Ahmed Khalidi does not give money to Interpal – which was eventually cleared of any wrongdoing – or the many other new Arab and Islamic welfare organisations, preferring other, older charities such as Medical Aid for Palestine. 'Muslims in Britain give money to charity and the funds are transferred to social projects; mosque X in Gaza or mosque Y in Kabul. But you don't always know where the line between social welfare and political activism begins and ends. If you feed a family of someone who will be sustaining a suicide bomber then you are sustaining him. The Israelis have put enormous pressure on Britain, but it's very difficult to prove that welfare organisations are engaged in military activities. I don't think they are. It's not a problem to get weapons there, you don't need to buy them in London and send them out. The Israeli policy is to uproot Hamas and go after the infrastructure, but it's wrong and unachievable because you cannot find the right legal mechanisms. The answer, as always, is political. Hamas is not a law and order problem and will not be resolved by military means. The answer is to provide Arafat with the means to co-opt Hamas into some form of co-existence with the peace process.'

For Palestinians such as Khalidi the next few years present a unique challenge, more than for any other Arab thinkers living abroad: that of building a new independent state, of Palestine. That challenge could yet see him return to his ancestral homeland as his people build up their own country. A signed photograph of PLO leader Yasser Arafat sits on the wall of Khalidi's study in his house in west London. His credentials are excellent for some future role in the new Palestine. Now editor-in-chief of the Arabic edition of the *Journal for Palestine Studies* quarterly, and a research fellow at the Royal Institute for International Affairs, Khalidi served as an advisor/delegate to the Palestinian delegation to the Washington Peace Talks in 1991–93 as well as senior advisor on security issues to the Taba/Cairo PLO–Israeli negotiations until December 1993.

'I have a very active relationship with the Palestinian leadership, I have been there five or six times since Oslo, I maintain close contact, but also some distance. I have access and I can talk to anyone I like. For the intellectual and political exiles the return will take much longer, but the Palestinians could be drawn in because of the state-building, a specific challenge. Personally, I'm fairly rooted here, I have children here and I would have to completely uproot, it's paradoxical because that's where I come from and I suppose you shouldn't think of yourself of being uprooted from exile.

'Living here I have a more sanguine view of both the West and the East. I can see both the advantages and disadvantages of western and eastern societies. The clear kind of difference relating to issues of personal freedom you learn and observe from being here, but you also learn the limitations, of democracies, which are not always immediately applicable to other types of societies. I have learnt a lot but I think I have maintained a kind of distance from the extremes of the Arab and the western one. Obviously the fact that you are here gives you a necessary and healthy distance from the Arab world and you can observe it much better from a distance, that is my job, to observe it. London is far enough and close enough at the same time.'

Yet while Khalidi, whose wife is Scottish, is well assimilated into London life, many Arab visitors and residents stay in their own business and social communities, and never really take the pulse of their new host country, although admittedly that's partly because of traditional British insularity.

'One of the regrettable aspects is that by and large many immigrants, not just the Arabs, tend to stick together. In England a foreigner is a foreigner, very clearly, unlike the United States where everybody is different. In some respects I am anglicised, I am not a religious Muslim, I don't spend too much time cultivating or addressing the Arab community. Culturally I consider myself a Muslim and if you ask me I would say I am a Muslim, but I haven't made very strong or conscious efforts to convince my children to be Muslim. I have a broad network of connections, some professional, some social. I haven't encountered much racism, or come across overt racism. The other day I was in a cab and I was giving him directions, when I got to the door he said: "It's good to hear an Englishman speak with such perfect diction." I just smiled to myself.'

Which is, of course, a perfectly English reaction.

The Satanic Verses *on fire in Bradford*

Chapter 5

BRITAIN: THE EMPIRE STRIKES BACK

'People are looking for an Islamic identity. You find someone called Muhammad, who grew up in western society, he concedes a lot so people accept him. He changes his name to Mike, he has a girlfriend, he drinks alcohol, he dances, he has sex, raves, rock and roll, then they say, "You are a Paki". After everything he gave up to be accepted, they tell him he is a bloody Arab, or a Paki.'
Omar Bakri Mohammad, leader of the radical group
Al-Muhajiroun

'We call ourselves British Muslims. Whether or not anybody acknowledges us or accepts us, we have decided that this is our country, this our home and that is where our children and grandchildren have decided to live.'
Sher Azam, Muslim community leader in Bradford

There is no spiritual hierarchy in Islam, for it has no ordained priesthood. Any male Muslim may lead the five daily prayers. Just as there is no ecclesiastical system of Muslim rank, there is no grade of mosque importance, or equivalent of the division between church or cathedral. The front-room of a small terraced house in Bradford, the grand mosques of Istanbul or Riyadh, wherever Muslims gather to pray five times a day, all are equal in the eyes of Allah.

But if one mosque in Britain is first among equals, it is the Central Mosque and Islamic Cultural Centre by Regent's Park in central London. The struggle for control of this complex has played a vital symbolic role in the evolution of British Islam, specifically in the battle for power and influence between the two main opposing currents of Muslim thought in Britain. In one corner, the outward-looking modernists, who want Islam to be engaged in a dynamic debate in British society, allied to Christians and Jews on issues of common concern such as morality, religion and education. These are thinkers and theologians such as the centre's former director Dr Zaki Badawi, who are trying to build a British Islam and develop a Muslim theology that responds to the needs of a minority community living in a late twentieth-century industrial society, and Pakistani-born Professor Akbar Ahmed, of Cambridge University. In the other, the strident radical Islamic activists, such as the late Dr Kalim Siddiqui, leader of the Muslim Parliament, arch-exponent of the *fatwa* on Salman Rushdie, and one of a vocal minority of Muslims who foment confrontation politics based on what they believe to be an unalterable enmity between the Muslim and non-Muslim world, an enmity that sustains them in their intellectual hostility to many western values. And somewhere in the middle, the majority of Britain's more than 1.5 million Muslims.

The Regent's Park Mosque is probably the most important site of British Islam, attracting thousands of worshippers on important Muslim festivals such as *Eid al-Fitr*, the feast which celebrates the end of the Ramadan fast. So, I thought, it was clearly the best place to start investigating the state of Islam in Britain today. I phoned the mosque to arrange a meeting, but the man to whom I eventually got through, Mr Bashir (he never would tell me his first name), one of the mosque's administrators, was not very encouraging. He was, it must be said, pretty rude at first, for he took it for granted that I was out to do a hatchet job on the mosque, on Muslims and on Islam in general, but eventually I persuaded him to meet me. He later warmed up and we parted, I hope, on good terms. He even gave me a rather beautiful copy of the Quran in Arabic and English, with a blue and gold cover and an inscription inside announcing it was a gift from King Fahd of Saudi Arabia. Mr Bashir himself is an educated and sophisticated man, who has studied at London University and is an industrial chemist. Born in Mauritius, he is now a British citizen. But like many Muslims he gives credence to the idea of a world-wide anti-Islamism, that is not exactly a co-ordinated campaign, but certainly a sequence of events against Islam and Muslims. Many more Muslims across the world think like he does than would disagree with his analysis.

Why, I ask him, are there so many misconceptions in the West about Islam?

'I would like to think that it's all just a coincidence, whenever you see so many bad things said about Islam and the Muslims. I honestly don't believe in a conspiracy because I don't think that anyone can harm Islam. But then again all these incidents took place in such a way that to put it down to just coincidence is a bit hard to accept as well.'

What are these incidents?

'Anything to do with Islam. Salman Rushdie. The Gulf War. Bosnia. You name it. Chechnya. Algeria. It's nothing new. The world is manipulated by certain individuals and certain countries for their own use . . . sadly.'

Which countries and which individuals?

'Primarily all the western countries. There is an order, the United States, UK, France, Germany, and Israel and all these. They are dictating events for their own benefit. They don't give a damn about the ordinary people . . . about their own people. Let alone show concern for anybody who is different from them, who is a different religion or a different colour to them.'

So what did he think of the claim that Islam and parliamentary democracy, as practised in Britain, France or Germany for example, are not compatible?

'I certainly can understand certain Muslims when they say that there is no place for democracy in Islam. What they mean by that is the western form of democracy that is being thrown at them throughout the colonial contact that they have with the western world and the western way of life. If you look at democracy, how many UN resolutions have been passed against Israel and never been upheld? When the UN resolutions were passed against Iraq and Saddam Hussein every single word of these hundreds of pages were adhered to. Now do you expect that an ordinary Muslim living in a remote region in an Arab country who for years has had his territory, his village, his house bombarded by the mighty Israeli gun power, will he say to you "Yes, I'm in for democracy?" You say democracy is a change of government by the ballot box. So why did the West suspend the Algerian elections?'

The western powers don't want an Islamic regime in Algeria, I reply.

'So that is why some people are saying that democracy as we term it has no place in Islam? I would have thought that this kind of democracy has no place whatsoever in any society, if you have this kind of conditional approach that you can select whoever you want as long as it's the one that we are all agreed on . . . If people want to decide for Islam then let them, that is their right. Why is it that we can decide any form of government except Islam?'

It's a hard point to argue against, and when Algeria is added to the list of Muslim grievances, from Bosnia to Palestine, via Chechnya and Kashmir, it's easy to see why many Muslims truly believe a small cabal of powerful countries dictates events across the globe.

We move on to life in Britain. How can Muslims keep their Islamic values in a society which so often seems diametrically opposed to those beliefs? By sticking to the precepts of Islam, says Bashir. 'I recently attended a meeting with the Chief Medical Advisor entitled "The health of the nation". They were concerned with certain aspects of young people's lifestyle causing concern. Drugs, marriage breakdown, things like that. Now you are saying to us, "How can Islam with all the things that we stand for find a place in that society?", yet the same government is trying its best to address these problems. But where we differ with them is they are reacting to the problems. We say that rather than reacting to the problems, which condemns them and us as well, we must proact to the problem and address it at its very core.'

At the core are Islam's moral precepts, of prohibitions against drugs and alcohol, and bans of pre-marital sex and homosexuality. There can be no compromise, he says. 'For example, the question of AIDS, the norm in British society is "a condom culture". This is the message that is being portrayed. But, no matter how careful you are, there are bound to be mistakes. So rather than saying to me, "Do whatever you like, but be careful", what we are saying, and what Islam is saying, is "No". Islam says there is no sex before marriage. The same thing with drugs and alcohol. Why do you have this fantastic campaign during Christmas, "Don't drink and drive"? Islam doesn't say "Don't drink and drive", Islam says "Don't drink at all" because there is no benefit whatsoever in drinking. You can only have problems. But western society is saying we cannot implement it so therefore forget it. Or we are dictated to by the big brewing and whisky companies.'

Like Islam everywhere, British Islam is a mosaic, a complicated pattern of different components. Mr Bashir represents one influential political strand, which takes a radical anti-western perspective on global issues. British Muslims are all united in their observance of their faith but they differ widely socially, culturally, economically and, perhaps most importantly, in their interpretation of Islamic theology.

British Muslims are Arabs and Iranian businessmen in London's Kensington and Bayswater. They are Turkish and Kurdish factory workers in the north of the capital and Bangladeshi snack-shop owners in the East End. They are Yemenis in Sheffield, Pakistanis in Bradford and Birmingham, Malaysians in Manchester and Somalis in Spitalfields. Most are Sunni, with small minorities of Iranian Shiites and Turkish Alevis. They speak half a dozen Arabic dialects as well as Farsi, Urdu, Bengali, Pashto, Dari, Hausa, Swahili, Turkish, Kurdish and, of course, English. But even among the Sunni majority there are deep divisions. The first division, political and ideological, is between the modernists and the radicals, but that distinction is useful only for evaluating the range of Muslim attitudes to life in a modern western secular society.

The theological gulf between the different traditions too is deep and wide, even among otherwise homogenous ethnic and national groupings. British Muslims with origins in the Indian sub-continent, for example, have brought with them en masse the ancient tension between the two groupings of Barelvis and Deobandis. Barelvis give the prophet Muhammad a superhuman status, seeing him as a figure who is everywhere at all times, who is not mortal flesh and blood, but light itself. Deobandis too revere the prophet Muhammad, but only as a mortal man, even though he was the *insan-i-kamil*, the perfect person. The emergence of the Barelvi tradition is an interesting example of how Islam, often wrongly viewed in the West as a monolithic homogenous faith, is in fact highly syncretistic and has evolved across the world by absorbing and adapting local traditions.

The Barelvis' particular veneration of Muhammad has its roots in the South Asian and Hindu tradition of myths and legends of fantastic gods who can perform miracles. Barelvi reverence for, and perception of, Muhammad, was shaped by the pre-existing polytheistic Hindu culture that accorded super-human status to religious figures. This might be just an interesting footnote in the development of Islam in Britain, except the Barelvis' veneration of Muhammad explains why they were so angered, hurt and infuriated by Salman Rushdie's *Satanic Verses*, why they led the campaigns against it and why, ultimately, they were driven to publicly burn the book. Many Arabs and Turks, for example, took a much more sanguine view of the whole affair, saying that far more insulting works about Islam than *The Satanic Verses* are available on the bookshelves of their own countries.

Several foreign Muslim political parties are also active in Britain, of which the two most influential are probably Hizb-ut-Tahrir and Jamaat-i-Islami. Both enjoy only minority support, but their presence as activists gives them a voice and a presence, and they set an agenda to which mainstream leaders must respond, if only to dismiss their claims and demands. Hizb-ut-Tahrir, about which more follows later in this chapter, was founded in east Jerusalem in the early 1950s. Jamaat-i-Islami was founded in Pakistan by Sayyid Abu'l Ala Maududi, one of the most important thinkers of modern Islam. 'Maududi's primary aim is to present Islam as an absolute and self-sufficient ideology, completely distinct from and opposed to both the western way of life and its eastern socialist equivalents,' writes Malise Ruthven in *Islam and the World*. Maududi's legacy of visceral anti-westernism can still be heard among Muslims everywhere from Mecca to Manchester. Jamaat-i-Islami does not enjoy wide support among the Pakistani electorate, but is influential in Britain.

Like Maududi, the majority of British Muslims have their roots in the Indian sub-continent, perhaps as many as two-thirds of the total, who were either born here or whose parents arrived in the great waves of immigration during the 1950s and 1960s.

They altered for ever the ethnic and social makeup of large swathes of London, as well as a string of cities across the country, but nowhere more so than the mill towns and cities of Yorkshire such as Bradford, Batley, Dewsbury and Leeds. Decades later it would be Yorkshire, rather than London, that would see the first outburst of Muslim rage, against *The Satanic Verses*, and one of the first examples of successful Muslim political activism, the campaign for the provision of *halal* meat in schools.

In fact Britain has played host to Muslim traders, travellers and envoys for centuries, and the southern port of Hastings was known to Arab cartographers. One of the first Muslim travellers to record his impressions of Britain was Mirza Abu Talib Khan, who was born in Lucknow in 1752 and was employed as a revenue officer by the British. He travelled in Europe between 1799 and 1803 and on one of his journeys he visited Dublin. He was not impressed, for in all the city he could find only two ramshackle bath-houses, and they were used solely by the very ill. In the summer, he writes, Dubliners wash their bodies in the sea, in the winter, they don't bother at all. The bath-house itself was a great disappointment, and lacked the specialist staff, such as a barber or masseur, that make a visit to a Muslim *hamam* such a relaxing and invigorating experience. But it seems there was no pleasing Abu Talib, for while the Dubliners didn't wash enough, the English gentlemen he met were too fussy about their appearance. Number six in his crotchety list of Englishmen's defects are 'their wasting much time on sleeping and dressing and arranging their hair and shaving their beards and suchlike'.

But it was the opening of the Suez canal in 1869 that marked the start of the first substantial Muslim presence. Indian, Yemeni and Somali seamen landed in Britain and Yemeni boarding houses opened in Cardiff and South Shields. Sufi Sheikhs arrived soon after, to meet their religious needs, opening *zawiyahs*, or centres which soon became the focus of those early Muslims' social and religious life. Many of the sailors married British women and stayed on, the start of Britain's first Muslim communities.

Unlike Christians and Jews, British Muslims have no organised national communal leadership, whether temporal or spiritual. Just as in France or Germany, they speak with many voices, which weakens their case on issues such as reform of the blasphemy law or the provision of halal meat in schools. Those same differences of ethnic origin, culture and theology that make Islam in Britain such a many-splendoured faith also mitigate against unity. Many Muslims argue for the establishment of a body along the lines of the Board of Deputies of British Jews, a quasi-parliament, representing different strands and factions, as the best solution, but so far the community has failed to agree on a framework.

But if there is no one voice of British Islam, British Muslims do have several leaders to represent them. Dr Zaki Badawi, now principal of the Libyan-funded Muslim College in west London, is often described as Britain's foremost Muslim cleric. Born in Egypt, Dr Badawi studied in London University, taught at Al-Azhar university in Cairo, the Muslim world's foremost theological academy, as well as in Malaysia and Nigeria before becoming Imam and director of Regent's Park Mosque. Dr Badawi took a strong stand against Ayatollah Khomeini's *fatwa* of death on Salman Rushdie, and was involved in the author's short-lived adoption of Islam, taking a brave public stance in favour of free speech which contrasted sharply with the public equivocation of many Muslim thinkers, who well knew how far back the *fatwa* would set relations between Islam and the West, but still kept quiet. It was a stand that then weakened Badawi's position in the community.

Demonstrations against the book were organised by Saudi Arabia, working through Jamaat-i-Islaami, he says. 'The presence of Jamaat al-Islami is one of our main disasters here . . . The Saudis were pouring money into them. I tried very hard to reconcile Salman with the Muslim community, despite the fact that I knew it was a thankless task. He wrote a document saying he was a Muslim, signed it and gave it to us. But he relented very quickly. He wanted to be a Muslim out of fear and I accept this is a legitimate approach, if you are under pressure you can become anything. I thought his opponents, if they had any sense, which they don't, should have encouraged him. He would have lost his liberal constituency very quickly. I thought he was an asset, he is a good writer and many of his opponents are ignorant illiterates. The Muslim community could gain a lot by having people of talent on their side.'

Bruised but not disheartened by the Rushdie affair, Dr Badawi is now among the vanguard of British Muslim modernists who believe that Islam must develop in line with Muslim life in a western secular society and evolve a new theology to meet those needs. His progressive beliefs have often brought him into conflict with the more obscurantist trends in British Islam, and their paymasters in Saudi Arabia and other Gulf States. Dr Badawi has been in the vanguard of inter-faith work, and Christian priests and ministers and Rabbis lecture at the Muslim College, but perhaps nothing he did was more controversial among Muslims than to introduce an Anglican priest and a Rabbi into Regent's Park Mosque to discuss religion and theology.

'We three thought that as religious leaders it's time we tried to make headway in bridging the gap between the communities. It was a risk because suspicion and hostility between them is always present . . . Muslims said, "How dare you bring a Rabbi and a priest into the mosque to speak?" Many Muslims then saw the mosque as their moral territory, something extra territorial, that we are not living in England, somehow when we enter the mosque it does not count under the British jurisdiction.'

But Dr Badawi persisted, and slowly the stereotypes, on all three sides, began to peel off and die. Those early meetings were perhaps the germ of an idea, now spreading in sections of both the Jewish and Muslim communities, that relations between Jews and Muslims should not be dependent solely on events in Israel/Palestine, when the two faiths have so many common concerns in 1990s Britain. 'At the beginning there was hostility, people tried to embarrass the Rabbi or the priest by saying, "You people hate Islam". But what was very heartening was the growing attendance and interest. And what really astonished many of them was how similar Jewish law is to Islamic law. So the people who came to be hostile and to embarrass the speakers gradually became accustomed to listening to other views. More importantly, for the first time Muslims frequented the synagogue and also the church. But the most important thing was people from different religions went to another place of worship, and were sitting there respectfully and participating politely in discussions.'

With the departure of Dr Badawi in 1981 from the Regent's

Park Mosque inter-faith work there has been much reduced, but the broader trend continues. The Jewish Board of Deputies, for example, does much work behind the scenes advising Muslim leaders on issues such as political lobbying, education and provision of *halal* meat. 'At the time this dialogue was sort of heretical, but suddenly it dawned on the Muslim community that it was really the way forward. It's really the convention that dialogue is not something to be accepted, particularly with the Jews, after Israel's creation. But the whole idea of dialogue, even with your greatest enemies, is something that is enshrined in Islam. But somehow people developed the idea that if you don't like someone then you don't talk with them. Ignore them. Don't learn about them. Don't even read about them. This is what I call the philosophy of ignorance, sanctified ignorance. If I don't like a book I burn it, which is a terrible thing. It is spreading.'

So where, I ask, does this originate from?

'From a contingent still living in medieval times in our local Muslim circles. We do not tolerate different opinions. We find someone who has a different opinion to be threatening and frightening. And rather than debate, we say, "You are no longer a Muslim" . . . Much of this comes from the Gulf states and their money. They exported a vision of Islam which was contrary to the development of what was taking place in Egypt, with a school of law and a school of thought that was tolerant, dynamic, and was moving forward, trying to adapt Islam to modern conditions and to incorporate modern science and technology. But its leaders were accused of being more or less heretical. That is what has happened, and now any new idea is regarded with suspicion.'

The answer, says Dr Badawi, is to deal with things as they are, and not as some Muslims would like them to be. 'To find a solution first, and then to theorise about it later. To live in the modern world and accept the modern world as it is. We cannot act like King Canute, demanding that science and technology must retreat at our command. It is there. No one lives without this, television, radio, the airplane. People can no longer escape the consequences of modernity.'

At the opposite end of the spectrum to the modernists such as

Zaki Badawi are the radical Islamists, who believe in, and work for, the eventual global triumph of Islam. They are not the same as the conservatives, described above by Dr Badawi, although they share their visceral hostility to western liberal values. In Britain the loudest voice for radicalism is that of Hizb-ut-Tahrir (Party of Liberation). Founded in east Jerusalem in the early 1950s by Sheikh Taqiuddeen al-Nabahani, the party calls for a new *Khilafah*, a pan-Islamic state based on the *Shariah*. Its members have been linked to several coup attempts in the Middle East, where it is banned but still maintains a secret cell network in most Arab countries as well as in Turkey, Pakistan and Sudan. It operates more or less openly in Britain. It is implacably opposed to the Arab–Israeli peace process and calls for the creation of an Islamic army to eliminate Israel, claiming that after the destruction of the Zionist state Jews will live in peace under Islam's guarantees. Hizb-ut-Tahrir publishes Arabic-language magazines across the Middle East as well as an English-language *Khilafah*, which is distributed at mosques in Britain and the United States. In Europe Hizb-ut-Tahrir has, or has had, bases in Denmark, Belgium, Germany and Austria.

Hizb-ut-Tahrir's growth is being watched with concern by a range of observers, including M15, which carefully monitors the activities of radical Islamists in Britain, especially those with connections to the volatile Middle East. Mainstream Muslim leaders fear, rightly, that the party's rhetoric of confrontation will further alienate middle-England and marginalise Muslim causes. Jewish organisations accuse Hizb-ut-Tahrir of virulent anti-Semitism. The National Union of Students accuses the party of intimidating Jewish, Hindu and gay and lesbian students, which it denies.

Judging from the many stories about Hizb-ut-Tahrir in the *Jewish Chronicle*, its student members seem ready enough to resort to verbal intimidation, although none has ever been arrested and charged with physical attacks. Either way Hizb-ut-Tahrir is also banned in dozens of British colleges and universities, where it now operates under cover names such as the 1924 committee, commemorating the year when Turkey's leader Kemal Atatürk abolished the Ottoman Caliphate. Hizb-ut-Tahrir supporters were accused of whipping up ethnic hatred in February 1994, after Ayotunde Obanubi, a twenty-year-old

African student at an east London college was stabbed to death in a clash between African and Muslim students. Four days before the killing, Omar Bakri Mohammad, then the party's leader in Britain, had spoken at a meeting for Muslim students.

The party's biggest triumph was the *Khilafah* Conference in August 1994, held at Wembley Arena. Thousands of Muslims jammed the aisles to hear the usual rhetoric against radical Islam's *bêtes noires*: Israel, the Middle East regimes – condemned as creations of imperialism – the World Bank, the United Nations, the International Monetary Fund, as the speakers verbally rounded up all the usual suspects. Despite the dire warnings from Jewish organisations of disaster if the *Khilafah* conference went ahead the main disturbance was the arrest of gay activist Peter Tatchell demonstrating on behalf of 'Queers Against Fundamentalism', as Islam forbids homosexuality.

But the question many, both Muslims and non-Muslims, were asking, was where had Hizb-ut-Tahrir suddenly appeared from, and more to the point, who was paying for extravaganzas like the *Khilafah* conference? 'Hizb-ut-Tahrir's presence is amazing, how they emerged so suddenly, they must have had some finances to appear so rapidly,' says Dr Badawi. 'When I was in Regent's Park Mosque in the early 1980s, they were not there. They appeared after the Gulf War, after the obvious and open delegitimisation of the regimes in the Gulf, which all appeared to be paper regimes, unable to defend themselves. Al-Mass'ari would never have happened without the Gulf War. I think Gulf money is behind Hizb-ut-Tahrir. Not the Saudis, they dislike them, almost all the governments dislike them, because they are revolutionaries, saying all the regimes are illegitimate. But just as all roads lead to Rome, all finances lead to the Gulf. There are people in the Gulf who will encourage any radical movement, because they hate the regimes and they want to get at them, but they cannot do that openly, so they sponsor any movement that is anti-establishment.'

Until January 1996 Hizb-ut-Tahrir's leader in Britain was Omar Bakri Mohammad. Bakri Mohammad, who arrived in London in 1985, was born in Syria in 1958 and describes himself as 'a leader on the ground', travelling across the country to speak at mosques everywhere from Barking to Bradford. Now based at the School of Shariah, a large scruffy house in Tottenham, north London, Bakri Mohammad has applied for British citizenship and receives welfare benefits. 'I am a very active person. I am not like Kalim Siddiqui, a nice sincere Muslim but who writes articles behind his desk, I am not like Zaki Badawi, who is funded by Libya, who has nobody to listen to him and doesn't give any talks, anywhere. There are people called leaders on the ground, there are many of them, we know each other very well.'

Hizb-ut-Tahrir's radical Islamic theology is the very antithesis of Marxist atheism, but it operates on classical Leninist lines. Core activists such as Bakri Mohammad travel the country, setting up closed cells in towns and universities, whose members are indoctrinated in the party's beliefs and worldview. The activists move on, but the new cadre-members are left behind, to infiltrate the Muslim establishment, such as established student Islamic Societies and try to take them over. For many young British Muslims Hizb-ut-Tahrir provides a set of ready-made answers to both political issues and the questions of personal identity that often draw alienated individuals to extremist organisations. And of anger and alienation, they have plenty. The party's millennarian message of proud self-reliance, rejection of the West, and unyielding Islamism has brought it growing support both in the backstreets of Bradford and Batley and across university and colleges campuses. 'If I was the Home Office I would be worried about the increasing anti-western sentiment among Muslim youth,' says one analyst watching developing trends among British Muslims. 'The idea of conspiracies against Islam, of Christian/Jewish/Zionist anti-Islamic cabals all working against Muslims are dealt with very casually, as accepted facts and this is a matter of great concern.' For some young British Muslims anti-western and anti-Semitic stereotypes are not myths, but facts.

Omar Bakri Mohammad has for several years been demonised as a hate figure in the British media, starting in the Gulf War when he was reported to have called for the assassination of John Major, which he strongly denies. He is a bearded, somewhat excitable man, who, like Arthur Scargill, has the slightly unnerving habit of referring to himself in the third person, and he plays a delicate game of working for the global

141

triumph of Islam, but fighting his battles verbally or in print, while his lawyer supporters advise him of the correct legal parameters in which he may operate. Bakri Mohammad explains how he set up Hizb-ut-Tahrir: 'Omar made a small group of brothers, I started to culture them, I moved with them from campus to campus. We divided the country into three areas: the community, society, universities, campuses or organisations. I played a role in all three of them. We started at the beginning moving from place to place. Wherever I go, I put one or two brothers there, and they convert other brothers. All the chocolate Muslims, they reject that. They don't want Islam to be introduced as a political or economic system.'

For Bakri Mohammad, 'chocolate Muslims' are Islamic Uncle Toms. Just as Uncle Tom is a term of abuse in the black community for someone who has black skin but is considered white inside, a 'chocolate Muslim' is someone who believes in Islam, but fails to live what Bakri Mohammad says is a truly Muslim life and struggle for its triumph. 'Definitely there is a struggle between western civilisation and Islam. The Muslims who say there is not are chocolate Muslims, who don't want to change reality. A chocolate Muslim is the one who does not believe Islam can change reality, who believes "I can let Islam fit with reality". But I believe that God sent a messenger, Muhammad, in order to change reality, not to find if he can live with existing reality, that is what Islam is about. Any Quranic text says God is the only creator, the only commander, that means sovereignty for God. Man-made law says sovereignty for Man. Remember I don't believe Islam is only a religion, the way you think about Islam, or the way many chocolate Muslims think about Islam. It is a system of life, therefore it is impossible to think like this mentality among certain intellectual Muslims, who say "Oh yes, we can live together", no, we cannot live together.'

I ask him what form this struggle between Islam and the West will take. 'It is an intellectual struggle. We do believe we must convince people about the new way of life. We struggle to change society . . . But when we have a debate on the campus they go to the NUS [National Union of Students] and say we are fundamentalists. If you reject the peace with Israel you are anti-Semitic, if you reject western civilisation you are a fundamentalist. If you reject the existing regimes in the Middle East you are an extremist. So they label you as they like. But we are living in the Muslim community and they know us. If I did this in the Middle East I would be arrested or killed. But here there is a facility. The environment of freedom of speech helped us a lot, to convey the message to the Muslim community.'

Many mainstream Muslim British leaders and thinkers are desperate to marginalise Bakri Mohammad and dismiss him as an irrelevance, for his rhetoric is a public-relations disaster for those Muslims who wish to work within British society. 'Kalim Siddiqui and Bakri Mohammad, what is their long-term impact?' asks Professor Akbar Ahmed, Pakistan-born author of several books on Islam, and fellow of Selwyn College, Cambridge. Just as Lenin, that arch-strategist once asked 'who benefits', when trying to gauge a rival's support and tactics, Professor Ahmed queries who is behind the sudden appearance of groups such as Hizb-ut-Tahrir, who, he believes, serve only to discredit Islam and British Muslims, most of whom, after all, merely seek to live their lives in peace and raise their families in the laws and culture of Islam.

'They have come from nowhere, and they go nowhere. Why is the British media encouraging them? Because they can damage Islam. The natural ally of the Muslims here would be the traditionally oppressed classes, such as the Jews, or the Irish. These are the very people who they [Hizb-ut-Tahrir] are targeting. They have alienated the traditional allies, the majority population, the mainstream Muslims, and who have they ended up with? This is an important question that Muslims don't ask in public, but there is not enough critical self-analysis in Muslim society.'

For Professor Ahmed, groups such as Hizb-ut-Tahrir have absorbed the worst of British yob culture. 'They have had their fifteen minutes of fame, which you have given them through the press, with the sensation of them threatening this, and threatening that. They are playing the game, for whatever reason, personal, to propagate their version of Islam, or to satisfy their ego. The damage they are doing to Islam, which is a very rational, balanced system, which encourages dialogue, and all sensible Muslims want this, a much more serious interchange of ideas, the damage is incalculable. People just assume that Muslims are mad fanatics.'

143

But for Bakri Mohammad, Professor Ahmed, seeker after tolerance and co-existence, is merely another Uncle Tom, courting the western establishment. 'Chocolate Muslims will say that. Go the community and tell me which chocolate Muslim is the leader of the community. We have an Islamic law school here with top barristers and lawyers. Why don't they go to the chocolate Muslims like Zaki Badawi? Or Akbar Ahmed, he admires western civilisation more than Islamic civilisation. There is no respect for him in the community. He is a sincere Muslim, but sincere is not enough.'

In fact Professor Ahmed – unlike Bakri Mohammad – enjoys widespread respect among many British Muslims, particularly those originating in the Indian subcontinent. He made headlines in May 1996 when he preached at evensong at Selwyn College, Cambridge, one of the first Muslims ever to address a Christian congregation. His address was a call for mutual understanding and tolerance between Christian and Muslim faiths and peoples, and he quoted the speech given by Muhammad at Arafat: 'God has made you brethren one to another, so be not divided. An Arab has no preference over a non-Arab, nor a non-Arab over an Arab; nor is a white one to be preferred to a dark one, nor a dark one to a white one.'

It was a speech not universally welcomed among British Muslims and reaction to it followed the traditional dividing lines: while the radicals denounced him, the liberals rallied around. Dr Ghayasuddin, the new leader of the Muslim Parliament – itself dismissed by many Muslims as the arm of Tehran in Britain – was quick to attack Professor Ahmed as an 'apologist' who sought the approval of the West. '. . . Muslims have always been oppressed by the West. And to try to merely sweet-talk to them in churches is neither here nor there. Such dialogues have hardly any implications on creating "understanding between religions", if that is the idea,' he told *The News*. That triggered an immediate salvo in Professor Ahmed's defence, from Dr Abduljalil Sajid, Imam of the Brighton Islamic Centre and Mosque, who was present at the lecture. He pointed out that Muhammad himself had invited Christians from Najran to Medina mosque and said Muslims who criticised Professor Ahmed were 'ignorant'.

As well as providing more ammunition in the intra-Muslim

war of words, the debate over Professor Ahmed's address illustrated that Britain's Muslims are a dynamic group, with different voices competing for the hearts of a maturing community. Professor Ahmed himself described his church speech as part of a much-needed 'jihad of understanding', a battle for heart and minds, both among Muslim and non-Muslims. 'For us Muslims this is happening on two fronts. Between Muslims, to understand the real nature of Islam, its message of peace and compassion, and at the same between Muslims and the West, so the West understands that Islam is not a religion of monsters and Muslims don't behave like monsters.'

All that said, I believe it is a mistake to dismiss Bakri Mohammad as an unrepresentative extremist. He has written over half a dozen books on Islam, such as *Political Struggle in Islam* and the *Hidden Force of the Muslim ummah* (world community). He is an intelligent and articulate public speaker who well knows how to play a Muslim audience and how to fashion and exploit the anger of much of Britain's unemployed and alienated Muslim youth. Talking with Bakri Mohammad, I was reminded of Dr al-Mass'ari, which is not surprising for the two are old friends, and al-Mass'ari too is a former member of Hizb-ut-Tahrir.

In his book *Postmodernism and Islam*, Professor Ahmed quotes the writer Fazlur Rahman on the 'postmodern fundamentalist' position, and points to Rahman's argument that this group concentrates on those components of Islam, such as the ban on bank interest, collection of *zakat* (charity) and so on, that will 'most *distinguish* Muslims from the West'. This, it seemed to me, was Bakri Mohammad's outlook. His book *Essential Fiqh* (jurisprudence) outlines the framework for living as a Muslim in 1990s Britain, with chapters covering the laws governing dress, buying meat from non-Muslims, imitating non-Muslims and the prohibition on bank interest, in short how to live in a society without being *of* that society.

So much for the theory. But the practice, say a range of groups from the NUS to the Jewish Board of Deputies, is one of abuse and intimidation. Not so, says Bakri Mohammad. 'We are in Britain. We are not in Zimbabwe. Britain has a government, has a police, authorities. Why has no one ever reported us to the police? I challenge them. It is against my intellectual dignity as a member of Hizb-ut-Tahrir to intimidate somebody. If he defeats me in an intellectual debate, I will lose everything. If I use my hands, or intimidate you, that means I failed.'

But it is Hizb-ut-Tahrir's public questioning of the Holocaust that has drawn the most attention, outraging Jewish groups and dismaying Muslim ones trying to build a common front to fight racism and discrimination. I asked Bakri Mohammad if he believed the Holocaust happened.

'I condemn the killing of women and children by the Nazis. The issue of the Holocaust, I have never seen it [and] I don't believe in something I did not see. But if somebody does not believe that Muslims were killed in Sabra and Chatila, I would never say they are racists . . . If the Holocaust means a political weapon to justify the existence of the Jews in Palestine, killing the Palestinians and the Muslims there, I do not believe in this Holocaust. If the Holocaust means the killing by the Nazis of innocent people of the Jews, women and children, of course we condemn it.'

Do you believe that six million Jews were killed by the Nazis?

'I don't believe so. Is it illegal if I don't believe that? It is not illegal. If somebody believes that a massacre happened to his own people, and someone else does not believe it, does that mean he is a racist? . . . Israeli leaders have said many times that their real defence weapon is not the nuclear bomb, it is the Holocaust and nobody should doubt its existence. So that put a lot of doubt in the minds of the Muslims in the Middle East. It has justified the killing and torturing of many people in Palestine. That's the way we think. Remember I am an Arab from the Middle East, I am talking about somebody growing up there, all my culture is from there . . . I condemn the killing of Jews because of their race, but if you use the issue of the Holocaust to justify something else we will never accept it.'

By the winter of 1995 the increasing media attention on Hizb-ut-Tahrir in Britain was causing disquiet among the party's underground leadership in the Middle East. Its aim, after all, was to rebuild the *Khilafah* in Islam's ancient lands, not engage in public bust-ups with Jewish and Hindu students in British universities. Bakri Mohammad was replaced. In January 1996 he launched Al-Muhajiroun, taking many members of

Hizb-ut-Tahrir with him, as well as former supporters of Jamaat-i-Islami and the Muslim Brotherhood. He won't say how many members the new group has but claims a presence in the United States, France, Pakistan, Malaysia and Mauritius. 'We don't want to talk about it, but we have a lot of members. Many people are Sufis, they can gather a million, but one million Sufis they just dance. But four or five people, politicians and statesmen, they can make more of a difference.'

That a Syrian political exile, a radical Islamist, is free to live in Britain, claim benefits, apply for citizenship, all the while working for the global triumph of Islam, enrages many. I asked Bakri Mohammad how he replied to the argument that, while here, he should shut up and keep his ideas to himself? 'If they said that, we will leave this country, we will never fight. But they have never said that. We political asylum seekers have never been told that our political activity is illegal. Where is the law – I have lawyers here – which says political refugees cannot run any political activity? There is not, and I am not breaking any laws. If they say I cannot, it is my duty to leave and I will leave the next day. I don't want to challenge you. I will leave.'

The Egyptian Dr Zaki Badawi and the Syrian Omar Bakri Mohammad. Both Arabs from the Middle East, both practising Muslims and influential leaders expounding their versions of Islam, but two men with radically different worldviews. Most British Muslims though are not from the Levant, but the Indian sub-continent. They live in the cities and mill towns strung out across the Dales and the Pennines, where even the first-generation immigrants have Yorkshire accents as flat as caps.

It was time, if I wanted to properly take the pulse of British Muslim life, to leave the metropolis. So I went to Bradford, home to one of Britain's biggest Muslim communities, nearly all from Pakistani Kashmir, northern India and Sylhet in Bangladesh. By the year 2000, Muslim leaders say, fifty per cent of Bradford school-leavers will be of Asian origin. Bradford, with its twenty-eight mosques and fourteen Islamic religion schools, is where British Muslims first found their political voice and discovered the power of the media. It was there that the image of a flaming copy of *The Satanic Verses*, nailed to a stick like a literary crucifixion, tongues of fire charring the pages as it

147

burned in front of a crowd of angry Muslim demonstrators, rebounded around the world. This is Muslim life in the post-modern age: where the Image – instant, powerful and permanent – defines the Message, a snapshot tableau that lives on for ever. And for many non-Muslims that picture symbolised every atavistic fear they had about Islam: that it was a faith irrevocably set on a collision course with the West, a religion of fury and rage, intolerance and censorship. The resonance with Nazi book-burning that triggered a deep shiver of repulsion among western Europeans struck no chord in Muslims from small villages in Sylhet and Kashmir.

Burning a book, any book, also horrified many Muslims. Not just because they knew what a public relations disaster it would be, setting back relations with non-Muslims by several years, but also because to burn a book is a deeply unIslamic act. To learn, to read, to study and perhaps most of all, to understand, is incumbent on every Muslim. The librarians of Cordoba, in medieval Spain, who guarded 400,000 volumes, or of ninth-century Baghdad, would have been aghast at setting the written word on fire. As Muhammad himself said: 'The ink of the scholar is more holy than the blood of the martyr.' But the spirit of enlightenment is sometimes absent in Bradford. It was there, that as a Jew writing about Islam, I encountered the single incident of anti-Semitism in the many months I spent travelling across Europe, Turkey and the United States. I was chatting happily with a taxi driver until he asked me what religion I was. I told him, and something hostile glittered in his eyes as he turned to look at me and the atmosphere changed. 'I think we have got a lot of problems then,' he said, as we bumped down the hill towards an almost-completed mosque. These 'problems' he could not well articulate, but they included, of course, Israel/Palestine. He was not exactly deliberately intimidating, but I was glad to arrive at my destination.

In Bradford, they would burn *The Satanic Verses* again, explains Sher Azam, former president of the Bradford Council of Mosques. 'We were quite clear that we do not want to upset anybody, or ridicule anybody, so why should somebody upset and insult us? We are in an environment where there is enough misunderstanding of Islam and when young people start studying this type of book, it will create further alienation and misunderstanding. We were asking for a ban. I supported that. I must make it clear that the Muslim community, and Islam, is as much a champion as anybody else, if not more so, of freedom of expression, but freedom of expression with responsibility. You are free to swing your arm around, as far as you can, as long as you don't touch my nose. Where your arm ends and my nose begins, that is the boundary.'

For many non-Muslims though *The Satanic Verses* came nowhere near Sher Azam's nose. Here two worlds, one western and secular, the other Muslim and religious, collided. The scars on both sides still run deep, still shaping perceptions, strengthening stereotypes and entrenching mutual acrimony. The Rushdie affair erupted over six years ago, but its legacy still poisons relations between Muslims and non-Muslims. There is no solution that could satisfy both sides, but if the long list of British Muslims' grievances had been considered with sympathy and understanding the Rushdie affair would never have become an Islamic standard behind which the community rallied.

'We are not powerful, we are a minority, a very powerless community,' says Azam. 'We hope that the people in power will understand our plight. You say we haven't succeeded, but we have at least brought it to the attention of the people, we have put forward the Muslim point of view and this has created a debate. You can burn pornographic material – we are not against books as books – when it is confiscated.' Before the book was burned Bradford's Muslims wrote to the publisher, the Home Secretary and the Prime Minister. They were, says Azam, fobbed off, or ignored. 'You could see the powerlessness of the community, that we were trying our best, without making an issue of it, and the number of complaints was growing day by day, people were worried, and angry and upset . . . If the circumstances which I have explained to you were similar or the same, we would do exactly what we did.'

Born in north-west Pakistan, Sher Azam, now in his middle-fifties, arrived in Britain in 1961. We met in his house in Bradford, where a Quranic inscription hangs on the wall, and the local paper is neatly folded by the sofa. The narrow streets of terraced houses resound to the sounds of Urdu as much as English. Shops signs are in both languages, the long curved consonants of that Asian language sloping down over more prosaic

English proclamations offering halal meat or basmati rice. Azam is a softly spoken man, much esteemed in the local community for his learning and stature as a community leader, dressed, like many Asians, in a baggy *shalwar-kameez*, long shirt and trousers. British Muslims are here to stay, he says, but they want their rights.

These include: state aid for Muslim schools; extension of race relations legislation to cover Muslims (there is no law outlawing prejudice against Muslims on religious, rather than ethnic origin); and the extension of the blasphemy laws to cover Islam. The Commission for Racial Equality (CRE) has so far maintained that religious discrimination falls outside its brief, which is based on enforcing the 1976 Race Relations Act. But after criticism from Muslim groups, the Commission has begun a series of consultations to begin countering religious discrimination. And groups working against racism such as the Runnymede Trust have appealed to the United Nations Committee on the Elimination of Racial Discrimination, based in Geneva, to press for the inclusion of Muslims in anti-racist legislation. Sher Azam gives an example of why new legislation is needed to protect Muslims: 'A precision engineering company in Bradford decided that their jobs were open to everyone except women or Muslims. The Commission for Racial Equality took them to court, and a very articulate Muslim presented our case, but we lost. The tribunal decided that the employer was found guilty on the first charge, but said that discrimination against Muslims was not an offence. This leaves the Muslim community very vulnerable and angry as well.'

But is there not a legitimate argument though that when an immigrant community comes to Britain it should conform to British values and not expect the host society to change? 'When you go to a place, you accept that country and you accept that land's laws. Muslims are expected and obliged to respect the law of Allah and of the land. For example, this is my house, and I have arranged the furniture according to my needs. If I accept you or someone else on an equal basis, it is a partnership. I expect you to pay the bills of the house, and contribute to the expenses. If this lampshade always catches you on the head, because you are a bit taller than I am, it's not unfair to ask me to raise it up a bit,' says Azam, gesturing towards the ceiling. 'I should do my best to accommodate you. That is what we are asking for, not special favours. The needs are different and it is not too much to ask if we, as an immigrant community, can be accommodated.'

Their campaign against *The Satanic Verses* failed, but Bradford's Muslims still learnt a lot about how to lobby, organise and campaign. 'I learnt a great deal about the misunderstandings we have about our perception of the community at large and the level of misunderstanding which does exist about the Muslim community. We had bags of letters, most of them were hostile. Our office was broken into and ransacked, nothing was taken, but a paper was left saying: "leave Rushdie alone", those sorts of things. I received hate mail, threatening phone calls. My little daughter answered the telephone and somebody said, "I'll cut your father's throat", I had to calm her down. You could see from these calls the hate that does exist in some people's minds. But we have learnt a lot about how British politics work.'

Beyond the media furore, Bradford is now home to a new and hardy plant, one putting down roots wherever Muslims live in Britain: British Islam. This is a hybrid but sturdy strain, born out of Mecca and Medina centuries ago, but now cultured in Manchester and Macclesfield. 'The trend is changing now towards a British system of Islam. The people who have grown up here and have been brought up in this environment, they understand the local language, the culture. They have not been brought up in an Islamic environment as we were. They have seen Islam practised by their parents or grandparents, they read about Islam, they have their own experiences together with the indigenous population. Western civilisation and psychology is more towards secularism and materialism than spiritualism, but they see that these are not the answers. So they are returning to the faith, a faith which is not my faith, but their perception of faith. A few years ago we were scared that our children would go away from faith, we were more worried about secularism. That feeling is not there now.'

Islam evolves in the environment in which it exists, says Azam. 'Islam is a tablet. In the Middle East, or the Far East, or eastern Europe, we have coated Islam with our own particular culture. British Muslims are coating that tablet with a British

Islam which is different from ours, very different. At the time of the Rushdie crisis the young people wanted to assimilate and were moving towards the local culture. That was really worrying, not because we don't want our children to be as British as anybody else, we want all the good habits of Britain to be in our children, but we don't want the bad habits. Parents were worried, but since the Rushdie crisis those people who were drifting away have come back. The Muslim booksellers said that they sold many times more books on Islam to young Muslims and non-Muslims after Rushdie.'

As well as spiritual problems, British Muslims, especially from the Indian sub-continent, face more mundane difficulties. Unemployment is over fifty per cent in some parts of Bradford, say Muslim leaders; racial attacks mean that some families are afraid to venture out, and young Muslims especially say they are regularly harassed by the police. Anger, alienation, racism, joblessness, an uncertain identity in a society that seems to have no place for them: all the ingredients in a literally explosive mixture. In the summer of 1995 it ignited. Bradford's Muslim youth – Hindus and Sikhs stayed at home – rioted and hurled petrol bombs at shops and cars, causing about £1 million worth of damage. The spark that lit the flame was reportedly rough handling of a Muslim woman by police in the course of two arrests. Muslim honour was defiled, Muslim pride was running high after a moderately successful campaign that spring to run local prostitutes out of town. So Muslim youth took to the streets, venting their fury.

And when Bradford's Muslim youth run into trouble, they call Aurangzeb Iqbal. Born in Mirpur, Pakistan, Iqbal, now in his early thirties, arrived in Britain in his early childhood. A flamboyant lawyer with a flair for snappy suits and controversial publicity – one campaign featured the Kama Sutra with the slogan 'We'll defend you in difficult positions' – Iqbal has set up his own firm of solicitors with offices in a former converted chapel. The waiting room is crowded with both British and Asian clients, asking for help with everything from divorce to criminal prosecutions. His clients are often young second- and third-generation Muslims, British citizens who lack the immigrants' deference that shaped their parents. Four of them, accused of threatening behaviour and assaulting the police during the Bradford riots, walked free after the prosecution case against them collapsed.

'British-born Muslims' aspirations are different. So far as they are concerned, the hunger that associated with the first generation, to take any menial job, is not there,' says Iqbal. 'They say why should we do that? We've gone to the same schools, colleges and universities, we are just as good, why aren't we getting the jobs? The authorities need to address this issue, there are parts of Bradford with sixty, seventy per cent unemployment in private residential areas, not council estates, and those kids have no channel, nowhere to go. With unemployment that high, one is asking for trouble. As far as I know not a penny has been spent to set something up for the kids, such as a leisure centre, job centre or training and skills centre.'

For Iqbal, a football fanatic, part of the answer is to bring young Muslims to sport. His greatest sporting achievement to date was helping organise the match between a UK Asian XI and Bradford City. The Asian XI lost 3–1, but that wasn't the point, says Iqbal. 'Muslims and Asians need to be involved in every sphere of activity. We don't just want Asians to be lawyers and dentists, we want them to become microbiologists and sportsmen. Sport has a unique ability to break down racial barriers. Ten years ago they said that black players will never make it because they will feel too cold in the winter. Now they say Muslim players will never make it because they will want to start praying in the middle of the match, when it's prayer time. I don't know where they get this from, I really don't. In five or ten years' time I want to see a situation with footballers wearing the names Khan and Abdul on their back. The Muslim community needs role models, to help bridge with the host community. Sports personalities can do that.'

I imagine that should he so desire, Iqbal could win a seat in Parliament in a few years. A graduate of Salford University and founder member of the Asian Business and Professionals Club in Yorkshire, he seems a natural spokesman for a community that surely needs young voices in 1990s Britain. Iqbal is in the vanguard of the British Islam that Sher Azam says is growing in Bradford. His religion is an integral part of his life. He prays five times a day, not always on time, for court cases wait for no man, not even Muslims, and reads the Quran every day. 'Islam gives you discipline, you never give up. You say your prayers, you don't say, "Oh, I don't feel like it today." No, you feel like it. You always look at the positive side, never the negative. The prophet Muhammad said never say "if", in your dealings in life. If something goes wrong you say it doesn't matter, I will learn from it and I will carry on. But Islam allows for flexibility, if you don't have time you read the Quran later. Because during the day I am involved in things, I pray in the morning before I come to work and the other four I say before I go to bed.

'If you see a Muslim you are supposed to have a very good idea that person is a Muslim, by the way he talks, the way he dresses, his demeanour, his manners, that shows he is a Muslim. So what good is it if somebody prays five times a day, wears tatty clothes, with rotting teeth. That is not living your life in the spirit of Islam. Islam tells you to dress smartly, to spend upon yourself, you are not supposed to hoard money. It's wrong to have a big bank balance or build a massive house in Pakistan that nobody lives in. That's wrong. Spend that money on your kids, on private education, don't hoard it. Have a nice house, invite your neighbours in regularly. Set yourself high standards, so that other people think, "I like that, I would like to be like that."'

But too many Muslims just sit back and complain, says Iqbal. 'The Muslim community needs more people like myself to get its views across. We can't always turn round to the local authority and say "it's your fault, your fault, your fault". The prophet Muhammad himself was a businessman, and there is no better example than him, his name was Amin, the truthful one. When any work was being done, he did it too. If it was building a mosque, he did it, if it was a battle he was involved in everything. Why is it up to the host community to provide us with jobs? To some extent we have done that. We employ thirty-two people, ten years ago you wouldn't have thought that a Muslim business could have so many people.'

British Muslims desperately need their own voice in the corridors of power, say Bradford community leaders such as Muhammad Ajeeb. When, in 1985, Ajeeb donned his ceremonial robes as Lord Mayor of Bradford, the first Asian – and

Muslim – to hold such a position, the city's Muslims glowed with pride, although many white locals were furious.

'I felt very honoured, very proud to achieve that, in a foreign country, I was from the first generation of immigrants,' says Ajeeb. 'Particularly coming from a country that was part of the British empire, where we were not regarded as equal human beings. Despite all these differences, prejudices, racist attitudes, I thought it was a remarkable achievement. Within the Muslim community and black community, they were elated. But within the white community there were mixed feelings. I got abusive letters, some very abusive.'

Now in his late fifties, Ajeeb arrived in Britain in 1957 from Kashmir. He trod the traditional immigrant's path, working for British Rail, and as a bus conductor, and has been an active Labour Party and union member for decades. He admits he has an eye on a possible seat in Parliament. 'When Muslims have their first MP it will be a watershed in British political history. It's vitally important the Muslim community gets a voice in British politics, every section of society should have representation. If you don't have any avenue to express your grievances or anxieties, other than lobbying, you feel disenchanted, left out, alienated, to the extent that you vent your frustrations on undesirable outlets such as riots. You become very bitter and the Muslims are no longer a tiny minority. They are the largest minority group within minorities. My major worry is that if marginalisation continues young Muslims will join Hizb-ut-Tahrir. I don't know who is advising the social planners up there in London, but in marginalising these minorities we shall be encouraging the extreme elements. My experience is that frustration ends in apathy or violence, there is no third outlet for it. We must not allow either.'

The Satanic Verses is still on sale. But Bradford's Muslims have notched up a significant political victory: over *halal* meat and Ray Honeyford, a Bradford headmaster who attacked the council's multi-cultural education policy and opposed the provision of *halal* meat for Muslim pupils. Honeyford went. *Halal* meat stayed on the menu. 'The *halal* campaign was a major victory. Muslims were ratepayers, tax-payers, they were entitled to some things they were not getting. They struggled for two or three years to get that right for their children. They couldn't understand why the council was not willing to provide the kind of meat that their children could eat. Any group of immigrants, such as the Jews, has to go through this struggle. Muslims are desperately trying to learn lessons from previous communities who have settled here in Britain.'

For now the buzz-words in Bradford and London, in mosques across the country, are 'British Muslims' and 'British Islam'. But Ajeeb, as always the far-sighted politician, is looking further ahead, to Europe. A common European home means one for Muslims as well, whether they are Bangladeshis in Bradford, Algerians in Amiens or Turks in Tubingen. 'The sense of being British, being English, is being eroded and once we become full members of Europe, that will become less and less significant and relevant. There will be new dimensions introduced into the concept of nationality, among those there will be Jews, Muslims, Hindus, Sikhs, so in that respect I don't think that the question of identifying truly with Britain will be inescapable,' he says.

'What we need to recognise as soon as possible, because the later we leave it the more problems we will face, is that Britain has become a multi-faith society. I use this word deliberately – it is irreversible – you cannot reverse it now. What are we going to do with these Muslims? Ethnic cleansing, like in Bosnia? We need to start from that recognition and then we can move on. That's why the Muslims are concerned. This is their country, they don't know any other country, if they want to fight for their rights they have to do it here, not in Palestine or Saudi Arabia.'

*French school students demonstrating for the right to
wear Islamic headscarves*

Chapter 6

FRANCE: THE DYSFUNCTIONAL FAMILY

'Liberté, Egalité, Fraternité' (Freedom, Equality, Brotherhood)
Slogan of the 1789 French Revolution

'France has two kinds of children, French children and Beurs [Arab] children but she doesn't want to recognise all her offspring, as though she had a bastard child.'
Nassera Sarah Oussekine, Paris-based feminist activist

Had I sat next to Mustapha Tougui on the Metro, as it snaked and rattled its way under Paris on the way to the city's main mosque, I would never have guessed that he was a lecturer there in Islamic theology. A sports teacher perhaps, or a musician, but not an expert on Islam, who has studied his faith everywhere from Saudi Arabia to Indonesia and translated the Quran. Tougui is quite pale-skinned, clean-shaven, eschewing the beard favoured by many Islamic radicals, and, indeed, looks like a slightly portly New York home-boy, with his blue baseball cap, white poloneck, furry thermal jacket and sports shoes. He is a very amiable man who smiles a lot and I imagine that he is a good teacher, for he has a pedagogue's gift for explanation. We met in the gardens of the Paris mosque, where twice a month on a Sunday morning he teaches the tenets of Islam to about 100 students, most of whom are from pure French, rather than north African, families, and who nearly all eventually convert to Islam.

A magnificent complex of mosque, school and courtyards that takes up most of a block in Paris's fifth *arrondissement*, the mosque was completed in the 1920s as a recognition of the contribution of the many Muslim, mainly north African, soldiers who fought in the French army during the First World War. It is a tranquil site, composed of several symmetrical courtyards with arched walkways that in the spring and summer are awash with verdant trees and shrubs, red and pink flowers and gently gushing fountains, a sort of mini-Alhambra. The ancient Islamic love of symmetry and geometry permeates the site, with its intricate arrangement of multi-coloured tiles that advance and retreat in patterns that are simultaneously rigid and linear, circular and flowing, symbolising and emphasising the unity of God. Just like the Alhambra, the Paris mosque complex is more ornate than the sparser mosques of the Arabian Gulf. It is a place of rest and worship, where both Muslims and those of other faiths can find a few moments' peace and tranquillity amid the bustle of modern-day Paris. Step inside the gate and you are transported to Fez or Marrakech or perhaps medieval Islamic Spain.

'All this work is by Moroccan artisans, they are specialists. If you go to Morocco it looks exactly like this,' Tougui explained as we examined the wooden cornices and inner coverings of the mosque and he pointed out the Quranic verses that have been carved into the panelling. The Paris mosque is built in the architectural tradition of Muslim north Africa, an area where several empires from Roman to Spanish have left their imprint. That mélange of cultures has produced something more intricate than the vast plains of the Arabian desert, says Tougui, with a different style of building. 'Morocco is a wet country, not a dry country like Arabia, it has water, trees and forests. The people there have a different style of life to the Bedouin people and you always behave like the place where you are. You have a look around you, at your surroundings. The mentality, the style of life, the mixture of people from the Arab world, Spain, the French, they all meet there.'

Talking to Tougui, who fluently expounds the doctrine of Islamic universalism, is an enjoyable and illuminating experience. 'The Quran is the synthesis of religion, not something new, a universal message for all people. You have absolute freedom to follow it or not. But as the Quran says, and the religions before it say, good deeds will bring their own rewards, and so will bad. Just as when you cross into a country, you have the right way and the wrong way, the long way and the short way, and you have very wide space to decide yourself which one to take,' he explained as we strolled through the gardens.

Born in Marrakech, Tougui, now 45, arrived in France over thirty years ago. Tougui is one of over four million Muslims in France, the largest Muslim minority in Europe, and the second biggest in the western world, after the United States. Unlike Germany, whose Turkish immigrants had no historical or cultural ties with their new homeland, most French Muslims were either born in France, or brought up in French colonial possessions in north and central Africa. They spoke French, understood French culture to a greater or lesser extent and felt ties of language and understanding. An Algerian who left Algiers for Paris after the war of independence in 1962, felt far more at home there than a Kurd transplanted from Diyabakir to Dortmund, or a Bengali who settled in Bradford.

In Algiers too, he could eat croissants for breakfast, read Albert Camus – himself born in Algeria – and drink *café au lait*, as well as thick Arabic coffee. France's Muslims felt an ambivalence, certainly, about the white European power that had ruled their countries, and helped itself to their riches, railroading their economies for its own benefit, but they also absorbed French ideas about liberty and social justice, the legacy of the 1789 revolution that still reverberates across the country today. But the immigrants' naive views of France were sorely tested when they arrived to work during the 1950s and 1960s. They were housed in ugly concrete developments on the edge of the great metropolises such as Paris and Marseilles, classic inner-city ghettos that soon deteriorated from neglect and isolation. The welcome that they hoped for did not materialise. Instead they found jibes and hostility, the sort of prejudice that would, in the late 1980s, boost the openly racist National Front. While Muslims in France have not suffered the physical attacks that Turks in Germany have endured, they must live in a country where the extreme right enjoys more electoral success than in any other developed European nation apart from Austria.

Tougui had been brought up a Francophile, in the tradition of that country's literature and culture, a land that claimed to be the home of human rights and human dignity. But it didn't seem that way when he arrived in 1966. 'It was the shock of my life when I came here for the first time. I was shocked at the reception. When I met French people in Morocco I received them very well, my family and my friends, like they were perfect people. When I came here they treated me like shit. It was everywhere. I can't give you an example because there are so many. They insulted you, completely insulted you. Now it's different, it's not better, I think it is worse, because they disguise it, before it was direct. They said all racist things, they call Arabs goats or rats. I was invited in by a family and the father couldn't resist from telling me that Morocco is full of rats. It meant I was one. The son, my friend, was very angry. He told me that his father was quite racist. These words give no dignity to the people who use them.'

As Tougui talked, an anger and resentment began to bubble to the surface and his voice tightened, the first manifestation of an almost-rage that I would sense or hear repeatedly as I travelled in France among its Muslims. It was the anger of a guest, in a house or a country, who is at first bemused, and then bitter that where he thought he would find a welcome, he encounters only hostility. 'I didn't come to France to get a job like a labourer. I have been to university, I can work everywhere. I chose this language from love, we came here for love. But we were very shocked when we came for the first time to see how people received us. They make a sort of advertisement in our country that Paris is the centre of the world, that French culture is most wonderful, just and perfect, democratic. But when you see the truth you are very disappointed, they talk in an opposite way on purpose about things at home. For example, about Islam, or customs in Morocco, when you come here the media turns everything upside down. In France they say when you are invited to a Moroccan family and you eat, you must burp. This is just rubbish. Or if they kill a lamb for you in your honour, they give you the eyes to eat, this is absolute rubbish, rubbish, a lot of rubbish. It is very harmful, because it's a little bit of rubbish here and a little bit there, and so it's more insidious.'

Now the media stereotypes – of bearded terrorist, or crazed fundamentalist – are more sinister than the former picture of eyeball-munching, burping Arabs, and their near-universal permeation across French society has disturbing implications for the lives of French Muslims. They affect everything from education and civil liberties to the simple freedom to build a mosque, the very idea of which drives provincial French politicians, who are happy for Muslims to carry out their town's dirty and menial jobs, into a collective frenzy. In Marseilles, for example, home to many hundreds of thousands of Muslims, there is not a single building that has been built as a mosque. Every Muslim place of worship there has been converted from a flat, shop or warehouse. Tougui, like most Muslims in France, is furious with the French media for their constant portrayal of Muslims as Islamic radicals who either wish to undermine the secular French state – that is based on the sacrosanct principle of separating church and state as a legacy of the 1789 revolution – or are planning a fresh campaign of bombings on the Paris Metro, such as the rash of attacks in autumn 1995.

That climate of hysteria, he believes, leads directly to the kind of physical attacks that several of his students have suffered. 'As you see I don't look very Arab or Muslim, so my life is not very miserable. But in the people who come to my lectures, three women have been stabbed on the Metro, just because they wear scarves, by normal people, not by thieves trying to rob them. They came with black eyes, people said to them you are Muslim and then boof! Another came with broken teeth. This is the media, the media who are responsible and the government as well. The media prints any rubbish, they say Muslims are the worst people in the world, Islam is the most horrible religion in the world, and you have stupid people with no conscience, no imagination, no instruction, no nothing and they think they are doing something good by attacking Muslims. I told them, when I saw these three women, I told them, it is the media who have stabbed you, not the others.'

At first I yawned inside when I heard this, for blaming the media for one's ills is an age-old past-time, and one that I, as a journalist, have heard many times and usually have little time for. I was told several times as I travelled in Paris and Marseilles that if I was a French journalist nobody would talk to me, particularly young people from the deprived urban satellite towns that are still home to many French Muslims of north African origin, who have suffered the brunt of misreporting. After just a few days in France I began to see why. Islam is an obsession for

France, an obsession stoked up by politicians and headline hungry reporters who feed off each other, creating a climate of intolerance and hysteria – it is no exaggeration to describe it thus – that I found nowhere else in Europe or America.

In the space of just a couple of weeks, in early 1996, I found the following special features on Islam on the newsstands. *Le Point*, a weekly news magazine, ran an eight-page story headlined '*Islam en France, Les intégristes sapent la République*' (Islam in France, the fundamentalists are sapping the Republic), with other headlines inside such as '*L'Islam radical est une idéologie de combat*' (Radical Islam is an ideology of combat). *Courrier international*, a digest of translated articles from around the world, ran a special issue entitled *Islam, Occident: L'Affrontement* (Islam, the West, the confrontation), its cover design overlaid on a background of Arabic calligraphy. *Le Nouvel Observateur*, another news magazine, also ran an eight-page feature entitled *Ce que veulent les Musulmans* (What the Muslims want), although some of the articles were quite balanced. *Enquête sur l'histoire*, a historical magazine, had *L'Europe et L'Islam, un conflit séculaire* (Europe and Islam, a secular conflict) as its main story. And so it goes on, day after day, week after week, an extraordinary obsession with the supposed Muslim menace.

But in a strange way, the media's obsession with Islam is that religion's best recruiting agent. In France, like everywhere, people don't believe everything they read or see on television. Many of Tougui's students wonder about the truth that lies behind the headlines. 'Everybody talks about Islam, the media of course are over the top, so they want to know. This bad propaganda in some way is the best help for Islam. Of course it makes your life a bit miserable, but it helps. The people who come here are looking for truth, they are serious people. They are the sort of people who want to know what Islam is, whether it is good or bad. Nearly ninety-nine per cent convert to Islam. They say it is incredible when they learn what Islam is, they thought it was something else. The media talks about Islam like Muslims are just crazy, we are terrorists, extremists, whatever. But as Giscard d'Estaing said, if you want authenticity you have to look at the origin, the source, which for us is the Quran.'

Tougui, like many Muslims, believes that France's Islamo-phobia is just a more subtle form of racism. Blatant anti-Arab prejudice is no longer acceptable, so now it's masked as an analysis of Islam – with the conclusion that Muslims are a menace to France – a perspective that gives those views a spurious intellectual gloss. 'In South Africa, racism is now frowned upon, in America they try to disguise it as well, the world now doesn't accept racism because it doesn't look good. So they make the same record, they just turn it over on the other side, and when they want to say the same thing about Arabs, they use Islam. If they say Arab, it's directly racist, but if they say Islam it's something intellectual, something different, they can pretend it is some kind of analysis. But of course not everybody is the same. I still love this country, the people who made the culture here and I still have very nice and lovely friends in this country. They feel sorry about what the majority says.'

Tuogui now spends half his time in London, in Earl's Court, a lively multi-cultural area in the west of the city that is home to a myriad of different cultures and nationalities. The difference in attitude is palpable, he says, for while Muslims in Britain certainly have their problems, there is an attitude in London of live and let live. 'There is a very obvious difference between France and England, how they talk about Islam. In England, I don't care if the people look cold, but they are all right, they are not nervous, they are quite fair. They are respectful of everything. You can feel it straight away at the airport, even at British Airways, where I have my silver executive club card, I always advertise for British Airways. As you get in the airplane you can feel the different atmosphere. Then when you come to England you can see straightaway that people are very calm, fair, you respect yourself and respect others, and you can live in peace. People here are very surprised when I tell them that in Britain if you work in the post office or somewhere else you can wear a headscarf, they think it's impossible. The Sikhs have their turbans, when I saw that I said I feel very proud of Britain. Here it's like a horrible dream. France is a lovely country, with a lovely culture but the people, the media and the government make it like a hell.'

The Paris mosque is something of a family sinecure, in the

hands of the Algerian Boubakeur family since the establishment of the Muslim society in Paris in 1921 that collected
subscriptions for the mosque's establishment. The mosque's first
director was Si Hamza Boubakeur, the second is his son Dalil.
Dr Boubakeur was formerly a medical practitioner and, sitting
in his office under a large painting commemorating the
mosque's establishment, he chooses a medical metaphor to
describe the state of Muslim life in France. 'Islam in part of its
life in France, is like a patient who has suffered and needs a
convalescence, in its spiritual life, in its practice and its way of
understanding with general society. The Islamic population has
grown too fast in a country which was not prepared to receive
such a great community in such a short time, but Islam has
roots, legitimacy in France for a long time.'

Like Mustapha Tougui, Dr Boubakeur says the waves of
Muslim immigrants into France, who arrived in France after
Algerian independence, expected to adapt quickly to life in
France. 'If an immigrant is from Sri Lanka, he does not know
the mentality of French people and the way they think. But not
in Algeria, so there was a natural installation. Then France
saw that these populations were fixed, and needed things to live
like schools and so on. So in the 1970s and 1980s a new problem arose when the immigrants became part of the French
population. They are Muslims, and we have a role in their spiritual, religious and Islamic life. Their children are Muslim, they
need mosques, religious leaders and spiritual guides, and that
needed an attitude from the political authorities, to take a decision, to clarify its attitude towards these new problems in
France.'

But French Muslims are still waiting for that decision, for a
framework for them to live and organise within. The mosque is
important, not just because of its location in the capital and
subsequent prestige, but because it is one of the few Islamic
institutions to be formally recognised by the French government. That lack of recognition has serious implications for
France's Muslims as they struggle to legalise and protect their
status along similar lines to the country's Jewish and Protestant
minorities. Islam is now France's second religion, but there is no
central Muslim organisation to run its affairs and negotiate with
the government, a legacy of French officialdom's unwillingness

to recognise the faith, but also of internal political and religious squabbles among French Muslims.

France's Muslims also want their religion to be officially recognised by the government with a formal concordat that would regularise its status and guarantee the civil rights of Muslims, says Dr Boubakeur. 'There are political reasons to make Islam something apart, there is a concordat with Jews and Protestants, but even now there is nothing with Islam. This mosque was built in 1922 under French law, there was a willingness then to give to Islam a structure. But that was 1922, and now it is 1996 and there has been no progress except this mosque. Many Muslims feel that Islam is considered not only the second religion, but as a secondary religion, and the authorities do not treat it equally. Take schools, there is an accord with other religions, to have schools, but with Islam there is nothing . . . In other countries in Europe, such as Britain or Germany, they have found an attitude of conviviality with Islam, with Muslims.'

But for all its pride of place as the public face of Islam in France, the Paris mosque is itself deeply controversial among France's Muslims. Many Muslims refuse to worship there because of the clashes of personality around its leadership and because of its strong north African emphasis. France's Turkish community, for example, more or less completely ignores the Paris mosque in favour of its own places of worship. Some Muslims decry Dr Boubakeur as a stooge of the French government, eager to do its bidding in return for power and prestige. Certainly he has his criticisms of the government, but he also has kind words to say about the right-wing minister of the interior, Charles Pasqua, the man whom many Muslims charge with infringing their civil rights and whipping up a nationwide anti-Muslim prejudice.

As the tempo of Algeria's civil war speeded up in 1994, and French casualties of the war lay dead next to Algerian ones, Pasqua ordered deportations, arrests and checkpoints to be set up in Muslim-dominated neighbourhoods in a police dragnet to catch suspected illegal immigrants. A whole community was labelled as potential terrorists, an image aided when French police proudly displayed a cache of arms they had discovered in a raid. Still, says Dr Boubakeur, Pasqua had an open mind when it came to the problems of France's Muslims. 'He was a man open to understanding of Islamic problems. With him there was a beginning of an understanding. He accepted an accord of an Islamic organisation in France, that was the first time that a Muslim organisation came to the ministry and got official recognition. But we have now returned to ancient considerations. Perhaps it is secularism and perhaps the period of last year that changed something in the consideration of Islam in France. That was a very difficult period, and now there is a convalescence. The Muslims did not accept the situation, or violence, absolutely not. But there was a change in people's social and psychological attitude. Muslims found it more difficult to find work and extremists can have a new influence against Muslims.'

Dr Boubakeur's view of Charles Pasqua, as a friend of French Muslims, is met with hoots of derision by one north African worshipper at the mosque, so much so that he rushes to tell his friend and they laugh out loud, before making a series of claims about strange connections between the government and the mosque's administration. 'You cannot say worse rubbish than to say Pasqua is good for Muslims. He is a bully boy. If Muslims have the possibility to pray somewhere else, they don't put their foot here, because of these connections. Many people hate this place, but you live and you have to do your prayer and you have to do what you can.'

Two dates define the lives of Muslims in France: 14 July 1789 and 1 July 1962. The Bastille, the prison in the centre of Paris, was stormed on the former, arguably the start of the French Revolution, and on the latter, Algeria declared its independence. Few, if any, Muslims raced with the mob on that summer's day over 200 years ago, but its demands to be freed from royal tyranny and to separate church and state are now enshrined in the constitution, and are wheeled out in defence of the status quo whenever Muslims ask for their own civil rights, such as the legal freedom to wear headscarves at school.

For many French Muslims, the headscarf affair, as it became known, was a defining moment, confirming them as perpetual outsiders in French society and pushing them towards a more radical Islam that is less accommodating with life in the late twentieth century. The headscarf affair began in 1989 in the Paris suburb of Creil, when three Muslim girls were expelled for refusing to uncover their heads, and its repercussions still reverberate across France today. Their headmaster argued that wearing a veil contradicted the principle of separation of church and state and what began as a local row ballooned into a nationwide debate, about how France can accommodate Islam, and how Muslims can adapt to life in a secular country. In the tolerance corner were the then Socialist education minister Lionel Jospin, who argued for a dialogue with the pupils to persuade them to uncover, but said they should not be expelled for refusing to do so. 'Society has changed, today it is more pluralistic,' he said. He was supported by Danielle Mitterrand, wife of the late President, who supported the pupils, saying: 'If today the principle of secularism proved itself incapable of welcoming all religions, this would be a significant step backwards.' In the intolerant corner, were, of course, National Front leader Jean-Marie Le Pen; other socialists; and a whole slew of French intellectuals and media pontificators such as Bernard-Henri Levy, who would later enjoy much acclaim as the self-appointed voice of Bosnia's Muslims in France when the war erupted a few years later. There are, it seems, fashionable Muslim causes, in Sarajevo, and less fashionable ones, in the suburbs of Paris.

But even more than the legacy of 1789, it is France's history in north Africa that still hinders French Muslims as they struggle to get acceptance in the country they now call home. It is a history, say many Muslims, with which France has yet to come to terms. Like all the great European powers, such as Britain, Spain and Germany, France had colonies and spheres of influence abroad, hers mainly concentrated in north and central Africa and the Levant, in countries such as Lebanon and Syria. But France has yet to make a gesture of reconciliation for its annexation of a slice of north Africa, and nor is it likely to. The blow to national pride is too great, the scars too fresh.

Even so, beneath the mutual resentment between former colonial master and subject, a web of strong historical bonds links France and north Africa for ever. There are Arab restaurants on virtually every street in France, serving cous-cous and kebabs, usually to a background of Arabic music. Admittedly, eating foreign delicacies does not necessarily trigger a broader interest in its country of origin – Britain's legions of beer-swilling and vindaloo-guzzling lager louts are rarely avid readers of Hindu vedas – but for those interested in other lands, food can be a gateway into other cultures. On a more serious level, there is still a strong Arabophile streak in French intellectual life, as witnessed by the construction of the Arab World Institute in Paris. The Institute, a few minutes' walk from Notre Dame Cathedral, is, in the words of its brochure: 'A symbol of the partnership between France and twenty-one Arab countries, and aims to generate a dialogue between cultures that have been in frequent contact for centuries.' It is a remarkable building, light and spacious, whose southern façade is composed of an intricate web of steel shutters that open and close according to how brightly the sun shines, to let in and shut out light. The Institute also houses a print and video library and hosts cultural exhibitions that can draw hundreds of thousands of visitors, testament to an enduring interest in France in Arabic culture.

That tangled relationship began in earnest with Napoleon's occupation of Egypt in 1798, the start of a serious French presence in north Africa, and the first military invasion of the Middle East since the Crusades. Its repercussions are with us still. Writing in *The Arabs in History*, Professor Bernard Lewis says:

> Their rule in Egypt was of brief duration, but profound significance. It began the period of direct western intervention in the Arab world, with great economic and social consequences. By the easy victory which they won the French shattered the illusion of the unchallengeable superiority of the Islamic world to the West, thus posing a profound problem of readjustment to a new relationship. The psychological disorders thus engendered have not yet been resolved.[1]

French claims on north Africa peaked with the repeated assertion in the 1950s, which seems ridiculous now, but was then taken very seriously indeed, that, in the words of the late President Mitterrand, 'Algeria is France'. All around the world empires were dissolving and former colonies gaining independence, but not Algeria. Those three words still colour French attitudes to its Muslim community, for while Britain had its empire in India and Pakistan, it was never government policy to claim that these lands were actually part of Britain. For France though, Algeria, unlike Morocco or Tunisia, was a *département*, or rather several *départements*, of France itself. This idea, that a part of another continent was as French as Calais or Bordeaux, was fantastic in its conception – and the native Algerians took a rather different view – and perhaps it may have worked better if they had been granted full political rights. But they were not and the legacy of bitterness on both sides lives on, poisoning relations between Muslims and non-Muslims in France. Countries, like people, are traumatised by events in their lives, and many Muslims believe that French national anger at the loss of that country still determines its current attitudes to Islam and its Muslim minority.

The Algerian war of independence began in November 1954, organised in part by Muslim veterans of the French army who had fought in the Second World War. Like the conflict in Bosnia, it was a semi-civil war, between two sides who both claimed a common land, and it was a brutal affair, marked by atrocities and torture on both sides. The National Liberation Front (FLN) won, and set up an authoritarian socialist regime that was corrupt and economically inefficient. The war was a disaster for France. It divided the country politically, over whether to grant independence, it was socially divisive and it brought down the Fourth Republic, bringing to power Charles de Gaulle, who became the first president of the Fifth Republic. De Gaulle reached an accord with the FLN in spring 1962 and independence followed soon after. The *pieds-noirs*, French settlers, flooded into France, angry and bitter at their betrayal, for many had lost everything they had spent years building up in Algeria. Algeria's large Jewish community left as well, together with thousands of Muslim Algerians, who saw little future under the FLN's Marxist-orientated regime.

France's domestic politics are still shaped by events in its former possessions. Mainstream French politicians are happy to play the race card, partly to head off the challenge from the National Front, and partly because they know their constituents and which notes to play to summon up ancient communal fears of Islam, which is why Charles Pasqua, the hard-line interior minister, was put in charge of France's Algerian policy. The civil war in Algeria started in earnest in 1992, after the government cancelled the elections that the Islamic Salvation Front (FIS) was sure to win. Ever since, French governments have backed the military regime against the will of the people. The nightmare scenario in Paris is a takeover by FIS and the mass exodus of hundreds of thousands of Algerians to France. Xenophobia would rocket, as would support for the National Front. But however events play out in Algeria, its fate is locked to France's, because it has nowhere else to turn, says Serge July, editor in chief of the daily newspaper *Libération*. Writing in *Newsweek* he argued that, 'France cannot break with Algeria. Algeria cannot turn for help to any other European power. The French and Algerians are condemned to co-exist . . . France can never escape the inconvenience, the side effects, indeed the direct impact of the Algerian tragedy.'

The most direct impact of the Algerian tragedy was the wave of bombings that terrorised France in the summer and autumn of 1995. Eight people were killed and 170 hurt in the first terrorist attacks in Europe blamed on radical Islamists. The Armed Islamic Group (GIA), a split-off of the FIS, reportedly claimed responsibility for the bombings. The communiqués attributed to it demanded that President Jacques Chirac convert to Islam and then vowed to continue the campaign of attacks until 'Islam has triumphed in France', a vow accompanied by a map showing the Eiffel Tower exploding. The attacks were greeted with horror by the overwhelming majority of French Muslims, both because they utterly opposed the loss of life and because they knew that the bombings were yet another setback in their struggle for acceptance. Not all believed that the GIA was responsible. Many Muslims saw the hand of the Algerian secret services, which maintain active networks among France's north Africans, in the attacks.

As one north African told me: 'These bombs originate from the special services. They manipulate people to do it. It's quite funny, but all these bombs went off just before the Algerian presidential election, and once the dictator was confirmed in power, they stopped, just like that. Every Muslim thinks this and even some of the media have said it. They need something like these bombs to justify their war against us, that they have to fight against the "Islamic monster".'

Certainly there are strange events occurring among France's Algerian community, such as the murder in a Paris mosque of Sheikh Abdelbaki Sahraoui, the 85-year-old co-founder of the FIS. And then there was the Kelkal affair. Khaled Kelkal, a 24-year-old Algerian, was shot dead by police in a village outside Lyon after a three-day manhunt. His fingerprints had been found on an unexploded bomb on the Paris–Lyon railway line. Police claimed that he opened fire first and so was killed by them in self-defence. Perhaps he did, but the shooting was filmed by the independent French television channel M6. The end of the crew's film showed Kelkal lying wounded on the ground, while an unidentified voice shouts, 'Finish him off, finish him off!' followed by Kelkal pointing his pistol. Shots ring out and a voice says: 'OK, it's good,' and the camera shows his dead body. That last part, that to many smelled of execution rather than arrest, was edited out of the segment broadcast on the news.

For France's Muslims all roads lead to Algeria, but I could not go there, as journalists are regularly targeted for killing in that bloodied land. So I went instead to the next best place, to Marseilles. I took the TGV train as it raced down the spine of France and transported me from chic metropolitan Paris to the ocean, to a city where the spirit of north Africa, its smells, tastes and languages hang heavy in the salty sea air. Marseilles is on the other side of the Mediterranean to Algeria, but the same ocean laps at the beaches of both countries, for the city is the doorway to both France and north Africa. Here, in tiny garrets on the steep, sloping backstreets, homesick Muslim exiles listen to Algerian radio and save for satellite dishes to watch the television news from home. In the villages nearby, their Christian counterparts, the *pieds-noirs*, do the same, except their dreams are of the businesses and estates that they lost and of the homes they were forced to flee when the FLN came to power.

Like all ports Marseilles is a tough town, where life is lived a little bit harder and a little bit faster than in the sleepy villages and holiday resorts nearby. It is home to a myriad of immigrant communities, from every former French possession from the

Antilles Islands to Algeria. Step off its main streets and into the Arab quarter and you are transported out of clean, antiseptic, ordered 1990s Europe, almost straight into the Middle East. And where there are Arabs taking a coffee, there are often policemen, stiff and unsmiling in their blue uniforms, standing and watching with their pistols at their sides. Halal butchers brandish cleavers as they hack at lamb carcasses, Arab music drifts from the tiny cafés and gaggles of men sip tiny cups of thick Arabic coffee. This is not the tourist's south of France, and the price of an Orangina on the beach at Nice will here get you a blow-out meal of cous-cous and lamb washed down with a few glasses of rough *vin de table*. Plush yachts bob up and down in the docks of the old port, but it is the creaking ships that ply the route from here to Algiers that are the city's lifeblood. The boat ride to Algiers from Marseilles takes twenty-four hours, the ferries ploughing across the ocean an aquatic link in the umbilical cord that even now binds Algeria to its former master.

For Ramzi Tadros, who works at the city's foreigner reception centre, Marseilles is the bridge from France to north Africa, and neighbouring countries. 'Marseilles is the gate between north Africa and Europe. If Marseilles cut its links with north Africa it would be a dead city.' The centre, which opened in 1977, is often the first port of call for bemused north Africans when they arrive in Marseilles. It provides legal help, a library and organises cultural events.

Its work is not always welcomed and it has been bombed several times, says Tadros, although nobody was charged with the attacks. Its clients' problems have changed over the years. 'Ten years ago it was to find somewhere to live, to learn French. Now we see the problems of the children of the first generation, who have the same difficulties as every young person in this town, to find work and to find their place in society, but that is very difficult. In most [French] people's minds to be an Arab is to be a Muslim, and to be a Muslim or a fundamentalist is the same thing. Everywhere young Muslims are told that they are fundamentalists, and they see that apparently the only people that are moving, or fighting, the West or the Christians are the fundamentalists. So it becomes a self-fulfilling prophecy. The problem is all the racist propaganda that has won plenty of people in France, but especially in this region.'

One reason for the high support for the National Front is the many *pieds-noirs* who live in Provence, says Tadros. 'They have long histories in Algeria, they had difficult lives there, they lost everything, so the National Front has easy support among them. They say, "The Algerians wanted us to leave Algeria, so we did, but now they want to come here because of the problems there, so let them stay where they are." But the idea that France is white and Catholic is not only a nonsense, it is using social and economic problems for political reasons, without knowing reality. They know that in the areas where there are social and economic problems it is very easy to say that the problem comes from these foreigners, and if we eliminate these foreigners it will eliminate the problems. It is easy and they use it. But in France one quarter of the population has somebody in his family, grandparents, that is foreign, Spanish, Italian.'

As a Palestinian, Ramzi understands the reality of exile and starting anew in a different land. He is a thoughtful man, who knows well the problems of the Muslim community in Marseilles, although he himself is an atheist, born into a non-Muslim Arab family. So what, I asked him, would he say to the *pied-noir* argument that Algerians have made their bed and should lie in it? 'To that I say that France stayed for one hundred and thirty years in Algeria. There are plenty of relations with Algeria and the Algerians, the language, the history, the economy, all these years and interventions. France transformed Algeria, in every way. If you go there you see the small villages were turned into French villages, the economy was turned to serve the French economy, producing wine and things that were used for France. The whole social and cultural structure of Algeria was adapted for France, more than in Tunisia or Morocco. There was never colonisation in other countries like in Algeria. I think it was an error to think this, to try to transform the whole country into a French *département*.'

Tadros, like many Arabs living in France I spoke to, argues that the very idea of a national, homogenous France is a relatively new idea, and one belied by the regional identities of areas such as Brittany in the north, and even Provence itself in the south, both of which still boast their own dialects and different cultural traditions. So France's Muslims are merely the latest tile in the ethnic mosaic, the argument goes. 'For a country that says it is the place where human rights were first established it's time now for France to build another type of human rights. This is a multi-cultural society, not only with Muslims and Arabs, but with Bretons and Basques. This must be the place to have new ways of thinking about citizenship. We are citizens of a world that is changing. When you accept people from north Africa to come and work in your country you must also accept them with their families, with their sons and daughters, with their culture and their religion.'

Acceptance yes, but France's Muslims want more than that. The key word that I heard repeatedly was 'respect'. Respect for their culture, respect for their religion and respect for them as human beings and not to be denied jobs, a flat, even a policeman's basic civilities because of their different religion. But there is not much respect for human values in places like Bassens in Marseilles, where Malik BenMessaoud grew up. The concrete jungle of Bassens was a dumping ground for the droves of Algerian immigrants brought to France during the 1950s and 1960s. Its inhabitants, a mixture of north African families and Gypsies, struggle to keep a sense of community, but that's difficult when the shops and school have closed and the city administration ignores the quarter's requests for aid, say locals.

BenMessaoud is a quiet, gently spoken 26-year-old, a street kid turned sculptor, but his body is hard and wiry and the muscles on his back and arms flex visibly as he moves. I would not like to see him angry, for there is rage as well as art in his soul. He was working in his studio in La Friche, an old tobacco factory turned arts centre, when we met, a cavernous site home to pop groups, theatre companies and musicians. He was building a model of a Citroën DS car, part of his project called Bassens Support Cité 1, its theme partly based on the mental and physical separation between Bassens and the rest of Marseilles. The choice of the Citroën DS represents the era of the Algerian war, he says, as well as the endemic car thefts that plagued Bassens when he was growing up there. 'The others in Marseilles have no idea about our lives. Bassens too has turned in on itself and made a closed community. We are not respected but we want to be able to express ourselves and to be listened to, to understand why things have happened this way. Now I have opportunities that I didn't have before and there are a lot of people whose potential is not being realised. Everyone has it, they are enraged and they have a message to get across.'

BenMessaoud is a believing Muslim, although not a very observant one. Islam for him is a set of ethical guidelines that governs the way he lives his life, rather than a set of strict rules. His girlfriend is Jewish, for unlike in Britain close friendships, love affairs, even marriages, are commonplace in France between Jews and Muslims. The film of life in a Parisian version of Bassens, *La Haine* (hate), which won best director award at Cannes in 1995, has three multi-racial heroes: Said, an Arab, Vinz, a Jew and Hubert, an African. *La Haine*, shot in black and white, is a gripping but deeply depressing slice of life in an inner-city ghetto where racism, violence, drugs and police brutality are daily occurrences. The events that might occur over a few weeks in real life are telescoped into one night for dramatic appeal, but most French Muslims I spoke to told me that its underlying perspective of despair, alienation and *La Haine* itself was accurately portrayed. Indeed, a report compiled by police intelligence units in 1995 warned that as many as 200 inner-city neighbourhoods suffered from 'extremely high levels of insecurity'.

BenMessaoud has found himself as a sculptor. As a child he dreamt of becoming an artist but never really believed it would be possible. *La Haine is* subsiding, slowly. 'Today, through this project, I am respected. Respect is very important. If I go somewhere to ask questions, I get answers.' Islam too, he says, gives him a moral and spiritual grounding as well a framework for his art. 'There is no sexual representation in my work, or portrayals of violence. I believe that sex is very personal and not something to be spread about. My religion gives me limits, culturally and morally. It's good because I have principles to work with and a framework to work in.'

A mile or so from the paint-spattered artists of La Friche lives Kamel Khelif, a friend of BenMessaoud. He is an artist who draws graphic novels. His work is dark and shadowy, vivid and detailed, but somehow gloomy, perhaps a bit like his life itself, I imagine, with pictures of men reading letters from home with lonely faces, smoking cheap Spanish cigarettes. 'Ah, yes, Malik, he has the rage,' says Khelif, as he makes a coffee in the tiny kitchen of his top-floor flat in Marseilles' Arab quarter. His flat is a sparsely furnished attic, an artist's residence, where drafts of future strips and finished illustrations are piled up on his desk. Critical acclaim is growing for Khelif's work, especially since the publication of his graphic novel *Homicide*, and he proudly shows me a letter from a professor at Miami University who wishes to meet him.

Together with his family Khelif arrived in Marseilles thirty years ago and, like BenMessaoud, he grew up in Bassens. 'There is a lot of hate on both sides in the suburbs, this hate is what they suffer, racism, rejection. A human being cannot see himself in the mirror and like himself because he has been rejected that much, so they become aggressive and start a war.'

At school Khelif's teachers told him he would never be an artist, so he trained instead to be a mechanic, before studying industrial design. 'For my parents, for the people in Bassens, being a graphic artist means nothing,' he said. 'When I decided to draw again, I drew the boat, it's my oldest memory. I will always remember the sentence that my mother whispered to me one night, "We are going". And then the voyage, what a voyage, I especially remember the smell of the ocean and the noise of the sheep and cows. My family and other families were together with livestock.'

Khelif is now working on a new project with a writer of Algerian origin, about the civil war there, but using symbolism instead of the television images of Islamic radicals and armed soldiers. 'There are two stories, one set in Algeria in 1968 and one in Marseilles in 1995, so it will be a double story. The main character imagines that the bees are coming out of the floor, a cockroach is walking on the wall, and it walks into his mouth. It's set in 1968 but this is what is happening in Algeria now, with the bombings and the extremists, that's the cockroach. It's telling in an indirect way, I don't draw people with beards and military uniforms, but I express it in a different way. It's very important for me what happens in Algeria.'

That same anger and rage that permeates the concrete suburbs of Paris and Marseilles, Lille and Lyon festers in the bazaars and tenements of Algiers, says Khelif, and is the best recruiting agent for radical Islam. 'The FIS is a long story and very complicated. But it's not the same story as in Iran or Saudi Arabia. The FIS was extremely popular in Algeria, because the leaders are not giving lectures in universities but in the street, like the National Front here, they are the ones who have contact with people. It's not just politics, it's also about hate. In both France and in Algeria young people feel hate. In Algeria 18- to 25-year-olds represent over two thirds of the population and when they have no hope of work and nothing else to occupy them, they live in a cultural *misère* [misery]. So they dream and you need to dream, not just to go to work at a factory, because we can do other things, we have other qualities. And all of this hate is expressed politically through the FIS, because the other parties are not representative, they are only interested in the élites.'

Khelif once said that he had never been to a chic place, unlike Sadia Ayata. I met her in a café in central Paris, a typically Parisian scene as we talked in the bustle of a busy morning with shoppers and office workers popping in for a quick coffee or beer. If there is an educated Muslim élite in France, then Ayata is part of it. Born in Algeria, she is a journalist who speaks fluent French, Arabic and English. For many years she returned with her family for two or three months.

'It was something inestimable for me to go there, because I understand something of my own past. I don't feel any rupture with my history, I know exactly who I am. When people talk about integration I feel very well integrated. The problem is the French people who don't want to integrate me. When I was young and we lived in the suburbs of Paris, it was mostly French people who didn't want to speak with my mother. I am a Muslim. I don't go to the mosque, but for me Islam is not only religion, it is part of a culture, but I feel Muslim anyway, for me it is part of my heritage, of my parents' roots, of a kind of *Weltanschauung*, a world-vision. My father was not religious, although my mother prayed, but both kept Ramadan, and for a long time I kept it as well. The way I feel, the way Islam influences my vision is how tolerant I am of other people, I accept them. At home in Algeria everyone was welcome. But here I always had the feeling that French people can never forget the past. They are trying to find something to prove that you are under them, that you are an Arab, or something.'

179

As well as knowing herself, Ayata also knows her civil rights, knowledge that was to prove extremely useful when she was arrested in a case that has since turned into a Muslim *cause célèbre*. For there are two Parises in France. The first is the tourists' metropolis of Robert Doisneau posters of kissing couples, the Eiffel Tower, and the *Mona Lisa*. The other is the Paris inhabited by north African Muslims. They may be physically the same streets and squares, but they live and work in a place where harassment by police is a daily reality and where racism and Islamophobia breed hate and hostility. On a hot summer's morning in August 1994, while waiting for a bus with her sister, Sadia Ayata found herself in the second Paris.

It was the time of Pasqua's crackdown, soon after several foreigners had been murdered in Algeria. Ayata had just returned from Berlin and was shocked at the number of policemen and women on Paris's streets. A few feet away policemen were checking a north African's identity. 'I started to speak with my sister about how the controls are always on north Africans, just because they have a darker face and are recognised. The control was finished but the policeman stayed and heard what I was saying to my sister, he told me to shut up. I told him that I was not speaking to him, that we were in a democracy and I was just expressing my opinion, I was not insulting him. It doesn't interest me to insult the police. Then he got nervous, pushed me and he wanted to search me. I told him to stop and said I preferred to be searched by his colleague, who was a woman. He called the police station and they sent a car. He asked me for my papers and I gave him my identity card. He showed me a card saying that the police can take anyone who is making a disturbance on the street.'

Ayata was arrested and taken to the police station in handcuffs. Her sister started calling French human rights organisations, who in turn telephoned the police station, driving the station chief into a fury. 'He went completely crazy with anger, asking me who I thought I was, and so on, he told me he was a *pied-noir* and he said, "I will take care of you personally".' Ayata groaned inside at this news, but held her ground when she was charged with 'committing an outrage against the police'. She demanded to see a doctor and was carted off with six policemen as an escort, again in handcuffs.

When she returned to the station she was interrogated about her religious beliefs. 'They asked me if I was a Muslim, if I was a practising Muslim, if I was for or against the Islamists. When he asked me if I was a Muslim I said: "If I say to you that I am a Muslim, you will think that I am an Islamist. If I say no then you will say that I don't even respect my own religion." He asked me if I put ham in my sandwich, I told him it was not his business what I have in my sandwich. When he asked me if I was for or against the Islamists, I told him that in the place where I was, I could only say I was against them. I kept my sense of humour, it's the only weapon you can keep.'

After six hours Ayata was released in a daze, at first unable to believe what had happened to her. 'I was really shocked. I didn't know which country I was living in any more. I wondered what was happening if they arrest me and hold me for six hours for just saying that, what will they do with others, someone from the suburbs? That was my great worry. I know how to speak French, how the system works, but what about the other young people who don't know their rights, who feel alienated, who feel uncomfortable in French society? Luckily I was not hurt, I didn't slip over in the police station. A lot of things happen like that. I can understand how some people come to hate the system.'

Ayata's case came to court on 15 November 1994. She was found guilty. 'It's a small crime, but it's a way to identify people, so when they pick up someone, they put the pressure on them and they have their name in the computer. At the trial Madame Mendès-France [widow of the former French president] came as a moral witness. I was fined 4,000 Francs, I appealed immediately. In June 1995 I had the appeal. I spoke at the stand, I said that a whole community could not be charged with events happening outside France, there are other means to fight against violence. I won the appeal on the basis of freedom of speech, in the universal declaration of human rights.'

But *La France* does not take defeats from young Algerian French women so easily and the police appealed against the appeal, setting the stage for a lengthy legal battle. Should Ayata win, then it will set a legal precedent and boost citizens' rights over police powers. 'If they find me guilty, they should not be surprised at all the shit that is going to happen. People will not respect the law and the police. I am adult enough to fight this case with papers, but if I was not I would become crazy, because what else can I have, what opportunity? I know the mind of the people in the suburbs, I know who they are in their minds, I know how despairing they are.'

For Ayata, her case has disturbing historical overtones. 'Fifty years ago the identity controls on Jews started like this, and nobody said anything. I didn't compare the French police to the Nazis, I just remember the facts. Anti-Semitism is not acceptable anymore, so the Muslims are the new scapegoat. People think that because I am a Muslim I am against Jews, but I have many Jewish friends, I am always interested in discussing with them how deep must be their wounds to have been so integrated in society and then start to be exterminated. I have a very good friend, when he was twelve he was in the subway with a yellow star on his jacket and he had to hide it, to go through the police controls. So now I wonder if there is the possibility to say to the police, "you are doing something wrong", if a citizen can interfere in this, not to insult them but say, "hey, come on, what are you doing, you are obeying rules that are not good". If there had been this possibility to intervene then, probably those things could not have happened and also those things today could not have happened.'

Sadia Ayata left the police station in one piece. Nassera Sarah Oussekine's brother did not survive his encounter with the Parisian forces of law and order. Malik Oussekine was beaten to death in December 1986 by three policemen after leaving a jazz club. He was 22, couldn't speak Arabic and he hoped to be a Jesuit priest, but his last, fatal, mistake was to be on the streets while students were demonstrating, for some French policemen always enjoy the chance to crack a few Arab heads.

'The cops were supposed to follow the demonstration on their motorbikes, but they left the route just to beat up some Arabs. Two of them were sent to prison for two years,' she says. Her brother's death changed her life, and her view of France, for ever. 'Until then I lived normally, I was working, I went home, I wanted to be happy, professionally, mentally, to have any kind of happiness. That day I realised that I was living in a racist country. We always talk about integration, but we are integrated, my roots are here in France, I was born in Paris, not in Algeria, so what more do they want? But they always remind us that we are different, that we are not from here, even if we are. In France racist crime is quite frequent, when the cops kill somebody from north Africa they are not really punished for it. It legitimises racist crimes.'

I met Oussekine in a municipal youth centre in the suburb of St Denis, in north-east Paris, where on streets nearby a soup kitchen doles out free meals to the poor and Arabic is spoken as frequently as French. She is one of the leaders of Nana Beurs, a feminist organisation for women of French-Arab background. Nana is slang for women, or chick, and Beur means Arab, of the second generation, born in France, and the word is a corruption of 'Arab' said backwards. French slang is now peppered with Arabic words, especially in the suburbs – *baraka* means luck; *fatma* a girl; and *clebs* a dog. Nana Beurs was giving a party, and we munched on slices of bread and pâté, washed down with kir or white wine.

The group was founded a decade ago by women who took part in the March of the Beurs, a mass demonstration for racial equality. But the question of sexual equality in the French-north African community had somehow been ignored. Their slogan, *Nana Beurs Il Faut Pas Sharia* (Nana Beurs don't need the *Shariah*) is a pun on the French saying *Il Faut Pas Charier*, meaning let's not go over the top. 'The needs of women were not taken into account during the march. The question of autonomy is very important for young Beur women who want to be independent. They are completely dependent on a residence card from a husband, and if a man gives it back, the woman has to as well. After the march of the Beurs, they obtained a ten-year visa and that was a relief because they can relax. Racism affects all of us, but sexism only the women, they have pressure from outside, racism, and inside the home, because of their sex. They are educated in a particular way, they are watched over constantly. This is a result of living in a patriarchal society. Islam does not have a monopoly on sexism, even Buddhism is the same.'

As she grew up Oussekine was inspired by the works of Simone de Beauvoir. 'She influenced me very much. There were many problems that I encountered in society and I didn't realise why, and then I read the *Second Sex*, but she deals with every class of women, bourgeois, workers, they all have the same problems as women. We have had hostility among the Beurs because we are secular, and we criticise religion in general and people don't like that and we push women's rights. Also we are from modest backgrounds, so of course there is always some kind of judgmental attitude. We criticise our religion and some aspects of Arab culture, not all of it, but everything in it related to women we strongly criticise. We are opposed to headscarves, because wearing one is not always a voluntary decision. The media are obsessed with this, they make it worse, they are always searching for a sensation.'

Whether they are artists or activists, journalists or teachers, France's Muslims are bringing an extra dimension to the country that they think of as home, but which has given them such a haphazard welcome, if welcome is the word. They ask to be considered humanely, as people, and for some understanding from the country that colonised Muslim lands for many decades, but now refuses even to guarantee the rights of French Muslim schoolgirls to cover their heads at school.

The headscarf affair was a turning point in the evolution of France's relationship with its Muslim minority. For many Muslims, it confirmed that they are forever destined to be *les étrangers* in French society, marginalised and discriminated against. It shook the very bedrock of French society, questioning the nature of the relationship between religion and the state, and bringing home, perhaps for the first time, the fact that France was home to several million Muslims, many of whom had different values and ideas to the Catholic majority.

The case was eventually referred to the Conseil d'Etat, France's highest court, which was asked to rule on whether the wearing of religious symbols was compatible with secularism. It fudged, saying that while 'in principle' it was not incompatible, anything which amounted to 'provocation, proselytism, or the undermining of the liberty or dignity of a pupil' could legitimately be banned. It was left to individual headmasters to decide. Finally, in the autumn of 1994, the right-wing then education minister, François Bayrou, issued a ruling that Muslims saw, quite understandably, as directly targeting them. Bayrou ruled that 'ostentatious' religious symbols were forbidden, but 'discreet' ones were permitted. Headscarves were ostentatious, crosses and skullcaps were discreet. Muslim organisations were quick to denounce the decision, and even the normally august newspaper *Le Monde* gave space to one of its columnists who asked, reasonably enough, 'Who is to decide what is discreet and what is ostentatious? What authority decides that the wearing of a cross, a *kippah* [skullcap], a Shield of David or a crescent is acceptable, but not a headscarf?'

I heard stories in the wake of the ban of north African schoolchildren refused entry to school for wearing T-shirts with the word 'Algeria' emblazoned on it in Arabic. Of course their peers, wearing clothes covered in logos proclaiming 'USA' have never had any problems. It seemed to me that if the very basis of the French republic was threatened by a few headscarfs – for only a tiny minority of schoolgirls wished to wear one anyway – then that mighty state was a very fragile edifice indeed.

Perhaps the last word should go to Ramzi Tadros, working on the front-line in Marseilles, where the two cultures meet and, sometimes, clash. I asked him if he was optimistic or pessimistic about the future of France's relationship with its Muslims. 'Like what is happening between Israel and the Palestinians, you cannot choose between optimism and pessimism. You must try and determine the outcome of things, and declare that another way can be found. We must do everything to get this good relationship, it will not be easy but we must do it. A good relationship is an equal one. Without saying that everything is wonderful, but to look at things like they are. If there is fundamentalism, there are reasons for it, let's talk to the Arabs and Muslims that are against it. Let's see what can be done, and not give lessons to other countries and people. If you want the second and third generation to respect French laws, you must show them that they are respected. This is the way to integrate. There is no other.'

Arrest of a Turkish demonstrator in Solingen, Germany

Chapter 7

GERMANY: THE RISE OF THE EURO-TURKS

'Hey ho house, we're neither liars, nor naturalised Germans, nor foreigners. We're just guest worker kids.'
Berlin Turkish rap band Islamic Force

'The Chancellor does not engage in condolence tourism.'
Helmut Kohl's spokesman, replying to calls for Kohl to visit the site of the Solingen arson attack, where five Turkish women were killed in May 1993

It was in a hot and smoky Kreuzberg bar in November 1988, a full year before the fall of the Berlin Wall, that I first heard the siren sound of pan-German nationalism. A clarion call for unity which, when it resounded across that then divided land, would have tragic consequences for Germany's Turkish community.

Now, as then, Kreuzberg, tucked away by the old border between east and west Berlin, is a bohemian mix of Istanbul and Islington, the trendy north London suburb. Veiled Turkish women queue for food next to grungy punks in the shops and small restaurants selling Turkish snacks. Smoky cafés are filled with Turkish men – there are no women inside – drinking tea and coffee and playing backgammon. The Wall is long gone and the nearby former border crossing for foreign trucks has become just another municipal crossroads, but the vibrant counter-culture that brought young rebels here from all over West Germany remains.

And from Turkey too, for as the immigrant community grew, the first political exiles arrived. Posters advertising a plethora of Turkish communist parties and leftist movements plaster Kreuzberg's walls, complete with stencilled pictures of Mao Tse-Tung or Lenin and clenched fists. Kreuzberg's Turkish communists still keep Marxist iconography alive. The two cultures – one German, western, green and tinged with anarchism; the other Muslim and mostly conservative – live tolerantly side by side, although separately. I have been to Berlin many times and have several friends in Kreuzberg, all young and liberal, but none of them ever introduced me to a Turkish speaker.

But back to that bar: we were chatting to Gerhard, a typical young left-wing Berliner, with a bottle of Beck's beer clasped in his hand, about the Wall and East Germany. Would it always be there? Suddenly his eyes opened wide and he declaimed: 'We were one country once. Just think what we could achieve if we were united again. Just think about it!' he said excitedly, waving his beer about.

I thought about it and I didn't like it. To me it sounded pretty ominous. If this was what they were dreaming of in Kreuzberg, capital of Germany's radical counterculture, then what were they hoping for in the small towns and villages of Bavaria and Saxony? A Germany free of foreigners, especially non-white or Muslim ones, it seemed over the next couple of years, as television screens filled with pictures of charred houses in Mölln and Solingen, weeping Turkish relatives of arson-attack victims, and the screaming faces, contorted with hate, of a German mob, as it attacked a foreigner's asylum in Rostock.

But for every action there is an equal and opposite reaction. The more the nationalists demanded that foreigners leave, the more young second-generation Turks said no, we are staying and we demand the same rights as you Germans. And now that's the message pounded out by Boe B, lead singer with the Berlin Turkish rap band Islamic Force. Boe B, whose real name is Bulent Ipek, says he wants to provoke people, and judging by the scars on his forehead he's succeeded several times. He's been in gang fights and had a violent tussle trading blows with the Berlin police. But Boe wants to put his sometimes violent past behind him and the only confrontation he seeks is an intellectual one, forcing Germans to accept the country's Turkish minority, almost two million strong, and grant them the same rights of citizenship and cultural freedom as their German peers.

Just as black rappers in America use pounding music as an expression of their identity, Berlin's Turkish youth have taken to the recording studio to get their message across: that, unlike their parents, they won't say 'please, sir' any more, because they are in Germany to stay. I met Boe at a party at Ypsilon Musix's recording studio in the Wedding district of West Berlin, celebrating the label's first birthday in June 1995. Back in the 1930s this workers' area was a stronghold of the Communist party, and was known as Red Wedding. But now, like much of inner-city Berlin, its dilapidated apartment blocks are home to Turks and their fellow guestworkers from a dozen countries, as well as plenty of gloomy Berliners dressed in the obligatory black.

The smell of spicy kebabs sizzling on the barbecue drifted through the studio as we spoke, and there were frequent offerings of ever more food and drink, for Turks, perhaps especially those living in Germany, are proud of their tradition of hospitality. The members of Islamic Force mix western and Turkish chords together, creating a new musical style that draws on its members' dual heritage as German-born, but of Turkish background. As an Alevi, whose version of Islam is far more relaxed than the Sunni one, Ipek seems an unlikely figurehead for a Muslim-inspired musical protest, but Islamic Force has more to do with rap music than mosques, he says. 'We are Turkish rappers, but we didn't want a name like the Turkish Posse or something like that because it sounds like a gang. Islamic Force comes from Afrika Bambaata's example, to show people where we are coming from. I'm an Alevi, I believe in Allah but I don't go to the mosque. I drink beer. The name aims to provoke people.'

Unlike many black American rappers, who extol a culture of violence and automatic weapons, Islamic Force calls for a peaceful change. 'Some other groups, they get on stage and rap in Turkish, they wear baggy trousers, but we don't live like that. They want to be black, and I don't understand why, those guys are shooting each other. They are like killers and gangsters and we are not killers or gangsters. That's what I hate. My mother listens to my music, she hears every new song that I make. I make songs that my mother can hear, I don't use words like "fuck".'

Born in Istanbul, Ipek arrived in Germany in 1979. The move from the Bosphorus, with its easy-going chaos and strong family life, to the Prussian orderliness of Berlin, and the formality of German culture, triggered a major culture shock, one that Ipek clearly has yet to come to terms with. 'It was hard coming here, because in Istanbul I had all my friends. The biggest problem was that it was a different atmosphere, the air, the people. When you are a kid in Istanbul and you come home from school and your parents aren't there, you sit on the steps of the house and the neighbours see you, they take you in, they feed you, and everything. When I was here, and I came home from school, I sat and waited and so many people passed by and nobody said anything. The Turkish neighbours were also like the Germans, because they learned it from them, to be closed. I think I will go back to Turkey. It's too cold here, not just the weather, the people, everything.'

Ipek has a son by a German woman, but says he barely gets to see him. 'I got a little kid, that's my son, but she don't want to show him to me because I was in trouble so much. That's what she can say against me, "Hey, he's a bad guy and I don't want him around my son", but that's not the real reason, the reason is she's cold, cold like the weather.'

But if Ipek wants to go home, Unal Yuksel, one of the founders of Ypsilon Musix record label, which has signed up Islamic Force, is definitely staying. For a start he was born in Berlin in 1969 and it's his hometown. Together with many young Turks, he is in the vanguard of creating something new and unique in Europe: a hybrid Turkish-German culture, a fitting evolution for a generation caught between Berlin and Istanbul, but neither wholly German nor Turkish. 'I take German culture from the street, and I take Turkish culture from the home, from my parents, so it's a mixed culture. From home I take religion, my character and my language, but character and language also from the streets, so this mix comes into the music as well, you make music that is mixed, it's music from a minority. But you need time for this, Turkish people have only been here for thirty years.'

The songs' lyrics make several simple points, says Yuksel. 'We are living here in Germany and I like Germany. My father has worked here for thirty years. My message to the Germans is that when you want to you can live with me, when you don't want to it's not my problem, it's your problem. The Turkish people have said to the Germans for thirty years: "We are also people, let us live together, please." But I don't want to say please. I am a human being here, I work here, I live here, I speak the same language. I feel something, although not the same as you [Germans] feel. I can't live in Turkey, so if I can't live in Germany, where can I live? But I don't say please. This is the message. The first generation cannot speak good German, but the second generation is more confident. They say we want this, this and this, and we don't want this. The first generation wants to make money and go back to Turkey. But that is not my problem, to make money and go back to Turkey. My problem is to stay here, make money and live here. What can I do in Turkey?'

What Yuksel can do in Berlin is run what he says is the first professional Turkish studio and record label in Germany, a venture promoting a new style of music that is getting increasing attention from major international companies such as Sony and PolyGram. 'There are differences between our music and Turkish music. The music that we make here is the music for the second generation and the problems of the second generation. The problems that we have here are not the same problems that they have in Turkey. I am twenty-five and people in Turkey who are twenty-five don't have the same problems as me. They can't understand me.'

For Yuksel, Boe B's angst reflects the alienation of many young Turks. 'He is in the second generation and he doesn't know how he wants to live, in Turkey, or Germany. He doesn't know and he is not alone, ninety per cent live like him. I think we must go on the street and say "I want to live here and I feel good here". Boe goes home and his parents say they want to go to Turkey. Then he goes back on the streets with his friends here. I have thought about this [identity] problem for a very long time and I analysed it. Now I think you have to choose to live either in Berlin, or in Turkey, but don't be in-between, because then you go crazy.'

First generation, second generation. He is from the first generation, she is from the second. These were the words that came up again and again as I travelled across Germany, speaking to

Turkish people. As with communities of immigrants everywhere, whether Turkish or Tasmanian, there are clear differences between father and son, but for young Turks especially, each of the two labels carried a distinct set of economic, cultural and political baggage. The first generation always knew they were Turkish and would never be anything else. They came to make money and then planned to go home, even though most have stayed. Their priorities were economic, rather than political. But the second has a different agenda: they are neither wholly Turkish or German, but a blend of both, with both the benefits and troubles that combination brings in its wake. The obsessions of many Germans, especially the conservatives in government – with their notions of race and blood, nation and order, punctuality and a rigid bourgeois lifestyle – mix ill with most Turks' relaxed Levantine approach to life, and that's before the issue of religion, of Islam and Christianity, is even considered.

It's hard to imagine that the German bureaucrats at the Ministry of Labour in Schleswig-Holstein had any idea what the eventual consequences of the agreement they signed in 1957 with the Turkish Ministry of Foreign Affairs would be. Almost forty years later debate rages over the very nature of German society and what it means to be German, as Germans struggle to adapt to the fact their country is now home, and will stay home, to one of Europe's largest Muslim minorities. Under the agreement's terms a dozen craftsmen arrived in Kiel, northern Germany, although there had been several private initiatives during the 1950s to bring in Turkish guestworkers. Although their Turkish qualifications were not recognised, immigration rules were relaxed then and the first wave of families began arriving. Many of the immigrants were from small villages and towns on the Anatolian plain, rather than the great metropolises of Istanbul or Ankara. Their concerns were local and parochial: their families and their work. Over the next few years the guestworkers' status became slowly regulated: government money went to start cultural associations and Turkish consulates opened in the many cities all over Germany where the Turks were settling.

But beneath the legal protection for Turkish workers, introduced under pressure from the German labour movement, there was always an assumption, whether spoken or unvoiced, that the foreign workers would not be staying. The very word itself: *gast-arbeiter,* or guestworker, said volumes more than any new piece of immigration legislation, with its message that: 'You Turks are welcome to come here and do the dirty work in reconstructing Germany, but then please go home.' This message is habitually reinforced by German politicians, such as Chancellor Helmut Kohl who repeatedly asserts that 'Germany is not an immigration country', as well as claiming that 'Germany is friendly to foreigners'. The statement, ritually invoked to reassure the *völkisch* right wing that the country of Goethe and Schiller will remain white, Christian and church-going, is clearly nonsense as a visit to any small German town will verify. But even among the first generation of Turkish immigrants, many have stayed on and are staying, such as Bilal Yuksel, father of Unal.

Formerly a builder, Yuksel senior, 51, now runs a stationery and toy shop in Kreuzberg with his wife Kevser. He came to Germany to make money and he succeeded, says his son Bilal. 'Now he has everything in Turkey, three or four homes and car. He could go back to Turkey now, but really he cannot. I am here, one of his three sons, one still goes to school. He will only return when he knows everything is OK with me and my brothers. I asked him why he doesn't go back? He said what can I do there? His mother is dead, his father is dead, he would live alone. Turkish people live with more than one person in one room, there is action everywhere. Nobody feels good alone. Nobody.'

Bilal Yuksel arrived at Munich's main station in 1966, after a two-day train ride from Istanbul. 'I have been in Germany for thirty years and that's a long time, it's a different situation for me now. I cannot go back to Turkey now, my wife has come and my sons are here. The thirty years went well, I learnt German. I have German friends, I never had any problems. We were one big family in the company, Turks, Germans and Arabs.'

At this claim Unal chips in, as we speak in the little kitchen behind the shop. 'This is the attitude of the first generation, this "please". Perhaps he had problems but he won't say. For all those years he went from home to work, from work to home. He doesn't have any conflicts, he doesn't go to the discotheque. He has his community.'

His father replies with dignity, smiling at his son: 'What is important is that every one of my sons can stay here on his own, can get money alone, can have a community. Then, when I see that everything is settled, I can go back to Turkey. I want my sons to have both cultures. I give my culture to my sons, the street and school gives the German mentality. This is something not German, not Turkish, but something new. When you want to live in another country, first you must learn the language, secondly you must integrate yourself into the culture. When you do both of these you won't have many problems. I achieved my aim, I did what I wanted to thirty years ago. These are the thoughts of a worker.'

The thoughts of the workers also exercise Safter Cinar, spokesman for the Turkish Union in Berlin. The Union's member organisations include groups for women, those promoting Turkish culture, for the elderly and even for hotel and restaurant workers. Its diverse membership is testament to an immigrant community that is now finding its voice. A youthful 49-year-old, Cinar came to Berlin to study economics over twenty years ago, stayed, and is now head of the German Trade Union Confederation's bureau that advises foreign workers. During the 1980s, when Turkey suffered a series of military coups, it was events in the homeland that exercised Turkish activists in Germany, but now community organisations are focusing on building a future in Germany, he says.

'Now we want to solve problems here in Germany, not like in the 1970s and '80s, when we made organisations to save Turkey. After the military coup in 1980 many immigrants, political ones, came to Germany. For a while everyone was only shouting about Turkey. But this produced a reaction, from the people who were born here or already living here who said we don't want people to come here from Turkey and tell us to fight for the problems in Turkey. This is a healthy and very important development.'

But the integration of the Turkish, and other minorities, must be a two-way process, says Cinar, and one that starts with German politicians recognising that their country *is* an immigration country and must adapt to the needs of its Muslim minority. 'Integration is happening, but it could be better if German politics were different. If they stopped making foreigner politics and began to make immigration politics, integration would be better and faster. The ideological position that we are not an immigration country is nonsense, every fact tells the opposite story. This is one of our problems, that politics is denying reality, with all its negative influences. If you say that seven million people here are not part of society, then you handle them as not being part of society.'

For all the politicians' rhetoric, German society is slowly changing under the influence of its immigrant communities. Germany's dull cuisine – one of the blandest in the world – is getting spiced up and evolving under Turkish and Asian influences. The country's legendary bureaucracy is even learning to bend some rules to cope with a massive influx of immigrants who don't always share Mr and Mrs Schmidt's legendary awe of authority. 'Not only the Turks but all the immigrants have had an influence here in daily life. In food, for example, the culture has changed. When I first came here in 1967, there were very few foreign restaurants, now it's hard to find a German one in Berlin. This influences what people cook at home and buy in the shops. Music has been influenced and also behaviour. This German idea of strong law and order in daily life, this is a little bit liberalised. I just read a book about hospital regulations that said in the 1960s and '70s you were only allowed to visit people two days a week, for only two hours. Now you can go every day, and the author says this is the influence of foreigners, because their families came every hour and so they changed the regulations. Such things cannot be measured, but there is a little liberalisation in daily life.'

German law is also evolving to cater for the beliefs of its Muslims, who are not just Turks, but also include north Africans, Arabs and Iranians. Judges have so far ruled that Muslim parents can refuse to allow their daughters to go swimming with boys and can wear headscarves to school. 'These court decisions also have a political aspect; they said, we are not only a Christian society now. It doesn't state that but this is the logic of it, the positive part. So to say that we are not an immigration country and we don't have to change anything is nonsense. But it is better if we have a general concept developing and we wouldn't have to keep going to court,' says Cinar.

But compared to, for example, Britain's Afro-Caribbean and

Asian communities, the Turks in Germany are substantially marginalised in public life. There are two German-Turkish MPs, one in the national parliament in Bonn and another in the Berlin state senate. But there are no Turkish news readers on German television, no senior Trade Union leaders – important in a country with a still powerful labour movement – no nationally acclaimed comedians or artists or musicians, although the German football team has one player, Mehmet Scholl, of Turkish background. The lack of a public face for Turkish people is the legacy of policies that deny Germany is an immigration country, says Cinar. 'This is because of factors such as the problem of the citizenship law, because to work in much of the official sector, you need German citizenship. But if you propagate only minority politics, that is how the majority sees you, only as a minority. For example, last year Berlin television invited us to discuss foreigners. I told them it would be good if they had a job training scheme for newcomers, and take people from ethnic minorities. Everyone said, yes that's a good idea, we could have better contacts with the Turks. I said it would be better the other way round, to take a Turkish person who has nothing to do with Turks, he specialises in traffic. They see you only as a Turk who only deals with Turks. And these are liberals.'

Not all Germans are liberals. Many are outspoken racists, such as the far-right Republikaner Party and its supporters and many more are tacitly prepared to condone, or even apparently tolerate – even exploit— racism for political advantage, such as Chancellor Kohl.

A very few Germans are killers, and the sites of their deeds, such as the city of Solingen, are seared into the consciousness of every foreigner in Germany. Durmus Genc, a Turkish immigrant, lost five members of his family – two daughters, two granddaughters and a niece – in an arson attack by neo-Nazis in May 1993. The youngest was Suyime, four years old. The Genc house at 81 Untere Wernerstrasse is no more, and the site is marked by a stone in commemoration of those who died there. A few days before the Solingen attack in May, the German parliament had passed a law limiting the right to political asylum in Germany. After the killings at Solingen, following a similar firebombing in Mölln in 1992 when Bahide Arslan and two young girls died, Turks invested in strong doors, dogs, alarms and security systems as they suddenly confronted the possibility of violent attacks by neo-Nazis.

The Solingen attacks also triggered four nights of riots by local young Turks and demonstrations across Germany. Chancellor Kohl's government expended much energy on denouncing the rioters, almost as much as it did condemning the firebombing. But while Kohl and his spokesmen dished out weasel words, Germany's President Richard von Weizsäcker seized the moral high ground when, speaking at a mosque in Cologne, he denounced the spreading anti-foreigner hatred. '[The attacks] spring from a climate generated by the extreme right. Even criminals alone do not emerge from nothing.' He parted company completely from Chancellor Kohl's government, though, on the question of citizenship and its refusal to amend restrictive laws that kept immigrants marginalised as permanent aliens. Many Germans, he said, 'speak too easily of "the Turks" . . . would it not be more honest and human to say "German citizens of Turkish heritage"?'

Yes, say the members of Germany's Turkish community. It would. Citizenship, more than any other issue, apart perhaps from physical safety, is the biggest concern for Germany's Turkish community. For many conservative Germans, the idea of granting German citizenship to millions of Muslim immigrants triggers a wave of horror at what they view as the inevitable subsequent pollution of the *Volk*'s blood-lines. Germany is one of the few countries whose citizenship laws are based on the idea of blood, rather than birthplace. Under the Nazis the obsession with pure blood-lines reached absurd and obscene heights as bureaucrats invented and carefully catalogued a maze of classes and sub-classes of Jews and part-Jews. The current rules, which state that citizenship is passed from parent to child, stretch back to the Imperial Naturalisation Act of 1913, and share a legacy of racism. In this case, originally, it was anti-Semitism, as the Act was passed to prevent granting citizenship to the many Polish and Jewish immigrants pouring into Germany. But now the rules are a wall against Muslims rather than Jews. Even *The Economist* described Germany's current citizenship statute as 'one of the oddest citizenship laws among large industrial countries'.

That 1913 act was inspired in part by the romantic nineteenth-century idea of Germany as a culturally uniform entity, of a *Volk* unified by blood and a common culture. That vision, still embedded deep into many Germans' psyche, does not encompass mosques or minarets. The current rise of Germany's intellectual new right, with its ideas of a homogenous German nation and the subsequent disturbing historical resonances these concepts stir up, has not helped the cause of extending German citizenship to its Muslims. The new right's buzzword is 'ethno-pluralism', which in the best Orwellian tradition may sound like a call for tolerance, but in reality is a mask for cultural apartheid. Many German intellectuals, instead of analysing how their country could integrate its Muslim minorities, are working out how to further marginalise them, drawing on the same old tired ideas of blood and race that obsessed Hitler.

For German nationalists the collapse of the Berlin Wall was a vindication of their *völkisch* ideas about the indivisibility of the German people. But for the Turkish community, indeed all non-white foreigners, unification has been a disaster in many ways, says Cinar. 'By the end of the 1980s we had reached a better acceptance of ethnic minorities. But now this problem is sidelined, because all discussions are about reunification. It cost a lot of money, and so there is less for immigrants social programmes. However, the murders in Solingen, Mölln and the riots in Rostock caused a stronger anti-racist discussion, because people who were indifferent to the problem were shocked and are getting active.'

The neo-Nazi murders have left a legacy of fear and uncertainty among all of Germany's immigrant communities. Racism, and the fear of a racist incident, from abuse to arson, is now a constant worry for foreigners, says Cinar. 'It is something there in daily life, not with something happening every minute, but there is the fear of something happening. Before the events of 1990 to 1993, if the children come home from school a bit late, the parents were just angry. Now everybody thinks maybe a racist incident has happened. This feeling is everywhere now, if you see a house burning, you think it was a racist attack. This fear is real and permanent. Factory workers say that there has been a change since the 1990s. Before, if a

Turk or non-German had a conflict with someone German, they shouted a bit and it was over. But now they say that the Germans say very quickly that you are a non-German. This is to do with the changes in the atmosphere, to be more German, more nationalistic. The skinheads are a problem because they are violent, but their numbers are small. The problem is what makes skinheads possible.'

But slowly, across the mainstream political spectrum, there is a growing realisation that the guestworkers are no longer guests, but residents. Even some on the conservative right believe that Germany's Muslims will have to be integrated into society, and that means granting full rights of citizenship. Debate continues over how to achieve this. Currently, immigrants can become German citizens if they have 10–15 years' residency and no criminal record or eight years of schooling. But, and it is a massive but, they must give up their original nationality. Many Turks, unnerved by the neo-Nazis and the opposition to granting them citizenship, understandably refuse to hand in their Turkish passports. As the wave of violence against foreigners grew in the early 1990s, many in the Turkish community began to compare themselves to the Jews in the 1930s. They had lost their German citizenship and many could find no other refuge, and died in the camps. Nobody in the Turkish community expected a second Holocaust, but many believed that what Germany gave, it could then take away. This feeling is widespread among the second generation, says Safter Cinar.

'There are two reasons for the problems over dual citizenship. First, the German policies that deny these people are part of society, which is a funny psychological position, although nobody says it outright. The second is that some [Turkish] people say, and the number is growing, "We don't know what will happen in Germany and we don't want to be in the position of the Jews in the 1930s who didn't have a country to go to." This is not yet a mainstream opinion, but if this situation of racism and violence continues to develop in Germany it could be stronger, not mainstream but it would be a big minority.'

The Turkish government has substantially changed its position on dual citizenship to help its nationals in Germany. Laws on property ownership and military service have all been altered to ease the guestworkers' position. In fact many Turks find a way around this, for they reapply for their Turkish citizenship, often even as they give it up, without telling the German authorities.

The twin devils of German society: the refusal to grant joint citizenship and the rise of neo-Nazism, are helping the rise of Islamic sentiment, especially among young Turkish people. Perpetually marginalised, subject to racism, marooned in a birth-place that rejects them, not surprisingly they are turning away from Germany, and towards Turkey and their religion instead.

Turkish Muslim groups, particularly the Islamic Welfare Party's (Refah) shadow set-up in Germany, provide a ready welcome, organisational infrastructure and perhaps most important, an immediate identity. The Islamic revival in Turkey, and growth in support there for Refah, is mirrored on the streets of Bonn and Cologne, Hamburg and Berlin. Almost everything that happens in Turkey reverberates through Germany, says Cinar, sometimes within hours. When Islamists torched a hotel in the Turkish city of Sivas, in protest at *The Satanic Verses*, there were demonstrations in Germany almost immediately afterwards.

Refah has no official presence in Berlin, but its supporters congregate at the Kreuzberg branch of National Sight, an Islamic club. A shop front, with a door onto the street, National Sight's walls are covered with posters expressing solidarity with Bosnia, the country which was once home to Nafiz Ucan's grandparents, who came originally from Tuzla. Quietly spoken and articulate, Ucan, 24, was born in Berlin and is studying law. He already delivers his opinions with a lawyer's careful use of language. 'I am a supporter of Refah, but not a member or representative. But objectively seen I will support the party's goals. I will support any decision that is rational and not violent. There is growing support for Refah among young Turkish people in Germany. People over fifty will have prejudices and heard all the propaganda that this is a radical Islamic party and is dictatorial, which it is not. Young people who objectively assess the information about Refah are very interested in it.'

Status, identity, the fundamental question of who they are, all these are driving young Turks in Germany into Refah's embrace, admits Ucan. 'The problem of identity that many young Turks have is definitely a reason why they are interested in Refah. Over the years Turks who lived here felt abandoned by their home country, there have always been many promises from Turkey, to support the community here, but none have ever been fulfilled. They feel hopeless, that they no longer have a homeland. This party has a mission and certain goals, and tells young people who live here that they will represent their interest both here in the community and in Turkey. It also offers strength among young people who live here, who want to live in peace in the community, and a philosophy to integrate and remain Turkish and Islamic.'

Refah's message to young Turks, that they can be proud of their Islamic heritage, both among Germans and among their fellow Turks, many of whom are fiercely secularist, also brings them support. 'Young people in the Turkish community don't feel recognised, they don't have an official status, they feel the pressure to assimilate but resist it, but they have nothing else to grasp at to defend their rights. Refah tells them we will support you in your attempts to form your own culture and not be assimilated. We will help you with lobbying for education, with the right to your own schools, with the right to be respected in this society. Young Turks will feel that this is very important. In Turkey French or British citizens have the right to protect their own culture, to go to school, or show films. But young Turks here don't have them, and Refah offers them the chance to keep a sense of identity and a sense of pride.'

The contrast between Islamic Force's Boe B, thumping out his rap lyrics about alienation and uncertainty, and Nafiz Ucan's fluent analysis of his situation and identity, could not be greater. It is the precepts of Islam that have brought him a peace that eludes many young Turks in Germany. 'First of all I am a Berliner, but one with a different cultural background. Berlin is a special place, I find it fascinating here that I can keep Turkish identity and still integrate in a foreign country. I can study here and be socially active, and lead the kind of life that I want. I am still attached to Turkey, but it's more cultural, whereas my attachment to Berlin is a realistic day-to-day one. This is my home city, where I feel first and foremost attached to.'

Beyond his love for Berlin, Ucan says he is both German and

Turkish simultaneously. 'I feel I have a double identity and I feel comfortable with that. I see myself as German, definitely, *and* Turkish. But beyond these questions of national belonging, there is something that comes from my religious beliefs, I feel first of all that I am a human being and must act in a humane way, and aspire to certain values, rather than a national identity. For me Islam is the idea of all people living together, to have a mixture of different cultures, religions and political tendencies. Religion is one way of attaining this, another is through my day-to-day life and how I live here in the city. That's what's so wonderful about Berlin. There is a great tolerance here, to live the life you want.'

Berlin is now home to forty mosques, and Islamisation has expanded so much that Turkish people can live an almost completely segregated Muslim lifestyle, says Safter Cinar, which, for him, is not a healthy development. 'This has been happening here, parallel to Turkey since the 1970s and '80s. They organise themselves here, not just religiously but economically, from birth to death. They have all their stores, groceries, everything you buy you can get in Islamic stores or travel agencies. Most of them are somehow connected to Refah. The main impulse comes from Turkey and other Islamic countries, the whole Islamic renaissance. They find good ground in Germany, because the people are feeling discriminated against, not accepted and Islam is something they can hold on to. I'm afraid this [parallel lifestyle] is becoming more popular.'

But why, I ask, is that a problem, for Turks to live an Islamic life?

'The problem is part of this Islamic community isolates itself from living here, it becomes more theologically undemocratic, this is the problem. This discussion will get more difficult unless Germany changes its politics and opens itself to these people. The normal German man and woman on the street see Islam as something from the last century, something undemocratic and so on. This is of course nonsense, but [this lifestyle] confirms their preconceptions.'

Several groups far more radical than Refah are active among Germany's Muslims. Rabah Kebir, the exiled leader of Algeria's Islamic Salvation Front (FIS) is under constant police surveillance at his home near Cologne. A report by the German counter-espionage service entitled 'Islamic extremism and its effects on the Federal Republic of Germany' identifies fourteen radical Muslim groups and over 22,000 active sympathisers across the country. German intelligence officers believe radical Islamists have set up a network stretching across eastern Europe and Germany to buy arms and detonators, which are then smuggled to Algeria, often via Hamburg. The reports says that: 'FIS sympathisers living in Germany have been involved in the dispatch of weapons and other technical material to Algeria.' Rabah Kebir though denies that FIS supports violence. 'We have chosen the peaceful path,' he has said.

Ismail Kosan, a Green MP in Berlin's senate, agrees that Germany's refusal to integrate Muslims into society is driving them to the mosque. Kosan, a well-groomed, paunchy and patronising radical, joined a Marxist party while at school in Turkey. His Marxism has faded away now, but he still shares the sense of self-worth and self-righteousness that is a hallmark of so many on the far left, constantly picking apart the wording of my questions. In fact he reminds me of nothing so much as the harder-line Islamic radicals I have met, who are far happier explaining what they believe is the true meaning of words such as democracy, rather than outlining the actual changes they would make in power. 'Do you know what an Alevi is?' he demands, when I mention the Muslim grouping, practising an extremely liberal form of Shiism, that forms about a third of Turkey's over 60 million population. And when I ask him if it is important for members of the Turkish community to be politically active in Germany he replies:

'It's important that all people become politically active, but what do you mean with regard to the Turks by politically active?' Oh dear. It's clear that we are not hitting it off, but I press on. As one of Germany's two elected Turkish politicians, his views carry a certain weight.

As for the Turkish community, that doesn't exist at all, he claims. At least in the sense that I use the word. 'There is no Turkish community here, but just people from Turkey. But there are people who speak Turkish but that does not necessarily make a Turkish community. When you talk about the Turkish community there are several different groups within it, Arabs,

Albanians, Kurds, Alevis. The European tendency of lumping together these different nationalities as Turks is completely false, because you cannot come to that simplistic conclusion, it's much more complex and there has never been proper research. It is not a homogenous community . . . The only thing in common they have is being a foreigner.'

Kosan's point is a fair one, although the members of all these groups have Turkish nationality, but even his allies in the Turkish community admit that his constantly hectoring tone does not make the best advertisement for his community. Whether or not it exists. He will agree though that Islamic sentiment is on the rise. 'Religious fundamentalism is becoming more popular everywhere. The Turkish-speaking community here does appear to be becoming more religious, for the simple reason that Europeans are the true fundamentalists, they are the ones who have rejected them, who pushed them out and marginalised them into having an Islamic identity. In the 1970s, very few women covered their heads and now you see them everywhere.

'This new religious identity has happened as a reaction, as a kind of self-defence, against European marginalisation, something completely different to what is happening in Turkey. Turkey has sent religious teachers here to Germany, to preach Islam and keep their Islamic identity. This fits in with the policy of integration which the German state has as well as the Social Democrats, that Turks should stay here and they should have a distinct Turkish and Muslim identity, when there are no distinctions made in terms of how heterogeneous everything really is.'

Germany's Turkish-speaking community may well be multi-ethnic, but religious Turkish Muslims are all united in their worship of Allah. At Friday afternoon prayers at the mosque in Cologne, there is the same sense of serenity and devotion that I have witnessed from Tirana to Washington DC. There is not enough room in the squat, square building for all the worshippers, who overflow into the courtyard outside, which, as is the Muslim custom, also boasts communal buildings and a kitchen. As the Imam recites the prayer the hundreds of worshippers fall as one to their knees, prostrating themselves before their God as they intone the verses from the Quran. There is an extraordinary sense here of spirituality, unity in worship and equality, for when all Muslims are kneeling on the floor facing Mecca, all are equal in the eyes of God. Unlike the many churches and synagogue services I have attended, nobody talks here during prayers, they come just to worship, and they chat and socialise later outside. For some reason it is at this mosque, rather than the glitzy new buildings at Regent's Park, or the plush Islamic centre of New York – for Cologne's mosque is certainly one of the plainest I have ever seen – that I begin to understand the true meaning of the word Islam: submission, surrender to a power greater than all of us, as I watch the rows of worshippers on the floor and sense the inner peace their faith brings.

A faith and a power that Ankara seeks to control, even at a distance. While Germany and Turkey argue about dual citizenship, they both agree on the need to control the rising Islamic sentiment in the Turkish community. Turks in Germany, like those in the mother country, reflect the diversity of Islamic beliefs. During the 1960s and 1970s some of the most active Turkish Muslim groups in Germany were part of the radical Islamic opposition, such as the Suleymanci movement. During the 1980s allies of Necmettin Erbakan, who in the summer of 1996 became prime minister of Turkey, were active, especially in Berlin. But even they were perceived as too liberal for followers of Cemaleddin Kaplan, who split off to form his own pro-Iran movement. Kaplan, nicknamed 'the Khomeini of Cologne' and 'the Dark Voice' for his aim of turning Turkey into an Islamic state, died in exile aged 69 in Cologne in May 1995.

For the majority of Turks groups such as Kaplan's and the Suleymancis had little appeal. They wanted to practise their religion and keep their cultural identity but had not come to Germany to make an Islamic revolution, either there or in Turkey. The answer, for Bonn and Ankara, both growing increasingly concerned at the Islamic radicals' activities – spurred on by the Islamic revolution in Iran – was firstly to control the influx of Turkish Imams into Germany. Bonn and Ankara signed an agreement that the only Imams and religious teachers that would be allowed to work in Germany would be licensed by Ankara. They would then have to return home after five years. (An immediate result of this agreement was a drive by radical Muslim groups in Turkey to infiltrate the new structures

and get their people posted, with official approval, to Germany.)
The two governments' next move against the Islamic radicals
was to take control of the growing network of mosques in
Germany, through setting up *Diyanet Isleri Turk-Islam Birligi*
(DITIB), a branch of the Turkish Department of Religious
Affairs.

By the end of the 1980s between half and two thirds of
mosques in Germany were controlled by DITIB. The organisation has also helped draw up educational syllabi for Muslim
schoolchildren in Germany, the classroom being another important arena for the battle between Turkey's Kemalists and
Islamists. Many secular Turks, such as Safter Cinar, oppose
DITIB's role in education. 'We are against this, and not just
because of the connection with Turkey. Even if we had a very
democratic Turkey this would be wrong. Firstly because of the
books, which should be written here because of the conditions
in Germany, and the teachers as well, must also live here to
understand the feelings and problems of the children. The
Turkish government saw that this was an opportunity to organise the religion of the immigrants in Europe under Turkish state
control, in competition with Refah.'

The DITIB office in Cologne, within the mosque complex, is the
first Turkish workplace I have seen to boast both a picture of
Kemal Atatürk and several verses of the Quran, inscribed in
ornate calligraphy, on the wall. The usual choice of decoration –
homage to the founder of the secular state, or verse from the
Quran – says everything about the dichotomy now cleaving
Turkey. As a virtual arm of the Turkish government, but one
dealing with Islam, DITIB carries out a delicate balancing act
among the Turkish immigrant community. The secretary, a jolly
man who I'll call Mehmet, and I, chat for a while about why I
am in Cologne, but Mehmet, 55, who arrived in Germany from
Tarsus in 1964, doesn't want to discuss controversial issues like
the rise of Refah or the influence of the Suleymancis. He's far
happier in his role as a sort of ambassador for Islam to the
many Germans who visit the mosque, curious about what it
means to be a Muslim. Just as in America and Britain, a growing number of Germans are becoming Muslim.

'Sometimes a Christian man or a woman comes to us and

says, I would like to become a Muslim. We say, all right, everybody is welcome for us. We write a paper, name, surname and address saying he was here today and wanted to become a Muslim. There is a chapter in the Quran, three sentences, "I believe and bear witness that there is only one God and Muhammad is his prophet." I say it, he says it, when he is finished, I say OK, you are a Muslim and we give him the paper. They need the paper if they want to marry a Muslim, woman or man. Perhaps because of a big love. Or perhaps he really believes and wants to be a Muslim. After this a few days later he may call us and say I changed my mind, I am not a Muslim and I need a new paper. But I say to him: "In the moment that you said you are not Muslim, you are not." You don't need a piece of paper saying it, we don't have it.

'We have school groups and university students, I go with them to the mosque and I explain what it is and what we are doing there. They have some wrong ideas, I correct them and say what the newspapers and television say, it's not true, because they don't know us or our religion. For example, they say why does the woman pray behind the men, is it discrimination? I say no, for God says this is not discrimination. Our prophet says that the key to paradise lies under the feet of the mother, that means you must respect a woman or a mother. If the Muslim religion discriminates against woman, he would not say it. This is not only a sentence, it is a book. You must think about it, what does it mean.'

But, I reply, a woman can never become an Imam. 'You know how Muslims pray, a man or the woman must go with their head to the floor, imagine such a woman doing the same thing in front of a man. Is it all right, is it praying if a woman lies before you? You don't think about praying if she lies on the carpet in front of you, can you think of God? There is no man in the world that would think of God while a woman lies in front of him. A Muslim wants to be in front of God but if the woman lies ahead he can't think about Allah. He thinks about the woman. That's why the woman cannot become a priest because she cannot lie in front.' That a woman might be distracted from her worship, for whatever reason, by the sight of rows of male posteriors doesn't seem to matter.

*

207

A few miles away from DITIB's bureau in Cologne is the Bonn office of Cem Ozdemir, Germany's first ethnic Turkish MP, elected for the Greens. A youthful 30, with a pierced ear, Ozdemir, who was born in Bad Urach, is positively buzzing with enthusiasm for his work. For many young Turks in Germany, whether Sunni Muslims, Alevis or Kurds, Ozdemir's election shows that even the corridors of power are opening to the Turkish community. 'Everybody in Bad Urach knows my parents and me, and they all say "our Cem", the Germans also say that, "our Cem, he's an MP".' Now of course, Ozdemir is a local hero, but growing up in Bad Urach his family's welcome was more ambiguous. 'I remember my teacher in the first class said, when I was seven, that there was no need to send me to the next class because we would go back to Turkey. Now when I see her in the street she says, "Of course I knew that you were going to be an important person." Unbelievable.'

Ozdemir became politically active in his teens, but while many of his Turkish compatriots in the early 1980s concentrated on opposing the then military junta, he threw his energies into the burgeoning Green movement. 'After Mölln, Solingen and Rostock I realised that it was important to be in Parliament and show the second- and third-generation immigrants that it was possible to change something here, and that we are a part of this society. I am not in the governing party so it is hard to change the law with a majority. But it is possible to use the media, to help some parts of the government to talk about things. I am from a workers' family, not an academic family. My mother works at a tailor's, my father works at a fire extinguisher factory. At the beginning some people said I sold out, but at meetings I tell people, look, I am not something different. I am one of you, nothing unusual and you can do it too.'

Now, says Ozdemir, all three main sections of the Turkish community – Sunnis, Alevis and Kurds – want to claim him as their own. His own background is Sunni, but his name, Cem, is the word for an Alevi prayer meeting. 'I was in Hamburg where I made a speech, and afterwards lots of young Turks said, "I've heard you are Alevi and I am so proud of you." I said I'm not Alevi, but I don't have to be one to think like one and understand the Alevis. Another one said you must be a Kurd, because only a Kurd can understand us. I said no, because I don't have to be a Kurd to understand what the Kurds are fighting for. This is a good sign that people say you must be this or that, because I can give the image that I work for the Turks and the Kurds, for Sunnis and Alevis.'

Ozdemir shares a different vision to the observant Muslims of the National Sight office in Berlin, but he agrees that Islam is on the rise among Turkish youth. 'These people who are not really integrated into society, who are in between, they are searching for another kind of identity because Germany never made it possible for them to feel part of the society. There are two possibilities here, either to become a nationalist, a real Turk, or become religious. Now a lot of people who never went into a mosque in Turkey started to be religious. This is very interesting, it's a sign that some things are not going in a good way. It gives those people feeling that they have something. They think: I am a Muslim, I have a mosque, community, security in this foreign world, foreign country. This is more than Refah, there is also a radicalisation in the classical Muslim organisations which are not near to Refah. In Turkey education is totally under the control of Islamic fundamentalists. That definitely has influence that spills over into Germany.'

The radical Muslims' recruiting drive is also fuelled by the graves of those killed in Solingen and Mölln and the squirming response of the Kohl government as it sought to balance enough outrage to appease international opinion and the votes of German nationalists. 'Mrs Genc in Solingen, she said we have to live together and she told me about the signs of friendship she has got from lots of Germans. But she also said this: that she can never forget what Kohl said and that he didn't even call her, or send her something, even a little token. Nothing, absolutely nothing.'

More positively, MPs from Kohl's Christian Democrat Party are now consulting Ozdemir about issues to do with their Turkish constituents, he says. 'They come to me and ask, "I have a mosque in my constituency, I don't know whether I should go there or not, what do you think?" Or they have been invited to something by the Turkish community. I feel like a kind of ambassador.'

Everything that happens in Turkey, the conflicts in Turkey

between Sunni and Alevi, Kurd and Turk, all these spill over onto the streets of Bonn and Cologne, Hamburg and Berlin, splitting the Turkish community on similar lines to their compatriots in Istanbul and Ankara. After an attack by gunmen on Alevi coffee shops in Istanbul, in March 1995, Alevis took to the streets across Germany and Europe to protest, from Cologne, Frankfurt and Hanover, to Innsbruck, Zurich and Paris. Kurds also regularly march to demand an independent homeland in the eastern part of Turkey. Coming back from my interview with Cem Ozdemir, I caught the tail end of a demonstration by tens of thousands of Kurds, marching across Bonn on an otherwise sleepy Saturday afternoon. There are about 400,000 Kurds in Germany, about 20 per cent of the total Turkish-speaking population. Some demonstrators waved the flag of the Kurdistan Workers' Party, the PKK, which is illegal in Germany and Turkey, the armed wing of the separatist movement. The PKK runs a nationwide underground movement across Germany, with cells in most major cities.

For Ahmet – not his real name – a Kurd living in Berlin, who is in contact with the PKK underground, the conflict between Turks and Kurds is part of his daily life in Germany. Ahmet fled Turkey after the military coup in 1980. As a member of a left-wing Kurdish political organisation he was an obvious target for the wave of repression that followed the generals' takeover. After being told initially by the German authorities that they considered him to be a terrorist he somehow acquired a French passport. He married a German woman, they divorced and he is now legalised in Germany, where he paints and writes.

'Relations between Kurds and Turks in Berlin are very tense. The atmosphere is terrible, they used to be one community when everyone first arrived, but increasingly as the conflict has developed militarily and socially, this has had consequences for the community here. Now it's very rare to find a television or newspaper in a [Turkish] café, because people try to remove any reason to heighten that conflict. To discuss it is to light up a powder keg,' he said when we met in a Greek restaurant in Kreuzberg. Most Turkish cafés are out of bounds for Ahmet, he says, as they would recognise him as a Kurd and he is scared to enter them.

'It changed three or four years ago, when it became especially bad. It started with the Gulf War then got worse. Now I would not dare to go into a Turkish café and order a tea, because I look Kurdish and I speak Turkish with a Kurdish accent. I don't feel comfortable. It's sad but it doesn't bother me that much. The PKK has a big presence in Berlin. It's not a question of noticing it, simply that they do, I know the people involved and I know what they do.'

What do they do, I ask?

He pauses. 'They look for help in every possible area, wherever they can for their cause. It's very strictly organised.'

Turks and Kurds are involved in the drugs trade out of proportion to their numbers, say German police officers. Certainly in Kreuzberg it doesn't seem hard to find heroin. A fresh harvest of syringes is scattered around the U-Bahn station at Kottbusser Tor every morning, a few feet from the Turkish snack stands. Heroin is the main narcotics problem in Berlin, says Chief Superintendent Rudiger Engler of the Berlin State Police. Of those involved in dealing in heroin, 40 per cent are foreigners and of those 49 per cent are Turkish, according to police statistics. 'There are known spots, in Kreuzberg at Kottbusser Tor, if you go there in the evening and open your eyes you can see the problem. A lot of the active street dealers are Turkish. Their parents came here to earn money, but they are dealing in drugs. Many are Kurds, from the cities and villages around Diyabakir, in the east.'

Other, Kurdish, sources say that while the PKK does not deal directly in Germany's heroin trade, it receives a percentage from many sales. Its members are suspected of several firebombings of Turkish businesses and assassination attempts against would-be defectors.

As the Berlin correspondent of the international edition of the Turkish daily *Hurriyet*, Ali Yumusak is a long-time observer of the Turkish community in Germany. He arrived in Germany in 1970 to study engineering in Augsberg, and began sending home stories about the lives of guestworkers. He has watched spreading Islamisation, the rise of the neo-Nazis (prompting him also to invest in new locks and security doors at home) and now the emergence of a new culture – the Euro-Turks, something new and unique.

Young Turks in Germany are distancing themselves from

German culture and regrouping amongst themselves, he says. 'The new generation is more self-confident and refuses to allow itself to be oppressed or discriminated against. They have reacted to the discrimination by forming their own networks. Young Turkish men used to date German women and even marry them. But now that trend is reversing, where Turks will only date and fraternise among each other. Now there are special Turkish discos, and some of the contact between the nationalities has been severed.'

Art, music and culture are building this new identity, he says, echoing the ideas of Ypsilon Musix's Unal Yuksel. Turkish-German writers such as Renan Demirkan and Emine Sevgi Ozdamar are fusing German and Turkish culture together to much critical acclaim. In Demirkan's first book, *Schwarzer Tee mit drei Stück Zucker* (black tea with three lumps of sugar), she told the story of a young Turkish woman's childhood and family in an Anatolian village as she waited to give birth in a German hospital. Some argued that the book received such praise because it presented the kind of picture of 'exotic' Turks in staid Germany that readers expected, and Demirkan has moved away from portrayals of a remembered Turkey in her newer work such as *Die Frau mit Bart* (the woman with a moustache), the tale of two German women's lives.

The first work of Emine Sevgi Ozdamar, winner of the Ingeborg Bachmann prize, the German equivalent of the Booker Prize, in 1991, was a play in broken German, inspired by a letter she received from a fellow Turk. 'He had typed his letter without knowing how to type, painfully finding the letters, just as I had to painfully find my words in German. This man had lost much of his Turkish; his mother tongue had become entangled with his newly acquired German and now not even his wife could understand him without difficulty. But this man never spoke out against Germany. He kept repeating: home is where you have a job,' she has said.

It was her prize-winning novel *Das Leben ist eine Karawanserei hat zwei Türe aus einer kam ich rein aus der anderen ging ich raus* (life is like a caravan trail: through one door you come in, through the other you go out), that propelled her into the serious writers' league. German critics saluted her achievement but remarked that the prize had gone to a foreign writer who had chosen German as her second language. 'I was accepted, but merely as a "guest-writer",' she said. In some ways her writing is similar to that of another writer who settled in a foreign country, the late Nobel Prize for Literature Laureate Isaac Bashevis Singer. The writings of both share an emphasis on the themes of relations between men and women, sex, are tinged with magical realism and shot through with a sense of psychological dislocation and the heavy baggage of exile.

Ozdamar's work reflects the fractured nature of not just the life of an immigrant, but of Turkey itself, as it juggles its secular system of government with its Islamic heritage. She uses a language that draws from both German and Turkish: 'We have no choice but to rebuild the tongue which we have lost with the tongue that we have found,' she has said.

As one of her characters proclaims: 'I screamed out poems on the anniversaries of Atatürk's death but he should not have forbidden the Arabic writing. This ban, it's as though half my head has been cut off. All the names in my family are Arabic: Fatma, Mustafa, Ali, Samra. Thank God I still belong to a generation that grew up with a good many Arabic words.'

The record producer Unal Yuksel, his father Bilal, the trade unionist Safter Cinar, the Refah-supporting law student Nafiz Ucan, the MPs Cem Ozdemir and Ismail Kosan, the Kurdish exile Ahmet, the writers Renan Demirkan and Emine Sevgi Ozdamar, all these, by their work and by their very lives in Germany are both changing that country's society and helping spawn something new, a Euro-Turkish hybrid lifestyle, that draws on both the norms of Bavaria and the Bosphorus, enriching Europe. And that doesn't necessarily mean an overlap in style or approach with the work of the French-Algerian artists of La Friche in Marseilles, or the books and films of Hanif Kureishi in Britain. Germany, like Britain and France, is spawning its own indigenous unique Muslim-local hybrid culture, as the Muslims of each European nation blend their own heritage with Islamic culture. Certainly there are common themes between the young Muslim artists and musicians of Berlin and Marseilles for example – of identity and alienation, say – but the work of all is shaped as much by their daily environment as by Muslim culture itself.

'This new generation is culturally richer than the first generation and its German peers, because they live in a mixture between two worlds, that is how they think and how they live,' says Ali Yumusak. 'They have adopted some of the German characteristics, of a good work ethic, they are disciplined and organised. But they think very pragmatically and practically, which they learnt from their parents who had to adapt to life here and learn how to integrate. They have kept some of their Mediterranean instincts, they are more mobile, more spontaneous and they can react more quickly. They also have a vision of the future, they see themselves between two cultures and they understand that they have two home countries. They can think in those terms, beyond national borders.'

Mme Matilt Manokyan, Turkish brothel owner

Chapter 8

TURKEY: THE VEILED
REVOLUTION

'Turkey is at a crossroads, either it will go on in this despotic way or they will listen to the voice of the people. There is no third way. The voice of the people is the voice of Islam, Islamic values and beliefs. The West is not the solution for Turkey, nor is the East, but Islam.'
Ahmet Hakan Coskun, news manager at the Islamic television station, Channel 7

'The Refah [Islamic Welfare] Party says if we come to power we will bring you equality, the world and everything you need. But Turkish people are not inclined to ask what is inside this everything.'
Izzettin Dogan, a community leader of Turkey's minority Alevi Muslims

The road some colleagues and I took to Istanbul followed a similar route to the one tramped along by Suleyman the Magnificent's soldiers over 400 years ago, as European city after city fell to the advancing Ottoman forces. He captured Belgrade in 1521, Rhodes in 1522 and wiped out the Magyar armies at Mohács, southern Hungary, in 1529, before taking Budapest. But while the Ottomans marched north, eventually reaching the gates of Vienna, we were headed south through the Balkans, from Budapest through Belgrade and Sofia to Istanbul, former capital of the Ottoman empire.

In the flatlands of Hungary, across much of Serbia, there are but few signs left that this whole swathe of Europe was once under Islamic rule. But while much of the Muslim contribution to the Balkans' cultural mosaic has vanished, the best traditions of Ottoman bureaucracy live on, as travellers are logged and registered. There are forms to be filled in and stamped, puddles of grubby-looking disinfectant to drive through and a succession of taxes and visa fees to be paid. As we drove across into Turkey from Bulgaria, our papers duly franked and authorised in the best traditions of Ottoman bureaucracy, a bank of duty-free shops glittered in the afternoon heat, their rows of shiny bottles stacked on the shelves. It was an unexpected welcome for an Islamic country, but one that perhaps symbolises the contradiction of a nation that forms the bridge between the Muslim Middle East and Christian Europe. Seeing the Budapest number plates on our car a shop-owner jumped forward excitedly, beckoning us over in the Magyar language, for this is a popular route for Hungarian shoppers. We asked if he had any whisky. 'Of course, of course, how many bottles do you want? Buy ten! Buy ten!' he exclaimed. One sufficed for a few nightcaps in Edirne.

The Thracian plain stretched away in front of us as we sped down the motorway towards Istanbul. Platoons of women worked in the fields by the side of the road, their heads covered, in accordance with Islamic custom, as they bent over the earth. The men, I learnt later, were busy in the coffee houses, smoking and chatting. The road was almost deserted until we approached the engaging chaos of Edirne when suddenly we were swept up in the shoals of ancient German cars which ebbed and flowed around us, while in front, high on a hill, the four minarets of the Selimiye mosque, Edirne's landmark, pointed skywards, to heaven.

Tucked away in the corner of north-west Turkey a few kilometres from the Greek and Bulgarian borders, this frontier town has been the stomping ground of a string of different empires' armies, from the Romans to the Ottomans. Murat I took the city for the Ottomans in 1361 and several decades later Mehmed the Conqueror garrisoned his forces here in preparation for the assault on Constantinople. The city was shunted back and forth over the next centuries between the Russians, Bulgaria and Greece, until Turkey regained its sovereignty in 1923 at the Treaty of Lausanne. Now Edirne is firmly Turkish, with a fine Ottoman heritage of mosques and markets. A small-scale

Istanbul, its streets are crowded with meandering pedestrians, cars hooting as they wander across roundabouts and junctions and everywhere stalls, shops and restaurants.

Towering over the city is the Selimiye mosque, whose minarets, the second tallest in the world after Mecca, are visible from kilometres away on the Thracian plain. The Selimiye mosque is the work of Mimar Sinan, court architect to three Sultans including Suleyman the Magnificent. Its dome is just a few centimetres bigger than that of the church-turned-mosque, Aya Sofya in Istanbul, a symbol in brick and mortar that the Islamic Ottoman empire would surpass its Byzantine predecessors. Centuries after the death of Sinan and the Sultans who commissioned him, Edirne is crowded with the heritage of urban Islamic architecture, for Islam at least in its Ottoman variety is an urban religion that demands schools, markets, bazaars and public kitchens as well as places of worship.

That the Ottomans could bring a rich and intricate civilisation in their wake, one that could teach medieval Europe plenty about religious tolerance and civic infrastructure, was unimaginable to Christendom's scribes. When the Ottomans captured Constantinople in 1453, for many in the West it was the darkest day in the history of the world and masses were said all over Europe. Here is the writer Kritovoulos on the capture of the city: 'This crowd made up of men from every race and nation, brought together by chance, like wild and ferocious beasts, leaped into the houses, driving them mercilessly, dragging, rending, forcing, hauling them disgracefully into the public highways and doing every evil thing.' His words sent a wave of horror across Christian Europe, just as they were intended to.

Blood ran in the gutters and a carpet of severed heads floated in the sea as the Sultan's soldiers took their three days of plunder, promised to them by their leader, Mehmed the Conqueror. Eyewitness reports from survivors describe an orgy of arson and destruction as children were raped and slaughtered while others were carted off to the slave market. Horrific, certainly, and the basis of a folk memory that still inspires fear to the cry of 'The Turks are coming', but standard practice for marauding armies, both then and now and certainly no different from the tactics of the Christian First Crusade, as they set off to rape and pillage their way across what is now Turkey.

When Pope Urban II's holy warriors set off in 1096 for Constantinople they never even reached the city, having expended most of their energy in the Rhine Valley where they set about slaughtering the region's Jews with great enthusiasm. When Constantinople was finally captured by the knights of the fourth crusade in 1204, Rome's warriors looted its Byzantine treasures in exactly the same way as the soldiers of Mehmed the Conqueror over two hundred years later. The Crusaders plundered so many valuables from Aya Sofya that they had to bring in mules to carry off their booty. For added entertainment they seated a prostitute on the Orthodox patriarch's throne.

The fall of Byzantine Constantinople on 29 May 1453 and its subsequent transformation into Muslim Istanbul was the Ottomans' greatest prize. Its importance was both strategic, for now Mehmed the Conqueror controlled the straits of the Bosphorus, gateway to both the Mediterranean and the Black Sea, and immensely symbolic, for the seat of Orthodoxy had been captured and Rome stood alone against Islam. Aya Sofya was converted into a mosque by Mehmed and stayed a house of worship until it opened as a museum in 1934.

Mehmed also built the first buildings of the Topkapi Palace, epicentre of the Ottoman empire for almost four hundred years. Now the *Bab-i Humayun* (sublime gate) is crowded with tourists and taxis, the general air of lax chaos in historical tune with its function as a space always open to the public, where citizens could petition the government. Virtually everything remains in the complex from the heyday of the Ottoman empire, including the Divan (council), from where the Sultan would listen in on its decisions, and the infamous cage, in fact a claustrophobic suite of tiny rooms. Tired of slaughtering their brothers on their ascension to the Ottoman throne the Sultans instead incarcerated them here for decades, complete with a retinue of deaf mutes and a harem. The Sultan's concubines had their ovaries removed to prevent them becoming pregnant, and giving birth to another potential rival to the throne. If they did conceive they were trussed up and drowned in the Bosphorus.

Not surprisingly most of those locked up in the cage went crazy. None more so than Deli Ibrahim (Mad Ibrahim) who emerged after twenty-two years to engage in unrivalled sexual excess. Egged on by his mother and his ministers, for the more time he spent in affairs of the bed the less he could devote to those of state, he designed his own orgy robes, lined with sable fur and studded with precious stones. His jaded tastes demanded ever fresh stimulation, and one day he decided that sexual pleasure was proportional to size. His messengers were sent out to find the fattest woman possible and located a gigantic Armenian woman who soon became his favourite. Sadly her appeal to the Sultan proved no insurance against the jealousy of her rivals, and she met the classic end of those who took a wrong turn in the intrigue-drenched corridors of the harem, and was strangled at a feast.

The institution of the harem is the source for centuries' worth of orientalist stereotypes about Turkey and 'lustful Turks', but its narrow corridors are more symbolic of a decaying empire, rotted by corruption and unable or unwilling to reform itself than any national sexual characteristics. Guarded by black eunuchs, who had suffered radical castration, the harem took its name from the Arabic word *haram*, meaning forbidden. As the Ottoman empire declined, the numbers of odalisques increased and by the reign of Sultan Abdulaziz (1861–76) they were over 800 strong. Many of the women, chosen for their beauty, were kidnapped from the outer reaches of the Ottoman empire, such as Georgia or Hungary. Only a few reached the Sultan's bed, when they would be given their own slaves. Many died of disease in the unhygienic conditions of the harem, their miserable end speeded up by boredom and homesickness.

From Topkapi Mehmed and his successors, particularly Suleyman the Magnificent, launched wars of conquest that saw large swathes of eastern and central Europe, Persia and the Arabian peninsula fall under Ottoman rule. At its furthest extent the Ottoman empire extended from the shore of north Africa, down the Saudi coastline, along the Black Sea to Daghestan in the Caucasus, north through the Crimea into Transylvania and northern Dalmatia. The international diplomatic, and sometimes military, consequences of its dissolution continues today in places as far apart as Sarajevo and Gaza.

So does the internal fall-out from the abolition of the Sultanate in 1922 and Turkey's transformation into a republic the following October. Turkey was among the losers of the First

World War, but compared to Germany or Hungary, she emerged with dignity, keeping reasonable natural borders in a state of manageable size and expelling the various foreign armies hoping for control of a slice of Turkish territory. In 1923 the capital was moved to the dreary central Anatolian town of Ankara, and Mustafa Kemal was designated head of state. The only undefeated general at the end of the war, Kemal had organised a campaign of guerrilla warfare against the plans to slice up Turkey between the victorious allies. An arch-westerniser, he was determined to drag Turkey into the twentieth century. In quick succession Kemal abolished the Caliphate, exiled the last Caliph Abdulmecid and all members of the royal house of Osman; closed the *medreses* (religious schools) and religious courts for the Jewish, Greek and Armenian minorities, and a new Ministry of Religious Affairs supervised the assets of the *vakifs* (religious endowments).

More was to come. The old lunar calendar was replaced by the Gregorian one, *seriat* (Islamic) law was abolished and a version of modern European law installed. This new corpus covered marriage and divorce which became civil matters, polygamy and the veil were outlawed and in 1930 women were given the vote. Surnames were made compulsory and Mustafa Kemal became Kemal Atatürk, meaning 'father of the Turks', for nobody could accuse him of modesty. The old Ottoman script, of Turkish words written in Arabic characters, was replaced with the Latin alphabet. Many of these measures appeared laudable decisions, by western liberal standards.

But Kemal's secularism was not of the western kind that decrees a separation of church and state and allows normal public expressions of religious belief, in communal structures or even dress. It was a planned anti-Islamism – in the sense of Islam providing the basis for a system of government – aimed at ripping out the old Ottoman society and replacing it with a secular government that drew on the 1920s ideas of nationalism and the corporate state that also inspired the early Benito Mussolini and the Soviet Communists. He tried, and succeeded, in creating a modern state out of a plethora of nationalities, for Turkey then, and now, is home to a plethora of minorities that include virtually every letter of the alphabet from Abkhazians to Turcomans, binding this ethnic mix together to form a new nation. Out in the villages in the vast stretches of Anatolia, his radical measures triggered anger and confusion. Atatürk was a great Gazi (fighter for Islam) who had liberated Turkey from the foreign infidels of the allies' armies and was revered by every Turk. But why did he change the alphabet and smash the power of the Imams in the drive for a one-party state, they asked?

Just as Stalin purged the old Bolsheviks in his drive for a completely new society, Atatürk jettisoned virtually every facet of the Ottoman empire that had centuries-old roots as he shaped his new republic with consequences that are even now sending ripples across Turkish society. By changing the alphabet, for example, he cut off many Turks from their own literary and cultural heritage. The legacy of his drastic attempts to kick-start a parochial, backward-looking Muslim country into a modern, secular state drove many Muslim clergymen underground, to form the nucleus of an Islamic opposition movement that, seventy years later, is gaining increasing political power in Turkey.

For Atatürk inflicted a trauma on the national psyche that only now is slowly surfacing, and helping build an Islamic revival, like a long-delayed case of national post-traumatic shock. By the time Atatürk took power the Ottoman turban had been replaced by the more modern Fez, but even that was not up to date enough, and that headgear was replaced by Atatürk's panama hat. A uniform western-style dress also helped bind his nation together, with its mix of national minorities. He wore his new hat for the first time in 1925 at Kastamonu, speaking in the heartlands of Islamic conservatism, records Andrew Wheatcroft, in his book *The Ottomans*, announcing that:

> A civilised international dress is worthy and appropriate for our nation and we will wear it. Boots and shoes on our feet, trousers on our legs, shirt and tie, jacket and waistcoat – and of course, to complete these, a cover with a brim on our heads. I want to make this clear. This head covering is called a HAT.[1]

Serpil Orcalan does not wear a hat. Her hair is covered instead with a shining grey, gold and green headscarf and she is dressed in a long skirt. An engaging young woman in her twenties, whose nose wrinkles when she smiles, Orcalan made history in July 1995 when she started work as a newsreader, reciting the day's events on television dressed in full Islamic garb, for the first time ever in Turkey's history.

Orcalan is a newsreader at Channel 7, a private Islamic television station set up in autumn 1994, with strong links to the Islamic Welfare Party (Refah). The station began with broadcasts to Istanbul and Ankara and plans satellite expansion to Europe, and the Turkic former Soviet republics of Azerbaijan and Turkmenistan. Refah's victory in municipal elections in 1994 gave it control of the city councils in both Istanbul and Ankara and led its supporters to proclaim that it was on the path to national political power. Its triumph was regarded with horror by Turkey's secular liberals and intellectuals, although the hysterical predictions of many that Turkey could become the next Iran are both premature and simplistic.

Like many young women who have now adopted Islamic dress, Orcalan is not from a religious background. Her family was split by her decision to cover her head, with her father unhappy and her mother supportive. 'The first time I put on the scarf I felt very happy and at peace, which I never found before in my life. I felt sheltered. I'm not from a religious background although my mother is religious. There was a lot of very powerful emotion at home about this. My father doesn't want me to wear it. He is a secularist, although now he accepts it. I was not religious when l grew up. I made a decision to dedicate myself to Islam. It was the search for truth. I was always searching for truth from childhood and what it was. When I was twenty I realised that Islam was the truth. I began to read the Quran during these years and that made me realise that Islam was the truth. I felt that God spoke to me, I felt his weight.'

For Orcalan, covering her head was more than an expression of piety. The choice, or lack of headgear, is a symbol of a desire for a whole regime: secular and democratic, as Turkey is now, or Islamic, with a form of government that would take shape only after a victory by Refah at the next general election. Donning Islamic dress is also a political and cultural statement, one that seeks to assert what observant Muslims see as traditional Turkish values and an Islamic heritage that is rapidly awakening after its eradication during the last seventy years of Kemalism, as Atatürk's doctrines are dubbed. 'I am not happy with living in a secular society. Secularism has oppressed Muslims. Secularism here is not the same as in the West, here it is a tool to oppress Muslim people. I and those like me underwent some bad experiences. Now there is a return to Islamic beliefs and values. It is a return to our origin, because Turkey has its origin in Islamic values and beliefs and now we are returning to the authentic ones.'

Like many young Turkish women who have turned to Islam Orcalan is keen to emphasise her modern credentials as an independent working woman, defying the western stereotype that Muslim women must be bound to the kitchen and the baby's bedroom. 'I am not against Muslim women working, this is a job among jobs. I am aware that there is such a wrong idea about Islam and women and Islam. I do not know the origin of this. They don't know us enough. Women are human and are equal in Islam. I've worked here for about two months. I don't have any previous experience but I always wanted to be a newsreader. Channel 7 is the only place where I can do this job without giving up my beliefs and wearing my Islamic clothes.'

Channel 7 broadcasts a varied menu of news, women's programmes, music shows and films. Some of the latter appear unexpected choices, at least to western eyes, for an Islamic channel. 'On Saturdays we broadcast classic films such as Laurel and Hardy, Tarkovsky and Charlie Chaplin. We find it normal to show these films. We don't show Sylvester Stallone films, because they are inhumane. That which is human is also Islamic. Charlie Chaplin's *Modern Times* is an Islamic film,' explains the station's news manager Ahmet Hakan Coskun, who is polite but cool towards me.

Dressed in a snappy check jacket and yellow shirt, with a neatly trimmed beard, Coskun works out of a modern house in the suburbs of Istanbul that is Channel 7's headquarters. It could be a private television station anywhere, full of state-of-the-art computers, editing equipment and shiny new desks. The only difference is that every woman in the bustling building is dressed in a long skirt and a headscarf. Channel 7's *raison d'être*, he says, is to make programmes based on genuine Turkish values, which are not those of the West.

'Our programmes at Channel 7 are liked by our grandmothers. The people themselves, our people, see themselves in Channel 7. In our news programmes we do not look at things through western glasses. For example if there is something happening in Algeria, the other channels are highly dependent on western sources in their news, but we rely on other sources. We underline the oppression of the Algerian government against Muslims. We depend on Muslim sources.'

As an Islamic television station, Channel 7 must carry out a difficult juggling act. 'If Muslims have made the discovery of television, it's because they think television as an instrument produced by man is in a sense inhumane, because it turns social life upside down and it poisons people. It is a western instrument. Being aware of all the dangers of television we try to develop a new model of television broadcasting. As far as possible we make coherent programmes. We are making a new philosophy of television. There is no propaganda for Islam in our programmes, but they support the view of Islam. When we make a programme of course we have Islamic women in it. They are there, but they don't say "women, come and be a Muslim".'

Suna Kili is a Muslim woman as well, one whose family can trace its Turkish roots back to the conquest of Constantinople in 1453, but she does not watch Channel 7 television. A loquacious professor at Bosphorus University, she worked in an office next to Tansu Ciller, formerly Turkey's prime minister, for seventeen years. Professor Kili is the author of several books on Atatürk and his legacy. But even among her students that appears to be crumbling, as more turn to Islam, to her increasing bafflement and concern. In part she blames the influence, and ready cash, of nearby Islamic states for what she sees as a retrogressive step, but like many intellectuals she also realises that a seismic change is transforming sections of Turkish society as her young charges exchange their jeans and mini-skirts for headscarves and long sleeves.

'Saudi Arabia helps them, and so does Iran. I had a student here, one who was wonderful, beautiful, could have been a model and in the middle of the term she started covering herself. She said to me, "I am the same one, I just decided to cover my head, it's a symbol."'

The move of many among Turkey's youth towards Islam is a potent symbol of a growing dichotomy about the very nature of the regime in Turkey. For those such as Professor Kili who believed that the principles of Kemalism were solid and immutable, these are troubling times. 'We have to reach a consensus on fundamentals about the nature of the regime we want to embrace. In Britain there are differences between the British Labour party and the Conservative party, on economic policies maybe, perhaps occasionally, it may be a difference of foreign policy. But you do not discuss the nature of the regime itself, in that freedoms are to be enjoyed by every party and by everybody, that the state should be secular and the way the press, the universities work, and so on, these are no longer a matter of discussion.'

Atatürk's revolutionary changes, such as the abolition of Islamic law, are protected by the constitution, but many intellectuals such as Professor Kili believe that the Islamists such as the Refah party aim to change the very nature of the regime itself. 'We have an identity crisis in Turkey. That identity crisis was solved in the Kemal era: you are a Turkish citizen and then you are a human being. Then we are back to saying you are foremost a Muslim. Turkey has become a more democratic society, but it did not become a more modern society. I don't think of modernity in terms of the number of factories or hotels, but in people's pattern of behaviour, if they are [Islamic] fatalists or not, but how society and the state function. From that point of view there is regression, and now the religious elements are contesting whether the nation-state was a good idea. The process of statehood has been gone through, but they want to change it.'

But Atatürk, argues Kili, did not move too fast. His radical changes were the only way of modernising the new Turkish republic. 'I am not someone who says draft the laws according to existing society. We couldn't move an inch. In a country like ours there should be an impetus. Aren't the Turkish people worthy of a twentieth-century government? Worthy of twentieth-century legislation? In the nineteenth century, the Young Turks aimed to establish constitutional government. We had a century and a half of concerted effort to political democracy. I can tell you unequivocally, it is not going to be possible in

Turkey to turn the clock backward, to establish a dictatorship of any kind, whether Marxist or a religious state like Khomeini.'

Istanbul is already a city of two parallel universes and the twain are meeting less and less as support for an Islamic revival grows. With its rolling landscaped gardens and view of the sea, Kili's Bosphorus University is like a Mediterranean version of Berkeley in California. Students who would look quite at home at Oxford or the Sorbonne loll about on the grass planning the evening's partying, while rock music carries over the verges. Student societies offer their wares, as well as a holiday company selling beach vacations. This is Turkey with its face turned firmly to the West. Downtown in bustling Taksim, shops offer western designer clothes and Swiss watches. REM blares from the bars, and the cafés are crowded with stylish young Turks sipping a beer or a coffee, the women in miniskirts, the men in sharp business suits.

But the snaking back-alleys of Fatih in central Istanbul are another world, where peddlers offer Islamic books and green banners among the dilapidated wooden houses and winding streets, the heartland of the Welfare Party's support. Women keep their heads covered and the otherwise ubiquitous portraits of Atatürk have been replaced with Arabic inscriptions and illustrations of Mecca and the Haj, the pilgrimage that every observant Muslim must make at least once. The cry of the muezzin calling the faithful to prayer drifts over the backstreets of Fatih. Observant Muslims gather outside the Suleymaniye mosque, masterpiece of Suleyman the Magnificent, and carefully perform their ablutions before praying.

For Welfare Party (Refah) activists such as writer Mehmet Metiner, now an adviser to Istanbul's Refah mayor Recep Tayyib Erdogan, the future of Turkey is already decided. 'The old systems are finished, all over the world and in Turkey. Kemalism is finished. In the hearts of the people the Kemalist system was finished when it was born. The public wants a system that fits with its religious beliefs. The public wants the state and the government to be like itself. The Welfare Party tries to give an answer to these demands of the public.'

In fact many attribute Refah's success more to political and economic factors than a genuine upsurge in enthusiasm for Islam *per se*. The city's population is growing by several hundred thousand every year as economic migrants pour in from Turkey's rural heartlands. Even university graduates struggle to find a job, so the waves of peasant arrivals fight to scrape a living as prices climb relentlessly. Neglected by the established parties, the rural influx into Istanbul and Ankara provides a steady source of recruits for the Welfare Party. 'The countryside people have flocked to the cities, now Istanbul has become a large shanty town. People are poor and they see all this ostentatious living and life is very hard. So religious people are really using that. This is very important. The fundamentalist leaders play on the fact that people are living in poverty, but they had to disclose their money, and Mr Erbakan, the head of the welfare party, is one of Turkey's wealthiest politicians,' says Professor Kili.

At the same time there is a growing malaise in the Turkish body politic. A feeling among many that the Kemalist project has run its course, the secular parties have run out of ideas, and now it is time for the country to return to its roots in Islam and the East. Writing in the English-language *Turkish Daily News*, Dr Nilufer Narli, author and sociologist, described 'a crisis of identity, a belief that we are in a state of moral decadence and political decline in a non-Islamic system, and the feeling of being lost'. Turkey's Islamic movement is a response to the anxiety of modernity, she writes. 'It has found fertile ground because of the shared belief that Turkey is in a state of decline – its cause, its departure from Islamic path; its cure, a return to Islam in private and public life to restore its Islamic identity and values.' This return, to the values of the Quran instead of Kemalism, would, Islamists believe, see Turkey return to the golden era of the Ottoman empire and take its rightful place as a world power.

Europe's at best guarded welcome for Turkey is boosting this Ottoman-nostalgia and Islamicism. With their country already a member of NATO, and a bulwark against the former Soviet Union during the Cold War, many Turks feel betrayed that Ankara is being fobbed off with plans for a customs union when they seek full EU membership. Turkey has joined the Organisation for Economic Cooperation and Development, the Council of Europe, the Organisation on Security and Cooperation in Europe and the Western European Union (as an associate member). Turkey has been an associate member of the EU for over thirty years. If there is a western international club, Turkey is in it. Ankara has also signed numerous international conventions in its drive to convince the world of its modern democratic credentials.

But deeds must follow words and serious question marks remain over Turkey's human rights record and its willingness to conform to European norms. Human rights activists say Ankara is in clear violation of the human rights conventions which it has signed. In 1993 the United Nations Committee Against Torture said that the 'existence of systematic torture in Turkey cannot be denied'. Gross human rights violations are being inflicted on civilians in the south-east of the country, says an Amnesty International Report released in February 1995. Much of these extra-judicial killings and torture take place in the ongoing war against the Kurdish separatists, but the situation is also deteriorating in the rest of the country. In the first ten months of 1994 there were fifty reported 'disappearances', nearly double that of 1993, with security forces blamed for the deaths and beatings.

Amnesty's report is a depressing and often grisly catalogue of torture, beatings, sexual assaults and apparent executions. Human rights activists and journalists have all been jailed for activities that would go unpunished in mainland Europe. In January 1996, Turkish journalist Metin Goktepe was found dead in Istanbul after witnesses said he was detained by police. He died from a brain hemorrhage brought on by beatings to the head, according to the official autopsy report.

But, and it's a big but, compared to its neighbours such as Syria and Iraq, Turkey, with its parliamentary democracy, reasonably free press and independent judiciary, is a beacon of light and progress. European concern at human rights issues in Turkey is seen as hypocritical by many Turks, who point to the West's inaction over the slaughter of Muslims in Bosnia and Chechnya. Disillusionment with the western values is growing, a process not confined to the slums and alleys of the bazaars and shanty-towns.

Here is Demir Inal, a manager at an Islamic bank, with a degree from a Canadian university, the sort of sophisticated figure who would be at home in a boardroom anywhere from

Zurich to New York. 'When Slovenia was attacked Germany came in and said "I don't want this" and the fighting was stopped in a week. But the same Germany does not do anything for Bosnia, or Chechnya. Things like this affect people's way of thinking. They keep seeing double standards. I would hope that the West does not think Muslim life is cheaper than Christian life, but it does sometimes seem like that. Look at Bosnia, 250,000 people dead and still dying. Look what America did to Libya when they killed two Americans. Imagine what they would do if it was an Iraqi or an Iranian who put the bomb in Oklahoma City, imagine what they would do to Iran or Iraq.'

The prospects of a Welfare Party government in a NATO member state, or a coalition with a powerful Welfare presence, has also caused nervous tremors in the West's corridors of power. Party officials, though, are careful to emphasise their commitment to democracy and their rejection of the Iranian and Saudi Arabian models of government. The *Refah Party and Truths*, the party's official publication, says: 'The allegation that the Welfare Party will develop policies based on enmity towards the West is not true because both East and West belong to Allah.' A statement whose intent can be interpreted in any number of ways, rather like the Refah Party itself.

The Refah Party has its roots in the National Order Party under the leadership of Necmettin Erbakan, founded in 1970. Declared illegal after the 1971 military coup it was reborn the following year as the National Salvation Party. Over the next twenty years its activists devoted themselves to increasing Islamic consciousness, especially among young Turks. Their target was not so much command of the state itself, but the building of an alternative Muslim vision of Turkey's future. Teachers, both in the state schools, where Sunni Muslim education is compulsory, and in private religious colleges, built up a solid bedrock for Refah's future Islamic edifice. In 1970 there were 72 *imam-hatip* private colleges, by 1992 their numbers had grown to 467.

Refah won 158 seats in Turkey's 450-member parliament after the December 1995 general election. By the following summer Necmettin Erbakan was prime minister of Turkey, after signing a coalition agreement with Tansu Ciller's True Path party. The agreement gave Turkey its first Islamist-majority government since Kemal Atatürk declared the country a secular republic in 1923. It made Erbakan prime minister for two years, after which he must hand over office to Mrs Ciller. There was much hysteria, both in Turkey and abroad about the victory of supposed 'Islamic fundmentalism', thanks in part to Erbakan's fiery Islamic rhetoric which included an array of unlikely and more or less unachievable promises. For example: 'We will set up an Islamic Common Market, an Islamic UN, a World Islamic Union and introduce an Islamic dinar . . . the Turkish lira is dead. Those who vote for us are good Muslims. Those who vote against us are unbelievers and atheists.' In fact only about 21 per cent of the electorate voted for Refah, and the vagaries of Turkish politics brought him to power, rather than any nation-wide lurch to radical Islam.

Erbakan's promises were enough to keep anyone busy for a while, but once again *realpolitik* triumphed over rhetoric and the reality of power proved radically different. As soon as his feet were under the Prime Minister's desk – and the Turkish financial markets began a depressed slide in anticipation of his rule – he immediately backtracked and promised to respect all international agreements, and seek close ties with the West. Even the military defence agreements with Israel, the subject of much harsh criticism among Refah's supporters, have survived. Instead Erbakan planned a visit to Iran, a smart move to keep the grassroots happy, while keeping steering the ship of Turkish foreign policy on a steady course. Why did he backtrack so rapidly? The answer is simple: had Erbakan attempted to change the fundamental nature of the Kemalist secular state the army would have almost certainly launched a military coup, leading to the nightmare 'Algerian' scenario. In one notable development Refah did open discussions with Kurdish leaders, and took a more conciliatory line towards that minority's aspirations.

Like any major political party, Refah is home to several factions and currents. It has its share of wild-eyed extremists who would force women back under the veil and introduce public amputations for theft, on the Saudi model. But these are a minority, and politicians such as its leader Erbakan who have swam in the dangerous straits of Turkish politics for decades are wily operators. He has survived the 'Bosniagate' scandal, when a parliamentary commission confirmed irregularities in the collection of millions of Deutschmarks from Turks in Germany by the Refah party. He was also little dented by revelations that he is the equivalent of a dollar millionaire in cash and property. In the autumn of 1994 he even visited the USA, reportedly to meet with think-tank analysts and intelligence officials.

But for all its anti-western election rhetoric, Refah's uniqueness is its remarkable and unprecedented synthesis of Islamic rhetoric and policies with political techniques of canvassing and recruitment that would be instantly recognisable by any Washington lobbyists. Recep Tayyib Erdogan and Melih Gokcek, the new young mayors of Istanbul and Ankara, wear suits, not turbans and robes, and appear little different from any western municipal leader. Refah party activists go door to door among the slums and alleys of Istanbul's shanty towns, offering help and a listening ear to the millions of Turks living in poverty.

Refah even formed its own tea-party brigades of female volunteers, from its women's wing, which claims almost one million members and whose activities were crucial to Refah's victory in Istanbul. Women Refah supporters were told to hold a small tea party and invite friends and relatives. Once the initial chat and courtesy greetings had been dispensed with, they moved onto politics. Videos of speeches by Erbakan were shown and calendars or perfumed handkerchiefs were handed out. It worked beautifully. No other party had bothered to court the hordes of bewildered female Anatolian immigrants who had poured into the big cities, to be completely bemused at the licentious ways of Istanbul and Ankara. Nobody ever asked them what they thought, except about the dinner they cooked every night for their husbands and families. When visitors came, who spoke their language, understood their concerns and gave them tea and attention, they were quickly converted to Refah's cause.

Battle has now commenced for the soul of Istanbul's shanty towns. Lined up against Refah's tea-party brigades are the secularist cadres composed of women such as Emine Sezer, who travel the slums for twelve hours a day and organise trips and summer schools for the children of the slums, as they seek to counteract what they see as Refah's brainwashing.

'Refah is putting pressure on, they give them money, they give them presents, and brainwash them, especially with the youngsters, and with Quran courses. About ninety per cent go on Quran courses and then they came to our summer school, so they saw the difference and they have to decide for themselves. I want to live integrated into western Europe, not into the Arabic countries. We tell this argument and they listen, because Refah hasn't told them this side of things. Refah is to denigrate western ideas and the western way of living. We tell them about the other story, the Arabic lifestyle, which is not something to admire, in my point of view. We are trying to show them two separate sorts of lives and they have to choose. I cannot dictate to people, you have to choose this sort of life or this sort of life, but if we educate them well enough they will be able to choose.'

For Sezer, part of Turkey's educated western élite, her work in the shanty towns has been a revelation. Her group should have started work there a decade ago, she says. 'Over the last few years there has been a gradual Islamisation of Turkish society and we were blind to it. That's the only way I can describe it, we talked, we had everything, we went on living grandly, and nobody ever thought of the poor. That was our biggest mistake. So we should be blamed, what we are trying to do now, we should have started ten years ago and fundamentalism would not have taken root in this country.

'But even if there isn't any danger of fundamentalism, I'd work like this. I have seen those people, they have nothing, absolutely nothing in their houses, in the way they live. Some of them have already been brainwashed, but they are open to you, they listen to you. But if I tell them don't cover yourself they will shut me out completely, they won't even listen to you, because they think it is their right to cover up themselves, so I have to respect them and I have to teach them to respect me. I keep telling them, don't give up, if you work hard and if your children are educated you can have a certain standard of living. You have to do something about it. We brought them birth control, they don't know anything about it, they have six or seven kids, they have no money, they were quite open to it. Now they keep asking us, what shall I do, please tell me, try to help me, and I am happy to see that.'

But while Erbakan, like every politician, is a prisoner of the global political and economic system on foreign policy issues, his victory has undoubtedly aided the cause of Islamicising Turkish society. Urban myth-style stories are already doing the rounds among the Istanbul intelligentsia, of armed gunmen threatening male and female swimmers for being 'unIslamic' by bathing together. Having Erbakan in the prime minister's office can only aid the cause of those who wish to steadily erode the Kemalist project. 'Refah members would like to introduce fundamentalism in Turkey. I'm scared of their group psychology. If one man says "attack the women whose heads are uncovered" the others will do it. I'm not so scared of Turkey becoming like Iran, as like Algeria,' says Sezer.

That seems unlikely at Istanbul's Refah-controlled municipal offices, where the hallways are crowded with bright multi-lingual young women, their heads covered and dressed in Islamic clothes, wielding mobile telephones. Here too, just as at Channel 7, the cliché of the cloistered Muslim female, condemned to a life of babies and cooking, has been proved wrong. 'In our country there is an idea that when you are living in an Islamic way, you must not be intelligent, you must be a fool. Our people always see us like this, as fools. I don't know why, they see the Islamic clothes as not modern,' said Isil Bulduk, an economics graduate from Istanbul University now working as secretary to the mayor. 'That view is wrong,' she says. 'I do everything the others do. I go to the cinema, theatres, I swim. There are places for us, of course, separately from men.' For many in Refah though, women have a strictly defined role. When Sibel Eraslan, Refah's women's leader in Istanbul, who many saw as a key to the party's victory for her organisational skills, demanded higher positions for women, she was sacked.

An intelligent woman in her twenties, who speaks English and German, Ms Bulduk said that other potential employers had rejected her on sight as soon as they saw her Islamic clothes. 'The others that finished the same school as me are working in very big companies. But when I went to business talks with them, with the head of the company, he looked at me, and said, "No, I don't need you." I said, "I know English, I know German, I finished university." But he said, "That doesn't matter, because your head is covered." I told him that the first thing is my job, if I do the job correctly, OK, if I make a mistake you can say everything to me.'

Bulduk became an object of media fascination after she started work as the mayor's secretary last year, to her consternation. 'They made a very big fuss when I got my job. They said, "How will you work in these clothes?" I wanted to tell them that I will answer the telephones and do all the other things that the mayor wants and I can do that with these clothes on. Before us in the past management, the secretaries working there only finished primary school. I said that I worked for four years and I am a university graduate and I have much more education. I said that they never took the others' photos but they took mine only because of my clothes. I felt very bad when this happened of course. But I got angry with them and I worked very well. Perhaps you won't believe me, but when I went to the toilet from my room, all the cameras were following me. I was like an animal in a zoo.' The unwillingness of many secularists to see the person beneath the scarf is substantially contributing to the growing polarisation of Turkish society as each reinforces the other's stereotype.

Married to an observant Muslim, Bulduk has found domestic happiness and career success. But she has lost many of her friends, who cut relations when she turned to Islam. 'My friends in university now don't talk with me. And they didn't come to my wedding ceremony. Of course I was upset, but I think why didn't they? I believe that they are not real friends. For example my old boyfriend from college came to me when I put these clothes on. He came to me and said: "I heard that you put your head under the cloth. I want to see you." He hugged me and he said, "You are very nice like that." That was a friendship, a real friendship.

'He said to me perhaps you won't be able to come to the bars with us, and you won't be able to drink alcoholic drinks with us, but I know your mind, and for me your mind is as important as your views. My college friends all supported me, but my friends in university, I don't know why, perhaps I didn't choose them very well, never accepted this. They thought that my husband made me so. He didn't, because putting these clothes on your head is a very difficult thing, because you give up things. I want to tell them that nobody can do this just because her husband wants it. And it is not a period for one or two years, it will continue for all your life.'

Modern Muslims such as Ms Bulduk reject the restrictions of regimes such as Saudi Arabia. They say they stand for a modern Islam, one that gives full rights to women. Turkey, they say, and on this many liberals agree with them, is a special case with its own history of tolerance and traditions. But it's hard to imagine a Welfare Party government ever joining – or being allowed to join – the European Union.

'Our face is turned to the East. Turkey has an Ottoman heritage. Our first target is to make good relations in politics, military and trade, with the countries that were once in the Ottoman empire. But that doesn't mean we are the enemy of the West, because we are not. We support political, military and trade relations with the West, but they must be just,' said Mehmet Metiner.

Other policy proposals may be less easy to implement. A system of differing religious courts for each faith, as existed under the Ottoman empire, will be introduced for each community, said Metiner. 'Everyone can live according to his laws. Christians and Jews have different legal systems. We don't want the government to insist on one legal system for all people, it is not democratic. That system [of different courts] worked very successfully in the Ottoman empire. We won't take that system into today's world but we will take it as a reference, and we will get some basic ideas from it and apply it to this world's concrete needs.'

Sorting out the more abstract needs of young Turkish women is the job of Leyla Melek, editor of Turkish *Cosmopolitan*. A softly spoken woman aged twenty-eight, Melek works out of the glass and steel extravaganza of the Sabah publishing house that also publishes several daily newspapers as well as *Penthouse* and *Playboy*. Like all of its international counterparts, Turkish *Cosmo* serves up a cocktail of sex, fashion and relationship advice, leavened with social and political features.

But there are limits to Turkish government tolerance. 'There are things we cannot print. We cannot show penises for example, if we do it will be censored. We do print explicit sex articles and we didn't have any problems until recently. We published a sex supplement, like all the women's magazines have done for the past year. Our March supplement was censored, but everything is so slow in Turkey so it was censored in April. I think they just had to pay a fine. The censorship is after it comes out. They inform us [if there is a problem] but usually that issue is already sold out and finished, so that nothing really happens. It was a very explicit educational book that we put out. *Penthouse* and *Playboy* don't seem to have many problems.'

Even here, in the heart of trendy medialand, Islam is making inroads, she says. 'I started working in this building in 1992, now it's 1995. Three years ago I would say maybe a third or a quarter of the 1,500 people working here used to fast during Ramadan, and this year it was more than half. Suddenly a lot of people started fasting. You think there's nothing wrong with that and nobody tried to beat me up or say something bad, but when people start staring at you if you are drinking a cup of coffee or some water it starts to become disturbing. It's not knowing what is going to happen in a few years that is disturbing. I'm trying not to think that it could be another Algeria.'

Islamic activists are focusing attention on young women, says Melek, and those who worked in western-oriented fields such as fashion are particularly prized. 'The Islamic section is very organised, probably the most organised in Turkey, where everything is very disorganised. In the high schools outsiders come to talk to the girls and try to convert them to Islam. It was effective over the years.

'Now they are converting models, because when someone who models clothes, or lingerie or swimwear, suddenly starts converting to Islam, covering their head, it's a very good image. westerners think it is disturbing, but a lot of Turkish people still believe that it's not disturbing and nothing will happen to the secular Turkish state. There is always the comparison with Algeria. I'm trying to be secular, everybody is free to believe in what they want. On the other hand I do believe that it is frightening and bad things may happen. I don't know yet, I haven't decided.'

For Matilt Manokyan, Turkey's richest woman, Refah's electoral triumph is bad news. Mrs Manokyan, an elderly Christian lady of Armenian origin in her late seventies, has built her multi-million-pound business empire on real estate and hotels, but she is best known for the string of legal brothels, somewhere between three and thirty-two, depending whom you believe, that she owns. Brothels are legal in Turkey, and her 'houses' as she calls them are supervised by doctors. Her palatial flats on the top floor of a block in downtown Istanbul are the most extraordinary residences I have ever seen. Lavish shrines to the cult of her personality, they are bedecked in portraits of Manokyan as a young girl, staring into the glorious future in classic socialist realist style. Every inch is taken up with press clippings, pictures, photographs and framed letters of thanks for the massive donations she makes to charity every year, including commendations from President Suleiman Demirel and Tansu Ciller. A battalion of servants bring a never-ending stream of tea and sweets and packets of opened cigarettes lay in ornate glass dishes all over the table. One of her officials cannot contain her excitement as we speak, and keeps jumping forward, proffering another letter of thanks from a charity to better illustrate the full extent of Manokyan's munificence.

Manokyan started in the sex business when she was working as a fashion designer and a client could not pay the bill. The client offered her a brothel instead and Madam Manokyan was born. 'The houses play an important role, they are necessary. In Europe relations between men and women are comfortable, if you like a girl you can go with her. But in Turkey it's not like that. We need these houses, because here a man cannot go out with a girl and just take her for a drink or something like that. I don't care if people criticise me for running the houses, everyone is free to speak in the way they want.' Such is her devotion to Kemal Atatürk that she has even written a poem in his memory, praising his merits as the founder of modern Turkey. 'Suleyman Demirel, Suleyman Demirel,' her assistant suddenly yells excitedly at the mention of Atatürk.

As we spoke, various men in suits, bodyguards and other functionaries, appeared and disappeared. Manokyan's wealth has earned her many enemies, and for all her charity work, not everyone wishes her well. In September 1995 her bodyguard and driver were killed in an explosion as she was getting out of her car in front of her apartment, and she was hospitalised with a broken leg.

*

But the question of a Muslim revival and its implications for Turkey and the West are more complicated than a simple dichotomy between the Sunni-Muslim-led Refah party and the pro-Atatürk secularists. There is a third force in Turkish society, firmly Muslim in its identity, but strongly aligned with Kemalism. These are the Alevi Muslims, about 20 million strong, including many Kurds, and they are now starting to flex their political muscle.

The Alevis have their own distinct version of Islam, one heavily influenced by the shamanistic roots of Turkic culture as it travelled to Anatolia from Central Asia. It is far more liberal, in western terms, than the majority orthodox Sunni Muslim faith. Like Sunnis, Alevis believe in Allah and his prophet Muhammad. However, they also follow Ali, one of the four Caliph, or successors, to the Prophet, after his death in AD 632, from whom they take their name, as do Iranian Shiites. But Turkish Alevis take a radically different approach to their religion than the Ayatollahs of Tehran. They pray together with women, include music and dance in their rituals and may drink alcohol. Alevis have a strong allegiance to the secularist ideology of the Turkish state, seeing it as a bulwark against a rising tide of Sunni Islam. This is not fully reciprocated: while Sunni Islam receives generous funding from the state, Alevis must finance their own religious and cultural activities. At the same time, religious education is compulsory in school, but the syllabus follows Sunni Islam, although community leaders believe this will soon change to include Alevi Islam as well.

That Alevis receive no financial help from the Ministry of Religious Affairs is in the tradition of the discrimination they have suffered for centuries from the Sunni majority, says Izzettin Dogan, an Alevi community leader and professor of international law. 'With the creation of the new republic [in 1923] Alevis found some kind of freedom to exercise their beliefs as they wish. But unfortunately this was special to the time when Atatürk was alive, and after his death the Alevis began to have difficulties. The Ministry for Religious Affairs is occupied only with the Sunnis and not with the Alevis.

'They have been pushed in a sense out of the social order of Turkey. They have to do something for the integration of this part of the Turkish community. The first step is to put in the programmes of education, beginning from next year, information about Alevi beliefs of Islam. If you are more than 20 million people you can influence the structure of your democracy. If they continue to ignore the Alevis it can cause some problems.' Sitting in his office in Taksim, Professor Dogan is clearly a major player in the Alevi community. A stream of visitors pours into his office to pay their respects and discuss the issues of the day, each greeting, and being greeted with, kisses on the cheek. In the background his black mini-dress-clad secretary bustles back and forth with trays of tea and coffee.

With the increasing support for Refah, which does not accept the legitimacy of the Kemalist project, Turkey's mainstream parties, and moderate Islamic groups, now see the Alevis as a useful ally in the struggle for the country's soul. 'They think they could bring the necessary freedoms to Alevis plus the other Sunni parties who believe the Atatürk model is still valid, and by the combination of these two parties they can overcome in a democratic way the danger of the Islamic parties coming to power. I don't believe that Refah is a danger to Turkey. But it can come to power. When Refah says "If we come to power we will bring you equality and the world and everything you need", Turkish people are not inclined to ask what is inside this everything. It creates, like Marxism or other ideologies, a kind of hope that a human being is intended to believe without analysing what is inside this ideology.'

While leaders of Refah are careful to cultivate good relations with the Alevis, both for reasons of PR and political expediency, both sides know that their visions of Turkey's future are radically different. Alevis will never transfer their allegiance to Refah, both because of politics and theology. 'The Alevi approach to Islam is quite different from the Sunni approach. It is that reason must reign in the state's affairs and not religious principles. Much of the Alevi way of life spread into the Ottoman empire in its first two hundred years, such as the tolerance of minorities, when the Ottoman empire accepted the refugees from Spain in 1492. But what is important to us is not to violate the rights of another, to respect everyone and not to injure them. We believe that all of us are very little pieces of God, the same God. So to injure a man, a human being, is the equivalent of injuring God himself. So you mustn't harm people, especially not to violate another's right. The rest of what we believe is not very important at all. This is something concerning man and God himself and not society.'

The differences between Sunni and Alevi Islam are also cultural, thanks to Alevi Islam's shamanistic and nomadic roots, which reject the rigid formalism of the Bedouin and Arabic lifestyle in favour of a relaxed humanistic vision. 'Sunni Muslims are based much more on the Arabic approach, which means what is important is the formalities of Islam, like the mosque, the way of prayer, fasting at Ramadan. This is very important for the Arab and the Arabic approach to Islam. We believe in the Quran and the Quran is the holy book of Islam and we are very attached to it. But we do not interpret it as the Arabs do. The Turkish approach is not the Arab approach. The Turks of central Asia did not accept Islam very easily. There were many wars between them and the Arabs before they accepted it. They did not cut their ties with their traditions and their past. They accepted Islam by making a mixture of Islam with their traditions with their Shamanistic traditions.'

As you might expect from a professor of international law Izzettin Dogan has a ready and fluent explanation as to why Alevis are permitted to drink wine. 'Alevis drink alcohol and they are very happy when they take it. It is not true that it says in the Quran that you must not drink alcohol. It is true that it says in the Quran that wine is not good to drink but we have translated that in another way. We believe that for the people who are going to be drunk if they drink then it is not good. But you know what Omar Khayyam says to God: "My God, if you have really forbidden to drink wine then why did you create grapes." For the people that are going to be drunk when they drink, then it is *haram* [forbidden]. But if not there is no problem and you can have it.'

Not surprisingly, for many hard-line orthodox Sunni Muslims, the Alevi approach is anathema. They consider the Alevis little more than heretics, who flout the five pillars of Islam. In March 1995 two radical Sunni Muslim groups claimed responsibility for an attack by gunmen on a coffee shop that killed three people in the Gaziosmanpasa district of Istanbul, home to many Alevis. Those deaths triggered mass riots by Alevis and between fifteen and thirty people were killed in the fighting against police and security forces that followed. It was an ominous sign of the tension that lies under the surface of Turkish society, provoking memories of the street-fighting that raged across Istanbul during the 1970s, when different groups controlled their own suburbs, turning some into no-go areas for the security forces, until the military coup of 1980.

The coffee shop attack was the second fatal assault on the Alevi community since 1993. In July of that year thirty-seven people died when a hotel in the city of Sivas, where a group of writers, poets and singers were celebrating an Alevi festival, was set on fire. Last year tension flared again when Istanbul's Refah party mayor Recep Erdogan tried to demolish Alevi mosques, used also as charity houses, triggering weeks of vigils in their defence, until he backed down. For Dogan, the violence is further proof that society must be based on modern democratic lines and not ruled by theology.

'It is true that if you like you can organise a society according to the Quran's principles. But it is not an obligation, the obligation that the Quran brings is the organisation of your entire life and your inner life, but other people's lives are not included. But there is another principle and many people don't give the right translation for this Sura: the Quran says that before everything I gave you reason. That means in all your approaches you have to ask your reason and if you put your reason before everything, before all your actions, then you can see that a state organisation must be based on rational principles and not religious ones. That's why I don't see any contradiction with a secular system. The source of a secular state system is also based in the Quran and in this Sura of the Quran . . . That's why I think the regimes that conform with the Quran are the democratic and secular systems, like in the West.'

From the activists of Refah, through the merchants in the bazaars, the generals who organised a string of coups, the tired politicians in the squabbling centrist parties and the nervous secularists, everyone agrees on one thing: after seventy years of Kemalism, an island of democracy in a sea of Islam, Turkey is at a crossroads.

Held at arm's length by Europe, nervous over increasing Russian influence in the Balkans, part of its former empire, and building pan-Turkic links with the former Soviet states of Central Asia, Turkey is searching for its place and role in the world. But wild predictions that as Refah's influence grows Turkey could turn into another Iran or Saudi Arabia are unlikely to come true, for the simple reason that Istanbul is not Tehran and Ankara is not Jeddah. Time and again Turks, of whatever political and religious hue, will proclaim exasperatedly that Turks are not Arabs or Iranians, but their own nation. And ultimately, the large population of educated women, the sizeable Alevi minority and, perhaps most of all, the guns of the strongly secularist army, will prevent a Turkish Islamic theocracy, where *Shariah* law rules. Turkey also has its own distinct Islamic Ottoman heritage, with a tradition of religious tolerance of Jews and Christians quite at odds with the present-day Muslim regimes of the Middle East.

Perhaps more likely is that Turkey could be the cradle for a new modern form of Islam that seeks to reconcile the dictates of the Quran with the twentieth century, where working Islamic women hold mobile telephones to their headscarves as they help run city governments, or edit the night's news on a computer screen. Refah itself is divided over the country's future and its proposed model of government, between the hardline Islamists and the pragamatists, who realise that even an Islamic Refah government cannot turn away from the modern world.

For Suna Kili, Turkish women could be the engine for this evolution. There is already a political battle inside Refah between its 60,000 women activists and the conservative old guard, who did not field a single woman candidate in the 1995 election. 'I am not expecting this reform from the men, I am expecting it from the women. I am expecting something to come out of all this religiosity in Turkey, from the educated Muslim women,' she says. Over seventy years of Kemalist modernism have left their mark on young Islamic women, such as the technicians and newsreaders of Channel 7, and they will not surrender so easily. Time, and the increasing age of their opponents within Refah are on their side, if nothing else.

'They are educated and they will not accept being thought of as second-class people. Today in my class on Bentham, the best interpretation was given by the girl who covered her head, in impeccable English. So what do you expect, they are driving cars, they speak languages, they own small businesses, they are running newspapers. I don't think religion in itself is that dogmatic. We wouldn't have so many sects if it were. If Islam is going to get a new liberal interpretation, it will come from Turkey, precisely because we have a different background, and it will come from Islamic women.'

*The Nation of Islam's million-man march,
Washington DC, 16 October 1995*

Chapter 9

THE UNITED STATES

'We Muslims will be the new communists. This country has always needed a bogeyman, we had the red threat and now it's the Islamic one . . . I'm a Muslim fundamentalist. That doesn't mean I'm ready to blow up a bridge. It means I strive to practise my religion in pristine purity.'
Abdel Hakim Mustapha, African-American Muslim, author and activist

'In many ways you are freer to be a Muslim in America than in many Muslim countries. I've lived and travelled in Muslim countries. You can't write a letter to the editor there, you can't hold a protest sign in front of an embassy, you can't do anything that is at variance with whatever government or dictator happens to be in power there.'
Ibrahim Hooper, of the Council on American–Islamic Relations

It is a busy winter morning at the Phoenician Art Gallery on 2580 Broadway and owner Hani Adawi, a Lebanese Palestinian, is juggling the demands of a stream of customers while explaining how he came to New York from the port city of Sidon, and has made a new life here for himself and his family. The snow lies thick on the ground outside and a bitter freezing wind is blowing up West 97th Street from the Hudson river, just a few blocks away. Framed prints and photographs line the walls as the prospective buyers wander in, steam floating above the ubiquitous cups of weak coffee that Americans seem to thrive on. Adawi's gallery is located on the Upper West Side, whose elegant brownstone buildings are home to an eclectic New York mix of Jewish professionals, students at nearby Columbia University and young singles who crowd onto the express subway every night to hit the bars and clubs in downtown Greenwich Village.

Adawi arrived in the United States in December 1982, the winter of the year that Israel invaded Lebanon, triggering a storm of international condemnation and a fresh flow of refugees from that battered country. 'I came here because of the war in Lebanon, and because of the lack of educational opportunities. I wanted to go to college but it wasn't safe, because of all the checkpoints. I still remember what happened, you can't forget these things, what the town and families went through. What the Israelis went through with the Nazis, they practised on us. Not everything. They didn't burn us, it wasn't that extreme, but it was psychological warfare, big time.'

He made it through the front-lines to Beirut to the American embassy, where he filled out the necessary forms, and, to his slight amazement, was quickly granted a student visa. He left soon after, to a country that seemed like another universe, free of checkpoints, warring militias and invading armies, although admittedly some of New York is considerably more dangerous than present-day Beirut. For the young Lebanese student America seemed like the promised land. 'It was a completely different world, like a paradise. In Lebanon it was like being in the darkness all the time. Here there were lights, traffic and buildings, everything was so quiet and peaceful.' Now married to Fatima, an Afghan woman from Kabul, Adawi studied in Oklahoma before starting work as a salesman for seven years. Then, like almost every other immigrant to the land of the free, he decided to go it alone and opened the gallery in the autumn of 1995. 'I decided to try the American dream and have my own business, and I opened it with my brother. We set it up with our own money.'

Like millions of immigrants before him, Adawi found that the United States, where almost everyone is descended from foreigners, whether recent arrivals or generations-old families, is still the land of opportunities. Unlike in Germany, the six million or so – nobody knows the exact numbers – Muslims in the United States are not perpetual outsiders, legally marginalised because of their skin colour and foreign culture or deprived of citizenship on racial grounds. Over half were born in the United States and for the immigrants there are no requirements to have the right pure blood-lines for gaining citizenship. Adawi plans to stay and become an American citizen. 'Eventually I will be like an Irish American, an Italian American, or an African American, but I'll be a Palestinian American. It means that I'll be an American citizen from Palestine, with my ancestors, my parents from Palestine. My parents come and visit me often. They are very happy that I'm living in America,' says Adawi. America is still the promised land across the Muslim world, even for many of the extreme Islamic radicals in the Middle East, who denounce the 'Great Satan' while wondering how to get a tourist visa or Green Card residence permit for the country they claim to see as a den of vice and debauchery.

Muslims are a comparatively new minority in the United States in this century, their numbers growing rapidly during the 1970s and 1980s as war and strife spread across Turkey, Afghanistan, the Levant and the Indian sub-continent and a major influx of immigrants arrived. Over half of American Muslims, 56 per cent, are immigrants, with the remainder born there. It's easy to meet Muslim immigrants – just hail a Yellow Cab. For just as the archetypal London cabbie is a Jewish man called Morrie, the New York version is a Muslim called Mustapha. Take a taxi there and it's odds-on that your driver will be Islamic, probably Arab or Pakistani, although not all immigrants from the Middle East worship at a mosque. I took a ride with a Coptic Christian from Egypt, who said he left Alexandria because of pressure from Islamic radicals, who were making life difficult for Christians. 'These Muslims, they want to ban everything, nothing is allowed, no dancing, no singing, no music, no drinking,' he exclaimed as we roared up the Hudson river embankment. 'All they know is Allah, Allah, Allah. That's OK, have God in your soul, all right, but not all day, every day,' which was as good a description I have ever heard of the puritanical, repressive streak that runs through one variety of Arab Islam.

But Islam is no stranger to America, as many of the millions of African slaves uprooted from West Africa and brought to America were Muslims. Yarrow Mamout, an African Muslim slave set free in 1807, may have lived to be over 128 years old, according to a pamphlet published by the American Muslim Council, and later became one of the first shareholders in America's second chartered bank, the Columbia Bank. His portrait now hangs at the Georgetown Public Library, Washington DC. Half a century later the United States cavalry hired his coreligionist Hajji Ali to raise camels in Arizona. In 1865, at the end of the Civil War, librarians at the University of Alabama saved one book from the advancing Yankee troops who would destroy the library. It was a translated copy of the Quran.

Islam is now America's fastest-growing religion, say Muslim organisations, and contrary to popular belief, only a small minority of American Muslims are Arabs. In fact most of America's Arab population of about 2.5 million are Christians. By far the largest group, over 40 per cent of American Muslims, are African-Americans, according to the American Muslim Council. Many are converts, descendants of those same slaves, who have turned to Islam as much for political reasons as religious ones, as a statement to express their separation from, and dissatisfaction with, the Anglo-Saxon Christianity of America's ruling élite. The next largest group, about 25 per cent, hails from the Indian sub-continent. Arabs such as Hani Adawi comprise about 12 per cent of America's Muslim population. But while Muslims can settle down and build up their communities, from California to Chicago, from Texas to Tennessee, they must battle against a tide of ignorance about their lives and beliefs, especially outside the major conurbations such as New York or Los Angeles, where Muslim immigrants tend to congregate.

Of all the western countries to host a substantial Muslim minority, it is the United States that offers the greatest challenge to those Muslims wishing to live fully as Americans, but maintain and cherish their Islamic heritage. Not because of any institutionalised anti-Islamism, but because the values and mores of much of contemporary America – widespread use of recreational drugs, alcohol, promiscuity, homosexuality, teenage dating, gun ownership, values that are ubiquitous across the media – clash completely with the demands of Islamic morality. As a parent of a two-year-old son, Abed al-Rahman, Adawi knows he will have to maintain a difficult juggling act to raise his children according to the values of Islam, while living in a consumer society that sells and markets sex, which for Muslims is a sanctification of marriage, as just another commodity. 'I believe all I can do is guide my kids the right way. There is a big difference between the Islamic way of living and the American way. Homosexuality, dating, you don't have to do this stuff to realise it is bad. Dating we believe is sinful.'

How do you know it is sinful, I ask?

'Because God said it's sinful, because it can lead to a lot of problems, sexual diseases, minor things to major things. If you don't have sex before marriage it rules out a lot of these things. It causes psychological problems too. There is too much freedom for teenagers. Eighteen-year-old kids live on their own and their own parents can't do anything. For us the family is extremely important. My brothers and I support our family. My father is retired, we support him. But we still listen to him, he can't stop us doing anything, he can't stop our money, because we support him, but we listen to him and we respect him. I don't think Americans have that much respect for their parents. They want to be independent, live on their own without family support. Young girls get pregnant and they don't know what to do. One girl I read about recently beat her kid, nearly killed it because it was crying and she didn't know what to do. There was no aunt to come round and chat, no mother to guide her. They want to be independent but there is a price for that.'

But these things could happen to Muslims, as well, I reply.

'But not as much. In Lebanon my neighbours are like my brothers and sisters. I lived in my apartment for three years and all I say is good morning to my neighbour. I don't even know her name. Most Americans are cold, everything is business, even friends are friends to a certain limit. You cannot say this is my best friend, I can rely on him. When I went back to Lebanon my friends were there, if I needed help I ask them and it is done. Here they say, "Well, I don't know. If it is convenient, OK." They put themselves first.' Hearing Adawi's description of America's lack of neighbourliness reminded me of an interview I once read with a Bosnian Muslim refugee who had gone to live in England and was amazed that even after months in his new flat, nobody had invited him in for coffee, that great lubricant of Muslim social life.

Adawi says that living in New York provides him with a challenge to live as a Muslim, one that brings its own rewards. He tries to observe the precepts of Islam, both religious and social, but to live as a Muslim in New York, without the support of family – for Islam is an intensely social and communal faith – is more difficult than in Sidon, although it's definitely easier in America's most cosmopolitan city than in Arkansas or Alabama. Adawi says he has encountered sporadic hostility, but that was probably based more on racism than any specific hatred of Muslims. More endemic, and sometimes more tiresome to deal with, is the inability of many Americans to understand why, for example, he doesn't drink alcohol, a refusal that is often met with baffled incomprehension. 'I don't drink and it is difficult when everyone around is drinking. They say, "Come on, have a drink, just a vodka, I'll mix it with orange juice." They don't understand. But I don't drink, I don't eat pork, I do my reading of the Quran, I say my prayers, I try to be a good person, to be good to my neighbours. I keep Ramadan, which is difficult to do here. Here there is no support like there is at home. You can look at it in a negative way and say how hard it is to do it here. But you can also look at it in a positive way and say to yourself: "I'm different. I am a Muslim but I can do it." Not because someone is telling me to, but because I want to do it. I'm doing it for myself.'

For Adawi, like many religious people, Muslim, Jewish or Christian or of whatever faith, there is a spiritual vacuum at the heart of the United States. The classified pages of newspapers such as the *Village Voice*, New York's weekly listings magazine, support his claim. They are filled with rows of advertisements promising instant fulfillment through everything from encounter therapy and self-exploration to Buddhism and even Sufi (Islamic mystic) dancing. Spirituality, like sex, is for sale to anyone with enough dollars, but that's only one half of the equation, for the peace that Islam or any of the great faiths brings their adherents comes at a price, of following a set of rules and requirements that do not always marry well with life in 1990s America. But in a society where more or less anything goes, many still yearn for some structure to their lives and a set of rules to live by, just as in western Europe an increasing number of Christian Americans are converting to Islam, finding in that faith the means to fill the spiritual vacuum in their lives. When I interviewed spokesmen and women in Washington for the American Muslim Council (AMC) and the Council on American–Islamic Relations (CAIR), both were white converts, who clearly found in Islam a faith and a cause around which to build their lives. But none of this is news to Muslims such as Adawi. He tells the story of one of his customers who brought him a picture of an Indian lady to be framed. 'I asked her, is it a relative? She said, "No, it is my mother, my spiritual mother, she's my healer, she gives me strength." I asked her what she meant. She said "She gives me an energy by her presence."'

He looks both incredulous and sympathetic as he talks. 'This woman, she's educated, beautiful, well dressed, not stupid. She asked me if I wanted to go and visit her spiritual mother with her. I said, no thanks. They are like a sponge, empty. They want something to fill them up, fill the emptiness. They absorb it all, there's nothing there so they need something. They want to belong, and when somebody talks to them sweetly they fall for it. They'll believe in anything. It's the way they are raised. They are empty. When I sat down and talked about religion at college they know only the basics. But they don't really know anything. When it is time to learn they are busy with girlfriends, sports, superficial things. But Islam answers my questions and satisfies my spiritual needs.'

Adawi prays at the Islamic Centre, on the other side of Central Park, a plush building financed by money from the Gulf states, popular with Muslim diplomats at the United Nations as well as local Muslims. On a quiet weekday afternoon a few dozen worshippers are gathered there for afternoon prayers, among them Abdel Hakim Mustapha, an African-American originally born Cedric McClester in Boston fifty years ago. Like many black Americans he has converted, or reverted – a term many Muslims prefer – to Islam.

Mustapha is keen to talk and so we retire to a nearby café where he tells his story over coffee and bagels. Born a Christian, he found it difficult to grasp the theological complications of a faith based on a trinity, rather than the unity of God. And like many African-Americans he has found an identity and a cause in Islam, with its message of universal brotherhood, where skin colour is irrelevant, that eluded him in white America. 'We see Christianity as something used to enslave us. Christianity said our pie would come from the sky, but Islam deals with the here and now. It challenges you, it's the perfect religion, always trying to improve you. Muslims are always reading, stressing education, constantly seeking knowledge. I was brought up a Seventh-Day Adventist, I always had a deep spiritual side. I challenged things at Sunday school, they couldn't describe the Holy Ghost or explain why prayers end in Jesus's name, not God's. I respected Jesus and I believed at one time that he was the son of God, but the son, not God. Islam is a thinking person's religion.'

Mustapha became a Muslim in 1983, after suffering a personal crisis involving drugs and some unpleasant people who were pursuing him, he says. A big man, he outlines his progress to Islam with an easy confidence, but like many black Americans, indeed a substantial number of Muslims all over the world, his views are tainted with anti-Semitism, seeing if not exactly a Jewish conspiracy, then at least a disproportionate Jewish influence in opinion-forming parts of society. 'The educators are Jewish, the media is prominently controlled by the Jews, but they have been made scapegoats and people play on that fact and it becomes a crutch. Yes, the Jews had some complicity with slavery, but where next, who do we hate next? It's more important what we do for ourselves, recognise the historical facts and then move on. We are in a corrupt environment, but I am one hundred per cent American, I love America and I don't want to live in any other country. They tell us money is the panacea for everything, but everyone is seeking answers because of corruption and moral decay. But Islam offers a moral compass, we understand there is another life when rewards and punishments will be handed out.'

The virtual collapse of communal life in part of the African-American community is also bringing many to the mosque. About one in four black men between 20 and 29 is either in prison, on parole or charged with a criminal offence and awaiting trial. As their society and values crumple under a hail of gunfire in the inner-city ghettos, many American blacks seek answers in religion. But which one?

Christianity is seen by many as the faith of the racists and enslavers, imposed on American blacks, and Judaism they view as the religion of the landlords and exploiters. So that leaves Islam, with its message of self-improvement and brotherhood. That many white Americans have a visceral distrust of Islam and Muslims only increases its appeal for many African-Americans.

'More people are turning to Islam because of the devastation in the black community. They see discipline, structure and a moral grounding that is now missing but once existed in the black community. Black rage is real. Even the most assimilated milk-toast black person has some seething rage. It comes from the little indignities, from white supremacy, whether it's subtle or blatant racism, people moving aside when you get into an elevator, or following you around in a store when you are shopping. I'm not a kid, I don't need an escort, I'm a middle-aged man. That's the appeal of the Nation of Islam, because we were subjected to the worst form of slavery.'

The idea that Jews controlled the slave trade is widespread among many African-Americans, whether Muslim or Christian. It shapes the black perception of a community that should be a common ally on which it shares many common values, and is a root cause of widespread anti-Semitism among many black Americans, despite the role that liberal Jews played in helping to organise and bankroll the civil rights movements of the 1960s. Their reward was a book published by the Nation of Islam, the fringe quasi-Islamic, but extremely influential, grouping led by Louis Farrakhan, in 1991, entitled *The Secret Relationship Between Blacks and Jews*, which accused Jews of being 'key operatives' in the slave trade. Although it was denounced and deconstructed across America by a legion of academics, the damage was done, and even mainstream orthodox African-American Muslims such as Abdel Hakim Mustapha have absorbed its ideas, helping sour black–Jewish relations for years, perhaps irrevocably. The book had over one thousand footnotes, but its basic premise is wrong on several counts. Firstly, there were no Jewish communities of note on the African coast where the slave trade was concentrated. Secondly, slavery was widespread in the Middle East before the advent of Islam, and Arabs, both pre-Muslim and Muslim, played a major role in transporting humans for sale. The Quran does not outlaw slavery, although it encourages slave owners to treat their human chattels fairly and, preferably, to set them free.

For most Americans, orthodox Islam, whether Shiah or Sunni, is synonymous – wrongly – with the Nation of Islam. The Nation was founded in July 1930 by Wallace Dodd Ford, who claimed to be descended from the prophet Muhammad. His assistant Robert J. Poole later took over, and changed his name to Elijah Muhammad, claiming that he was divinely inspired by Allah. In his book *Message to the Blackman in America* he set out the sect's core beliefs, which include a claim that white people were devils created in an experiment by a wizard called Yakub 6,000 years ago. The Nation's bizarre creed, combined with its message of communal and economic self-help, brought it growing support among African-Americans. It came to national prominence in the 1960s, through the voice of its then spokesman Malcolm X. But Malcolm X broke with the Nation after completing the *Haj*, the pilgrimage to Mecca that is one of the five pillars of Islam. That was his epiphany, where he saw that racism against whites was both un-Islamic and unproductive. This is part of a letter he wrote to his assistants in Harlem from Saudi Arabia:

> There were tens of thousands of pilgrims, from all over the world. They were of all colors, from blue-eyed blonds to black-skinned Africans. But we were all participating in the same ritual, displaying a spirit of unity and brotherhood that my experiences in America had led me to believe could never exist between the white and non-white.
>
> America needs to understand Islam, because this is the one religion that erases from its society the race problem . . . I have never before seen sincere and true brotherhood practiced by all colors together irrespective of their color.

It was a moving personal testimony, but was greeted with fury by the radical separatists around Elijah Muhammad, including Louis Farrakhan. It was also, perhaps, his death warrant. Malcolm X was shot and killed while speaking at a rally in Harlem in February 1965. After a series of disputes that saw many of the Nation's followers adopt orthodox Islam, Farrakhan refounded the Nation of Islam in all its hard-line bigotry. Here he is, tape-recorded in a Chicago mosque, outlining his views on the Holocaust:

> Little Jews died while big Jews made money. Little Jews being turned into soap while big Jews washed themselves with it. Jews, playing violin, Jews playing music, while other Jews [were] marching into the gas chambers. We wasn't there. We had nothing to do with it. But why all of a sudden am I anti-Semitic? Tell me!

Farrakhan's most recent triumph was the 'million march', when hundreds of thousands of American males took to the streets of Washington DC. His ideology of racial separation – apartheid by any other name – is increasingly influential among mainstream black organisations. Despite his record of virulent racism, anti-Semitism and sexism – for women are often banned from his meetings – his is the loudest and most-listened to voice in black America, and he uses Islam as a platform to proclaim his beliefs, to the anger of many mainstream orthodox Muslims. In fact, by even the most heterodox standards, members of the Nation of Islam – whose members number about 10,000, less than 0.2 per cent of America's Muslims – do not qualify as mainstream Muslims. For starters, the belief of the Nation's second leader, Elijah Muhammad, that he was divinely inspired by Allah, marks him as a major heretic. The essence of Islam is the *shahadah*, the belief that 'there is no God but Allah and Muhammad is his prophet', and this definitely does not include the claims of Elijah Muhammad that he was Allah's messenger, or his 'discovery' that white people were created by Yakub 6,000 years ago. The nation's creed also diverts from orthodox Islam on several other important counts. Nation of Islam members are not required to go on the *Haj*, and fasting from dawn to dusk during the month of Ramadan is not compulsory. Instead they fast in December, when the days are shorter.

Although Farrakhan's beliefs exclude him from mainstream Islam, he found a ready welcome among the leaders of Islamic countries such as Iran, Libya and Sudan when he toured the Middle East in early 1996. Even Necmettin Erbakan, leader of

Turkey's Islamic Welfare Party, found time to meet Farrakhan and eat *iftar*, the fast-breaking meal at the end of Ramadan, with him. Compare the warm welcome extended to Farrakhan in the Islamic Republic of Iran, with the treatment of Iranian Bahais there. Both are heretics by orthodox Muslim standards. Like Farrakhan, Bahais claim that another prophet followed Muhammad, in their case Bahuallah, a nineteenth-century Persian. Hundreds of Bahais have been executed or imprisoned for their beliefs. Nobody in Tehran threatened to charge Farrakhan with heresy. But then even in an Islamic theocracy politics can still triumph over theology.

For all its small numbers, the Nation of Islam also exercises considerable influence over mainstream orthodox African-American Muslims, many of whom, like Abdel Hakim Mustapha, are sympathetic to at least parts of its message. 'The Nation of Islam was the first to introduce us to the concept of Islam, because we had lost a lot of that through the Christian heritage. They stopped us from eating pork, they stopped us from smoking, they stopped us from fornicating. They reconstructed lives, and made many of us ready for Allah. Many of the Black Muslims, followers of Elijah Muhammad, are orthodox Muslims, and they are orthodox Muslims because they are exposed to a concept of Islam. If we were to judge him on orthodox Islamic standards we would find a lot lacking, however, if we judge the work of the man there must have been some divine inspiration there for him to put us on the path.'

Thankfully though, the Nation's creed of hate, racism and separatism still puts it beyond the Islamic pale for Mustapha. 'I had contact most of my life with Black Muslims. I debated with them, my problem with them was racism, this talk of the blue-eyed devil I couldn't accept. I'm not prepared to hate, it doesn't do anything for my creed. I'm not prepared to use racism as a crutch, because that's just emotional masturbation. I'm more results oriented. We don't make nationalistic separations in Islam, we are all one nation, whether white or black, that's why the term Black Muslim is anathema to Islam.'

Play the game of word association with most white Americans, saying the terms 'Muslim' or 'Islam' and their immediate reaction will probably be 'terrorist' or 'Oil Sheikh', says James

Zogby of the Arab American Institute in Washington DC. 'I have taken cartoons from Nazi Germany and Tsarist Russia about Jews and compared them with American cartoons about Arabs at that time. You could see a definite similarity, about "the alien in the midst and beyond who threatens us". Both Muslims and Jews were feared because they were perceived as threatening the values of various states and their peoples. The way both the Jew and the Arab were presented was they lusted after women, their wealth was illicit by definition. This was still going on in the '70s and '80s, this idea that, "they have our oil" or were involved in terrorism.' The AAI even puts out a leaflet outlining the contribution Arab-Americans have made to politics, business, arts and culture. It is a proud little document, yet I found it somehow sad that such a pamphlet needs to be published at all.

But the stereotypes about oil money are wrong, as only a small percentage of American Muslims are Arabs, and most of those hail from the Levant – Syria, Lebanon, Palestine and Egypt – rather than the Gulf. As for terrorism, the vast majority of American Muslims are law-abiding, peaceful citizens. Muslims have been involved in or carried out two acts of terrorism on American soil in recent years: the murder in 1990 of the extreme right-wing Rabbi Meir Kahane in New York, and the bombing of the New York World Trade Center in February 1993. El Sayyid A. Nosair was jailed on gun charges for the death of Rabbi Kahane, although he was acquitted of murder. To put this in perspective, James Zogby rapidly quotes FBI statistics that there were over sixty-five terrorist acts on US soil between 1981 to 1992, of which just two were carried out by Arabs and Muslims.

The World Trade Center bombing, in which six people died and hundreds were injured, was an unmitigated disaster for Muslims living in the United States as well as the families of those killed or injured. The actions of a tiny clutch of fanatics, based around the radical Egyptian Sheikh Omar Abdul-Rahman, now imprisoned on conspiracy charges, tarred a whole community with the terrorist's bloody brush. When the bomb went off Hani Adawi was working for a company which had an office in the World Trade Center. His colleagues reacted, not so much with hostility, as sudden interest in his Palestinian heritage, assuming that he would be privy to all sorts of inside information about the attack. 'If you tell them you are a Palestinian, a Muslim, they think you are a terrorist and you know about hijackings and stuff like that. They asked me what's going to happen next. They ask stupid questions as if I'm the expert or I am one of them. I don't know. I'm not an expert. I may understand why it happens. But not the details. When the World Trade Center bomb happened we had an office in there. People at work started asking me about it, an army reservist asked me what kind of bomb did I think it was, he said, "All Palestinians are terrorists, they have experience." At first I got upset, then I got used to it. But I realised that they know nothing about us.'

At the other end of Manhattan to Adawi's gallery is the office of Daniel Felber, former defence lawyer for Siddig Ibrahim Siddig Ali, one of the defendants in the New York conspiracy bombing trial. Felber works out of Wall Street in the heart of New York's financial district, encased in a glass-fronted office, twenty-one floors up in one of the skyscrapers that soar up dizzily from the sidewalks. There were two trials in the aftermath of the bombing. The first was over the bombing itself, the second, in which Felber acted, over a conspiracy to organise a campaign of urban terrorism across New York. The conspiracy trial defendants, almost all of whom were Arab Muslims represented by Jewish lawyers, were accused of planning what would have been one of the deadliest campaigns of urban terrorism in modern times, that, if carried out, would have resulted in hundreds of deaths and many thousands of injuries.

Not to mention a massive national backlash against their Muslim co-religionists across the United States, that judging from the aftermath of the Oklahoma bombing – for which no Muslims were charged – would have triggered a wave of assaults and attacks on Muslim community centres and mosques. (A report issued by the Council on American–Islamic Relations on the aftermath of the Oklahoma bomb catalogues a nationwide series of arson attacks, fake bomb threats, death threats and various forms of harassment and intimidation against American Muslims.) New York prosecutors said the conspirators' planned attacks included: bombing the United Nations building in New York; bombing a string of bridges and tunnels that connect Manhattan with the rest of New York and New Jersey; planning to assassinate Egyptian president Hosni Mubarak and, for good measure, organising the murder of Rabbi Kahane.

At the centre of the plot was the frail figure of the blind Egyptian radical Islamist Sheikh Omar Abdul-Rahman, who was accused of being the plotters' spiritual leader and guiding spirit. Muslim and Arab leaders are now quick to marginalise Sheikh Abdul-Rahman, claiming that he was way out on the extreme fringe, but his bloodthirsty sermons at the mosques where he preached were always crowd pullers. Abdel Hakim Mustapha listened to the Sheikh speak several times and did not always like what he heard. 'He was a fire and brimstone speaker, not the jolliest fellow you could hope to meet. I heard him five times in Brooklyn and Manhattan. I didn't like his anti-West bent. I have difficulty with people who come here to tell me about the Great Satan, and who left a Muslim country to do it. If they were in heaven then why did they come to hell? I also prayed with Siddig Siddig Ali. I would have thought he was the most religious person in the world. If he truly believed that he was acting in the name of Islam his faith must have been weak.'

Felber's office is lined with hundreds of hours of transcripts of FBI wiretaps, taped conversations of the conspirators' telephone recordings and secret video-recordings of the house in Queens, New York city, where the operations were planned. The conspirators also practised shooting and combat at a firing range, provided to them occasionally by a sympathiser. Although Abdul-Rahman was found guilty of the charges, there are no tapes of him ordering the attacks against the target list. Much of the prosecution's case rested on testimony from Emad Salem, a former Egyptian army officer who was an FBI agent and had infiltrated the group, and who received more than $1 million for his evidence, according to Reuters. Defence lawyers built some of their argument on the claim that Salem was not so much an agent as an *agent provocateur*.

The transcripts of the secret FBI video, telephone and wire-taps on the conspirators make truly bizarre reading, a lengthy and perverse mix of domestic banality as the group has lunch at McDonald's (that ultimate symbol of the America they plan to strike so many blows against) and then plans to obliterate hundreds of innocent bystanders and passers-by in the name of their version of Islam. The conspirators had their own code language in a failed attempt to evade detection: salt = fertiliser; food oil = diesel; balls, small balls = hand grenades; chicks from the eggs = grenade detonators; groceries = bomb supplies; box = timer and toy, or children's toy = machine gun. Certainly Emad Salem seems more together than the rest of the group, whose members sometimes appear extremely amateurish in their organisation. Here is an extract of Videotape V18, recorded in the Queens safehouse on the evening of 29 May 1993, just under a month before a posse of FBI agents burst in on 24 June. Salem is Emad Salem, and Siddig is Siddig Ibrahim Siddig Ali, Daniel Felber's former client. The Sheikh is Sheikh Abdul-Rahman. Amir is Amir Abdelgani, another in the group. The comments in brackets appear as they did in the full transcripts. This tape is written up in the third person, rather than as a direct transcript.

> . . . Salem then says the Sheikh is very enthusiastic about night missions and that he wants to see strikes against the American army.
> Siddig then asks why (or how) did Salem talk to the Sheikh in this way. Salem explains that he whispered in the Sheikh's ear.
> Salem picks up his secret identity disguise wig from the table.

After a while the discussion turns to the need for a camera, or video-camera to plan one of the attacks.

> Siddig says he doesn't know how to use a camera. Salem says this is not a problem, he will teach him. He promises to obtain a very sophisticated camcorder and states that 'we are now in the training stage' of the mission.
> Salem takes a small white plastic unit out of his briefcase (probably one of the timers) and places it on the bench.
> He then places his wig inside the briefcase and shuts it. He places the timer on the far, right-hand corner of the bench.
> Amir places plastic wrap over the cake on the table in order to keep it fresh.

Even would-be bombers don't like stale cakes. By the evening of

23 June the plans had advanced considerably and work on making the bombs had started, with Salem again playing a major role, especially on the more technical matters. The mixing of the chemicals is underway, as this extract from tape V5-V8 shows, again recorded at the safehouse. Emad Salem is giving directions to someone to get to the house, although he's not sure of the way.

> Salem: . . . nitrogen can be dissolved completely. (Amir is still mixing. Salem's beeper went off, he made a phone call.) May peace be upon you yes, (beeper noise) uh huh, did you page me twice, okay, uh huh, hello, yes, yes, okay, yes, eh? No problem, listen take the PATH train, yeah we are spending the whole night here, we have a lot of work, you know what I mean, yeah, yes, please, because we have a lot of, because there is new stuff come over. Thank you, brother, I don't know, wait one second (asking Amir). Do you know how you get here from New Jersey? Which train to come?

The discussion moves on to plans for an attack, involving stalled cars that will be left in place while another picks up the would-be bombers. And so it goes on, for page after page, tape after tape. It is depressing but fascinating reading, providing an illuminating and detailed insight into the day-to-day workings of a would-be terrorist group, for the FBI's cameras and tape-recorders have captured everything, from the most banal domesticity to the conspirators' mocking each other's accents to their plans to turn downtown New York into a killing field. Everything, that is, except a direct order from the Sheikh to carry out the group's planned attacks. Many Muslims believe that he was imprisoned because he knew too much about American operations in Afghanistan and needed to be removed from the public eye, and public access.

Certainly his conviction highlighted the American government's murky relationship with Islamic radicals, who are prepared to use violence to achieve their goals. That the American intelligence services armed, funded and helped train radical Muslim fighters in Afghanistan to help drive out the Soviets has been well documented. The Sheikh himself had reportedly been involved in the United States' secret *jihad* in Afghanistan. He had been charged, but acquitted, of involvement in the 1981 assassination of then Egyptian president Anwar Sadat, the first Arab leader to sign a peace treaty with Israel. In his interrogation statement to the Egyptian police he said that the killers had twice asked him to be the *emir*, or leader of the group, but he had refused. A few years later the Sheikh and his followers might have accepted help from the Great Satan in Afghanistan, but they had their own agenda. Like Dr Frankenstein before them, the CIA found that the monster they had created – in this case to kill Russians – cut loose from its masters' plans and soon span out of control.

The Sheikh's conviction on conspiracy charges raised several questions about the links between the CIA and Islamic radicals. He preached in mosques across New York and in New Jersey, fiery, inflammatory sermons that seemed to if not advocate violence, then at least condone it. These sermons, indeed probably his every waking and sleeping moment, were watched with interest by the FBI, the law enforcement office combating terrorism on US soil, but many agents asked how he had got into the country in the first place, his entry suggesting some degree of government collusion in his activities. The Sheikh's name was on the State Department's list of possible terrorists for some time, but he still managed to obtain an American visa in Sudan in 1990, although he was later charged with falsifying his application form. Even so, as any visitor arriving at New York's JFK airport, or any other international American airport, can testify, immigration officers run every new arrival's name through a computer before allowing them in. It's impossible to believe that those machines do not contain the State Department's list of suspected terrorists. The Sheikh freely entered and left the United States several times. The State Department reportedly said there had been a computer error about his status, a true 1990s catch-all evasion.

For Daniel Felber, the trial raised serious and worrying questions about the murky web of links between the American government's intelligence services, their operations in Afghanistan and its support for extreme Islamic radicals. 'This is my pet peeve. Our foreign relations are often wrong-headed and hypocritical. There is no question in my mind that when the Sheikh's followers were fighting the communists in Afghanistan we put money, personnel, armaments, you name it behind those Muslims. What we did in the process in order to fight the red enemy was arm and validate, comfort and give courage to a fanatical sect of Muslims. When the communists fell and the cold war was no longer a war we had to contend with them because they now became a threat to our national security all over the world.'

Felber met Sheikh Abdul-Rahman several times, both at the trial and in prison. He appeared sick and devout, says Felber. 'He was not well. He is diabetic and blind. He is a very pious and devout man. I have listened to many of his lectures and speeches as government evidence. He represents a sect of Muslims that are not opposed to using violence to achieve their aims.'

What, I ask, are their aims?

'There are short-term ends and long-term ends. I don't want to speculate on what the longest-term end could be, but in the short term they would like to have the ability to eradicate prejudice against Muslims, and to kill their sworn enemies, anyone who doesn't believe in their beliefs.'

Did he sense at the trial the involvement of American intelligence operatives, or feel that undercurrents of collusion were flowing through the courtroom?

'The court itself was an armed camp. It reminded me of the Eichmann trial in Jerusalem, because there was a glass booth, and it was that kind of spectacle. There was a ceremonial courtroom with a fifty-foot vaulted ceiling, there were marshals ringing the courtroom, and there were people with walkie-talkies and earpieces all over the place. The security was as tight as I imagine it is in Belfast. So who was there it was impossible to tell. You have to also understand that defence attorneys are treated by the government just one notch higher than the persons accused of the crimes. We are not given any confidences so I don't have any information about that.'

The conspiracy trial started in January 1995. On February 6 Siddig Ibrahim Siddig Ali entered a guilty plea and Felber's involvement ended. But that move raised more questions than it answered. Siddig has not been sentenced, even though, by his own admission, he helped plan a series of terrorist attacks across New York. Many believe that he turned state's evidence in exchange for, what? A shorter sentence, followed by a new life somewhere perhaps, under the witness protection scheme. Felber, I am sure, knows more than he can or will say about this strange case, for he is still bound by the laws of client confidentiality. But he is an engaging and helpful man, so I tried my own attempt at courtroom-style questioning, in the case of the Pushy Journalist *v.* the Not Necessarily Unforthcoming Lawyer.

'Where is Siddig at the moment?' I ask.

'Unknown.'

'This is a bit strange, isn't it?'

'You know, I can't speculate as to what is going on and what is not going on because I no longer represent him. For security reasons or whatever reasons, I'm not made privy to his whereabouts. I imagine that he is incarcerated somewhere, but I have no way of knowing where he is.'

'He's not freely wandering the streets?'

'I don't think so.'

'So he entered a guilty plea but hasn't been sentenced yet?'

'That's right.'

'Was he the one that turned state's evidence?'

'I can't speculate. I don't know.'

'But he was the one that was suspected to have?'

'I believe that he was.'

'What kind of man was he?'

'He was an intelligent man, a handsome man. He was tall with a certain muscular build. He was a weight-lifter. He had olive skin, black hair and brown eyes. He spoke English fluently. He was from the Sudan.'

'Did he seem like someone who was working on a fundamentally rational level or did he seem like a fanatic?'

'He certainly harboured at certain times in his life fanatical beliefs about fundamental Islam . . . I've listened to hours and hours of taped conversations both surreptitiously taken from his telephone and secretly recorded by a government informant. And in those conversations he says things that are very fanatical.'

'Can you remember any of those things?'

'There were certain instances when he advocates violence on American soil for its policies of supporting Zionist policies. He volunteered to fight in Bosnia on behalf of the Muslims in the Balkans. Certainly there is videotape and audiotape of him in possession of weapons. There were training camps in Pennsylvania and Long Island where they trained in armed warfare, so I would categorise him at certain times in his life as having those fanatical beliefs.'

'Why did he plead guilty?'

'He is a former client now, but even so I cannot reveal confidences. I can't talk about what he said to me and what his confidences are. I can only speculate that he believed that he could do better for himself by admitting his guilt, showing his remorse, and not making the government prove his guilt.'

For many observers, especially outside the United States, the news that radical Muslims were planning a series of bombings was perhaps less surprising than the revelation that the conspirators were free to practise shooting and engage in target practice whenever they felt like popping off a few rounds. The firing ranges were under FBI observation, but apparently were not in themselves illegal. Videotape V18 records how one day Siddig was unable to work as Sheikh Abdul-Rahman's interpreter as he was going to the Pennsylvania camp instead. The Sheikh was scheduled for an immigration hearing to legalise his status – more grist to the conspiracy theorists' mill – but Siddig could not accompany him. This is an extract from the tape, a conversation between Siddig and Emad Salem:

Siddig says he did not tell the Sheikh that the reason he was unavailable was because he was going for training in Pennsylvania. Siddig is emphatic with Salem that the training at this camp was all legal, but nonetheless the FBI seems to be shadowing the group. Siddig then says that the fact that their training was legal won't prevent the FBI from arresting them.

As indeed it didn't, eventually. It's hard for a British person, indeed any European, to understand the depth of feeling in America about the constitutional right to bear arms. Even foreign Islamic radicals can practise their version of the *jihad* out in the forests and parks of Pennsylvania, or anyone else, as long as they have the permits. I asked Felber how it was possible that the conspirators could have such easy access to weapons. 'Americans, as you know, have an obsession with weapons. I'm sure that there are militias training with arms as we speak. It's not an unusual thing to have shooting ranges and that type of thing. They are all around the States, and providing that the weapons are licensed and not concealable, which as rifles they are usually not, and providing that the land is not being trespassed upon and providing that they carry out the other regulations you need to comply with, then there's no harm in arming and training.'

Felber, like many involved in the two bombing trials, does not believe that the threat of further terrorist attacks has ended with the imprisonment of the conspirators. 'I don't think it's been rolled up. But I do think that our country is very vigilant in connection with trying to observe, and surveille and infiltrate Muslims of fanatical persuasion because they have egg on their face allowing the World Trade Center to explode like that. And there is the knowledge that we have now, that they had various hints in advance and did nothing to prevent it . . . There is a videotape that totally appalled me at the time, because I was defending my clients.'

The videotape that so appalled Daniel Felber, indeed almost anyone who saw it, was *Jihad in America*, reported by journalist Steve Emerson. Emerson was kind enough to send me a copy of his film, which was aired across the United States on the PBS network in November 1994. The programme is extremely disturbing viewing. It purports to show the existence of a nationwide network of armed Muslim terrorists across the United States, who are prepared to kill their enemies, especially Jews and supporters of Israel. Emerson summed up his claims in an article published in *New Republic* in June 1995:

> The investigation . . . revealed that these groups have established elaborate political, financial and, in some cases, operational infrastructures in the United States. Because of the scope of this movement (it ranges from New York to Oklahoma to California), its bellicose rhetoric (it calls for a holy war or 'jihad' against the US and other governments) and its advocacy of terror to achieve its ends (it

trains recruits in the use of car bombs), it's no wonder the FBI has made this network a top priority.

Although there is no evidence that these myriad Islamist groups are centrally coordinated, it does appear that they collaborate and cross-fertilise . . . In addition, Sheikh Omar Abdul-Rahman . . . had been hosted or sponsored in the US by at least half a dozen mosques and innocent-sounding Islamist 'charitable' and 'religious' organisations.

Coming in the wake of the World Trade Center bombing, the video drove Muslim and Arab-American organisations into – it is no exaggeration to describe it thus – a frenzy of denials and accusations. I have a thick file of cuttings, of claim and counter-claim from both sides about the programme. Emerson's film, which is based in part on information received from Israeli intel-ligence, was a watershed in American Muslim history. It played a major role in shaping many Americans' perception of the growing Muslim community. It sharpened divisions on both sides; between the Muslims who believe that prejudice against Islam is widespread and unalterable and there is a Jewish/Israeli conspiracy to smear Muslims and Islam; and among those many non-Muslims who see a potential terrorist behind every Islamic name. Time after time, when speaking to Muslim groups and individuals, Emerson's name came up, accompanied by angry denunciations. The Council on American–Islamic Relations, sin-gled out by Emerson in the above article as promoting 'the interests of militant Islam in the US', issued a twelve-page rebuttal of the film's claims.

For me, *Jihad in America* included some valuable material. It showed that in a small section of the Arab Muslim community in the United States there is considerable loathing of Israel, Jews and the West in general, and that a few Imams in American mosques often resort to the rhetoric of hatred – fuelling what the *New Republic* has called 'a culture of incitement' to vio-lence. However, the film's essential premise, that there is a widespread Islamic terrorist network operating in the United States, failed to convince me. As a journalist I know how easy it is to shape an argument in any direction one chooses, using original source material, of which Emerson had plenty. But if there is a nationwide network of Islamic terrorists, then what are they doing all day? Apart from the World Trade Center bombing, there haven't been any other terrorist attacks carried out by Muslims. Instead there have been a stream of hate-crimes against Muslims, ranging from abuse and violence to daubings of obscene graffiti.

What there does seem to be, says Daniel Felber, is a network of fundraising organisations that raise money for charitable causes in the Middle East, some of which could be diverted to land in the bank accounts of organisations like Hamas, which is responsible for several terrorist outrages in Israel. In January 1993 two Arab Americans were arrested in Israel and accused of transporting funds there for Hamas. Israeli officials went into full information offensive mode about the pair, pouring out streams of reports and press releases about what they claimed was an international Islamic terrorist network. Three years later President Clinton signed an executive order that bans fundraising for twelve organisations opposed to the Middle East peace process, including Hamas, Abu Nidal Organisation, two factions of the Popular Front for the Liberation of Palestine and two extreme Jewish groups, Kach and Kahane Chai (Kahane Lives). American Arab and Muslim organisations, who strongly deny that cash raised in mosques goes to fund bombers, now fear that this order will be used indiscriminately against their charitable and fundraising work in the Middle East.

'If Emerson's film is to be believed, and it is a very frightening tape, there is a very organised network of groups that solicit large sums of money for educational purposes and then this money ends up in the hands of terrorists bent on destruction,' says Felber. 'But is there an organised terror network? I don't think so. Is there an organised network of Muslim groups, some of which do good things like creating schools and help? Yes, I believe that is true. And I believe that there is probably a portion of that money that goes back to fund fundamentalism. The problem is with the Muslims that don't believe it. You listen to some of the speeches when they talk about death to blank, death to blank. There were videotapes taken that were used at the trial, where there were these individuals that had their faces masked by headdresses and who were shooting automatic weapons in the air. That kind of thing does give you pause.'

*

270

I took a pause then, to travel to Washington DC and meet the Muslim lobbyists at the Council on American–Islamic Relations, whom Steve Emerson accused of keeping such dubious company. If New York is where America goes to party, then Washington DC is where it does its politicking. Washington DC is not so much a city as a collection of government offices and museums, surrounded by some of the most dangerous inner-city ghettos to be found anywhere in America. The switch from one to the other is quick and total, and if you take a wrong turn at the traffic lights you could be looking down the barrel of a gun. Washington DC, like New York, also has a luxurious Islamic Center, derided by many Muslims as the 'Ambassador's Mosque'.

At lunchtime prayers on a Friday the worshippers spilled out onto the streets, under the wary eye of an armed policeman. They were a cosmopolitan group, including Arabs, Africans, African-Americans, South-east Asians and Turks. After prayers the worshippers were served a free meal of spicy lamb, rice and salad, cooked in the adjoining kitchen. It was good food, also much enjoyed by the several homeless people who wandered in for a serving before dozing off in the library. Just as in New York this Islamic Center is often the first point of call for curious non-Muslims. Americans interested in Islam telephone or write with queries or may attend one of the tours of the mosque. A minority eventually convert, and their names – overwhelmingly of European origin – are listed outside on the wall.

Visitors to the Council on American–Islamic Relations in downtown Washington are greeted with tea or coffee and a choice of bagels, that Jewish snack, ubiquitous across America, that Muslims also enjoy. CAIR is, in the true sense of the phrase, a public relations organisation.

Ibrahim Hooper, a white convert, or revert, as he prefers to be called – because Muslims believe everyone is born into their faith, only moving away as they grow up – greeted me. 'For me the explicit appeal of Islam was the anti-racism, the egalitarianism, there's no priest, no ministers. Just you and God,' he says.

With a Master's degree in television journalism, Hooper was a natural choice for the job of National Communications Director at CAIR. 'Our mission statement is that we are dedicated to presenting an Islamic perspective on issues of importance to the American public. Our main goal is to have the Muslim community in a principled engagement with the larger society. We don't want assimilation and we don't want isolation. We want Muslim school board members, city council members, Muslim mayors, Muslim views taken into account, whenever large decisions are taken in society, that they are seen as a constituency that must be served,' he explained.

And while radical Muslims in the Middle East denounce America as a den of iniquity and fornication, Hooper says it's easier to live as a Muslim in the United States than in Turkey or Algeria. 'In many ways you are freer to be a Muslim in America than in many Muslim countries. I've lived and travelled in them and you can't write a letter to the editor, you can't hold a protest sign in front of an embassy, you can't do anything that is at variance with whatever government or dictator happens to be in power there.'

CAIR is one of several Muslim lobbying and communal organisations that have started work in the last few years, many of which are based in Washington DC, because that is where America's decision-makers live and work. As well as CAIR there is the American Muslim Council, an umbrella grouping for a range of communal organisations and the Arab American Institute. In its information pack the AMC includes leaflets entitled: *Making Our Voice Heard, Educating the Media*; *Islam and Muslims: An American Stylebooklet*; *Islamic Vision on Moral Issues* and *Political Empowerment for the Muslim Community*. All these organisations have drawn heavily on the experience of the Jewish community which runs a powerful and effective political lobby, influencing American policy on Israel as well as domestic issues. Their existence is testimony to a minority that is growing in confidence as it finds and articulates a voice in America's ethnic and religious mosaic. The long-term goal of many Muslim activists is to turn the United States into a society where an Islamic perspective and the desires and needs of the Muslim community are automatically considered on issues as diverse as policy towards Bosnia and anti hate-crimes legislation.

Hooper puts a tape in the VCR, CAIR's go-getting video promotional film. When the need arises, a voice explains, for Muslims to take action, CAIR is there. 'Within the hour an emergency call to action is faxed to everyone of the 1,500

Muslim community centres and mosques in the US, uniting them into a formidable community.' CAIR also leads the charge for Muslims to reach out and be an active part of their communities, sponsoring workshops nationwide on dealing with the media and publishes a guide showing mosques and community centres how to become an indispensable part of their neighbourhood.

'Even though the Muslim community is growing by leaps and bounds it has not been able to impact the system like other minority groups have, so there is a real need for organisations like ours, to be a facilitator for political and social activism,' says Hooper. 'Probably our biggest achievement is to give Muslims a new attitude on what they approach and show that they can be successful. For example a Muslim woman will be fired because she wants to wear a *hejab* [headscarf], and she will take it and say, "Oh well, there's not much I can do, I am in this society, it is not a Muslim society, that's life." Now they are discovering through our campaigns that there are things that can be done and be done quite effectively.'

CAIR's newsletter is full of its successes in combating ignorance about, and prejudice against, Islam. Hooper is proud of how quickly the sports shoe manufacturer Nike took down a billboard in Los Angeles, after its intervention. The billboard, captioned 'They called him Allah' was intended to celebrate a local basketball hero. A local Muslim e-mailed CAIR to alert them. 'Many of our achievements take place within hours: someone called us, says there is a billboard up in LA comparing Nike to Allah, within hours Nike is saying we will take it down. We got some e-mail setting out the details, I contacted Nike headquarters, I said, "Here's the situation, here's why we think it is offensive," and they said, "We agree."'

CAIR also had an advertisement for Budweiser which featured a scantily clad model with an Islamic phrase across her T-shirt taken off air, and a greeting card removed from sale. The card, which featured a veiled Muslim woman on the cover, read inside: 'So you're feeling like Shiite. Don't Mecca a big deal out of it.' Both were offensive certainly to Muslims, particularly the juxtaposition of words from the Quran and beer, which is forbidden to Muslims. But aren't these advertisements an integral part of the freedoms that allow Hooper to live an Islamic lifestyle in peace and freedom? Isn't CAIR restricting free speech, I ask him?

'We never say to restrict free speech. When we did our campaign against *Jihad in America*, we never said take it off the air, we said PBS should follow its own guidelines, on objectivity, accuracy and fairness,' he exclaims.

Ah yes, *Jihad in America*. 'It constantly pops up, in the editorials, in the writings of those who are against Islamic activism. It's a horrible, horrible film, it was biased, inaccurate . . . If they came to us and said there is some Muslim underground terrorist network in America, and it were true, we would say go and get them. But it's not true, aside from the World Trade Center.'

So, I ask, is money being raised for Hamas through mosques here?

'Most mosques can barely support themselves, let alone send massive amounts of money overseas. They raise money to build the mosque, first of all, to build schools, to build all the things that any other community does. Believe me, I wish we had all this money that Mr Emerson was claiming, we would be a lot better off.'

Where does CAIR get its money from?

'From its members. Sometimes ten dollars, sometimes a thousand.'

Does CAIR raise money for Hamas?

'No, absolutely not.'

Either way, of all the countries I visited while researching this book on the lives of Muslims in the West, America was definitely one of the most complex and fascinating. A superpower in a state of flux, often uncertain about its own identity, it is now, perhaps slightly to its own surprise, home to the biggest Muslim minority in the western world. Four strands emerged on my travels there that define the Muslim community: the immigrant experience; the way that black Americans are turning to Islam as a revolutionary anti-white statement; the furious controversy about terrorism and some American Muslims' alleged links to terrorists; and, finally, perhaps most significantly, the slow emergence of an American Islamic lobby in Washington DC's corridors of power.

Instead of always reacting to their host society, American

Muslims have decided to try to change it to suit them. In a country based on immigration and the rights of free speech and assembly – and even with the widespread ignorance about, and prejudice against, Muslims and Islam – they have a fair chance of succeeding in some of their aims. Their ultimate strategy is to turn the United States into a society based on Judeo-Christian-Islamic values. But that will be a long struggle, and as the voice of American Muslims becomes louder, more effective and articulate, it will produce a counterweight to argue against their aims. Many Americans will already see, for example, CAIR's successes in getting advertisements that offend Muslims withdrawn, as conflicting with free speech, enshrined in the constitution. But it is in the offices of groups like CAIR and the AMC, as well as the network of mosques and community centres across the fifty states, that the debate over Muslims in America, and the sometimes uneasy marriage of Islamic and American values, is being shaped. It is a new chapter in American history, one of which only the first few pages have been written.

(Andrew Ward)

Brick Lane, London, 1996

Chapter 10

THE FUTURE

---

*'To Allah belongeth the East and the West; whither-
soever Ye Turn, there is Allah's Face. For Allah is
All-Embracing, All-Knowing.'*
The Quran, Surah 2: verse 115

*'There is an Islamic storm blowing. It is an Islamic wind
which is blowing around everything in the world.'*
Izzettin Dogan, a community leader of Turkey's
minority Alevi Muslims

Think of Islam not as a rock, rigid and unyielding, but as an
ocean. As a vast expanse of water flowing this way and that,
whose many currents, cultural, political and theological, swirl
and eddy through history and across the globe. Its tides have
advanced and retreated through the centuries, as the great
Muslim empires, ruled from Baghdad and Granada, Cairo and
Istanbul, Isfahan and Delhi, rose and fell. Each one absorbed
local culture and traditions on the way, whether Jewish, Hindu
or Shamanistic, for like every religion, Islam evolves in the environment in which it exists.

Medieval Muslim Spain, for example, was the zenith of
Islamic history in the western hemisphere, but some *Shariah*
laws were often breached as much as observed. It was an intricate and sophisticated civilisation, whose great cities were home
to great libraries and universities, fabulous architecture, public
parks and baths, a thriving culture of study and literature, all
ruled over by a spirit of intellectual inquiry and religious tolerance for Jews and Christians. Quite an apt model in fact for a
united Europe in the 1990s as it struggles to find a framework
for minority rights. In Granada and Cordoba an educated

Muslim élite led a life of opulence and hedonism, and their poets often praised the joys of wine-drinking, strictly forbidden for Muslims. Other poets, such as Ibn Arabi, Spain's most famous Sufi mystic, used the imagery of Judaism and Christianity, even polytheism – the ultimate abomination for Muslims – to express Islam's universalism:

> My heart has adopted every shape; it has become a pasture
>   for gazelles and a convent for Christian monks,
> A temple for idols and a pilgrim's Kabah, the tables of a
>   Torah and the pages of a Quran.
> I follow the religion of Love; wherever Love's camels turn,
>   there Love is my religion and my faith.[1]

Muslim Spanish poets also celebrated the joys of sensual love and erotic passion, a literature firmly in the traditions of Islam, which enjoins both husband and wife to sexually satisfy each other. There is no monasticism in Islam, said Muhammad. Even now in some Islamic countries a woman can initiate divorce if her husband fails to have sex with her at least once in four months. 'Almighty God created sexual desire in ten parts; then he gave nine parts to women and one to men,' said Ali, husband of Muhammad's daughter Fatima, and the founder of Shiah Islam, faith of the Iranian revolution. Islam is not a puritanical faith and sex – within marriage – is something to be enjoyed and treasured. *The Tales from the Thousand and One Nights*, the famed collection of bawdy popular Arabic folk stories, are crowded with a cast of fornicating characters, or more often would-be fornicators, tales as entertaining now as they were in ninth-century Baghdad and Basra.

Take just one issue in Islam – women and sex – and you will find a myriad of opinions, each densely supported by quotations from the Quran and the *hadith*. If *The Tales from the One Thousand and One Nights* represent one extreme, of vibrant sexuality, female circumcision is its other. Many Muslim women are for ever denied the full pleasures of the marital bed, as their genitals are mutilated, usually with some form of partial or total clitorodectomy. One in five Muslim girls lives today in a community that sanction some type of genital mutilation, writes Geraldine Brooks in her book *Nine Parts of Desire*. The practice probably began in Africa, before spreading into Egypt and then moving across the Islamic world, probably as a means of defending male family honour by preventing daughters and sisters from having sex before marriage. There is absolutely no Islamic justification for mutilating a woman's genitals, and the practice is nowhere advocated in the Quran. In fact self-mutilation is not approved of in Islam, as this *hadith* shows:

A man once said to Muhammad, 'O Messenger of God, permit me to become a Eunuch.' He said, 'That person is not of me who maketh another a eunuch, or becometh so himself; because the manner in which my followers become eunuchs is by fasting and abstinence.' The man said, 'Permit me to retire from society, and to abandon the delights of the world.' He said, 'The retirement that becometh my followers is to live in the world and yet to sit in the corner of a mosque in expectation of prayers.'[2]

Many Muslims are well aware that there is no Islamic justification for female circumcision, but few speak out against it and many Islamic leaders still advocate the practice. A clitorodectomy is a barbaric and terrible operation, often performed without anaesthetic, but most girls survive it.

The Pakistani village women set on fire by their husbands in 'honour killings' often do not. In Pakistan, still a traditional, patriarchal society, Islam has been influenced in several ways by Hinduism, some theological, others deeply misogynistic. Hindu traditions of polytheism have helped shape the South Asian Barelvi Muslim view of Muhammad as someone super-human whose presence is ubiquitous, not flesh, but light itself. Until it was abolished in 1829 by the British, Hinduism also had a history of *suttee*. The word *suttee* comes from the Sanskrit *sati*, faithful wife, and a faithful Hindu widow was one who immolated herself on her husband's funeral pyre, ostensibly to join her beloved husband in the afterlife, but in reality to prevent her having a share of his property. This traditional Hindu view of wives, as people not worthy of living after the death of their husband, has helped shape some South Asian Muslims' perception of women. Even now some Pakistani hospital doctors must treat the wives of Muslim villagers who have been set on fire by their husbands, for some or other imagined transgression of male honour, suffering horrific burns which many do not survive.

In Africa women have their genitals mutilated in the name of Islam. In Pakistan they can be immolated, although both these practices are nothing to do with genuine Islam. But in Pakistan and Turkey women can also rise to the office of Prime Minister, like Benazir Bhutto and Tansu Ciller. There seem to be major contradictions here. No issue causes as much misunderstanding in the West as the position of women under Islam, and few questions exercise contemporary Muslim scholars more than the interpretation of women's rights. It is also a vital question for western Muslim women. Will they rally to the western feminist standard and fight for equal rights, or meekly submit to a life of domesticity? The latter seems unlikely. Muslim women in Europe and America have been raised in an environment of equal rights in the workplace and public life. Those victories, won after long struggles, are totally compatible with a modern European or American Islam, say western Muslim women.

The Quranic view of women's rights was centuries ahead of Europe and the rest of the world. Muslim women can own property and are entitled to their own share of inheritance. They can get a divorce, are entitled to sexual satisfaction in marriage, can conduct business and study. Muhammad's wives and daughters played vital roles in establishing the new faith. They were pillars of the new Muslim community. And what westerners see as restrictions, such as headscarves, many Muslim women see as protection from prying, lustful eyes. Women must dress modestly, but so must men. What is modest to one Muslim, though, is still licentious to another. Many modern young Muslim women, in Turkey for example, wear a brightly coloured *hijab* (headscarf) and cover their arms and legs with fashionable blouses and long skirts. Arab women in the Gulf states often dress in the *chador*, the shapeless black garment that covers them from head to toe. Many Muslim women though argue that this drab dress stems not from Islamic civilisation, but Christian and pre-Islamic Iranian traditions. What is Islamic is in the eye of the beholder, or rather in the avoiding eye of the non-beholder.

Either way it is a fact that the Quran enshrines certain rights for women. In a seventh-century tribal society these were truly revolutionary concepts. They must be considered in the context in which they appeared, say modern Muslims, and not judged by the standards of 1990s Europe. It was only in 1870, after all, that the British law that put a woman's wealth under her husband's control on marriage was abolished. But the scales are still tipped towards Muslim men. A man may take several wives, but only if he is able to support them (and this provision was partly a way of allowing widows to re-enter society and be provided for). A woman can only have one husband. A woman cannot lead the prayers in a mosque, for fear of distracting male worshippers. A woman's testimony in court is worth half that of a man. Even so, the Quran provides the basis for women's participation in society. The problem for Muslim women who seek a wider role are the rulings of the male scholars of the four main schools of *Shariah* law, who interpret the Quran and the *hadith* and until now have decided how, in practical terms, Islamic law is to be implemented.

For Kareema Altamore, a former activist in the American trade union and women's movement, wearing a *hijab* is a means of defining her own role in the world, and setting out her space within it.

I met Altamore at her workplace at the American Muslim Council in Washington where she works on legal issues. She makes a forceful case in favour of the *hijab*. 'I really wanted to try another way of life, and this is part of the Muslim way of life. It has become very enjoyable. In this society, which is predominantly secular and very permissive, it allows me to make a statement, that some people choose to live their life in a religious framework. It identifies me as someone who has made that choice, which I'm proud and happy to. In a society which has blended the public and the private, it allows me to say I have decided what is private, my hair is private, this is for in my house and for my husband. I can decide that and maintain that. Our dress is very modest because we have that line. It's very healthy. I enjoy demonstrating that it is possible to draw this line.'

Like many American Muslim activists Altamore is a convert, who became a Muslim six years ago after long study, finding in the faith the concern for social justice that drew her to progressive causes. For her there is no conflict between Islam and feminism. At the same time Islam is a religion that sanctifies sexuality. 'I studied the basis of Islam, the Quran, and the principles of Islamic jurisprudence. I made a decision based on that study, it was conversion by conviction, that I was convinced that this is a balanced, just system . . . Islam celebrates sexuality, in marriage, very much so and it is very beautiful, and again very balanced. Sexuality is also for bonding and for affection, both the male and female have the right to be sexually satisfied so to me it is like a natural order.'

I asked her what brought her to Islam. 'That's always a complicated question, because usually it's a path that you have been on for a long time without being aware that you were on it. It culminates with the conscious choice to become a Muslim. My work in the women's movement gives me an interesting perspective, because I see things that were taken to extremes, and perhaps backfired and were detrimental. That gives me a special value for the balancing aspects of Islam. I see Islam as a way of life, not only worship, importance of balance and justice. I was a grandmother before I was a Muslim, so that gives me an idea of the importance of balance, of growth of justice. Islam highly emphasises the importance of justice, based on the idea that if Moses brought the law, Jesus brought compassion, then Islam is the balance between law and compassion.'

The concept of balance, of giving importance to *dunya*, matters of the world, as well as *din*, matters of religion, is a vital part of Islam. Islam does not reject the world, in fact it is incumbent on a Muslim to try to propagate the values of Islam. But when it comes to practically implementing Islamic laws, many Muslims cannot agree what form they should take.

In many ways Iran, site of the 1979 Islamic Revolution, is the best example of the clash over women's rights under Islam. For many in the West the triumph of the Ayatollahs was the ultimate victory of an alien and obscurantist faith, that sought to force women off the streets and back to the kitchen. Certainly there were plenty among the Shiah clergy who would have women swathed in black and kept in the house by their husbands. But the pictures of the *chador*-clad Islamic women revolutionaries marching and shouting down the boulevards of Tehran belied the western myth that Iranian women would lose all the rights they had gained under the deposed Pahlavi dynasty. Not just Ayatollahs, but women too tasted political power in the street battles of the revolution.

A defender of women's rights emerged in the – to western eyes – unlikely form of the leader of the revolution, the Ayatollah Khomeini. Where it said in the Quran that the prophet's wives were to remain in their houses, the conservative clergy extrapolated that to mean that all women should do so. Khomeini, the stern-eyed puritanical cleric, disagreed. He read the verse literally, and encouraged women Islamic revolutionaries to fight side by side with the men. After the revolution the imposition of Islamic law meant that unmarried men and women could not be allowed to be alone together, touch each other unless in a medical situation and women had to cover themselves in a *hijab*, or headscarf. The sexes were separated, to the dismay of many Iranians and the incomprehension of many westerners. But at the same time those new laws suddenly opened a myriad of new work opportunities for women: as television journalists covering women's sports events; as driving instructors, gym instructors and hairdressers. Even so, the new jobs did not compensate for the way fervent young revolutionary guards harassed women for wearing make-up or leaving their heads uncovered.

I did not travel to Iran, as this book's concern is the lives of Muslims in Europe, but I still wanted to ask some questions about the way that country juggles the demands of statehood and *Shariah*. In both Bosnia and Britain, Iran competes with Saudi Arabia for influence. The way Islam develops in the Muslim academies of the Shiah holy city of Qum eventually reverberates across Bradford and Birmingham. In London the voice of Tehran could often be heard behind the pronouncements of the late Kalim Siddiqui, leader of the self-proclaimed Muslim Parliament. Siddiqui was a frequent visitor to Tehran, although most mainstream Muslims dismiss him as an unrepresentative figure whose only strategy for British Muslim advancement was to call for the implementation of the *fatwa* on Salman Rushdie, about which he reportedly advised the Ayatollah Khomeini.

In London I met an Iranian government official, tieless of

course, because for many Muslims neckties are a symbol of western imperialism. He was a liberal and sophisticated man, keen to show that Iran too is evolving to find a balance between the demands of its interpretation of Islam and those of the international community. Even so, he wouldn't let me use his name in print, or tape-record our interview. 'That way I can deny what I told you,' he laughed.

Iran was born in revolution and street battles but by now the logic of statehood has imposed itself. In Iran, just like Moscow in the early 1920s, revolutionary rhetoric has been muted to meet the demands of diplomatic reality, but juggling the demands of diplomacy and the ideologues at home is a difficult balancing act. For example, Iranian officials have said that while the *fatwa* of death on Salman Rushdie cannot be withdrawn, as it was a religious rather than governmental edict, the country will not send assassins to seek the author out and kill him. The Islamic Republic seeks to take its place in the international community, although American and Israeli officials repeatedly accuse Iran of sponsoring international terrorism, which it denies. Most analysts believe that some elements of the hard-line Islamic clergy though are involved in supplying funds and manpower to extremist groups in the Middle East. At the same time hundreds of anti-government Iranian exiles have been killed or wounded since the Islamic Revolution. In March 1996 Belgian customs officials found a mortar-launcher and shells hidden in food containers on board the Iranian freighter *Kohladooz* in Antwerp. Officials in the Paris-based Iranian opposition group, the National Council of Resistance, believed their headquarters was a likely target for a mortar attack.

In London though, my Iranian friend is presenting a more conciliatory line. 'Like all revolutions, the Iranian one has had to adapt to the world outside and the international community. The revolutionary rhetoric has been toned down, we have compromised,' he explains over tea and biscuits. Iranian thinking is evolving not just on foreign policy, but also on domestic questions, such as weddings and the amount of celebration and ostentation that is advisable, he says. 'People used to get married in very simple circumstances in small mosques, without a big feast, and without spending a lot of money, or showing off, because that was the correct interpretation at the start of the revolution. But recently you can see the same families who had those simple weddings spending a lot of money on celebrating, because we only get married once. Now they are saying that Islam is not against people enjoying themselves. Things are changing.'

Satellite dishes are illegal in Iran, to protect its Muslim population from what the Ayatollahs believe are the corrupting influences of western culture. The ban is not very strictly enforced, though, says the official. 'People use them. Sky News came to Iran and broadcast a film of an Armenian Christian wedding, with wine, drinking and dancing, as Christians can drink alcohol. We had complaints from non-Iranian Muslims, asking how we can allow this, when we are supposed to be a Muslim state? I have met Muslims in the West who don't think we are good Muslims, who think we have forgotten Muslim principles.'

The role of women too, is changing, and they are playing a greater part in public life. 'We have women MPs, a woman minister in the Health Ministry, women working in the judicial system, and we are passing new laws to give them a more important role,' he says, beckoning me over to the television in the corner.

We tune in to Iranian morning television, and watch a few minutes of an Iranian soap opera, with a group of middle-aged women, their heads all covered. That women even appear on television angers the conservative clergy, he says, gesturing at the television. 'Ayatollah Khomeini was in favour of increasing the role of women, but the conservatives are against women on television, or women newsreaders. Some of the Ayatollahs don't like to see ladies laughing or entertaining themselves on television, they want women to be covered from head to toe in a black *chador*. We have women producers and directors who have won international awards, but they say this is un-Islamic.'

The western media tends to categorise any Muslim who stands up for his or her rights as a 'fundamentalist'. I have generally avoided using this term for two reasons. Firstly because several Muslims, mainstream ones, such as Abdel Hakim Mustapha in New York, told me that they were fundamentalists, in that they were believers in the fundamentals of Islam, as any Muslim must be, which does not imply a radical anti-western agenda, let alone support for violence or terrorism. Muslims also object to the word, not just because of its sub-text of violence and hostility, but because it was, they say, originally coined to deal with Christian fundamentalism. More to the point, terms such as 'fundamentalist', even 'moderate' are often meaningless labels, loaded with value judgments.

A more useful distinction is between those who seek to impose Islam, radical Islamists, and those mainstream Muslims who merely seek provision for the needs of Europe's Muslim minorities, and even between these two poles there is a sliding scale of opinion. Hizb-ut-Tahrir for example, are radical Islamists, in that the party is working to topple established regimes in the Middle East and impose a new *Khilafah*, a pan-Islamic state. At the other end of the scale is Dr Zaki Badawi, principal of the Muslim College in west London, who is working to develop a modern Islamic theology for life in 1990s Europe. But this process of recognising the world as it is, and dealing with reality instead of a demonised opponent, must also extend to the Muslim world. When thousands of demonstrating Iranians yell 'Death to America, death to Israel' it serves no purpose except to rally support for an uncertain regime that is riven by factional infighting over its future direction. America will not die, and neither will Israel, especially not through death by shouting.

This realisation, that the West, like the Muslim world, is a complicated mosaic that needs to be analysed and understood, not rejected outright, is slowly permeating through to Islamic leaders in the Middle East. For Raschid al-Ghannouchi, leader of the Tunisian An-Nahda (Renaissance) movement, currently living in exile in Britain, part of the blame for Islam's negative image lies with Muslims themselves. 'We ourselves are partly responsible for this. Even today, much of what passes for Islamic thought is very reactionary,' he said in an interview with the British Muslim magazine *Trends*. 'We have given ammunition to the opponents of Islam through our own actions, hijacking, hostage taking et cetera. We must change all this image. Islam is a call to all mankind. It calls for *ukhuwwah* [brotherhood], *'adl* [justice], *hubb* [love], *wahdat insanniyah* [unity of mankind], *huquq insan* [human rights]. We want to help mankind, not to hurt mankind.' That said, members of An-Nahda have been blamed for a terrorist attack in Tunisia in which a young girl was injured.

Muslims should not engage in blanket condemnation of the West, he said. 'Some Muslim groups have been telling the Muslims that the entire western civilisation is *kufr* [land of disbelief]. Was this the methodology of the Blessed Prophet Muhammad? No! The prophet did not set out to destroy all the values of the pagan Arabs . . . He said, "The best of you in *jahiliyah* [ignorance/pre-Islamic] will also be the best in Islam." So how can we just say "forget everything"?'

At the Muslim College in west London Dr Zaki Badawi is honing Islamic theology to grapple with the moral and theological dilemmas sparked by the scientific advances of the 1990s. Scholars such as Dr Badawi are developing a new European Islam. Part of that process is returning to the spirit of Islam, and stripping away all the cultural accretions that have coated and distorted the lives of Muslim believers. 'There is a great failure on the part of a large number of Muslims to understand the spirit of Islam altogether. What happens is that certain cultural traditions were drafted over Islam and given justification in terms of Islam. Like almost every society it tries to justify itself in terms of the overall embracing doctrines. If you are working in a communist society then you justify your actions in terms of communism. In a Christian society in terms of Christianity. In a nationalist society in terms of nationalism. So in Islam it's the same thing.'

I ask Dr Badawi which sort of customs he is referring to.

'For example the habits and customs regarding outsiders. Christians, Jews, Buddhists, Hindus, depending on the circumstances. The attitude towards women. Some want to make them invisible altogether, while others say that they've got to participate. It depends on your cultural heritage. Islam can be used by any culture. Islam has a system of law with which you can justify virtually anything. And because it is a universal religion it has readily been adapted by various countries. That is why I coin the phrase that we are developing a British Islam. That is quite legitimate in my view. If you want to compare societies in Muslim countries and you go from Indonesia to Nigeria across the whole sub-continent of India through Iran, Afghanistan, Iraq and the Arab world, passing the Balkans into north Africa, then going down to Nigeria, every time you cross the border there is a different culture but they are still Muslims.'

Developing a modern Islam – the religious principles of which can be relevant for all Muslims in the West – demands a theology of the minority. In most Muslim countries Muslims are either in the majority or are substantial minorities. But not in Britain and Europe, and so Islam must evolve according to its new circumstances, says Dr Badawi. 'It has to adapt and accept the new conditions and consequences. Most of our laws, like the Jewish ones, were written in the middle ages. But Islamic law had a particular characteristic. It is the law or theology of the majority. Very seldom do you find the theoreticians writing about the law of the minority, and how to operate as a minority. The Jews adapted to this because they were a minority for so many centuries. We are still grappling with how to live as minorities, develop a theology of the minority.'

The family is the cornerstone of Muslim life, but Muslim families too break up in acrimony. Divorce is allowed in Islam, but is not encouraged. The thing which is lawful, but disliked by God, is divorce, as one of Muhammad's *hadith* says. The problem in Britain for Muslim women is when their former husbands refuse to grant them a religious divorce after the civil one. There are parallels here with the situation of orthodox Jewish women, who cannot remarry in synagogue, says Dr Badawi. 'When a woman seeks and obtains a divorce in the civil court, she remains married according to Muslim law. So when she wants to re-marry there can be a problem for her if she goes to the husband and asks for a divorce but he refuses. Sometimes the woman can have an agreement before the marriage to have the right to divorce herself,' he says.

Now the Muslim Law Council, which Dr Badawi helped found, has formed a court of Islamic law to arbitrate in bitter divorce cases. It has also developed a standard marriage contract, to be developed to all mosques in Britain. It may not be accepted by every Muslim, but it is a step forward in dealing with conflicts between Muslim and British civil law, says Dr Badawi. 'If the husband will not give her a divorce, she can be blackmailed. So we decided to act as a Muslim court when there is such a case. We write to the husband and ask if he can give a good reason why there should be no divorce. If he doesn't give sensible reasons then we consider that he is acting against her interests just to spite her. Then the law of Islam permits us under these circumstances to declare a separation between them. In other words to cancel the marriage. This solved a huge problem, and now we have created a marriage contract in which the woman herself has the right to divorce.'

Three Muslim lawyers also sit on the council, to ensure that its deliberations conform with British law. 'This is part of the whole development of having a law for the minority because the people also ask us about *fatwas* [Islamic legal judgments] for various issues and we give them *fatwas*. The latest *fatwa* that has made some headlines was the one about organ transplants, which we have discussed and debated for eighteen months. We brought all the schools of *Shariah* law together and we arrived at the answer that this would be permissible. We have dialogue with everybody because you cannot live in isolation.'

It's not just community leaders such as Dr Badawi who are juggling the demands of Islam and life in a modern secular society. Every fortnight Dr Syed Mutawalli ad-Darsh has a page in the weekly Muslim newspaper *Q News*, in which he answers queries from readers. Some – usually about finance, as it is generally forbidden for Muslims to pay or receive interest – necessitate complicated responses. A whole industry of Islamic banking has developed in Muslim countries to cope with the difficulties of trading and financing without interest. Others are more straightforward. One, presumably from a woman reader, asks, in a household where both man and woman are working, shouldn't a man do his share of the housework? Yes, he should, says the paper's columnist. If housework was good enough for Muhammad, who shared domestic duties, it is good enough for every Muslim male. 'It is the husband's duty to care for his family and home, not just sit in front of the television for hours, while his wife does the cooking and looks after the children. This is simply unfair,' writes Dr Mutawalli ad-Darsh.

These are the thinkers revitalising Islam, developing the theology of the minority for the needs of Muslims in the West. But how is the interplay between science, modernism and Islam developing in the Middle East? For any new theological ideas born there will soon spread to Berlin and Bradford. I decided to ask Dr Muhammad al-Mass'ari, who is part of a trend of modern and modernising Islamists engaged on an *Ijtihad*, or intellectual struggle, honing their version of Islam as a faith for the 1990s.

A discussion with such an eminent scholar seems an excellent opportunity to raise many of my own questions about Islam, specifically on the compatibility of Islam with pluralism and the separation of religion and state that is a key element of western parliamentary democracy; the rights of women and minorities under Islam. But first, the *Ijtihad*, or *jihad* of the brain, that al-Mass'ari hopes will revitalise Islam. 'Any new *Ijtihad* will bring many buried jewels, such as questions of the rights of women, on the millet system [of rights for religious minorities], on the authority of the nation, on political power that everyone is equal. If you jump over the sectarian and party divides, you will find something which is very solid and could establish a stable society quite reasonably,' he says.

Dr al-Mass'ari, he says, wants to see an 'Islamic democracy' in Saudi Arabia. What exactly is one of those, I ask? 'I use it in the procedural sense, that all powers will emerge from the nation, the *ummah*, which everyone, man and woman has exactly the same stake, all those regulations and acts must be decided by the appropriate state authorities as decided in the constitution, such as a National Assembly, or president or a combination or both. This means that the clergy do not have any more say than anyone else. This is different from the Iranian model. . . . Generally my feeling is that the American model would be more tasty, or something with a President and prime minister. The President could choose to head the meetings of the council of ministers, or could leave it to the prime minister, head the National Security Council or give it to the prime minister. The word Islamic means that the people and the assembly will go back to the Quran as the source of Islam to produce laws. So if someone elects a communist or someone who is opposed to Islam, fine, there is no more Islamic state, then it will be a secular democratic state.'

This would be unlikely though, as secular parties will be outlawed, their members and supporters arrested. 'The state, as long as it is Islamic, will throw non-Islamic parties in front of the court, like the German model of dealing with non-democratic parties, which was applied to the communists in the 1950s. The only way to change will be to have a revolution, which is possible if the people decide.'

However much he talks about democracy, it would be a serious error of judgment to see Dr al-Mass'ari as a sort of Muslim Tony Blair, propagating a 'New Islam', with all the contentious – for non-Muslim westerners – parts excised. He may be opposed to the Saudi royal family, and planning to abolish the repressive religious police that enforce the demands of *Shariah*, but the CDLR's vision of Saudi Arabia as an Islamic democracy does not mean that the desert kingdom will be transformed into a Gulf version of Germany. The massive, perhaps unbridgeable gulf between the western liberal pluralistic agenda and the Islamist one, will remain. There will be a universal franchise yes, and women will be given the vote, but only Islamic parties will be able to stand for parliament. All others will be banned. Public beheadings and lashings will remain.

Our discussion moves on to the West's favourite *bête noire* in Saudi Arabia, the weekly public executions and floggings of those who transgress various part of the *Shariah*. In 1994, according to Amnesty International, at least 53 people were executed in Saudi Arabia, many publicly beheaded. These customs, barbaric to us, but rational to many Muslims, will remain, says Dr al-Mass'ari. 'If we abolish all penalties we will not be able to control criminality. Human beings need to be terrified and deterred by certain punishments. In Islamic history we have a great aversion against prisons, more than against lashes. We feel that if you insult someone, calling a woman an adulteress for example, a nice beating of eighty lashes is very painful, and very disgraceful. But after a few weeks he is over it, his life can go forward, and when he remembers it, it was very painful and he can learn from it. There is also a spiritual part, that if you accept the punishment and you are sorry, it leaves you free from all sins. Calling a woman an adulteress is a major crime in Islam, like murder. So if I put you in jail for this how long will it be? Six months will not be long enough. Five years? Five years, for heaven's sake, it is very demoralising and character changing and you have the problem of supporting the family.'

There are prisoners in Saudi jails, who endure a living hell, sentenced to hundreds, or even thousands of lashes, which are administered in batches over a period of weeks or months. Each time their bruises fade, and their scars begin to heal, they are taken out and flogged again. It is un-Islamic to administer more than a hundred lashes, says Dr al-Mass'ari. 'I have the feeling that lashes, purely rationally because this is not coming from a revelation, are not as bad a punishment as jail. One hundred lashes is the most you can get in Islam. If someone cannot take it it will be administered in parts, or even symbolically. One hundred lashes can be given to a sick man, who is found guilty of adultery, he gets one hit with a bundle of one hundred sticks. There is a lot of possibility for mercy.'

And so to executions, and the various methods thereof. 'Execution is a deterrent, I think it works. There are three points, the principle, the method of execution and the publicity. It seems to me, and God knows best, because nobody came back to tell us how painful it is, that cutting off the head is the fastest way of despatching a human. It is not very nice to imagine, but it is definitely superior to the electric chair. With beheading you cut the spinal cord, the blood pressure drops to zero in a few seconds and the brain itself doesn't feel pain. And the cut part, the lower part, does not feel pain anymore, because it is not connected to the brain.'

Executions are carried out in public in Saudi Arabia, usually on a Friday. Crowds are encouraged, foreigners pushed up to the front to witness Islamic justice. For most of us it is an abhorrent spectacle, although considering the overwhelming popular support for capital punishment among the British public, similar spectacles in London would probably be just as crowded. 'The publicity, it is the most terrifying, because of the blood and everything. So it combines two positive aspects – we have assumed its deterrent value. Once we decided that he does not deserve to live, we do it in the most gentle and clean way. We have to be rational, if you kill then kill in the most beautiful way. But I am inclined to think that if some people see the execution they will tell the others and it will be a deterrent.'

As for the other great bone of contention, women's rights under Islam, here Dr al-Mass'ari takes a fairly liberal position, at least when speaking to foreign journalists, arguing, as always on the basis of his interpretation of the Quran and *hadith*, particularly the later ones. Saudi women should be able to both drive and vote, he says. I ask his view of the legal situation in Pakistan, for example, where a woman's testimony in court is worth only half that of a man's. 'There are various schools on this, some say it is universal for every female witness, some say only when dealing with a written document dealing with debts and indebtedness, and some argue that later *hadith* which call for witnesses do not mention males or females and so abrogate the previous answer, and this is my point of view. Before Islam a woman could not be accepted as a witness, the first step was to abrogate that and make it half, the second was to abolish it altogether. But a woman being a scholar of Islam, that is well established, the wives of the prophets were major scholars, responsible for many *fatwas* and judgments.'

All religious minorities will have full rights, he says. 'My point of view, for which I have solid evidence in the Quran and is a point of view which will defeat any other – the reason it was not the most popular in the past is that most points of view emerged early in Islamic history before all the *hadith* were collected and the scholars were just working with the materials they had – later on, after a lot of scrutiny and weeding out, the evidence is very clear. All minorities have the same rights – the only specialisation with Christians and Jews is that there are regulations about intermarriage. All minorities will have full rights, there has been a very good model developed by the Ottomans, of the millet system, which gave the minorities very wide rights, which is a good system for other minorities.'

Dr al-Mass'ari though is a man of many voices, and the article in his newsletter *Al-Huquq* that called for the 'elimination of the Jews' – whether only in Israel or across the world – leaves room for much doubt about his real vision of a multi-faith society.

That the Muslim world can embrace so many differing perspectives shows how simplistic is the view of the supposed inevitable clash between Islam and the West. This view, popular among former Cold War warriors looking for a new enemy to justify their huge budgets and military-industrial empires, has acquired a spurious academic gloss with the publication of lengthy articles in specialist academic publications, such as Samuel

Huntingdon's influential piece in *Foreign Affairs*, published in 1993. Huntingdon argued that the next global conflict will be between western nations and peoples who do not share their values such as Islamic and Confucian civilisations. 'The clash of civilisations will dominate global politics. The fault lines between civilisations will be the battle lines of the future,' he wrote. 'This centuries-old military interaction between the West and Islam is unlikely to decline. It could become more virulent,' he cheerfully adds later.

Admittedly this view does strike an atavistic chord across Christian Europe, stirring up ancient memories of Muslim hordes besieging Vienna, and Islamic armies advancing up southern Italy towards the final prize: Rome. Muslims counter that argument with the Crusades, and the western domination of Muslim lands for centuries, a process culminating in the Gulf War. But many non-Muslim long-term observers of the Islamic world strongly disagree with the Huntingdon thesis. 'There is too much emphasis placed on this aspect of a clash of civilisations, that there is a Muslim enemy out there. We are used to dealing with enemies, so now Communism is dead, we need a new "ism" to deal with, even if there isn't one, otherwise why do we have such massive defence establishments?' argues one Washington-based analyst with long experience of the Muslim world.

And just as Islam is not a monolithic bloc, neither is the West. American experts take a different perspective on the implications of large Muslim minorities now living in the West to their European counterparts. This analyst believes that European governments are using the idea of the supposed 'Muslim menace' as a means of compensating for their own uncertainties about their continent's future. 'The European Union, which is looking for a way to define itself, could find this idea attractive. Human identity celebrates not just what you are, but also what you are not. This is ingrained into the European psyche, the Ottoman hordes at the gates of Vienna, the crusades, the old concept of Christendom against Islam, people have inherited all of these.

'The future though depends on the way Muslims integrate. Here the Muslim immigrants are not drawn from the lumpen, they are doctors and lawyers, integrating into American life.

But in Germany whole Turkish villages have moved across, from the blacksmith to the village idiot. Muslims in America are not directly associated with unemployment, they are not people on the street. In Europe a lot of Muslim immigrants have nothing to offer a modern economy, so there you have two problems in one: immigration together with a great religious antipathy.'

New thinking is needed to deal with a new situation, he argues, and old stereotypes must be jettisoned. 'There is a danger of overplaying Islam. Everybody focuses on the "danger" of Islam, as though there was a masterplan to get back at the West. This theory is falling on post-Cold War ears, of people who are used to a world of grand alliances. But you need to put this in perspective and look at the history of Christian Europe. Everybody fighting each other in the Second World War was a Christian. They have been killing each other since the Middle Ages. All the idea of Christendom means is that Christians killed Muslims instead. The Thirty Years War was a conflict between Christians, and nationalism and other factors were much more important.'

There is a rage, fueled by hatred, against the West among many Muslims. It is the anger of the slums of Algiers, of the refugee camps of Gaza and the West Bank, of the back streets of Karachi and the remains of Kabul. It is the fury of the poor, living in countries that have for decades seen their resources diverted to the West, while their own poverty increases. It is the anger of those living under corrupt and despotic regimes such as the Algerian military junta, that would have been voted out of existence had elections not been cancelled, and is propped up by western aid. It lives too, in a muted form, in the terraced houses of Bradford and Besançon, wherever Muslims live, and struggle to find work and build their lives.

Islam's message of social justice, and its holistic unity that combines religious belief with the necessity for action is rattling the old regimes, says Izzettin Dogan, a leader of Turkey's Alevi community. Like many Alevis, moderate and liberal Muslims who wish to keep Atatürk's legacy of state secularism in Turkey, Dogan is worried about the future. Islam has replaced Marxism as the ideology of the dispossessed and unless the West remedies global economic injustices, the rage will spread, he argues.

'There is a gap between the north and the south of this world. After seventy years of the practice of Marxism we saw that Marxism hasn't given any response to the problems of poor people. Islam could have this possibility, have these principles by saying that all human beings have to share their wealth and it is happening even in Turkey. All the people that were voting for the socialist parties now vote for the Islamic party, Refah, this is very typical. The western powers have to do something. They cannot continue to be selfish and if they do they will only reign by power and force because they have superiority of technology and weapons. But this will not bring peace to the world. They have to divide their revenues and bring some hope so there will be a more peaceful Islam in the world.'

Bosnia, Palestine, Chechnya, Kashmir, this is the roll call of Muslim grievances. Muslims believe that the life of their co-religionists is cheap for the West, contrasting the response to the invasion of Kuwait to the late and tawdry action to stop the war in Bosnia. They have a point. The war in Christian Croatia, the result of the same supposedly endemic ancient Balkan hatreds that prevented intervention in Bosnia, was stopped after just a few months. The war in Bosnia lasted over three years. 'Many Turks are disappointed with the West and that is normal,' says Dogan. 'I myself am very disappointed to see the weaknesses at that point of the West, their failure to bring international order and justice. They say that all that they do is on behalf of justice and human rights, as they have done in Kuwait, intervention against Saddam. But what we saw in Bosnia was a double standard and nobody can be convinced that western people do something just for humanity and justice, for the sake of humanity. Nobody believes that, because the only power that can bring order, justice and peace is the West.'

Far away from Dogan's office in downtown Istanbul, Bradford community leader Mohammad Ajeeb echoes his arguments. The Islamic resurgence is triggered by global economics, he says. 'The Islamic resurgence, the revival of fundamentalism, I think, is simply a reaction, to the western domination of the Third World, including the Islamic countries. They have no power to decide about their economies, about their future. People in the Islamic world are saying we are a billion people and still we have no power. Our fate and our destiny is being decided by Washington, London and Paris. In that type of situation you use religion as a defence mechanism. When the western forces attacked Iraq, Saddam Hussein, who was closely allied with the Soviet Union, and belonged to a party which claimed to be socialist, suddenly he says "*jihad* is the only way". He knew that was the last weapon he could use politically, to unite and boost his people's morale.'

Islam gives a voice to the streets and the bazaars, not the politicians and the diplomats, he says. 'It's the masses who are questioning why they have no control of their resources, not the rulers. They are saying to the West, "You haven't given us anything, you have taken away everything." Islam is making them realise who they are, it's a revival of identity, they think they have been cheated for too long. People there are suffering from a multiplicity of grievances, social, economic and political. Many of them regard Islam as a panacea for all their grievances. But I think it is going to be hard to find all answers in Islam in the modern age. If you want to adopt Islam, you have to do it according to today's requirements.'

Both the Muslim world – and the western countries that are now home to millions of Muslims – must realise that their fates are locked together, I believe. A mosque now stands on the global village's high street, as well as a church, a synagogue and a Hindu temple, and nearly all of their worshippers can receive CNN.

The days of countries, or even empires, as fortresses, closed bastions against the 'other', whatever that might be, Marxist or Muslim, are finished, thanks in part to satellite television. And religious and ideological questions aside, such states do not work economically. Albania and Romania under their warped version of one-nation communism were economic disaster areas. Even Iran, one of the last countries in the world to proclaim foreign and domestic policies driven by ideological concerns – in its case Shiah Islam – needs foreign trade and to be part of the international community. Islam and the West must co-exist, even coalesce, because there is no other option. I believe that Europe must jettison the tired old stereotypes about Muslims, and begin to meet their justifiable needs, within the framework of a democratic pluralistic society. The earth will not spin backwards on its axis and the French Republic collapse if a Muslim schoolgirl wears a headscarf to school. In Britain there will still be crumpets for tea if the Race Relations Act is amended to include outlawing discrimination against Muslims.

'Many of the traditional concepts that have held for a century or two are now being challenged or disintegrating, such as the notion of the nation-state, or the Cold War,' says Professor Akbar Ahmed. 'That is plunging many Muslim societies into a completely new era. The most important part of this for Muslims is the emergence and power of the global media. The global media – meaning the western media – is so dominant and so intrusive, that for the first time in history the Muslim household is no longer safe from the world media. I mean safe in a traditional sense, that a Muslim felt he was lord and master of his house, however small and humble. But now however remote a Muslim is, if he has a television, he has access to the images shown in any part of the world.

'I don't want to trivialise this as something in the West versus the East. This is a problem for traditional societies, not as a challenge between Islam and the West. This is global. The question is how traditional societies can maintain their sense of identity, even integrity, in the face of this emerging global culture. What this means for Islam is that because Islam emphasises these values so much, the sense of family, with a structure of defined roles, of a cosmological order, with man as God's finest creation, when the base of Islam is being affected, something is going terribly wrong. That is the core of the problem, this lack of balance.'

There is a whole industry in London and Washington of think-tanks and research centres, staffed by full-time professionals and experts whose job it is to ponder weighty issues such as the future of relations between Islam and the West, and the implication for Europe of its new Muslim minorities.

My reporting was almost over and now it was time to draw some conclusions, so I went to Chatham House, also known as the Royal Institute of International Affairs, to meet Dr Rosemary Hollis, head of the Middle East Programme. For Dr Hollis, the Huntingdon thesis of the inevitable clash of civilisations is too crude a mechanism with which to analyse the relationship between Islam and the West. The multi-cultural society is already a reality. 'There is an element of truth in it [Huntingdon's argument], in that when we are in a state of crisis, which we have every reason to be, in materialist and resource terms, these things will come to the fore, as justification for a scrap. There is a sense that there is not enough to go around and we have to have what we have at the expense of the others. But one of its biggest flaws is that it ignores the fact that we have now got mixed cultures all around the Mediterranean and Indian Ocean. The Muslims already live in western Europe, and they are already part of western European culture.'

This process, she argues, works both ways, and western culture and political ideas are spreading among educated Muslims, many of whom went to university in the West. 'The influence of western culture has permeated the élites across the board in Muslim countries. For example, I see an attitude towards religion that it should be the responsibility of the individual, and let's have a society in which the majority of people are Muslims, but the way they practise their religion doesn't infringe on those who don't wish to practise it to the same extent. I realise that is part of the problem because westernisation is identified with collusion with the West, but I don't think you can draw a neat split between what is a Muslim society and what is a Christian society.'

At the same time the struggle between Saudi Arabia and Iran for global influence, and ultimately, leadership of the Muslim world means that there is no unified Muslim enemy. 'Saudi Arabia is in competition with Iran to be the better Muslim. If you look at Central Asia, Afghanistan, Bosnia, it's almost as if as soon as you tell the Saudis the Iranians are spending money there, they must as well. That's another reason why it doesn't make sense to say they are the enemies of the West, because they are so busy competing with each other.'

For Dr Hollis, the focus on the supposed Islamic threat has exposed the spiritual vacuum at the heart of capitalism and modern industrial society. 'The timing is right for a focus on religious differences, because of this search for identity which has come in the wake of the collapse of the Cold War. It shows that capitalism is hardly something to live by, all by itself. It doesn't actually provide a sense of identity whatsoever, so there is a quest for identity, with religion right back in there at the centre. Some of the kinds of stereotypes that Muslims in developing countries have of the West are valid, in the sense that morality has gone to pot there. There is no regard for the extended family, there is no support for the elderly, sex and drugs are freely available.'

Life in a global village means that Europe must concern itself with events in north Africa. 'We in western Europe, including Britain, have to care about what happens in north Africa, and the viability of the systems and the economies there,' she says. The rebound effect is not confined to the Muslim communities, but has implications for the whole of the host society. Emigrants leave north Africa to find work in Europe because of political instability, and their presence here is exploited by right-wing extremists. The far-right sets a political agenda, to which the mainstream parties must respond if they are to keep up their electoral support. Which usually means police harassment of immigrants or their children, which leads to riots, which leads to increased support for the extreme-right, and so the depressing circle goes on.

'The mood has turned in western Europe against immigration, there is this association between foreign immigrants and scarce jobs, even though the reality is that the vast majority of the unemployed in France are of north African origin. Things turn ugly in Europe, right-wing racist elements in France know they can stir up trouble by aggravating the immigrant community, the majority of which would prefer to keep a low profile.'

More than that, there are implications for the very nature of civil society in late 1990s Europe. Sadia Ayata, the young French-Algerian woman I met in Paris who was arrested for criticising police identity checks on north Africans, knows all about that. 'This kind of friction could lead to a change in the nature of European society and the nature of civil rights if it boils up, with more draconian police powers, powers of arrest and search. We saw some of this when the French launched a clamp-down. It changes the nature of civil society inside Europe itself. But the connection is already there, it is not a question of kicking something out, we are already interconnected,' says Dr Hollis.

*

Indeed, we are all connected, not just by television, but by new global media channels such as computer modems. Like many Islamists, Dr al-Mass'ari spreads his message through cyberspace, using e-mail and a page of the World Wide Web, part of the Internet that globally links millions of computers. Muslims are great exponents of cyberspace, the term for the electronic media that communicate by telephones and modems, for two reasons. Computer technology is firmly in the spirit of Islamic scientific discovery, part of a chain of historical development that stretches back to ninth-century Baghdad and the academies of Caliph al-Ma'mun. More importantly, for the first time since the death of the Ottoman empire, Muslims can consider themselves part of a re-born *ummah*, or Islamic brotherhood that spans the globe, even if it only exists in cyberspace. It's hard for a non-Muslim to understand the depth of feeling that believing Muslims have for their co-religionists around the world, especially in places such as Bosnia and Chechnya, but for them the concept of the *ummah*, that all Muslims are linked in one community of faith and belief is an integral part of Islam.

The Internet is crowded with discussion and news groups about Islam, and some of the finest sites on the World Wide Web, which combine graphics, words, pictures and downloadable sound clips, are Islamic. Using my laptop computer I can take a trip through cyberspace across the world, to visit a mosque in Kazan for example, where I can examine its architecture in detail, or step back and see its whole glory. If I am feeling a little mystical perhaps, I can discuss Sufism with Muslims in Paris or Berlin. I can download verses of the Quran, soundclips that my computer will play. I can visit the Cyber Muslim Information Collective or the Azhar Mosque of the Internet. I can even download the whole Quran and keep it on my hard drive. 'Everyone is into cyberspace now and that is another aspect of the democratic debate that is happening in the Middle East,' says Dr Ahmad Khalidi, London-based advisor to the PLO leadership. 'Not that all these people are democrats, but there is an open debate.'

For repressive regimes like the Saudi one, the Internet is bad news. It cannot be fully controlled, either domestically or internationally, and the House of Saud expends much time and money on controlling information about the Middle East and its own activities. 'They want control and they won't allow, as far as they can, criticism of their own regime to go beyond a certain limit,' says Dr Khalidi. The Internet though, cannot be bought, or realistically controlled. 'There are a few outfits, but not many, which can resist this. The Internet is one way around this, because so far it is free and accessible to all and I presume it would be very difficult for the Saudis or any other government to impose the kind of regulations they want. It will be very interesting to see what kind of impact this will have and how the Saudis will try.'

More constructive than restricting information is its free exchange, and an equal dialogue between East and West, says Professor Akbar Ahmed. 'For the first time we are seeing serious attempts at dialogue and interchange of ideas and some mutual respect, which is important. On one side you had the Prince of Wales and his speech at Oxford, the Pope saying that Islam is a respected world civilisation, that contributed a great deal. In the Muslim world, you have people like King Hussein, Mahathir in Malaysia, people who are constantly making an effort to come to the West, to universities, to think-tanks, to dispel the idea that we are all mad fanatics.'

Muslims want to contribute an Islamic perspective to life in a modern industrial society, says Dr Dalil Boubakeur, rector of the Paris Mosque. The alternative, of continued prejudice and marginalisation, will only help spawn an angry Islamic underclass, that will turn increasingly to a radical, confrontational Islam and polarise further the divisions between the Wests Muslims and their host countries. Better by far then to let Muslims in the West introduce a new approach – or rather a much older one – founded in spiritual values, rather material ones,' says Dr Boubakeur.

'Muslims who are living in another country than their original one, have a new way to discuss life, religion, nationality, philosophy, which are not only about the problems of life in their own countries. Islamic life can be part of communal life in France, and Muslim values can help provide answers to problems of modern life, to make an *Ijtihad* [intellectual renewal]. For instance, on artificial insemination, that brings a lot of theological problems in Islam, on abortion, on genetic engineering, problems of malformed foetuses.'

There is a word in French, *misère*, which does not translate exactly into English. It means misery, depression, angst and alienation, the symptoms of life in depressed inner city ghettos across Europe and the United States. Islam can help solve *misère*, says Dr Boubakeur. 'Islam gives an explanation for ethical questions. We also have an attitude towards *misère*, on racism, and on the problems of the world, such as Bosnia. The Muslims there have their civilisation, their humanity, their spirituality, they are not without origin, people coming from the sky. Muslim people want to participate in the civilisation from which they were excluded for centuries, centuries. Now, it is difficult, I know. Muslims want to remember their past civilisation, which is not dead, absolutely not. We say that civilisation is not a fact only of one part of the world, of one country, the American way of life, or the European one, but civilisation is for all men in the world. We are all brothers, there is not a humanity of first-class and second-class people. That is the message of Islam.'

There is something more that Islam can bring to Europe, immeasurable, intangible, but nonetheless vital, say believers in all religions: God and spirituality. The missing part of the jigsaw of life in the late twentieth century. 'Islam offers a high spiritual life, one very open to God and unity of mankind and to philosophical insights into life, both spiritual and rational. Islam recognises the unity of God and a consideration that reason is not all, in our knowledge of life and the world and the cosmos. It leaves a part to mystery,' he says.

There is mystery, of God, religious belief and spirituality, and there is reality, of dealing with the world as it is, as Dr Zaki Badawi said, and not as some Muslims would like it to be. The challenge for the Islamic world is learning to deal with the West. To understand, that just like the Muslim nations, the Judeo-Christian half of the globe is a complex grouping of different states, nationalities and outlooks.

That necessary, subtle and intuitive interplay is already happening, as the two cultures are blending and fusing, often in unexpected places. Much of this book was written to the background of music by Natacha Atlas, and Nusrat Fateh Ali Khan. Atlas, who sings in Arabic, was born in Belgium of mixed English, Jewish and Arab ancestors. Her music blends Arab song, hip-hop, dub and dance, a new sound for a generation whose hearts are turned East, but who live in the West. Nusrat Fateh Ali Khan – another favourite of World Music fans – sings Sufi-influenced Pakistani devotional songs of such atmospheric rhythmic power that his voice can make you feel as though your soul is being turned inside out. I play both on a CD, the computer technology within it the latest manifestation of the mathematical advances that began over 1,000 years ago in the Islamic scientific academies of Baghdad. There is a continuum of knowledge, science and culture that stretches across the centuries, and across the globe. The West's new Muslim communities too, are part of that advance.

In mosques and Muslim academies across Europe and America, Islamic scholars are defining a new Islamic theology for the 1990s. In Bosnia, Islam has been reborn, harder and less starry-eyed about western ideas of human rights. In Britain, France and Germany, artists, musicians and writers have blended pictures, music and words to create a Euro-Muslim culture that fuses the Islamic traditions of north Africa and Turkey with that of the host countries. In Turkey the struggle continues for the soul of the Islamic Welfare Party, between the young women activists, influenced by western ideas of women's liberation and the sexist, conservative old guard. In Albania, Muslim activists have revitalised Islamic life, tolerant and heterogeneous, firmly in the Ottoman tradition. In the United States a Muslim political lobby is competing for influence in Washington and the White House.

I think they are all succeeding, slowly. I hope so, for as the Quran says:

Oh Mankind! We created you from a single pair of a male and female, and made you into nations and tribes, that you may know each other, not that you may despise each other.

Surah 49: verse 13

ISLAM: A SHORT GUIDE

'There is no God but Allah and Muhammad is the prophet of Allah.'
The *shahadah*, or core Islamic creed

Islam is the religion of Muslims. The word 'Islam' itself is Arabic and best translates into English as 'submission'. It comes from the Arabic word *salaam*, which shares the same semitic root as the Hebrew word *shalom*, both of which mean peace; peace not just in the absence of conflict, but a deep, inherent tranquillity. A less literal but more intuitive translation of the word 'Islam' would be something like 'submission to the peace that comes from following in the path of Allah (God)'. What this means in practice can perhaps best be understood by visiting a mosque when dozens, or hundreds of Muslims prostrate themselves as one, intoning their prayers, united in their faith and belief.

The semantic roots of the word 'Islam' are important, because for believing Muslims, Islam is more than just a matter of weekly attendance at a mosque. Rather it is a faith that provides a holistic unity, that balances matters of faith, *din*, with those of *dunya*, matters of the world. For Muslims the peace of submission ensues from following the path of Islam in every matter, from brushing one's teeth (Muhammad brushed his five times a day) to the implementation of the correct provision of civil rights for religious minorities under Islamic rule.

Muslims believe that Allah revealed his divine truths to Muhammad, who was born in Mecca about the year AD 570 in the Christian calendar. Unlike Christians who believe Jesus was the son of God, Muslims do not revere Muhammad as a divine figure. He was the *insan-i-kamil*, the perfect man, but a man nonetheless. Muhammad specifically warned Muslims against worshipping him, rather than Allah.

Muhammad became a prophet in the year 610 at the age of forty. He worked as a trader in Mecca, with the caravans that criss-crossed the deserts of Arabia and the Levant. He was a well respected local figure, known as *Al-Amin*, the one who is trusted. A contemplative man, he liked to retire to a cave on Mount Hira, where he would meditate and pray. It was there, during the night that Muslims now call 'the Night of Power and Excellence', that he received his first revelation, when the archangel Gabriel commanded him to 'recite'. The revelations, memorised by many Muslims during the time of Muhammad, continued for twenty-two years, and were written down as the Quran.

The Quran, divided into *surahs* (chapters) and verses, is the primary source for Muslim law and beliefs, supplemented by the *hadith* (sayings of Muhammad), which have been collected and systematised. Both the Quran and the *hadith* are firmly in the tradition of Jewish prophets such as Amos and Jeremiah, calling for social justice, protection of the poor, widows and orphans, and threatening the power and legitimacy of the unjust oligarchies who oppress their subjects. Muhammad knew the consequences of what he preached and he was at first overwhelmed with humility, but nevertheless continued with the role God had thrust upon him. He encountered stiff resistance from the Meccan élite to his message of egalitarianism, monotheism and social justice, but persisted before his faith triumphed in Arabia.

That was in the seventh century of the Christian era, although Muslims use a different calendar, based on a lunar cycle. Today there are over one billion Muslims in the world, mainly spread throughout the Middle East, Africa, the Indian sub-continent, south-east Asia, Europe, and the United States. Most people associate Islam with Arabs, as the land now known as Saudi Arabia was the birthplace of Islam, specifically the cities of Mecca and Medina. In fact Indonesia is the world's most populous Muslim country, and many Arabs, especially those now living in the United States, are Christian.

Although Muslims believe in the idea of the *ummah*, the international community of faith and brotherhood uniting every believer, they are differentiated by nationality, culture, politics, and theology. Over three-quarters of all Muslims in the world are Sunni, who take their name from the *Sunnah*, the customs and practices of Muhammad. Within Sunni Islam there are four main schools of law: *Maliki, Shafi, Hanafi,* and *Hanbali,* which reflect varying perspectives of Islamic interpretation and traditions from flexible and liberal to more conservative and literal.

The main split in the Muslim world is between Sunni and Shiah Islam, with between ten and twenty per cent of Muslims across the world professing allegiance to mainstream Shiah Islam, or one of its variants. Most are concentrated in the Middle East, particularly Iran and Iraq. The Sunni/Shiah split originated in a dispute over the succession to Muhammad after his death. Shiahs believe that Ali, the first young male Muslim and Muhammad's son-in-law, should have been the first *Caliph* (or successor to Muhammad after his death) rather than the fourth. Some Shiahs do not recognise the first three Caliphs, although Sunnis do recognise Ali as a Caliph. After Ali and his son Hassan were assassinated, Ali's other son Hussain led an army against the fighters of the Caliph Yazid at the battle of Karbala in 680.

A few dozen Shiah (partisans) faced hopeless odds against an army of thousands. They were slaughtered and Hussain's body was beheaded. Over 1,300 years later the Battle of Karbala is still of deep and profound significance for present-day Muslims, whether Sunni or Shiah. The motifs of martyrdom, revolution and uprising – as well as the disregard for death that can characterise Shiah Islam – could all be seen in the events of both the Iranian revolution and the Iran–Iraq war.

But whether Sunni or Shiah, all Muslims are united in their adherence to the five pillars of Islam, which are the core of their faith. These are:

1. The Creed (*shahadah*)
The statement that 'There is no God but Allah and Muhammad is the prophet of Allah' is known as the *shahadah*. This is recited during prayer, religious ceremonies and as an assertion of Islamic faith. The *shahadah* is the core tenet of Islam and by reciting it in front of two Muslim witnesses, a non-believer converts to Islam. The *shahadah* itself is split into two parts. The first half is known as the *tawhid*, the expression of God's unity. He is omnipresent, omniscient, without beginning, or end, forever indivisible. The Christine doctrine of the Holy Trinity is

totally alien to Muslims, who believe God can never be divided or manifest himself as a divine being in a human body.

The second half of the *shahadah* is called *risallah*, the acceptance of Muhammad and all the previous prophets as messengers of God. Islam recognises the prophets of Judaism and Christianity such as Abraham, Isaac and Jacob, as well as Jesus and Mary, as conduits for divine revelation. The virgin birth, the symbolic miracles and the Gospels are confirmed in the Quran. But the idea that God had a son is specifically denied.

To describe Islam, as some non-Muslims do, as 'Muhammadanism' is not only deeply offensive, but also a nonsense. Muslims revere Muhammad as God's messenger but they do not worship him, only Allah. Implicit in the *shahadah* is the core Muslim belief that Muhammad was the *last* messenger of God, perfecting the recognised truths of Judaism and Christianity in the final version of God's one true religion. As the Quran says (Surah 5: verse 3):

This day I have perfected your religion for you, completed my favour upon you and have chosen for you Islam as your way of life.

Those who claim divine revelation for messengers after Muhammad are committing *apostasy*, denial of the faith – a crime which can be punishable by death for Muslims.

2. Prayer (*salat*)

A Muslim must pray five times a day, every day: before sunrise (*al-fajr*), noon (*az-zuhr*), mid-afternoon (*al-asr*), after sunset (*al-maghreb*) and before midnight (*al-isha*). Ideally a Muslim should pray in a mosque, but if for practical reasons of work or travel this is impossible, then a Muslim may pray in any quiet, clean place, facing Mecca, often on a rug kept especially for praying. There is no priesthood in Islam, and any believing Muslim, who knows the Quran and is of high moral character, may lead the prayers. Prayers are usually led by the *Imam*, the mosque's resident scholar.

The way the mosque has evolved across the world illustrates how Islam absorbs the culture of the local environment in which

308

it grows. Prayers can be offered in any space that is in a state of ritual purity, free of impurities such as excrement. In Bradford Muslims pray in terraced houses, in central Africa mosques are built from mud and tree trunks. A mosque must include a *mihrab* (an arch facing Mecca) and a supply of clean, drinkable water for worshippers to perform their ablutions. The *minaret*, the tall tower from which the *muezzin* (caller) summons the faithful to prayer, is an Ottoman addition.

3. Poor tax (*zakat*)

Zakat is a tax on all Muslims living above the poverty line, who have amassed some capital, for the economic relief of poor Muslims. It is not charity, which is an additional duty. *Zakat* is paid on wealth according to the amount of property and goods that a Muslim owns, at 2.5 per cent. However, it is forbidden to establish any kind of bureaucracy to monitor or collect *zakat*, rather it should be paid out of an individual's conscience. *Zakat* is part of Islam's message of egalitarianism, that all men are equal in front of Allah. *Zakat* also illustrates how Islam is a religion of the community, that Muslims have an obligation to their less well-off co-religionists. Usury, the practice of giving or receiving interest, is forbidden for Muslims, and hoarding wealth is frowned on. Money should be used instead to benefit the community, either through *zakat*, charity, or honest commerce.

4. Fasting (*sawm*)

Every Muslim who has reached puberty must fast between dawn and dusk for the thirty days of Ramadan, the ninth month of the Islamic year. *Sawm* is how Muslims show their obedience to Allah and ability to do without food, drink, tobacco or sex. Muslims prepare for the day of abstinence by getting up an hour before dawn and having a large breakfast. The timing of Ramadan varies from year to year as Islam operates by the lunar calendar. Nursing mothers, the sick, and travellers are exempt from *sawm*, but they must make up for lost days by fasting later in the year. *Sawm* can be problematic for Muslims living in countries far to the north or the south, such as Britain or the United States, when the time between dawn and dusk can be over twelve or even sixteen hours. The end of Ramadan is marked by a feast to break the fast, *Eid-al-Fitr*, a time of great happiness and celebration.

5. Pilgrimage (*Haj*)

All Muslims must attempt the *Haj*, the pilgrimage to the Great Mosque at Mecca that houses the *Ka'bah*, Islam's holiest site, at least once during their lifetime. Muslims believe that the *Ka'bah* was built by the Judaic prophet Abraham (known to Muslims as Ibrahim). But by the time of Muhammad's lifetime the *Ka'bah* was a shrine to polytheism, filled with idols, until Muhammad destroyed them and began to preach the message of Islam. The core ritual of the *Haj* is *tawaf*, circling the *Ka'bah* seven times anticlockwise, when the *hajji* (pilgrims) attempt to touch or kiss the black stone which Muhammad put in the wall of the Ka'bah.

Until the advent of international jet travel going on the *Haj* was the most difficult journey a Muslim would have to make. Now its spiritual importance remains, although getting to Mecca from Bangladesh or Birmingham is a lot easier. A Muslim man travelling on the *Haj* must be healthy, sane, solvent and provide for his family while he is away. Muslim women may also make the *Haj*, although if they are unmarried they must be accompanied by a male chaperon whom they cannot marry, such as a father or a brother. Non-Muslims are strictly forbidden from joining the *Haj*.

Hajji start preparing for their journey at the end of Ramadan, by praying extensively and observing meticulous personal hygiene. *Hajji* wear a special garment of two unsewn pieces of white cloth, with sandals on their feet, while they observe a series of extensive rituals to complete the *Haj*.

There can be as many as two million people on the *Haj*, from all over the world, in any one year, and Muslims who have completed the *Haj* say the experience is one of the defining moments in their life.

Steven Barboza, author of *American Jihad: Islam after Malcolm X*, describes going on the *Haj* thus:

'For me it was wonderful. When I first saw the Great Mosque of Mecca I looked out the window of the bus, I saw the people on the plaza in front of it, all dressed in long robes, wearing white or pale blue, and it looked like everybody was dressed to go to their own wedding. I saw this marvellous structure with a greenish-grayish tint to it. When the people assembled to pray, when I saw the vast throng of people kneeling, and the complete silence and serenity of it, all with two million people, it was awe-inspiring. It inspired a feeling of peacefulness. It was a spiritual trip. It's about taking a journey into yourself; even though you're among two million people, you're alone. They call it a practice session for the Day of Judgment. You go through all these motions and emotions with two million people and yet you're alone. You do exactly what I said, you journey within yourself. And I found that to be wonderful. Afterwards I felt more inspired to live the religion by the letter. It brought some amount of solace to me. It made me see as an eyewitness that Islam is practised by people around the world.'

The *Haj* ends with the Feast of Sacrifice (*Eid al-Adha*), also known as the Great Feast, where the *hajji* sacrifice animals such as cattle, camels, sheep or goats, just as Ibrahim was permitted by God to sacrifice a ram instead of his son Ismail. The *hajji* can eat the meat, but most should be given away to the poor and the needy. A Muslim who has completed the *Haj* can preface his or her name with the word '*Hajji*'.

These, then, are the five pillars of Islam, but they are merely the foundation of a Muslim's faith. For readers interested in learning more about the details of Islam I recommend the following works: *Living Islam* by Professor Akbar S. Ahmed (Penguin); *What is Islam* by Chris Horrie and Peter Chippindale (Virgin) and *Islam: The Straight Path* by Professor John L. Esposito (Oxford University Press).

NOTES

Chapter 3 – Europe's Forgotten Islamic Heritage
1 Quoted in Bernard Lewis, *The Muslim Discovery of Europe*, p. 155.
2 Quoted in H.T. Norris, *Islam in the Balkans*, p. 29.
3 Bernard Lewis, ibid, p. 227.
4 Fouad Ajami, 'The Other 1492: Jews and Muslims in Columbus's Spain' in the *New Republic*, vol 206, no 14, p. 22.
5 Bernard Lewis (ed), *The World of Islam*, p. 181.
6 Bernard Lewis, *The Muslim Discovery of Europe*, p. 222.

Chapter 6 – France: The Dysfunctional Family
1 Bernard Lewis, *The Arabs in History*, p. 166.

Chapter 8 – Turkey: The Veiled Revolution
1 Andrew Wheatcroft, *The Ottomans*, p. 21.

Chapter 10 – The Future
1 Poem by Ibn Arabi, in Akbar S. Ahmed, *Living Islam*, p. 72.
2 Sir Abdullah Suhrawardy, 'The Sayings of the Prophet Muhammad'.

BIBLIOGRAPHY

Books

Aburish, Saïd K. *The Rise, Corruption and Coming Fall of the House of Saud.* London: Bloomsbury, 1995.

Ahmed, Akbar S. *Living Islam.* London: Penguin/BBC Books, 1995.

——. *Postmodernism and Islam.* London: Routledge, 1992.

Almond, Mark. *Europe's Backyard War.* London: Mandarin, 1994.

Andrić, Ivo. *The Bridge on the Drina.* Chicago: University of Chicago Press, 1977.

——. *The Days of the Consuls.* London: Forest Books, 1992.

Ayliffe, Rosie. *Turkey: The Rough Guide.* London: Rough Guides, 1994.

Barboza, Steven. *American Jihad.* New York: Doubleday, 1994.

Bousfield, Jonathan, and Richardson, Dan. *Bulgaria: The Rough Guide.* London: Rough Guides, 1994.

Brooks, Geraldine. *Nine Parts of Desire.* London: Hamish Hamilton, 1995.

Chippindale, Peter, and Horrie, Chris. *What is Islam?* London: Virgin Books, 1993.

Coughlin, Con. *Hostage.* London: Warner Books, 1993.

Council on American–Islamic Relations. *A Rush to Judgement.* Washington DC, 1995.

Dent, Bob. *Blue Guide Hungary.* London: A & C Black, 1991.

Dunsford, Martin; Holland, Jack; Bousfield, Jonathan, and Lee, Phil. *Yugoslavia: The Rough Guide.* London: Rough Guides, 1990.

Esposito, John L. *The Islamic Threat.* Oxford: OUP, 1995.

Glenny, Misha. *The Fall of Yugoslavia.* London: Penguin, 1993.

Haque, M. Atiqul. *Muslim Heroes of the World.* London: Ta-Ha Publishers Ltd, 1990.

Helsinki Watch. *War Crimes in Bosnia-Hercegovina*, Volumes 1 and 2. New York, 1993.

Johnson, Paul. *A History of the Jews.* London: Phoenix, 1995.

Lari, Sayid Mujtaba Rukni Musawi. *Western Civilisation Through Muslim Eyes.* Houston, USA: Free Islamic Literatures, Inc.

Lewis, Bernard. *The Arabs in History.* London: Hutchinson, 1970.

——. *Islam and the West.* Oxford: OUP, 1993.

——. *The Muslim Discovery of Europe.* London: Phoenix, 1994.

Lewis, Bernard (ed). *The World of Islam.* London: Thames & Hudson, 1994.

Lewis, Bernard, and Schnapper, Dominique (eds). *Muslims in Europe.* London: Pinter, 1994.

Maass, Peter. *Love Thy Neighbour.* New York: Knopf, 1996.

Malcolm, Noel. *Bosnia: A Short History.* London: Papermac, 1994.

Mohammad, Omar Bakri. *Essential Fiqh* [laws]. London: Islamic Book Company.

Nielsen, Jørgen S. *Muslims in Western Europe.* Edinburgh: Edinburgh University Press, 1995.

Norris, H.T. *Islam in the Balkans.* London: Hurst and Company, 1993.

Richardson, Dan, and Hebbert, Charles. *Hungary: The Rough Guide.* London: Rough Guides, 1995.

Ruthven, Malise. *Islam in the World.* London: Penguin, 1991.

Said, Edward. *Orientalism.* London: Penguin, 1995.

Silber, Laura, and Little, Alan. *The Death of Yugoslavia.* London: Penguin/BBC Books, 1995.

Simpson, John. *Behind Iranian Lines.* London: Fontana, 1988.

Simpson, John, and Shubart, Tira. *Lifting the Veil.* London: Coronet, 1995.

Thompson, Mark. *A Paper House.* London: Vintage, 1992.

Watt, W. Montgomery. *Muhammad.* Oxford: OUP, 1978.

Wheatcroft, Andrew. *The Ottomans.* London: Viking, 1993.

Internet
Islamic Computing Centre
www.islam.org/iqra/index.htm
Islamic Resources on the Net
www.utexas.edu/students/msa/links/other.html

Cyber Muslim Information Collection
www.uoknor.edu/cybermuslim

The Wisdom Fund
www.twf.org

The Truth About Islam
www.twf.org/truth.html

The Sayings of the Prophet
www.twf.org/sayings.html#jihad

Azhar Mosque of the Internet
http://thales.nmia

INDEX

Note: references in *italics* indicate illustrations.